# Studies in Systems, Decision and Control

Volume 237

The series "Studies in Systems, Decision and Control" (SSDC) covers both new developments and advances, as well as the state of the art, in the various areas of broadly perceived systems, decision making and control–quickly, up to date and with a high quality. The intent is to cover the theory, applications, and perspectives on the state of the art and future developments relevant to systems, decision making, control, complex processes and related areas, as embedded in the fields of engineering, computer science, physics, economics, social and life sciences, as well as the paradigms and methodologies behind them. The series contains monographs, textbooks, lecture notes and edited volumes in systems, decision making and control spanning the areas of Cyber-Physical Systems, Autonomous Systems, Sensor Networks, Control Systems, Energy Systems, Automotive Systems, Biological Systems, Vehicular Networking and Connected Vehicles, Aerospace Systems, Automation, Manufacturing, Smart Grids, Nonlinear Systems, Power Systems, Robotics, Social Systems, Economic Systems and other. Of particular value to both the contributors and the readership are the short publication timeframe and the world-wide distribution and exposure which enable both a wide and rapid dissemination of research output.

Indexed by SCOPUS, DBLP, WTI Frankfurt eG, zbMATH, SCImago.

All books published in the series are submitted for consideration in Web of Science.

Muneer Al Mubarak
Editor

# Sustainable Digital Technology and Ethics in an Ever-Changing Environment

Volume 2

Springer

*Editor*
Muneer Al Mubarak
Management, Marketing, and MIS
Ahlia University
Manama, Bahrain

ISSN 2198-4182 ISSN 2198-4190 (electronic)
Studies in Systems, Decision and Control
ISBN 978-3-031-86707-1 ISBN 978-3-031-86708-8 (eBook)
https://doi.org/10.1007/978-3-031-86708-8



This Springer imprint is published by the registered company Springer Nature Switzerland AG
The registered company address is: Gewerbestrasse 11, 6330 Cham, Switzerland

# Preface

Welcome to the second volume of the book titled *Sustainable Digital Technology and Ethics in an Ever-Changing Environment*. This volume represents a culmination of rigorous academic scholarship, peer review, and careful curation by the editor.

This book, with its meticulously selected chapters, delves into the intersection of business ethics, sustainability, and technology, catering to both academic and practical realms. It aims to spark discussions, prompt critical thinking, and provide insights into the ethical dimensions of contemporary technological advancements.

Comprising a total of 54 chapters, the book is divided into four parts:

Business Ethics and Environment Sustainable Development.
AI, Business and Education.
Technology and Sustainability in Marketing, Business and Hospitality.
Technological Innovation and Business, Environment, Economics Sustainability.

Each part offers a unique perspective on the intricate relationship between ethics, sustainability, and technology, offering a comprehensive exploration of these critical themes.

In closing, I extend my heartfelt gratitude to all the contributors, authors, and reviewers who have played a pivotal role in shaping this significant volume.

Manama, Bahrain | Prof. Muneer Al Mubarak
December 2024

# Contents

**Business Ethics and Environment Sustainable Development**

**Business Ethics in E-commerce: Ethical Concerns and Safeguarding Consumer Trust and Loyalty** ..... 3
Hamad Adeeb Al-Mutawa, Ebrahim Khalil Sowailem, and Muneer Al Mubarak

**The Role of Digital Technologies in the Economic, Social and Environment Sustainable Development** ..... 21
Sara Hameed and Muneer Al Mubarak

**An Assessment of the Impact of Technical Analysis and AI on the Effectiveness of Intraday Trading and the Behavior of Traders** ..... 37
A. Aravind, S. Sumayya, Arya Ajith, and Ananthi Rajayya

**Barriers to Sustainable Practices in the Construction Industry—A Bibliometric Analysis and Thematic Classification** ..... 47
Angela Benham and Lijeesh Pullot

**Students Perception on Corporate Social Responsibility in Coimbatore District** ..... 59
S. Hemaprasanna, J. Karthikeyani, K. Suresh Babu, R. Muthukumar, and M. Janani

**Masculinity, Emotional Regulation and Suicidal Ideation Among Young Adult Men** ..... 71
Celeste Sam, Ekta Singh, Fathimath Sahala, Sana Hyder Ali, and Soumya Simon

**Ethical Challenges in Technology Development** ..... 79
Noof Alsaud, Sara Saleh, and Muneer Al Mubarak

**Ethical Ramifications and Remedial Approaches on Bias in Artificial Intelligence** ........ 95
Shaima Isa Ali, Shahd Faisal Albadoo, and Muneer Al Mubarak

**Sustainable Employee Retention Strategies in Dynamic Environment of Higher Educational Institution** ........ 109
K. Rajeswari, Niranchana Shri Viswanathan, Katya, and S. Jayashree

**To Study the Impact of Government Subsidies, Financial Literacy, and Policy Awareness in Consumer Adoption of Green Finance with the Mediating Effect of Perceived Financial Benefits—SEM Analysis** ........ 119
Neerupa Chauhan, Reenu Mohan, Pandurangangagouda Honnali, S. Gokilavani, and Mallikarjun Biradar

**Nutrient-Rich Elixir the Health and Wellness Benefits of Tropical Fruit Cocos Nucifera** ........ 133
Tinto Tom, E. Eswara Reddy, Jerrin Jose, S. J. G. Preethi, Uma Maheswary, and Akhila Variyar

**Green Versus Greenwashing: How Consumers Differentiate Authentic Green Marketing from Deceptive Practices** ........ 145
Shilpa Chadichal, Prasanna Mohan Raj, and Sanjeev Padashetty

**Gender-Inclusive Circular Economy Policies: Assessing the Effectiveness of Current Policies in Supporting Women Scrap Collectors in Karnataka Specifically Bangalore** ........ 167
Madhu Druva Kumar, Cherian Thomas, M. Lokanadha Reddy, K. Lavanya, S. Varshini, and N. T. Jeevitha

**AI, Business and Education**

**The Ethics of ChatGPT in Academia** ........ 177
Husain Mohamed Tarish and Muneer Al Mubarak

**Unleashing the Power of Big Data and AI: Innovative Education for Tomorrow's Learners** ........ 191
Hassan Nader Hussain and Muneer Al Mubarak

**The Intersection of Culture and Tourism in Fashion Experiences** ........ 209
Richa Gupta and Rafaela De Angelis Barros

**Impact of AI Tool-Chatbot on Customer Experience in Singapore** ........ 219
Divya Tiwari and Easwarmoorthy Rangaswamy

**Resilience and Perceived Stress—A Comparative Study Among Performing Arts Versus Non-performing Art Students in India** ........ 231
P. V. Ankita, Mehak Mehreen, U. P. Riddhi Raj, S. Subasree, and Soumya Simon

**Unraveling It Sophistication: Driving AIS Success in Medan's Food and Beverage MSMEs** ........ 241
Risanty, Dede Ryansyah, Sambas Ade Kesuma, Adi Syah Putra, and Yuni Annastia Suherman

**Does Financial Literacy Sway Decision-Making in Personal Investments?** ........ 251
P. A. Mary Auxilia, K. R. Sowmya, S. Rajkumar, K. Jagannathan, J. Chandrakhanthan, and D. Ravindran

**An Analytical Review on the Protection of Children from Sexual Offences Act (POCSO), 2012 in India** ........ 265
S. Muruganandhan, S. V. Akshayareena, G. Jayashree, P. A. J. Aujita, V. Vinothina, and K. Sivasubramanian

**Revolutionizing Waste Management in Developing Countries: Challenges and Opportunities** ........ 271
Tinto Tom, Jerrin Jose, Akhila Variyar, Cherian Thomas, S. J. G. Preethi, Uma Maheswary, and Pradeep S. Raj

**Analysis of Backward and Forward Linkages Associated with Migrant Street Fish Vendors from Goa** ........ 285
John Xavier D'Souza and Jeronimo Monteiro

**The Role of Digital Technologies in Supply Chain Sustainability** ........ 293
Mohamed Ali Kadhem and Muneer Al Mubarak

**Sustainable Waste Management in Business-to-Business** ........ 313
Mohamad Osama Nesrallah and Muneer Al Mubarak

**Technology and Sustainability in Marketing, Business and Hospitality**

**Evaluating Service Quality in Private Hospitals: A Comprehensive Framework** ........ 331
S. Razeena and P. Shareena

**Municipal Solid Sustainable Waste Management in the Kingdom of Bahrain** ........ 347
Fatema Alaauddin Showaiter and Muneer Al Mubarak

**Currency Dynamics in Determining the Prices of Precious Metals in the Domestic and International Markets: A 52-Year Analysis** ........ 369
B. G. Shobha, K. U. Jayalakshmi, B. Vidhya, and M. Manisha

**Informal Sector Employment: An Analytical Review on Trends, Challenges and Avenues for Development in India** ........ 383
S. Muruganandhan, K. Udhaya Kumar, C. V. Kiran Raj, R. Thamilmani, V. Vinothina, and K. Sivasubramanian

**Ecological Transition Potential of Tourists in Mountainous Adjara** ..... 391
Merab Putkaradze, George Abuselidze, and Landa Putkaradze

**Mental Health Literacy and Depression Literacy Among Young Adults** ..... 401
Mushfirah Sayed and Anjana Sinha

**Sustainability Practices in Bajawa Coffee Cultivation: A Case Study in Beiwali and Wajamala Villages, Ngada Regency** ..... 411
Rudy Pramono, Juliana, Yosep Dudedes Timba, Diena M. Lemy, and Amelda Pramezwary

**Measuring the Role of Foreign Direct Investment in Generating Employment Opportunities: An Analysis in India** ..... 425
S. Muruganandhan, A. R. Sanjeev Kumar, P. B. Shrithik, Amita Biju, V. Vinothina, and K. Sivasubramanian

**Impact of Freebies in Electoral Politics and the Role of Dravidian Government in Tamil Nadu** ..... 435
S. Muruganandhan, Y. G. Gokul, S. Abishek, R. Thamilmani, K. Sivasubramanian, and B. Haripriya

**Gender Discrimination and Empowerment of LGBT Community in India: A Systematic Review** ..... 443
S. Muruganandhan, A. R. Sanjeev Kumar, R. Saiesakkisakthivel, M. Abishek, K. Sivasubramanian, and Mathun Muinao

**Potential for Tourism Based on Sustainable Livelihood in the Ngarai Sianok Geopark Area** ..... 451
Rosianna Sianipar, Juliana, Nova B. Sitorus, Diena M. Lemy, and Amelda Pramezwary

**Sustainable Presence of Women in Company Management: The Role of Gender Composition and Strategic Management** ..... 459
Vildan Bayram and Ayten Nahide Korkmaz

**An Analysis of Service Convenience's Effect on Online Ticket Performance: A Perception of Accounting Students on Faculty of Economics and Business Universitas Sumatera Utara** ..... 473
Fahmi Natigor Nasution, Yasara Ulfa, Risanty, Sambas A. Kesuma, and Alyssa Nadia

**An Analytical Review of Poverty Alleviation and Impact of Globalisation Policy in India** ..... 483
S. Muruganandhan, J. Yazhini, S. Isai Yazhini, R. Thamilmani, K. Sivasubramanian, and S. Guhan

**The Influence of Social Media on Perceived Body Image and Self-esteem Among Young and Middle-Aged Adults** .............. 491
Hrishita Lobo, Rekha Devi Chauhan, Candida Antonetta D'souza, Marissa Ruth Furtado, and Soumya Simon

**Assessment of Patients' Perceived and Expected Health Care Service Quality** .............. 499
J. Rajalakshmi and L. Jayanthi

**Technological Innovation and Business, Environment, Economics Sustainability**

**Consumer Perception and Behaviour Towards Organic Personal Care Products** .............. 513
L. Jayanthi, J. Rajalakshmi, and K. P. Jaheer Mukthar

**From Waste to Wellness: Circular Economy Strategies for the Indian Healthcare Industry** .............. 523
Jean Maria George, U. S. Syama, Geril Scaria George, M. M. Nirmala, and K. Vinoda Devi

**The Impact of Economically Weaker Section Reservation in India: A Study on the Syro Malabar Community in Kerala, India** .............. 533
Subin Thomas, Smita Satapathy, and Sheen Thankalayam

**Sustainability in the Context of Remote Work in IT Sector** .............. 547
Tisha Tomy and R. Sheeja

**Evaluating the Impact of Financial Inclusion on the Effectiveness of the Pradhan Mantri Fasal Bima Yojana** .............. 561
Ashwitha, K. A. Thanuja, K. Vinodhadevi, and Priyanka Ghosh

**Harnessing the Potential of Artificial Intelligence: Transforming Industries and Workforce Dynamics** .............. 573
Bijin Philip, Geethu Anna Mathew, Jinu Mathew, Roshen Therese Sebastian, Mary Jacob, and M. Jeeva

**Navigating Corporate Governance's Role in Health Insurance Performance: A Bibliometric Approach** .............. 583
N. Rakesh and Vishal Srivastava

**A Systematic Review of the Impact of Artificial Intelligence on Digital Art** .............. 593
Diego Pozú-López, Johan Peralta-Rodríguez, and Orlando Iparraguirre-Villanueva

**Social Entrepreneurship as Catalyst for Social Change: A Mediation Approach** .............. 605
Debolina Gupta, R. Mallieswari, Jayashree Kowtal, Niharika Mishra, and S. Jayashree

**Analysing the Status of Agricultural Finance in Rural India in Promoting Agriculture** ........ 615
Roopa Adarsh, E. Jothi, K. Sivasubramanian, and R. Madhusudhanan

**Challenges on Wheels: A Study of Food Delivery Executives and Influencing Factors in Bangalore** ........ 623
Bijin Philip and Soumya Simon

**The Devastating Impact of Greenhouse Gas Emissions: A Looming Threat to Sustainable Development in India** ........ 635
P. Remmiya Rajan and K. B. Rishikesh

**Factors Affecting the Performance of Women Entrepreneurs, Kothanallur Taluk of Thiruvarur District, Tamilnadu** ........ 647
M. Rani Therasa and K. V. R. Rajandran

# Business Ethics and Environment Sustainable Development

# Business Ethics in E-commerce: Ethical Concerns and Safeguarding Consumer Trust and Loyalty

**Hamad Adeeb Al-Mutawa, Ebrahim Khalil Sowailem, and Muneer Al Mubarak**

**Abstract** This Research focuses on the topic of business ethical concerns and safeguarding consumer trust and loyalty, with a specific emphasis on its implementation in the context of e-commerce. The study aims to explore how ethical considerations are integrated into consumer trust and loyalty. The paper emphasizes ethical implementation in e-commerce, focusing on key issues (privacy, security, environment, intellectual property). It examines the impact of ethical practices on consumer trust and loyalty. The findings suggest that to address ethical concerns in e-commerce, corporate leaders should implement efficient strategies, establish long-term management protocols, develop robust security policies, and implement training programs for sustainable success and customer satisfaction. This can provide a stable success and foster customer satisfaction and loyalty. It is important to note that different companies and fields within E-commerce may have unique security problems, and thus require tailor-made strategies. Loyal and trusted customers are crucial for long-term growth and profitability. Promoting a culture of ethical leadership requires a top-down approach that begins with senior leaders setting a strong example of ethical conduct. This article explores the significance of loyalty and trust as essential elements for success in e-commerce. It examines the ethical considerations and strategies that e-commerce businesses can employ to build strong relationships with their customers. The article highlights the benefits of identifying these relationships for researchers, practitioners, and individuals seeking a deeper understanding, as well as for e-commerce businesses to navigate effectively.

**Keywords** Business ethics · E-commerce · Privacy · Security · Fraud · Fair competition · Consumer protection

---

H. A. Al-Mutawa · E. K. Sowailem · M. Al Mubarak (✉)
Ahlia University, Manama, Bahrain
e-mail: malmubarak@ahlia.edu.bh

 
M. Al Mubarak (ed.), *Sustainable Digital Technology and Ethics in an Ever-Changing Environment*, Studies in Systems, Decision and Control 237,
https://doi.org/10.1007/978-3-031-86708-8_1

## 1 Introduction

E-commerce has revolutionized the way we shop, providing convenience and accessibility to consumers worldwide. However, with the rapid growth of online transactions, security issues have become a pressing concern [17]. This paper explores the ethical concerns surrounding e-commerce security and highlights the importance of safeguarding consumer trust in the digital marketplace. As online shopping continues to gain momentum, maintaining consumer trust and loyalty has become a paramount objective for businesses operating in the digital marketplace [9]. This paper delves into the ethical concerns that arise in e-commerce, ranging from privacy and security. It highlights the significance of safeguarding consumer trust and loyalty by implementing robust ethical frameworks and practices. This research aims to shed light on the key ethical considerations and strategies that e-commerce businesses can employ to navigate the evolving landscape of online commerce while fostering a strong bond with their customers.

The theoretical framework covers various aspects (ethical concerns in e-commerce, implementing ethical practices in e-commerce, safeguarding consumer trust and loyalty practices in e-commerce). Additionally, the fundamental concepts of Business ethics and E-commerce, such as its definition and main issues overview. The research problems addressed in identifying the ethical issues that organizations encounter in online commerce, understanding the challenges they face in terms of ethics, and examining how they affect consumer trust and loyalty. The main objective of the study is to explore how ethical considerations are integrated into consumer trust and loyalty. The paper emphasizes that ethical concerns in e-commerce are not only matters of legal compliance but also crucial for building long-term relationships with customers. It highlights the ethical challenges e-commerce businesses face and explores various aspects of ethical considerations in the online marketplace [15].

One key area of focus is privacy and security. With the increasing prevalence of data breaches and privacy concerns, customers are becoming more conscious about how their personal information is collected, stored, and used. The research delves into the ethical responsibilities of e-commerce businesses to protect customer data from unauthorized access and outlines best practices for data security, encryption, and transparency in data handling. In addition, Combating Deceptive Practices and Misrepresentation is an essential and ongoing endeavor aimed at safeguarding consumers' rights and promoting fair and transparent business practices. In today's complex and interconnected world, deceptive practices and misrepresentation can take various forms, ranging from false advertising and misleading product claims to fraudulent financial schemes and online scams. Although, Fair competition in e-commerce is a critical aspect of the digital marketplace, where businesses operate and compete online. It revolves around the principles of equal opportunity, transparency, and non-discriminatory practices among online sellers. In the realm of e-commerce, fair competition ensures that all participants, regardless of their size or resources, have a level playing field to showcase their products or services and engage with consumers. It involves preventing anti-competitive practices such as predatory pricing, monopolistic behavior, and deceptive advertising that could hinder fair market access and choice for consumers [8].

Moreover, fair pricing and responsible advertising are vital for fostering consumer trust. The article examines the importance of transparent pricing, avoiding deceptive practices, and ensuring that customers are not misled by false advertising or hidden charges. By adhering to ethical pricing and advertising principles, businesses can establish credibility and maintain long-term relationships with their customers. Also, Counterfeit products and the protection of intellectual property rights are significant challenges in the realm of e-commerce. With the widespread accessibility of online platforms, the sale and distribution of counterfeit goods have become more prevalent, posing serious threats to both businesses and consumers. Intellectual property rights, including trademarks, copyrights, and patents, play a crucial role in safeguarding the creative works, inventions, and brand identities of individuals and businesses [4].

The research also delves into the significance of inclusivity and accessibility in the e-commerce space. It explores the ethical obligations of businesses to ensure that their websites and digital interfaces are designed to accommodate individuals with disabilities, providing equal access to products and services. By prioritizing accessibility standards, such as providing alternative text for images and ensuring intuitive navigation, e-commerce businesses can foster a sense of inclusivity and demonstrate their commitment to ethical practices. More aspects to include are social engineering and phishing attacks are critical aspects of e-commerce that significantly impact the security and integrity of online transactions. In today's interconnected digital landscape, where individuals and businesses rely heavily on online platforms for various purposes, it is essential to understand the risks associated with these concepts. Social engineering involves manipulating individuals to gain unauthorized access to sensitive information, while phishing attacks exploit human vulnerabilities through deceptive tactics. Furthermore, ensuring vendors, supply chain and sustainability ethics is crucial in maintaining trust and transparency in e-commerce, as it directly influences customer confidence and the overall integrity of the online marketplace. Safeguarding against these risks and promoting ethical practices is paramount for the sustainable growth and success of e-commerce in the modern era. Because Sustainability emerges as another critical ethical consideration in e-commerce. It reflects the importance of minimizing environmental impact by adopting eco-friendly practices. This includes reducing packaging waste, optimizing shipping routes to minimize carbon emissions, and offering sustainable product options. By integrating sustainability into their operations, e-commerce businesses can align with growing consumer expectations and contribute to a more environmentally responsible digital marketplace [4].

In fact, it is crucial to establish effective avenues for consumer complaints and redress to build a strong trust relationship. With the increasing popularity of online shopping, consumers face unique challenges, such as product quality issues, delivery problems, fraudulent activities, and unsatisfactory customer service. To address these concerns, e-commerce platforms and regulatory bodies have recognized the importance of providing accessible and efficient channels for consumers to voice their complaints and seek redress. These avenues aim to protect consumer rights, promote fair business practices, and maintain trust in the online marketplace. By establishing robust mechanisms for complaint resolution, e-commerce platforms can enhance customer satisfaction, foster long-term relationships, and contribute to the growth and sustainability of the digital economy [6].

Overall, the research delves deeply into the multifaceted landscape of ethical concerns within the realm of e-commerce. It provides an extensive and thorough exploration of the various ethical dilemmas and challenges that businesses operating in the online marketplace prioritize ethical practices to safeguard consumer trust and loyalty. By highlighting the need for businesses to prioritize ethical practices, the research underscores the dynamic role that ethics play in maintaining and fostering consumer trust and loyalty. In today's digital age, where consumers have a substantial option in their reach, trust becomes a pivotal factor in their decision-making process. When businesses demonstrate a genuine commitment to ethical conduct, consumers are more likely to place their trust in those companies and remain loyal to their brands. E-commerce businesses can not only ensure legal compliance but also differentiate themselves in an increasingly competitive landscape, fostering long-term relationships with customers and building a reputation as a trusted and responsible online retailer. This article explores the significance of loyalty and trust as fundamental pillars of success in e-commerce is explored in this article. The article examines the crucial ethical considerations and strategies that e-commerce businesses can utilize to foster strong relationships with their customers. It highlights that the identification of these relationships can be beneficial not only to other researchers, practitioners, and individuals seeking a deeper understanding but also to e-commerce businesses, aiding them in effectively navigating the e-commerce landscape.

## *1.1 Scope of the Study*

E-commerce has revolutionized the way we shop, providing convenience and accessibility to consumers worldwide. However, with the rapid growth of online transactions, security issues have become a pressing concern. This paper builds particular emphasis on the ethical implementation of E-commerce, focusing on major ethical issues faced by companies in e-commerce include (privacy, security, environmental concerns, and intellectual property rights) and their consequences on consumer trust and loyalty and highlights the importance of safeguarding consumer trust in the digital marketplace.

## 2 Literature Review

### 2.1 *Business Ethics*

Business ethics refers to the principles and standards of conduct that guide a company's actions when making decisions and taking actions. These principles are based on moral concepts of right and wrong and are designed to ensure that businesses operate in a responsible and ethical manner. One aspect of business ethics is corporate governance, which involves the way in which a company is managed and controlled. This includes the responsibilities of the board of directors, the CEO, and other senior executives, as well as the company's disclosure practices and its compliance with laws and regulations. Social responsibility is another important aspect of business ethics, referring to a company's obligations to its stakeholders, including its employees, customers, suppliers, and the community. This includes issues such as environmental sustainability, labor practices, and philanthropy. Sustainability is also a key consideration in business ethics, focusing on a company's ability to meet the needs of the present without compromising the ability of future generations to meet their own needs. This encompasses issues such as environmental protection, social responsibility, and economic sustainability. Fairness and transparency are integral to business ethics as well, emphasizing a company's commitment to treating all stakeholders fairly and being open and transparent about its activities. This includes issues such as non-discrimination, anti-corruption measures, and the disclosure of information to investors and the public. The importance of business ethics lies in the various benefits it brings to businesses. By operating in an ethical manner, businesses can build and maintain trust with their stakeholders, including employees, customers, suppliers, and the community. This trust can lead to increased sales, improved employee morale, and reduced risk of litigation. Business ethics also help companies make better decisions that are in the best interests of the company and its stakeholders. For instance, a company committed to environmental sustainability is more likely to make decisions that protect the environment and reduce its carbon footprint. Compliance with laws and regulations is another advantage of business ethics. Ethical conduct helps businesses adhere to legal requirements, avoiding costly fines, penalties, and reputational damage. Furthermore, businesses seen as ethical are more likely to attract and retain top talent, as employees want to work for companies, they believe are doing the right thing. In summary, business ethics plays a vital role in guiding responsible and ethical conduct within companies. It encompasses corporate governance, social responsibility, sustainability, fairness, and transparency. By adhering to ethical principles, businesses can build trust, make better decisions, comply with laws, attract top talent, and maintain a positive reputation, ultimately contributing to their long-term success and the well-being of stakeholders [10].

To foster a culture of ethical conduct within their organizations, businesses can implement a comprehensive approach that encompasses establishing a clear code of ethics, providing comprehensive ethics training, implementing a robust reporting system, and holding employees accountable for their actions. A well-crafted code of ethics serves as a guiding document that outlines the company's values and expectations for employee conduct, providing a framework for ethical decision-making. Ethics training programs help employees understand the nuances of ethical dilemmas and equip them with the skills to navigate these situations with integrity. Establishing a confidential reporting system empowers employees to raise concerns about suspected unethical behavior without fear of reprisal, fostering a culture of transparency and accountability. Finally, holding employees accountable for their actions through appropriate disciplinary measures reinforces the importance of ethical behavior and sends a clear message that ethical conduct is a fundamental expectation of employment. By implementing these strategies, businesses can cultivate a workplace environment that promotes ethical behavior, protects the company's reputation, and enhances stakeholder trust. By promoting ethical behavior, businesses can create a more positive and productive work environment, and they can build a stronger reputation with their stakeholders Purwaningsih et al. [20].

## 2.2 E-commerce

E-commerce, or electronic commerce, is the buying and selling of goods and services over the internet. It encompasses a wide range of activities, from browsing and shopping for products online to making payments and tracking orders. E-commerce has become increasingly popular in recent years, as it offers a convenient and affordable way for consumers to shop from the comfort of their own homes [16]. Taher [26] provides a comprehensive categorization of businesses operating within this domain. Retailers, exemplified by industry giants like Amazon, Walmart, and Target, engage in direct selling of products to consumers. Marketplaces, such as eBay and Etsy, serve as platforms where various businesses can showcase and sell their goods. Service providers, on the other hand, offer a range of online services like web hosting, search engine optimization (SEO), and social media marketing. The successful operation of e-commerce businesses hinges on the utilization of various technologies Websites serves as the primary interface for consumers, necessitating user-friendly design, easy navigation, and secure payment processing capabilities. Payment gateways play a crucial role in securely processing online transactions, providing consumers with a safe and protected environment to input their credit card or other payment details. The shipping and logistics sector comes into play by facilitating the transportation of goods from e-commerce businesses to end consumers. Reliable and efficient shipping services, alongside a diverse range of shipping options, are essential to meet consumer expectations. E-commerce boasts a multitude of advantages over traditional brick-and-mortar retail. Perhaps the most prominent benefit is convenience, as consumers can effortlessly shop from the comfort of their own homes, at any time, even outside regular business hours.

Moreover, e-commerce businesses typically offer an extensive selection of products and services, surpassing the limitations of physical store space. Price competitiveness is another advantage, with e-commerce entities often able to offer lower prices due to reduced overhead costs compared to traditional retail establishments. nevertheless, e-commerce also presents certain disadvantages. One drawback is the lack of personal interaction, as consumers cannot physically interact with products before making a purchase decision. The process of returning products purchased online can be more intricate and time-consuming compared to the relative ease of returning items to a traditional brick-and-mortar store. Additionally, the online shopping landscape carries inherent security risks, including the potential for fraud and identity theft. Despite these limitations, e-commerce continues to flourish as an industry and is predicted to sustain its growth trajectory in the years to come. Ongoing technological advancements and the increasing comfort of consumers with online shopping are expected to solidify e-commerce's position as the dominant form of retail in the foreseeable future [26].

## *2.3 Ethical Concerns in E-commerce*

Ethical concerns in e-commerce encompass a range of issues related to privacy, data protection, security, trust, consumer protection, intellectual property rights, fair competition, environmental impact. These concerns arise from practices such as inadequate data protection, security breaches, false advertising, intellectual property infringement, unfair pricing, unsustainable practices, poor labor conditions, and unethical supply chain management. Addressing these concerns requires the implementation of robust privacy and security measures, transparent and fair business practices, responsible supply chain management, and adherence to relevant laws and regulations.

### 2.3.1 Privacy: Protecting Personal Information in the Digital Age

One of the primary ethical concerns in e-commerce is the vulnerability of consumer data to breaches. Data breaches can expose sensitive personal and financial information, leading to identity theft, fraud, and other malicious activities. Major e-commerce platforms have fallen victim to such attacks, compromising the trust consumers place in online transactions. To address privacy concerns, e-commerce businesses must prioritize implementing robust security measures. This includes encryption protocols, secure payment gateways, and regular security audits to minimize the risk of data breaches and protect customer information [27].

In the realm of e-commerce, privacy stands as a paramount ethical concern. Online businesses collect a vast trove of personal data about their customers, including names, addresses, email addresses, browsing history, and purchase patterns [11]. This information, while valuable for targeted marketing and customer profiling, also carries inherent risks if not adequately safeguarded. Data breaches, unauthorized access, and the potential for misuse of personal information pose significant threats to consumer privacy [5]. To address these privacy concerns, e-commerce businesses must prioritize transparency and user control over data collection and usage practices [19]. Clear and concise privacy policies should inform consumers about the types of data collected, the purposes for which it is used, and the measures taken to protect it [11]. Additionally, businesses should provide options for consumers to opt out of data collection or restrict its use and establish robust data security measures to prevent unauthorized access and breaches [22].

### 2.3.2 Safeguarding Transactions and Preventing Cybercrime (Payment Fraud)

Payment fraud is a significant issue affecting e-commerce, posing ethical concerns for both businesses and consumers. Cybercriminals employ various techniques such as phishing, card skimming, and identity theft to exploit vulnerabilities in online payment systems [22]. This not only results in financial losses but also erodes consumer trust in e-commerce platforms. To combat payment fraud, e-commerce businesses should adopt multi-factor authentication, tokenization, and real-time transaction monitoring. Additionally, educating consumers about safe online shopping practices, including the importance of strong passwords and secure payment methods, can help minimize the risk of fraud [1].

The security of online transactions is another critical ethical issue in e-commerce. E-commerce platforms must implement robust security measures to protect sensitive financial data, such as credit card numbers and bank account information, from cybercriminals. Phishing scams, malware attacks, and vulnerabilities in payment gateways can lead to financial losses and identity theft, undermining consumer trust in e-commerce [14]. E-commerce businesses should adopt industry-standard security protocols, such as encryption and secure payment gateways, to protect financial data during transmission and storage [19]. Regular security audits and vulnerability testing are essential to identify and address potential weaknesses in systems. Additionally, educating consumers about online security threats and promoting safe practices, such as creating strong passwords and avoiding suspicious websites, can contribute to a more secure e-commerce environment [22].

### 2.3.3 Combating Deceptive Practices and Misrepresentation

Fraud is a pervasive problem in e-commerce, encompassing various forms of deception and misrepresentation [14]. Fake websites, counterfeit products, misleading product descriptions, and hidden fees can all lead to financial losses and consumer dissatisfaction. Businesses must prioritize transparency, authenticity, and fair pricing practices to maintain consumer trust and minimize the occurrence of fraudulent activities. E-commerce businesses should adopt clear and transparent pricing practices, including disclosing all fees and charges upfront, and providing accurate and detailed product descriptions [19]. Authenticity can be enhanced through partnerships with trusted suppliers, product verification processes, and clear labeling of products.

Additionally, businesses should establish effective fraud detection and prevention mechanisms and collaborate with law enforcement agencies to prosecute fraudulent activities.

Overall, Governments, regulatory bodies, and consumer protection agencies should work diligently to develop and enforce laws and regulations that hold businesses accountable for their actions, ensuring that consumers are well-informed, protected, and empowered to make informed decisions. Through proactive monitoring, enforcement actions, and public awareness campaigns, the fight against deceptive practices and misrepresentation strives to foster trust, integrity, and ethical conduct in the marketplace, benefiting both consumers and businesses alike [2].

### 2.3.4 Ensuring a Level Playing Field (Fair Competition)

E-commerce has revolutionized the competitive landscape, introducing new players and challenging traditional business models. However, this competitive environment also raises ethical concerns about fairness and equity [11]. Anti-competitive practices, such as price discrimination, product bundling, and exclusionary deals, can disadvantage smaller businesses and limit consumer choice. To promote fair competition in e-commerce, regulatory bodies should establish clear guidelines and enforce anti-trust laws to prevent anti-competitive practices [5]. Additionally, fostering a supportive ecosystem for small and medium-sized enterprises (SMEs) can help them compete effectively in the e-commerce space. This includes providing access to training, resources, and mentorship programs to help SMEs develop their e-commerce capabilities. By promoting fair competition in e-commerce, societies can foster innovation, consumer trust, and a vibrant online marketplace that benefits both businesses and consumers alike [2].

### 2.3.5 Transparency in Pricing and Advertising

Transparency in pricing and advertising is an important ethical consideration in e-commerce. Misleading pricing practices, such as hidden fees, inflated discounts, or false advertising, can deceive consumers and erode their trust in online retailers. E-commerce businesses should adhere to fair pricing practices and provide accurate product information to ensure transparency. Regulatory bodies can play a crucial role in enforcing ethical standards in e-commerce, ensuring that businesses adhere to advertising guidelines and accurately represent their products or services [2]. Consumer advocacy organizations and online review platforms can also hold businesses accountable and contribute to promoting transparency. Consumer protection is a crucial ethical imperative in e-commerce. Consumers should have access to accurate and transparent information about products, services, and their rights as buyers [19]. Businesses must adhere to fair advertising practices, provide clear return policies, and establish effective grievance redressal mechanisms to protect consumer interests. E-commerce businesses should provide clear and accurate product descriptions, including specifications, warranty information, and pricing Hasan & Rahman, [11]. Fair advertising practices should avoid misleading claims, hidden conditions, and deceptive.

### 2.3.6 Counterfeit Products and Intellectual Property Rights

E-commerce platforms often face ethical dilemmas regarding counterfeit products and intellectual property rights infringement [17]. Unscrupulous sellers take advantage of the anonymity provided by online marketplaces to sell counterfeit or pirated goods. This not only deceives consumers but also harms legitimate businesses and undermines intellectual property rights. To address this issue, e-commerce platforms must implement stringent verification processes to authenticate sellers and ensure the legitimacy of products. Collaborating with intellectual property rights holders and utilizing automated monitoring systems can help identify and remove infringing listings promptly. Ensuring the protection of these rights in e-commerce is essential to foster innovation, maintain consumer trust, and support legitimate businesses. By addressing the issue of counterfeit products and enforcing intellectual property rights in the digital marketplace, societies can create an environment that promotes fair competition, encourages creativity, and upholds the integrity of the online economy [7].

### 2.3.7 Ethical Use of Consumer Data

The collection and use of consumer data in e-commerce raises ethical concerns related to privacy and transparency. Businesses must be transparent about the data they collect, how it is used, and provide consumers with control over their personal information. Failure to do so can result in breaches of trust and potential misuse of sensitive data [21]. To prioritize ethical data practices, e-commerce businesses should adopt clear privacy policies, obtain explicit consent for data collection, and allow consumers to opt-out of data sharing. Implementing secure data storage and anonymization techniques can also minimize the risk of data exposure and protect consumer privacy [17].

### 2.3.8 Accessibility and Digital Divide

Another ethical concern in e-commerce is the digital divide, where certain groups may face barriers to accessing online shopping opportunities. This includes individuals with limited internet access, disabilities, or language barriers. E-commerce platforms must ensure inclusivity and accessibility for all consumers, regardless of their background or circumstances. To address this concern, businesses should focus on providing user-friendly interfaces, multilingual support, and assistive technologies to accommodate diverse user needs. Collaborating with organizations that advocate for digital inclusion can help bridge the accessibility gap and ensure equal opportunities for all consumers Sánchez, [25].

### 2.3.9 Social Engineering and Phishing Attacks

Social engineering techniques, such as phishing attacks, are a significant security concern in e-commerce. Cybercriminals use deceptive tactics to trick users into revealing sensitive information like passwords or credit card details [22]. This unethical practice not only compromises individual users but also undermines the trust consumers have in online transactions. To combat social engineering attacks, e-commerce businesses should educate their customers about recognizing and avoiding phishing attempts. Implementing email authentication protocols, such as DMARC (Domain-based Message Authentication, Reporting, and Conformance), can help prevent fraudulent emails from reaching consumers' inboxes [3].

### 2.3.10 Environmental Impact and Sustainability

While not directly related to security, the environmental impact of e-commerce raises ethical concerns. The growth of online shopping has led to increased packaging waste, carbon emissions from transportation, and energy consumption in data centers. E-commerce businesses should consider implementing sustainable practices, such as optimizing packaging materials, utilizing eco-friendly transportation options, and adopting renewable energy sources. Consumers are increasingly conscious of environmental issues, and businesses that prioritize sustainability can build trust and loyalty among eco-conscious shoppers. Certifications such as carbon–neutral shipping or partnerships with environmental organizations can demonstrate a commitment to ethical and sustainable practices [18]. E-commerce has a significant environmental impact. The packaging, shipping, and returns of e-commerce products all contribute to greenhouse gas emissions and other environmental problems [19]. Businesses need to take steps to reduce their environmental impact.

### 2.3.11 Vendor and Supply Chain Ethics

Sustainable Supply Chain Management (SSCM) offers a comprehensive framework for addressing sustainability in e-commerce supply chains. It encompasses various aspects such as the environmental impact of e-commerce activities, including energy usage, packaging waste, and transportation emissions. Additionally, it recognizes the social dimension of sustainability, emphasizing fair working conditions, ethical sourcing practices, and responsible consumption. By embracing the principles of SSCM, organizations can align their operations with sustainable practices and contribute to the overall sustainability of e-commerce supply chains. In the e-commerce realm, platforms often serve as intermediaries that connect consumers with vendors. Upholding ethical standards throughout the e-commerce ecosystem necessitates a strong emphasis on vendor and supply chain ethics. It is crucial to address ethical concerns such as exploitative labor practices, inadequate working conditions, and the sourcing of products from unethical suppliers. E-commerce enterprises must exercise careful due diligence when selecting and onboarding vendors, thoroughly evaluating their adherence to ethical standards. The implementation of vendor code of conduct policies, along with regular audits, can effectively enforce ethical practices. Collaboration with industry organizations and non-governmental organizations (NGOs) specializing in supply chain ethics can provide valuable guidance and support, enabling businesses to actively promote responsible sourcing and contribute to the advancement of the industry [24].

#### 2.3.12 Avenues for Consumer Complaints and Redress

To address ethical concerns in e-commerce, providing avenues for consumer complaints and redress is essential. E-commerce businesses should have clear and accessible channels for consumers to report issues, seek assistance, or request refunds. Prompt and satisfactory resolution of consumer complaints demonstrates commitment to customer satisfaction and helps build trust. Implementing customer support systems, such as live chat, email, or dedicated helplines, can facilitate effective communication and problem resolution. E-commerce platforms should also have transparent return and refund policies that protect consumers' rights and ensure a fair and ethical shopping experience. Loyalty and trust are two fundamental pillars of success in e-commerce. Building and maintaining these qualities with customers is crucial for long-term growth and profitability. In the absence of physical interactions and face-to-face encounters, e-commerce businesses must rely on digital means to establish trust and foster customer loyalty [21].

### *2.4 Implementing Ethical Practices in E-commerce*

From the previous chapter we conclude that we must robust privacy and security measures, transparent and fair business practices, responsible supply chain management, and adherence to relevant laws and regulations to implementing ethical practices in e-commerce. Moreover, implementing ethical practices in e-commerce involves a comprehensive approach that encompasses establishing a clear code of ethics, providing ongoing ethics training, establishing grievance procedures, conducting regular audits and reviews, and promoting a culture of ethical leadership. These strategies work together to cultivate a workplace environment that values integrity, transparency, and responsible decision-making. A well-defined code of ethics serves as a guiding framework for ethical behavior, outlining the company's values and expectations for employees at all levels. This code should be easily accessible to all employees, regularly reviewed and updated, and effectively communicated throughout the organization. Regular ethics training is crucial for reinforcing the principles outlined in the code of ethics and providing employees with practical guidance on applying these principles in their daily work. Training should be tailored to the specific roles and responsibilities of employees, using real-world scenarios and case studies to illustrate ethical dilemmas and encourage thoughtful decision-making [10].

Establishing clear and accessible grievance procedures empowers employees and customers to raise concerns about unethical behavior without fear of retaliation or reprisal. These procedures should be transparent, ensuring that complaints are handled promptly, impartially, and confidentially. Addressing grievances effectively demonstrates the company's commitment to ethical conduct and helps to prevent potential problems from escalating. Regular audits and reviews are essential for assessing the effectiveness of the company's ethical practices and identifying areas for improvement. These reviews should involve a comprehensive evaluation of policies, procedures, training programs, and complaint handling mechanisms. The findings of these reviews should be used to inform ongoing efforts to strengthen the company's ethical culture [12].

Promoting a culture of ethical leadership requires a top-down approach that begins with senior leaders setting a strong example of ethical conduct. Leaders should consistently demonstrate their commitment to ethical values in their decision-making, interactions with employees, and external communications. This sets the tone for the entire organization and encourages employees to follow suit. By implementing these strategies holistically, e-commerce businesses can foster a workplace culture that prioritizes ethical behavior, builds trust with customers, and contributes to long-term success [9].

## *2.5 The Future of E-commerce*

The e-commerce industry is undergoing rapid transformations, and as it continues to evolve, it is expected that ethical dilemmas within this sector will become increasingly complex. It is vital for businesses to proactively address these challenges and establish ethical practices that will contribute to a sustainable and equitable future for e-commerce. According to [22], several trends are set to shape the future of e-commerce. Firstly, the rise of mobile commerce driven by the widespread usage of smartphones and tablets necessitates businesses to optimize their online stores for mobile devices, ensuring a seamless shopping experience for customers on the move. Secondly, the integration of artificial intelligence (AI) is revolutionizing e-commerce by enhancing customer experiences, providing personalized product recommendations, and combating fraud. By leveraging AI, businesses can offer tailored product suggestions, automate customer service interactions, and identify fraudulent transactions. Thirdly, the expansion of social commerce highlights the increasing significance of social media platforms as channels for product sales and marketing. Businesses can utilize social media to showcase their products, engage with potential customers, and drive traffic to their online stores.

Additionally, the adoption of augmented reality (AR) is transforming the shopping experience, enabling customers to virtually try on products before making a purchase. Implementing AR technology allows businesses to enhance product visualization, create more immersive shopping experiences, and reduce return rates. Lastly, the emergence of new business models, including subscription services, direct-to-consumer (DTC) brands, and social media marketplaces, underscores the ever-evolving nature of e-commerce. To thrive in this dynamic and exciting industry, businesses must stay updated on these trends and adapt their strategies accordingly while upholding ethical practices. By embracing the latest trends, responding to shifting consumer behaviors, and adhering to ethical principles, businesses can position themselves for success in the future.

## 2.6 *Safeguarding Consumer Trust and Loyalty*

### 2.6.1 Trust in E-commerce

Trust is the foundation of any successful relationship, and it holds even greater significance in the digital realm. E-commerce customers must have unwavering confidence in the security of their personal information, financial details, and transactions. They must also trust that the products or services they purchase will meet their expectations and that the company will promptly and professionally address any issues or concerns. Nguyen [17] identified several factors that influence trust in e-commerce, including security, transparent practices, reputation, and reliability. Customers must be assured that their personal and financial data is protected from unauthorized access or theft. This necessitates robust security measures like encryption, secure payment gateways, and clear privacy policies. Transparency is crucial in building trust. E-commerce businesses should be open about their policies, procedures, and practices. This includes providing clear information about product descriptions, pricing, shipping, returns, and customer service [23].

A strong reputation and positive customer reviews play a significant role in fostering trust. Encourage customers to leave feedback and address negative reviews promptly and professionally. Delivering on promises and maintaining consistent performance reinforce trust. This entails providing accurate product descriptions, ensuring timely deliveries, and efficiently handling customer inquiries. To bolster trust in e-commerce, prominently display security seals and certifications from reputable organizations, providing reassurance about data protection. Clearly articulate privacy policies and data handling practices in an easily accessible format. Partner with trustworthy payment processors to guarantee secure transactions and safeguard sensitive financial information. Encourage customers to leave reviews and testimonials on the website and social media platforms. Respond to negative reviews professionally and take steps to resolve any issues or concerns [13].

### 2.6.2 Loyalty in E-commerce

Customer loyalty is the cornerstone of e-commerce success, representing the willingness of customers to repeatedly purchase from the same business. This valuable asset drives long-term revenue and profitability, making it essential for e-commerce businesses to cultivate and nurture customer loyalty. Several factors contribute to customer loyalty in e-commerce, including positive customer experiences, reward programs, personalized interactions, and community building. A positive customer experience encompasses the entire purchase journey, from browsing and ordering to payment and customer service interactions. It involves providing user-friendly websites, efficient ordering processes, seamless payment options, and prompt and helpful customer service. Reward programs play a significant role in incentivizing repeat purchases and fostering loyalty. Points-based or tier-based loyalty programs offer exclusive benefits to loyal customers, rewarding their continued patronage. These benefits can include discounts, early access to new products, personalized recommendations, and special promotions. Personalized interactions further enhance customer loyalty. By utilizing customer data, e-commerce businesses can tailor marketing messages, product recommendations, and special offers to individual customers' preferences and purchase history. This personalization creates a sense of recognition and value, strengthening the customer-brand relationship. Community building fosters a sense of belonging and shared interest among customers, deepening their connection to the brand. E-commerce businesses can engage with customers through social media, forums, and loyalty programs, creating a vibrant community that extends beyond transactional interactions [23].

To effectively build customer loyalty, e-commerce businesses should implement a comprehensive strategy that encompasses these factors. Implementing a loyalty program that rewards repeat purchases and provides exclusive benefits is a crucial step. Additionally, segmenting customer data and tailoring marketing communications based on preferences and purchase history enhances personalization. Providing excellent customer service by responding promptly to inquiries, resolving issues effectively, and exceeding expectations further strengthens loyalty. Actively seeking customer feedback and encouraging positive reviews reinforces trust and attracts new customers. By cultivating trust and fostering customer loyalty, e-commerce businesses establish a solid foundation for long-term success [9]. Remember, satisfied and loyal customers are the lifeblood of any thriving e-commerce venture.

# 3 Conclusion

By implementing these strategies holistically, e-commerce businesses can foster a workplace culture that prioritizes ethical behavior, builds trust with customers, and contributes to long-term success. Security issues in e-commerce and the associated ethical concerns require ongoing attention and proactive measures from businesses, regulators, and consumers. By addressing data breaches, payment fraud, counterfeit products, and privacy concerns, e-commerce platforms can foster a safer and more trustworthy digital marketplace. Upholding ethical standards, promoting transparency, and prioritizing sustainability will contribute to the long-term success and sustainability of the e-commerce industry while ensuring consumer confidence and satisfaction. While e-commerce offers numerous benefits, addressing security issues and ethical concerns is crucial to maintaining consumer trust. By prioritizing robust security measures, protecting consumer privacy, combating counterfeit products, and promoting inclusivity, e-commerce businesses can create a safer and more ethical digital marketplace. Ultimately, safeguarding consumer trust will foster long-term success and sustainable growth in the e-commerce industry.

Implementing ethical practices in e-commerce involves a comprehensive approach that encompasses establishing a clear code of ethics, providing ongoing ethics training, establishing grievance procedures, conducting regular audits and reviews, and promoting a culture of ethical leadership. These strategies work together to cultivate a workplace environment that values integrity, transparency, and responsible decision-making. Loyalty and trust are two fundamental pillars of success in e-commerce. Building and maintaining these qualities with customers is crucial for long-term growth and profitability. In the absence of physical interactions and face-to-face encounters, e-commerce businesses must rely on digital means to establish trust and foster customer loyalty. By cultivating trust and fostering customer loyalty, e-commerce businesses establish a solid foundation for long-term success. Remember, satisfied and loyal customers are the lifeblood of any thriving e-commerce venture.

## *3.1 Theory Implications*

Loyalty and trust are two fundamental pillars of success in e-commerce. Building and maintaining these qualities with customers is crucial for long-term growth and profitability. In the absence of physical interactions and face-to-face encounters, e-commerce businesses must rely on digital means to establish trust and foster customer loyalty. With a focus on fostering a strong relationship with customers, this research explores the critical ethical considerations and strategies that e-commerce businesses can leverage. The goal is to help these businesses navigate the constantly changing world of online commerce effectively.

### 3.2 Practice Implications

Promoting a culture of ethical leadership requires a top-down approach that begins with senior leaders setting a strong example of ethical conduct. Leaders should consistently demonstrate their commitment to ethical values in their decision-making, interactions with employees, and external communications. This sets the tone for the entire organization and encourages employees to follow suit. By cultivating trust and fostering customer loyalty, e-commerce businesses establish a solid foundation for long-term success. Remember, satisfied and loyal customers are the lifeblood of any thriving e-commerce venture.

### 3.3 Study Limitations and Future Research

It is important to note that different companies and fields within E-commerce may have unique security problems, and thus require tailor-made strategies. However, these strategies must always be legal, feasible, effective, and innovative. E-commerce is a dynamic and exciting industry with immense potential for growth. Businesses that can embrace the latest trends, adapt to changing consumer behavior, and adhere to ethical practices will be well-positioned for success in the future. This literature review provides a concise summary of the key points about it can be expanded further by incorporating additional research studies, empirical evidence, and real-world case examples. The study primarily relies on existing published data and information as the primary source, without directly collecting primary data from original sources.

## References


1. Abdallah, A., Maarof, M.A., Zainal, A.: Fraud detection system: a survey. J. Netw. Comput. Appl. **68**, 90–113 (2016)
2. Adam, R.: Predatory pricing for e-commerce businesses from a business competition law perspective. J. Law Sustain. Dev. **11**(8), e1438–e1438 (2023)
3. Adu-Manu, K.S., Ahiable, R.K., Appati, J.K., Mensah, E.E.: Phishing attacks in social engineering: a review. J. Cyber Secur. **4**(4), 239–267 (2022)
4. Arief, M., Mustikowati, R. I., Chrismardani, Y.: Why customers buy an online product? The effects of advertising attractiveness, influencer marketing and online customer reviews. LBS J. Manag. Res., (2023)
5. Badu, A.A., Abdul-Rahim, A.A., Bakar, S.H.A.: The ethics of e-commerce: a literature review. Int. J. Bus. Ethics **26**(1), 1–22 (2019)
6. Brogan, P.: Consumer redress through online dispute resolution: the role of online dispute resolution in facilitating consumer access to justice in e-commerce. Arbitr. Law Rev. **12**(1), 12 (2020)
7. Cai, L.: A cross-border e-commerce intellectual property rights. J. Financ. Res., **4**(1), (2020)
8. Donmaz, A.: Privacy policy and security issues in e-commerce for eliminating the ethical concerns. Regul. Appl. Ethics Bus. Pract. 153–164, (2018)

9. Eid, M.I.: Determinants of e-commerce customer satisfaction, trust, and loyalty in Saudi Arabia. J. Electron. Commer. Res. **12**(1), 78 (2011)
10. Ghillyer, A.: Business ethics now. McGraw-Hill (2018)
11. Hassan, S.A., Rahman, M.A.: The impact of e-commerce on consumer behavior: an ethical perspective. J. Bus. Ethics **39**(2), 217–232 (2016)
12. Jones, M.J.: Creative accounting, fraud and international accounting scandals. Wiley & Sons (2010)
13. Lin, X., Wang, X., Hajli, N.: Building e-commerce satisfaction and boosting sales: the role of social commerce trust and its antecedents. Int. J. Electron. Commer. **23**(3), 328–363 (2019)
14. Mabokangwe, L.V., Egbe, F.O.: The ethical implications of e-commerce in developing countries. J. Bus. Ethics **127**(2), 395–406 (2015)
15. Mainardes, E.W., Coutinho, A.R.S., Alves, H.M.B.: The influence of the ethics of E-retailers on online customer experience and customer satisfaction. J. Retail. Consum. Serv. **70**, 103171 (2023)
16. Nanehkaran, Y.A.: An introduction to electronic commerce. Int. J. Sci. Technol. Res. **2**(4), 190–193 (2013)
17. Nguyen K.: Business ethics in E-commerce. (Parboteeah & Cullen 2013), (2016)
18. Oláh, J., Popp, J., Khan, M.A., Kitukutha, N.: Sustainable e-commerce and environmental impact on sustainability. Econ. Sociol. **16**(1), 85–105 (2023)
19. Peterson, R.A.: Ethical considerations for e-commerce businesses. J. Bus. Ethics, **44**(1), (2017)
20. Purwaningsih, P.M., Siwiyanti, L., Nursiva, H.: The Importance of Applying Business Ethics in E-Commerce. Nusant. Sci. Technol. Proc., 38–46, (2022)
21. Saikrishnan, S.: Consumer perception and awareness on e-commerce. Int. J. Sci. Dev. Res. **8**(5), 1867–1870 (2023)
22. Santanam, R., Sethumadhavan, M., Virendra, M. (Eds.). (2010). Cyber Security, cyber crime and cyber forensics: applications and perspectives: applications and perspectives. IGI Global
23. Sharma, G., Lijuan, W.: Ethical perspectives on e-commerce: an empirical investigation. Internet Res. **24**(4), 414–435 (2014)
24. Sindakis, S., Showkat, S., Su, J.: Unveiling the influence: exploring the impact of interrelationships among e-commerce supply chain members on supply chain sustainability. Sustainability **15**(24), 16642 (2023)
25. Sánchez-Torres, J.A.: Moderating effect of the digital divide of e-commerce. Int. J. Soc. Econ. **46**(12), 1387–1400 (2019)
26. Taher, G.: E-commerce: advantages and limitations. Int. J. Acad. Res. Account. Financ. Manag. Sci. **11**(1), 153–165 (2021)
27. Zhong, G.: E-Commerce consumer privacy protection based on differential privacy. J. Phys.: Conf. Ser., **1168**(3), 0–8, (2019)

# The Role of Digital Technologies in the Economic, Social and Environment Sustainable Development

**Sara Hameed and Muneer Al Mubarak**

**Abstract** This paper explains on the relationship between two subject topics; digital technologies and sustainable development considering growing interest of organizations in how digital technologies contribute to sustainable development. In this new era of technological transformation, digital technologies are the new way of steering businesses, which significantly improved organizations operational effectiveness. Since the way businesses get done has an economic, social, and environmental footprint, digital technologies affect the sustainability aspects of doing businesses, which is crucial for the organizations since consumers, and stakeholders have visibility to businesses impacts on the organization's revenues and valuation. This paper illustrates a study of "Digital Sustainability" over a thorough analysis of various journals with the goals of; (1) bringing together the existing research findings, (2) manifesting how the various studies are related to each other, and (3) identifying breaks in the conducted research to advance the improvement of this research. There are important implications, theoretical and practical, i.e., the call for support in comprehending the vital role of utilizing digital technologies in multiple sustainable development objectives is being addressed herein, consultants, and policymakers are just a few of the practitioners who shall be impacted by this research paper. This paper reveals and concludes that digital technologies can be well thought-out as a predecessor to sustainable development, and both should be a key component of any organizational strategy. Markets are changing, new business paradigms are emerging, and organizations now have the tools necessary to address sustainability-related issues.

**Keywords** Digital technologies · Sustainable development · Economic · Social · Environment · Sustainable development goals (SDGs) · Open innovation

S. Hameed
The George Washington University, Washington D.C., USA

M. Al Mubarak (✉)
Ahlia University, Manama, Bahrain
e-mail: malmubarak@ahlia.edu.bh



M. Al Mubarak (ed.), *Sustainable Digital Technology and Ethics in an Ever-Changing Environment*, Studies in Systems, Decision and Control 237,
https://doi.org/10.1007/978-3-031-86708-8_2

# 1 Introduction

The use of digital technologies can assist with the achievement of many goals. Nowadays, the main topic of discussion among leading organizations is how to take advantage of and accelerate the digital technologies and sustainable development processes. The scientific community is looking for conceptualizations of digital technologies and their connections to and relationships with other concepts. While the practical applications and advantages of digital technologies at businesses, are of interest to experts. The provision of new services, products, and processes is made easier by digital technologies, which also enhance existing ones. Wu et al. [39] to encounter market's rapidly fluctuating needs, in order to ensure that innovation is a driver to the sustainable development. As digital technologies spread quickly, businesses are being compelled to find new methods of operating. Digital technologies are considered disruptive processes that organizations adopt as the business environment changes [37]. Digital technologies foster innovation because it calls for the development of new business models, new forms of industry and organizational collaboration, the acquisition of new knowledge and skills, and the sustainable use of organizational resources. Additional transformations are also sparked by digital technologies, including changes to the governance, HR, and IT departments of organizations.

Digital technologies are fundamentally changing how businesses operate, develop, produce, market, and sell their goods and services by their very nature. As a result, both ideas become related and interconnected, which is encouraged by innovation. Furthermore, innovation is a collection of modest but important perspectives for ongoing development that help businesses become more successful and long-lasting. This study links the ongoing debate about sustainable innovation, and in particular, how to balance sustainability issues with innovation methods [7], Orlando et al. [26], [32], developing it further while giving Digital technologies important consideration. A recent analysis of how digital technologies are being used to accomplish the Sustainable Development Goals (SDG) of the UN. In particular, taking into account increased access to new data sources, and collaboration among digital ecosystems, their paper demonstrates growing expectations for the role that digital technologies play in achieving the SDGs. However, a variety of scholarly publications and journals are available today. (e.g. [35] and [22]) have specifically targeted the SDGs and largely based their research on academic sources. The development of this research paper was greatly influenced by the need for understanding and evaluating scholars' contributions to the field. In an effort to comprehend the comprehensive use of digital technologies for Sustainability goals, this work analyzes the existing academic literature on the relationship between digital technologies and sustainability. The following research question serves as the basis for the analysis:

*How Can Digital Technologies Contribute to Sustainable Development?*

## 2 Methodology

Digital technologies subject research has been approached deductively. Then, inductively few sub themes have been derived from several journals and articles related to digital technologies and sustainable development. Since systematic literature reviews serve as the basis for this research paper, like other business management researchers, hence, guidelines and procedure for systematic literature reviews have followed. Accordingly, this research paper follows a systematic review which consists of 3-steps. First, data collection which included defining the research paper scope of study. Second, data analysis which involved extensive and comprehensive analysis of the collected data based on the earlier defined scope in the first step. Finally, synthesis to which findings, conclusions and results have been demonstrated according to the analyzed data.

## 3 Research Design

In this research paper the approach that has been used is qualitative characterization based on content analysis which follows five structured steps, as follows:

(1) Create the main topic category which serves the purpose of this research paper.
(2) Develop sub-categories of the main topic category which demonstrate multiple aspects of the main topic category and serve the primary objective of this research paper.
(3) Interpret and discuss the results.

The above-mentioned process supported the approach of highlighting and linking the sub-categories with the main category of the research paper, which all ultimately interpret and discuss the primary objective of this paper, that led to detect four main themes which are the following:

(1) The correlation between digital technologies and sustainable development.
(2) Digitalization strategies for sustainability purposes.
(3) The catalyst for sustainability and open innovation is the digital technologies.
(4) Digital technologies as a contributor in the economic, environmental and social sustainability.

# 4 Literature Review

## *4.1 The Correlation Between Digital Technologies and Sustainable Development*

Digital technologies and sustainable development are the trends that are shaping both, the society and economy. Therefore, they require major evolutions. Nevertheless, digital technologies and sustainable development seem to be two unrelated concepts at first. Some authors refer to how their opposition to one another causes a substantial move in both, ecological and social systems. Yet, others, note that these concepts are being the concepts that will change the game, yet they are strategic imperatives that are capable of bringing about significant revolutions. It is challenging to define the two concepts due to their broad nature. Many other authors claim that sustainability concept has frequently been misused, incorporating various points of view that have diminished its original significance. In fact, as the term is multi- and trans-disciplinary that affects socio-economic organizations at all levels through procedures, resolutions, and actions, therefore, the domain of "Sustainability" is challenging to define. For example, the decision-making process for Sustainability frequently displays cultural prejudices and differing viewpoints on open-mindedness, investment behavior, and risk assessment. This is true for a variety of stakeholders such as, owners, decision-makers, government bodies, and consumers. Finally, stakeholders' participation of the co-creation of shared value is significant for sustainable development. Consequently, the definitions argument is far from being closed [6]. Insert here: Technologies like artificial intelligence (AI), the Internet of Things (IoT), financial technology (FinTech), and blockchain have significantly improved the world. Their applications span various fields, including healthcare, education, and business, enhancing efficiency and innovation across these sectors [40].

The debate can be enriched by the presence of multiple different perceptions on sustainability, possibly increasing its applicability to particular stakeholders or industry sectors. 'Urban Sustainability' is one example of a specification in in different context, or 'Sustainable Manufacturing'. Furthermore, professional bodies such as the Chartered Institute of Procurement and Supply (CIPS) developed its own definitions, such as "Sustainable Procurement" as "the act of adopting social, economic and environmental factors alongside the typical price and quality considerations into the organizations' handling of procurement processes and procedures". Additionally, different stakeholders from a wide range of sectors, such as governmental bodies and non-governmental organizations have varying perceptions of what sustainability entails [27]. Sustainable development. Consideration of minority perspectives, it is said, may alternatively result in considerable integration as well as novel notions of socio-technical transformation, defined by a more equitable and natural understanding of Sustainability. Yet, one of the key breaks in the implementation of sustainable development is the availability of many stakeholders' perspectives and each and every attempt to extend the development of a universal definition and concept could make the contentious discussion last longer.

The consensus on a term for "Digitalization" and its descendants, including "Digital technologies" and is of a similar complexity. The phrases "Digitalization," and "Digital technologies" are frequently interchanged in literature. For instance, according to Gartner [13], "Digitalization" refers to both the course of action to convert to a digital business and the utilization of digital technologies to provide novel opportunities for earning value. This is not far from Hanelt et al. [17] definition of "Digital technologies" as the alteration in organizational structure brought upon and modified by the pervasive of Digital technologies. Digitization is currently the most significant force behind innovation and entrepreneurship, regardless of terminology Berger et al. [8]. The term "Digital Sustainability" was created in the professional world to acknowledge the correlation and convergence between "Sustainability" and "Digital technologies". In this sense, the accomplishment of Sustainable Development Goals (SDGs) is stressed by both practice and academia. Blockchain is one of the prominent digital technologies. It functions as a decentralized ledger, recording all digital events that have been executed and shared among participants. Its versatility makes it widely applicable across various domains, including supply chain management [4].

However, digital sustainability is "the means by which digitalization, as a key part of the fourth industrial revolution, can deliver on the global sustainability goals" for the Cybercom Group (2021). Similarly, [14] describes digital sustainability as "the organizational activities that seek to advance the Sustainable development goals through creative deployment of technologies that create, use, transmit, or source electronic data". As cited by Markman et al. [24], it is being seen frequently that sustainability is as in opposition to other objectives or missions. In this instance, "Digital Sustainability" has the potential to bridge the gap between the both strategic goals of sustainability and digital technologies in order to spur beneficial societal and environmental improvements as opposed to merely minimizing them. The potential of various forms of digital technologies to improve sustainable development is well acknowledged in many literature reviews.

The notion of "Digital Sustainability" suggests or implies that there is a positive association between the two concepts of Sustainability and Digital technologies, yet this is not a given. Innovation challenges include not just the potential for economic gain but also the potential for societal challenges and the impact these challenges may have on Sustainability's social and environmental facets. As such, normative innovation development is a requirement of Sustainable development. It has been noted by one of the researchers that only with full participation from all levels of the socio-economic organizations, sustainable innovation is conceivable. However, it has been mentioned by one of the researchers that there is no proof that a company's performance improves when digital technologies and Sustainability are combined. Although the opportunities that digital technologies present for sustainability are generally seen as positive, it is an imperative necessity to retain the high levels of awareness that digital technologies can be a disturbing power that, if unintentionally, unrestrained, or undervalued, may harmfully affect sustainability and its development

[11]. Therefore, urge for the establishment of sustainable governance frameworks that can lessen the conflicts between digital technologies and sustainability and increase the potential. In order to clarify the situation, examines the various perspectives on digital technologies in connection to the sustainability in various aspects such as, ecological, economic, and social aspects.

## 4.2 Digitalization Strategies for Sustainability Purposes

One of the most common themes when thinking about the vital and leadership role that digital technologies play to improve sustainable development is the SDGs. The 17 SDGs, also known as the UN Agenda 2030, cover a broad span of subjects and build on the developments in the sustainable development area over the previous 30 years. In this regard, many authors and researchers have emphasized on the various ways in which digital technologies might help attain the SDGs. Nevertheless, this research paper shows a bias in favor of two traits. The first concerns the emphasis on policy-making. The second is that, despite the assertion that the SDGs have a broad, global scope, the published articles have localized and sectorial boundaries. The contribution that businesses make to sustainability and the SDGs through digital technologies is under-discussed and appears to be ignored by academic production, which only briefly and barely touches on this subject. However, some publications expressly explore digital potential for sustainability as part of a corporate strategic plan in a new era of scientific and technological growth. Business models are recommended as a framework for strategic application in various publications. Even in this situation, most publications are concentrated on a certain sector or job function, such as modern architecture or quality management.

Digital technologies are sometimes incorporated and integrated to support sustainable businesses. For instance, some researchers approach the subject with a focus on the value creation, while others look at how to predict how to minimize the digital transformation risks. The fact that many studies investigating the strategic relationship between sustainability and digital technologies have a limited geographical emphasis, is also considerable. Sometimes the title emphasizes this (e.g., but typically comes from the content. Some authors, for example, use cases from the Italian industrial sector to illustrate how digital technologies servitization affects sustainability. The findings from Finnish SMEs are the basis of a study on the influence of sustainability strategy on the correlation between business digital strategy and financial performance. The research also demonstrates a propensity among academics towards particular sectors or industries. Among them, the use of digital technologies in agriculture is a common theme or to rural communities [36]. Understanding the usage, access, and the implications of digital technologies politically, environmentally, and socially for agriculture is a focus research area.

Several publications examine "Industry 4.0," stressing the concepts and tendencies it refers to in terms of sustainability. Manufacturing and supply chain are given special consideration [22, 30]. It is noted that articles become far more strategic than informative at the industry or sector level. Among them, there are few who are specifically acknowledge that digital manufacturing is an important topic in manufacturing area. The authors contend that digital technologies have a positive impact on sustainable development if challenges associated with the technological social changes are tackled. Several articles about libraries are also highlighted by the search, notably those that discuss the importance of digital information and the causes and progression of digital learning. The potential of digitalization for urban Sustainable development is also explored in articles, especially considering trends and responses to events. Several outputs of literature also touch on the impact of digitization on healthcare Sustainability and tourism. Themes of digital technologies that are related to the sustainable production and consumption are another recurring issue or related to the energy sector [38].

## 4.3 The Catalyst for Sustainability and Open Innovation is the Digital Technologies

Sustainable development is a key priority for global organizations, as it addresses critical challenges. It offers significant social, environmental, and economic advantages to firms, fostering long-term growth and resilience [3]. Organizations should continually advance their performance, productivity, and efficiency and provide their clients with new customer experience in order to be sustainable. To put it another way, they must be innovative. Innovation in digital technologies is how businesses work, reassure that their stakeholders are satisfied, and provide value proposition and social welfare. Process, a change, and an impact are all perceived concepts of digital technologies. Digital technologies assist in the organizations' economic societies and development through an optimal and effective utilization of resources, waste elimination, opportunities provision to streamline and improve organization's operations that lead to lowering costs. Furthermore, digital technologies assist in creating new ecosystems and business models, such as the circular economy and cross-industry cooperation, which enables businesses to respond more quickly to the demands of rapidly shifting consumer demands, boosts income, value-added to the economy, and encourages the growth of an innovation culture (Robertsone et al. [31]).

Digital technologies are significant to improve an organization's performance from social and environmental aspects and employees' welfare and happiness. Digital solutions for effective and efficient energy monitoring, cost-effective water and resource use, and the development of new manufacturing processes are just a few examples of the potential of digital technologies for environmental sustainability. They are based on cutting-edge technologies that are more environmentally friendly are a few examples of the innovative potential of digital technologies for environmental sustainability. Digital technologies offer numerous potentials in the social domain to expand the accessibility and affordability of various services for larger populations, despite their location or socioeconomic status, and at lower costs, such as public health and education services. With the help of digital technologies enabling technologies, digital technologies provide a wide range of options for sustainable practices that will foster open innovation, sustainable growth, and a brighter future for both humanity and the globe. Digital transformation is a continuous process enabled by digital technology tools that lead to evolution and developments, i.e., a change by itself that transforms an entire organization and its business; the ways in which value is established and transferred. Three viewpoints are perceived on Digital technologies: process, change and impact Robertsone et al. [31].

To survive and ensure sustainable development in the operational environment of today, which is rapidly changing, organizations and businesses must continuously improve their effectiveness, quality, application of innovation in digital technologies and solutions, improvement of their procedures and cultures, and existing products, services, and customer experiences. This is innovation. Closed innovation has lost its place in the modern world. By utilizing knowledge sharing between parties, open innovation accelerates the release of new products, services, and customer experiences to market and creates new business opportunities. Organizations need outside resources and expertise to generate ideas. Organizations must share and exchange their ideas via current digital platforms or newly emerging ones in order to acquire this expertise and knowledge [12] and other technologies that encourages the introduction of new digital ecosystems [19]. Digital technologies tools are required for o to build up their capacities. Open innovation promotes the development of new technologies and collaborative digital platforms that serve as the foundation for and enabler of digital technologies [18, 28]. Digital innovations are being adopted and implemented as a response to the difficulties that the organizations are facing [29], that, in turn, leads to digital technologies that transform all organizational operational aspects, which in turn spurs the development of new goods, services, programs, systems, and business models that encourage further innovation.

Technological innovation has invented the digital technology revolution [3] that boosts the sustainability of the organizations [16] yet, being pushed by it. Innovations that are driven by technologies, are factors that contribute to the economy advancement [3]. The enablement of digital technologies is through several technologies that have different effects on the sustainable development. On the other side, some technologies use a lot of energy, leave a carbon imprint, and have recycling restrictions. [4]. Despite the fact that technology is a key enabler of digital technologies and a common component across all digital technologies' concepts and frameworks, the authors come to the conclusion that adopting change across all facets of businesses, organizations, and society—including strategy and culture and value proposition—leads to the creation of new business models. By focusing on innovation and efficiency, new business models promote digitization and transformation even more. Sustainability is made possible by the process of digital technologies. The digital technologies promote and supports open innovation. Organizations must undergo digital technologies in order to survive and advance. New business models are working within and amongst new ecosystems, are brought about by digital technologies, and these effects foster Sustainability and encourage open innovation. The connections between open innovation, Sustainability, and the three views of Digital technologies.

### *4.4 Digital Technologies as a Contributor in the Economic, Environmental and Social Sustainability*

The effect of digital technologies on the economic and environmentally sustainable development is substantially greater and significant when it is compared to the social sustainability. This can be explained by the fact that consumer wants and expectations are the primary forces behind digital change [4] and competition. The expectations of the final consumers, who desire a better and more individualized experience, are the primary force behind the change of businesses everywhere. Utilizing modern tools makes that experience possible. Consumers are also conscious of how items affect the environment, due to their interest in sustainable consumption and shopping, people are willing to spend more for organic and ethically produced goods. The use of digital technologies helps make materials and resources, especially energy resources, more efficient, favorably affecting the sustainability of the economy and environment. Digital technologies contribute to improving operational and production effectiveness [25], reduce costs, increase productivity [4], and enhance the products and services quality [25] as well as the effectiveness of a product development process [20]. Digital technology, with the help of its enabling technology, helps businesses balance supply and demand and avoid overproduction [33]. By the implementation of innovation, organizations can become more competitive [4, 15]. Digital technologies also facilitate cooperation, enhance communication with the stakeholders, and integrate them into corporate activities, which create value. By doing so, we contribute to corporate sustainability and enable open innovation [4]. Overall organizations' improved performance drives sustainable development [28].

Other advantages include increased employee knowledge and abilities, which make them more adaptable to labor market changes. Big Data and AI improves the decision-making process [4, 19], enable businesses to respond more quickly to changes and challenges, thereby improving innovation [5], and assist with the social and environmental aspects of the businesses. These technologies also help companies better understand their customers'. The importance and significance of digital technologies in attaining the Sustainable Development Goals (SDGs) is also emphasized by earlier study. Reducing waste, and integrating renewable energy from manufacturing and digital technologies play a leadership role in achieving environmentally sustainable development. With regards to the optimization of supply chain, AI and IoT technologies are frequently deployed. Environmental performance is improved by simulation and modeling technology. AI is also used to optimize energy use. Traditional methods of environmental management are changing as a result of cloud, social network, mobile technologies, big dana and IoT [28].

Digital technologies contribution to the social sustainability is developing, and there is not enough study on this topic. According to [10], pleasant working circumstances, maintaining a healthy work-life balance, and performance enhancements all contribute to employees' well-being and have a beneficial effect on their satisfaction and loyalty. digital technologies have a tremendous effect on the growth of the economic and social communities by promoting employee wellbeing and improving environmental performance [10]. The quality of life is generally improved by digital technologies, to which it has ensured approachability to the public services like health care, and education [4]. Despite the numerous advantages that digital technologies offer to businesses and society, research shows that it has a negative influence on Sustainability. More investigation and focus are needed to address worries with regards to the high rates of consuming energy brought on by some digital technologies, a significant carbon consumption, a lack of recycling options, and e-waste [4]. Several academics agree that the correlation between digital technologies and sustainability is ambiguous [4] and that digital technologies might have both, good and bad effects on the environmental and social sustainability. Economic and social progress may be hampered by the digital divide, dependence on digital systems, cyber-security, internet addiction, and information manipulation. The Sustainability of established organizations and their business practices is threatened by the rise of new business models and market players brought on by the digital technologies [34]. The development of new complementary technologies, products, and services, accessibility, and the opening of new markets with partners are all made possible by digital technologies. This is because it supports and promotes partnerships and integration amongst and within the suppliers, consumers, and internal and external partners.

At the same time, industry 4.0 technologies [12] and digital technologies are moving innovations outside of the organization's borders into cooperative networks and driving international business expansion of companies. Due to a lack of in-house digital proficiencies, businesses are forced to interact with external organizations, go outside for specialists with the requisite knowledge and skills, and hire them. The authors have also discovered evidence that open innovation is what is driving the digital technologies in the literature they have analyzed. First and foremost, working with outside partners encourages the development of new digital cooperation tools [12], as well as other digital resolutions, findings, and inventions. These tools, in turn, drive digital technologies [4]. The main forces behind the digital technologies are intellectual and IT potentials that businesses acquire as a result of the open innovation process.

The results of the current study also show how open innovation benefits Sustainability. Although the interrelationships between open innovation and Sustainability were not the main focus of the research, which was initially intended to investigate the influence of digital technologies on open innovation and sustainability, the authors still think these findings are significant. Open innovation unquestionably strengthens a company's ability to compete by boosting innovation, which is a crucial Sustainability driver [9], supporting market expansion, and boosting the desirability of the products as consumers praise the products from larger users and manufacturers' networks [12]. [9, 23] and other authors argue that adopting new business models, expanding chances to join new markets, and enabling risk- and resource-sharing are all ways that open innovation helps corporate Sustainability initiatives. Data sharing is made possible through collaboration and partnership, which benefits society and the economy [21].

## 5 Conclusion

Digital technologies and Sustainability play primary and vital role in the concept of Environmental, Social, and Governance (ESG) aspects. We live in an era where technology has brought remarkable changes to all aspects of our lives. Furthermore, recently the topic and discussion about climate revolution has speeded up at an extraordinary rate, which poses a significant challenge for organizations and businesses to operate and change into more sustainable models. Sustainability-minded investors are pressing businesses to address their internal governance, social, and commercial challenges in a targeted and open way. Environmental, social, and governance (ESG) aspects come into play in this situation. ESG refers to a set of criteria and standards used to assess aspects of sustainable development and the social effect of businesses. ESG, which originally referred to a focused reportage scheme for financial stockholders, is now a general phrase for how businesses evaluate the influence of these variables on their products, services, processes, and employees.

Environmental elements take into account aspects like climate revolution, energy, carbon footprints, management of natural resources, and animals well-being. However, social elements evaluate the business in terms of business networks, stakeholder demands, and potentials in areas like health, education and safety. Finally, governance assesses the use of accounting principles, transparency, and shareholders' right to vote on significant issues, as well as the handling of conflicts of interest. Comprehensive digital technologies allow businesses to alter the perception of their problem-solving approaches, as well as it allows them to function more rapidly and efficiently. Likewise, ESG helps businesses to become more attractive and mature stronger.

Presently, sustainability and digital technologies are becoming the prevailing concepts in the modern economy. Businesses will be helped by digital technologies in their efforts to generate and develop ESG data sets methodically and to make sustainable investment decisions. In the long run, the revelation of the organization's ESG data will result in great results amongst its rivals. Reduced capital costs and the risk of the business, improved situation of the stakeholder, chances to the entrance of long-term capital, and enhanced the effectiveness of organization's businesses and operations and organization status are all advantages of effective ESG implementation. The financial sector, and specifically the strategies of banks, are likewise exhibiting and advancing this tendency.

As a matter of fact, in addition to the foundations that the technology provides for the investment decisions, it can also assist with raising awareness amongst consumers about the repercussions of their consumption on the environment and social aspects, which will encourage them to make more sustainable choices. Consumers are now more conscious of green products and services and approach their purchases differently. Future consumers will readily support and choose green brands if technology advances sufficiently and there aren't too many features that differ across service or product providers. As a result, businesses should think about utilizing green approach as a long-term scheme to gain competitive advantages.

Financial institutions can help consumers choose goods and services that fit their personalities and environmental preferences by giving them a breakdown of their daily spending through fin-tech applications. In Europe more than the half of the customers are ready to pay extra for sustainable services and products, according to research by EY teams. Two-thirds of millennial shoppers in the US believe they are ready to pay more on a product that has received a sustainability certification. The majority—75%—believe that businesses should focus more on serving the community than just increasing profits. More precisely, the pandemic's effects on the economy have strengthened the movement toward sustainable development. Additionally, investors are now more likely to make their decisions on ESG evaluation criteria than on conventional financial factors.

Global ESG assets may exceed $41 trillion for the entire year of 2022, according to Bloomberg Intelligence, and reach $50 trillion by 2025. With a 40% increase over the previous two years, the United States is a major contributor to this growth. According to the Institute of International Finance, during the global sell-off, 80% of the stock indexes of ESG-compliant funds beaten comparable indexes of non-compliant funds. With substantial contributions from Southeast Asian nations, green bonds, the biggest part in the sustainable commercial market, were issued globally in 2021 for a total of more than $620 billion.

Sustainability and digital strategy are becoming more and more entwined. 40% of 400 executives from different regions and industries who participated in a recent survey by Bain & Company and the World Economic Forum said they already see a positive impact from digital technologies on their sustainability goals. A more circular economy is now possible thanks to digital technologies, which are also used to measure and track sustainability progress, optimize resource use, reduce greenhouse gas emissions, and reduce greenhouse gas emissions. But collaboration and creativity are also made possible by digital technologies. Strong tools like digital twins, additive manufacturing, and design-based artificial intelligence (AI) will enable the next wave of climate change solutions. Data-sharing platforms, IoT-enabled sensors, block chain-based authentication, gamified apps, and other technological examples encourage collaboration along the entire value chain and align participants' metrics and goals. Digital has some drawbacks in addition to all of its advantages. 10% of respondents to the survey think that digital technologies endanger sustainability. They are particularly concerned about the effects on diversity and inclusion, data privacy, mental health and wellness, and future-ready skills. It is crucial to address these problems immediately. Before we could fully respond to the impacts of the First Industrial Revolution on our environment, 200 years had passed. The consequences of the Fourth Industrial Revolution will manifest much more quickly.

## 6 Implications

### 6.1 Practical Implications

Besides to the implications of the above study, this research paper also provides practical perceptions related to organizational change and strategy. Subjects with regards to digital technologies, they are not only academic, but also a core issue of organizational change and strategy. It is also more practically relevant because it provides insight into how organizations match, where and how they compete, etc. In this respect, the multidimensional framework of digital technology not only provides individuals, organizations, and businesses with an all-inclusive comprehension of the phenomenon, yet it can be used as a high-level list combined with the survey matrix as well. More importantly, the findings of this research paper provide individuals, organizations, and businesses with valuable insight into the digital technologies changing aspects of and suggest the key leadership skills needed to master them. As mentioned earlier, digital technologies entail a shift towards adaptive organizational approach that allow constant customization on the one hand and stormy digital business ecosystems on the other.

### 6.2 Theoretical Implications

The results show that this change requires three key leadership competencies. Awareness of digital technology, acceleration of digital technology, harmonization of digital technology. Each skill contains specific qualities and capabilities and is connected to a multidimensional outline of digital technology. First, individuals, organizations, and businesses need to be aware of digital technology as it relates to the process and content of their strategies, as contextual situations change intermittently with sudden explosions. Not only do we need to monitor and respond to change, but we also need to be cognizant of the volume and types of data and new digital technologies and their characteristics. As our review reveals, it is important to recognize that these factors are evolving rapidly and that they are interwove with aspects relevant to each market and consumer characteristics. To consider this complexity when evaluating appropriate conditions, managers should use data analysis. Existing and upcoming digital technologies research on technology impacts and systemic change is best well-matched to inform consciousness of digital technologies.

Second, in relation to the innovation mechanism in the multidimensional framework, individuals, organizations, and companies need the accelerating capabilities of digital technologies in implementing strategies. This is the mental capacity to continuously conceptualize new digital products and processes using available resources, as well as the readiness to devote management time and money during brief spurts. Above all, keep in mind that quick implementation and investigation to learn in the market outperform upfront preparation and examination due to the dynamics of the digital business ecosystem. It is crucial to note that research in the area of holistic coevolution can inform and advance this skill.

Third, people need to have the ability to harmonize digital technologies. Harmonizing digital technologies skills are those that allow people to incorporate new digital products into their current goods and services. This includes integrating between "physical" and "digital" goods and services, identifying areas of overlap and conflict, settling disagreements, and taking into account cultural characteristics through balancing and mixing. In this situation, people must develop specific perception of what to maintain and what to change as a result of digital technologies. This perception must be based on both the general opportunities and threats and the context of the relevant industry, as we identified in the change's nature. Studies on segmented adaption provide inspiration specifically for integrating digital technologies. It is suggested that those businesses and organizations are prepared to encounter the challenges of digital technologies by the utilization of the above mentioned leadership skills. It is proposed that businesses and organizations are best equipped to handle the difficulties of digital technologies if they make sure the aforementioned abilities are present and maintained in their organization.

### 6.3 Study Limitations and Future Research

The steps of the research procedure can be attributed to the following mentioned constraints of this research study. First, some studies may have been missed during data collection because this study only used one database, particularly those published on the Internet particularly: Google Scholar. Second, due to the choices used in the cut-off selections (such as those relating to timeframe), the filters used in data analysis may have missed certain pertinent studies. Third, there may be worthwhile substitutes to an "input-process-output" in terms of synthesis. Given these constraints, it is suggested that more researchers to build upon and improve the findings by using different data or analytical techniques. Fourth, in terms of the conclusion, the ideas were built on the difference between intermittent and constant change, as it indicates its importance for comprehending digital technologies. The motors of change are one such framework that distinguishes between different change conceptions, as was mentioned above. Future research should be done to examine how these frameworks relate to digital technologies. The aforementioned observations demonstrate that few publications at this time are able to provide findings that are applicable to a variety of situations. To correct the misalignment between the digital sustainability theory and practical aspects, it is also important to address the lack of attention paid to the correlation and connection between digital technologies and sustainable development.

## References


1 Al Mubarak, M.: Sustainable development through five senses of effective corporate social responsibility strategy. Soc. Respons. J. **17**(6), 829–839
2. Alalwi B., Mazzuchi T., Hamdan A., Al Mubarak M.: Blockchain technology implications on supply chain management: A review of the literature. In: Hamdan A., Hassanien A.E., Khamis R., Alareeni B., Razzaque A., Awwad B. (eds) Applications of Artificial Intelligence in Business, Education and Healthcare. Studies in Computational Intelligence, vol 954. Springer, Cham, pp. 23–38 (2021). https://doi.org/10.1007/978-3-030-72080-3_2
3. Bican, P.M., Brem, A.: Digital business model, digital transformation, digital entrepreneurship: is there a sustainable "digital"? Sustainability **12**(13), 5239 (2020)
4. Camarinha-Matos, L.M., Fornasiero, R., Ramezani, J., Ferrada, F.: Collaborative networks: a pillar of digital transformation. Appl. Sci. **9**(24), 5431 (2019)
5. Chu, Y., Chi, M., Wang, W., Luo, B.: The impact of information technology capabilities of manufacturing enterprises on innovation performance: evidences from SEM and FSQCA. Sustainability **11**(21), 5946 (2019)
6. Ci, S., He, H., Kang, C., Yang, Y.: Building digital battery system via energy digitization for sustainable 5G power feeding. IEEE Wirel. Commun. **27**(5), 148–154 (2020)
7. Cillo, V., Petruzzelli, A.M., Ardito, L., Giudice, M.D.: Understanding sustainable innovation: a systematic literature review. Corp. Soc. Responsib. Environ. Manag. **26**(5), 1012–1025 (2019)
8. Berger, E.S.C., von Briel, F., Davidsson, P., Kuckertz, A.: Digital or Not—The future of entrepreneurship and innovation: introduction to the special issue. J. Bus. Res., 125:436–442, (2021)
9. El Hilali, W., El Manouar, A., Janati Idrissi, M.A.: Reaching sustainability during a digital transformation: a PIS approach. Int. J. Innov. Sci. **12**(1), 52–79 (2020)
10. Farías, A., Cancino, C.A.: Digital transformation in the chilean lodging sector: opportunities for sustainable businesses. Sustainability **13**(14), 8097 (2021)
11. Flyverbom, M., Deibert, R., Matten, D.: The governance of digital technology, big data, and the internet: new roles and responsibilities for business. Bus. Soc. **58**(1), 3–19 (2019)
12. Fukawa, N., Zhang, Y., Erevelles, S.: Dynamic capability and open-source strategy in the age of digital transformation. J. Open. Innov. Technol. Mark. Complex. **7**(3), 175 (2021)
13. Gartner: Information technology gartner glossary. Available at: https://www.gartner.com/en/information-technology/glossary/digitalization, Accessed 30 April 2023
14. George, G., Merrill, R.K., Schillebeeckx, S.J.D.: Digital sustainability and entrepreneurship: how digital innovations are helping tackle climate change and sustainable development. Entrep. Theory. Pract. **45**(5), (2020)
15. Gomez-Trujillo, A.M., Gonzalez-Perez, M.A.: Digital transformation as a strategy to reach sustainability. Smart Sustain. Built Environ. **11**(4), 1137–1162 (2021)
16. Hajishirzi, R., Costa, C.J., Aparicio, M.: Boosting sustainability through digital transformation's domains and resilience. Sustainability **14**(3), 1822 (2022)
17. Hanelt, A., Bohnsack, R., David, M., Marante, C.A.: A systematic review of the literature on digital transformation: insights and implications for strategy and organizational change. J. Manage. Stud. **58**(5), 1159–1197 (2021)
18. Hinings, B., Gegenhuber, T., Greenwood, R.: Digital innovation and transformation: an institutional perspective. Inf. Organ. **28**(1), 52–61 (2018)
19. Jeong, H., Shin, K., Kim, E., Kim, S.: Does open innovation enhance a large firm's financial sustainability? a case of the korean food industry. J. Open. Innov. Technol. Mark. Complex. **6**(4), 101 (2020)

20. Junge, A.L.: Digital transformation technologies as an enabler for sustainable logistics and supply chain processes—an exploratory framework. Braz. J. Oper. Prod. Manag. **16**(3), 462–472 (2019)
21. Kasperovica, L., Lace, N.: Factors influencing companies' positive financial performance in digital age: a meta-analysis. Entrep. Sustain. Issues, 8(4):291-311, (2021)
22. Lafferty, C.: Sustainable internet-of-things-based manufacturing systems: industry 4.0 wireless networks, advanced digitalization, and big data-driven smart production. Econ. Manag. Financ. Mark. **14**(4), 16 (2019)
23. Mahdad, M., Hasanov, M., Isakhanyan, G., Dolfsma, W.: A smart web of firms, farms and internet of things (iot): enabling collaboration-based business models in the agri-food industry. Br. Food. J. **124**(6), 1857–1874 (2022)
24. Markman, G.D., Russo, M., Lumpkin, G.T., Jennings, P.D., Mair, J.: Entrepreneurship as a platform for pursuing multiple goals: a special issue on sustainability, ethics, and entrepreneurship. J. Manag. Stud., **53**(5):673–694, (2016)
25. Mu, R., Wang, H.: A systematic literature review of open innovation in the public sector: comparing barriers and governance strategies of digital and non-digital open innovation. Public Manag. Rev. **24**(4), 489–511 (2020)
26. Orlando, B., Ballestra, L.V., Scuotto, V., Pironti, M., Giudice, M.D.: The impact of R&D investments on eco-innovation: a cross-cultural perspective of green technology management. IEEE Trans. Eng. Manage. **69**(5), 2275–2284 (2022)
27. Pepe, F., Paternostro, S., Monfardini, P.: Sustainability standard setting as local government matter: an italian experience. Public Manag. Rev. **20**(1), 176–200 (2018)
28. Pînzaru, F., Dima, A.M., Zbuchea, A., Vereş, Z.: Adopting sustainability and digital transformation in business in romania: a multifaceted approach in the context of the just transition. Amfiteatru. Econ. **24**(59), 28–45 (2022)
29. Pînzaru, F., Zbuchea, A., Viţelar, A.: Digital transformation trends reshaping companies. The 17th international conference on business excellence, **13**(1):635–646, (2019)
30. Plumpton, D.: Cyber-physical systems, internet of things, and big data in industry 4.0: digital manufacturing technologies, business process optimization, and sustainable organizational performance. Econ. Manag. Financ. Mark. **14**(3), 23 (2019)
31. Robertsone, G., Lapiņa, I.: Digital transformation as a catalyst for sustainability and open innovation. J Open Innov: Technol, Mark, Complex, **9**(1) Article 100017, (2023)
32. Shahzad, M., Qu, Y., Zafar, A.U., Rehman, S.U., Islam, T.: Exploring the influence of knowledge management process on corporate sustainable performance through green innovation. J. Knowl. Manag. **24**(9), 2079–2106 (2020)
33. Skordoulis, M., Ntanos, S., Kyriakopoulos, G.L., Arabatzis, G., Galatsidas, S., Chalikias, M.: Environmental innovation, open innovation dynamics and competitive advantage of medium and large-sized firms. J. Open. Innov. Technol. Mark. Complex. **6**(4), 195 (2020)
34. Sraml Gonzalez, J., Gulbrandsen, M.: Innovation in established industries undergoing digital transformation: the role of collective identity and public values. Innov. Organ. Manag. **24**(1):201–230, (2021)
35. Stuermer, M., Abu-Tayeh, G., Myrach, T.: Digital sustainability: basic conditions for sustainable digital artifacts and their ecosystems. Sustain. Sci. **12**(2), 247–262 (2017)
36. Tim, Y., Cui, L., Sheng, Z.: Digital resilience: how rural communities leapfrogged into sustainable development. Inf. Syst. J. **31**(2), 323–345 (2021)
37. Vial, G.: Understanding digital transformation: a review and a research agenda. J. Strateg. Inf. Syst. **28**(2), 118–144 (2019)
38. Vlasov, A.I., Shakhnov, V.A., Filin, S.S., Krivoshein, A.I.: Sustainable energy systems in the digital economy: concept of smart machines. Entrep. Sustain. Issues **6**(4), 1975–1986 (2019)
39. Wu, L., Lou, B., Hitt, L.: Data analytics supports decentralized innovation. Manage. Sci. **65**(10), 4863–4877 (2019)
40. Al Mubarak, M., Hamdan, A.: Sustainable competitive advantage through technological innovation: An introduction. In: Al Mubarak, M., Hamdan, A. (eds) Technological Sustainability and Business Competitive Advantage. Internet of Things. Springer, Cham. (2023). https://doi.org/10.1007/978-3-031-35525-7_1

# An Assessment of the Impact of Technical Analysis and AI on the Effectiveness of Intraday Trading and the Behavior of Traders

**A. Aravind, S. Sumayya, Arya Ajith, and Ananthi Rajayya**

**Abstract** Intraday trading is the act of buying and selling a financial instrument within the same day or even multiple times over a single day. It applies a distinct range of tools to analyse data, detect patterns, implement algorithms, interpret sentiment, and handle risks. It covers key technical ideas like Dow Theory, Elliot Wave Theory, candlestick charts, support and resistance points, and tools such as moving averages and RSI. We gathered data through surveys and deep talks with traders, plus info from published sources. We then used MS Excel and Trading View to break down this data. Results show that most traders think technical analysis helps, but mental factors play a big role in how they trade. Good risk control and varied plans that use both technical and basic analysis are key. The study points out that traders need to learn about technical analysis and risk control. It also suggests getting more women into trading. We recommend personal learning plans for better ways to test past data, practice trading, and try a mix of methods. This full study aims to give tips and plans for new and skilled traders alike. It stresses how important mental smarts technical know-how, and strong risk control are to do well in intraday trading.

**Keywords** Intraday trading · Technical analysis · Algorithmic trading

## 1 Introduction

Intraday trading is the type of trading that is characterized by effecting purchases [1] and resale of different financial assets to capture the price change within the modeling session [6]. However, such traders should always stay in the loop and in movement to enhance their returns and employ techniques [2, 23]. This process is enhanced

A. Aravind · S. Sumayya (✉) · A. Ajith · A. Rajayya
Amrita School of Business, Amrita Vishwa Vidyapeetham, Amritapuri, Kerala, India
e-mail: sumayyasajeev99@gmail.com

A. Rajayya
e-mail: ananthir@am.amrita.edu

 
M. Al Mubarak (ed.), *Sustainable Digital Technology and Ethics in an Ever-Changing Environment*, Studies in Systems, Decision and Control 237,
https://doi.org/10.1007/978-3-031-86708-8_3

with the introduction of Artificial Intelligence (AI) which assists in analyzing data better, recognizing patterns, making trades, gauging sentiment, and controlling risk [7].

To effectively participate in trades, the trader needs to be aware of some principles of technical analysis [22] such as Dow Theory, Elliot Wave Theory, and candle charting. In addition to that, they employ support and resistance levels [20, 21], trend reversals, and other indicators such as Moving Averages and Relative Strength Index (RSI) in their practices [8]. Effective risk and money management strategies are crucial, such as setting stop-loss orders and optimizing position sizes. Platforms like Meta Trader 5 and Quant Connect offer advanced tools and data for comprehensive analysis and trading strategy development [4, 9].

In this paper, Sect. 1.1 discusses the objectives to achieve, and Sect. 2 on the methodology which consists of the Research Design (2.1) and Research Methodology (2.2). Section 3 discusses the literature review, Sect. 4 on data analysis, Sect. 5 on findings and recommendations of the survey, and finally Sect. 6 discuss on conclusion.

### *1.1 Objectives*

- Evaluate the technical analysis indicators' precision and dependability in pointing out the best intraday trades.
- To explore the trading approaches and the strategies utilized by the intraday traders.
- To Explore the behavioural and decision-making patterns of intraday traders.
- To provide recommendations to new and aspiring intraday traders.

## 2 Methodology

### *2.1 Research Design*

The research design for this project is descriptive. It is the framework or preparation of collecting and analyzing the data for a study. Intraday trader's data is collected, evaluated, and used to conclude. This would be helpful for beginner traders to improve their method of trading and develop and improve their strategies [17].

### *2.2 Research Methodology*

Sources of Data Collection

Primary Data: This data was collected through a study that was designed in a unique way for the study. A systematic questionnaire was used to collect the data for this study. Also, in-depth interview was conducted with experienced traders for valuable insights into the behaviour of the traders, their strategies, and their experiences. Secondary Data: The secondary data for this study is taken from published research articles, relevant books, interactions with company officials, and websites related to trading.

**Sampling Method**: Convenience Sampling.
**Sample frame**: Intraday traders
**Sample Size**: 100
**Tools and Techniques used**: MS Excel, Google Forms, Technical Analysis (Trading View platform)

## 3 Literature Review

### *3.1 Technical Analysis in the Financial Markets*

Technical analysis is a widely adopted approach in the financial markets. Achelis (2013) and Murphy (1999) provide comprehensive guides to technical analysis methods and applications [18]. They discuss various comprehensive guides to the technical analysis methods and applications. They discuss various indicators, charts patterns, and strategies traders use to make trading decisions [19]. Lo, Mamaysky, and Wang (2000) delve into the computational algorithms and statistical inference behind the technical analysis and explore its empirical implementation in the financial markets [9].

### *3.2 The Effectiveness of Technical Analysis*

The researchers have examined the effectiveness of technical analysis in different contexts. Menkhoff (2012) focuses on currency momentum strategies, highlighting the potential profitability of technical analysis in foreign exchange markets. The presence of positive returns is detected. Currency momentum exists in positive returns. Further Statman (2002) explains how investment analysts, or in this case, technical analysts are influenced by behavioral economics and thus support the notion that there are psychological issues involved in trading decision-making [10].

### *3.3 Trader's Behavior and Psychology*

Regarding a trader's activity, the distinctive psychological aspects that influence such activities have been documented in the area of behavioral finance. For example, Nofsinger (2005) studies the connection between social mood and financial economics and focuses on the emotional aspect of the trader's decision-making processes [15]. In the detailed work of Tversky and Kahneman (1974), the concept of Heuristic approaches in the financial market has been explained where several biases are attached to the trader while making decisions [16].

### 3.4 Volume of Trade and the Actions of Investors

Odean [4] examined the relationship between trading volume and investor activity and found out that investors, because of great engagement in the trading activity, incur great costs and results are poor. The study addresses the attitude of a trader to discipline and self-control which is very critical in excessive trading.

### 3.5 Insights from Successful Traders

Schwager (1995) presents interesting and instructive interviews with outstanding market participants from the book "Market Wizards", where one can find extremely useful information [3]. These interviews provide insight into the trading methods and risk-reduction tactics used by profitable traders. They can impart useful knowledge and lessons about intraday trading from their experience [5].

## 4 Data Analysis

### 4.1 Commonly Used AI Tools

**MetaTrader 5**

MetaTrader 5 (MT5) is a free trading platform that allows users to perform technical analysis and execute trades in the Forex and stock markets [11]. It features real-time quotes, financial news, and charts, and offers options for advanced trading strategies. The platform is user-friendly and provides access to demo accounts for practice. Users are encouraged to double-check any crucial information regarding its features and trading conditions [12].

**QuantConnect**

QuantConnect is an algorithmic trading platform that leverages the cloud to allow you to create, back-test, and deploy trading strategies across asset classes like equities, FX, futures, options, derivatives, and cryptocurrencies. Driven with an open-source engine called Lean, it makes research and live trading very easy [13].

**Table 1** Long entry intraday trade

| | |
|---|---|
| Quantity | 100 shares |
| Entry price | 2496 |
| Stop-loss | 2482 |
| Target price | 2518 |
| Period of trade | 13th June 2023 |
| Risk to risk-to-reward ratio | 1:2 |

### Intraday Trade on Reliance Industries: Long Entry

Here in this long entry intraday trade, the technical indicators used are:

1. Central Pivot Range (CPR)
2. Exponential Moving Average (200 Days)
3. Volume Bar [14].

On the date 13th of June, the reliance chart was in a bullish trend 0.100 shares of Reliance Industries were traded. Following are the details regarding the trade taken:

As shown in Table 1, the target is 22 points above and the risk is 14 points down. The target is achieved in this long entry. Profit Calculation*:*

Profit = Target points * No of shares.
Therefore,
Profit = 22*100 = 2200.

### Intraday Trade on Reliance Industries: Short Entry Stop-Loss Hit

Here in this intraday trade, the technical indicators used are:

1. Central Pivot Range (CPR)
2. Exponential Moving Average (EMA)
3. Volume Bar.

On the date 26th of April, the reliance chart was on a Bearish trend. 100 shares of Reliance Industries were traded.

Following are the details regarding the trade taken:

As shown in Table 2, the target is 18 points below and the risk is 9 points above. The target is not achieved the trade fails and hit the stop loss is hit in the short entry. Loss Calculation:

Loss = Risk points * No of shares.
Therefore,
Loss = 9 *100 = 900.

**Table 2** Short entry stop-loss hit intraday trade

| | |
|---|---|
| Quantity | 100 shares |
| Entry price | 2357 |
| Stop-loss | 2366 |
| Target price | 2339 |
| Period of trade | 26th April 2023 |
| Risk to risk-to-reward ratio | 1:2 |

**Table 3** Short entry target hit intraday trade

| | |
|---|---|
| Quantity | 100 shares |
| Entry price | 2485 |
| Stop-loss | 2495 |
| Target price | 2455 |
| Period of trade | 16th May 2023 |
| Risk to risk-to-reward ratio | 1:3 |

### Intraday Trade on Reliance Industries: Short Entry Target Hit

Here in this trade, the technical indicators used are:

1. Central Pivot Range
2. Exponential Moving Average (EMA)
3. Volume Bar.

On the date 16th May 2023, the reliance chart was on a bearish trend 0.100 shares of reliance industries were traded.

Following are the details regarding the trades taken.

As shown in Table 3, the target is 30 points below and the risk is 10 points above the target achieved in this trade. Profit Calculation:

Profit = Target points * No of shares.
Therefore,
Profit = 30*100 = 3000.

## 5 Findings and Recommendations of the Survey

The findings of this study are.

- Gender: The majority of the participants were male i.e.85%. This created a gender imbalance in the sample.
- The 25–34 age group was the largest (53%), it is followed by 18–24 (36%). So, the young and the middle-aged individuals are showing a significant interest in intraday trading.

- Most of the respondents had a Bachelor's degree (56%), reflecting a higher level of education among the traders.
- Occupation: Traders constituted the largest group, with a diverse range of occupations, including students, assistant managers, and business owners.
- Technical Analysis: Participants demonstrated a belief in the effectiveness of technical analysis in guiding their intraday trading decisions.
- Psychological Factors: Emotions and mindset are crucial in intraday trading, influencing decision-making and performance.
- Risk Management: Effective risk management strategies were not explicitly mentioned, but they are essential for successful intraday trading.
- News Impact: The study explained that news has a huge impact on making trading decisions (especially after the recent Lok Sabha (election results)
- Wide Range of Occupations: The participants represented various professions, indicating diverse individuals interested in intraday trading.
- Younger Participant Engagement: The study revealed a substantial presence of students, suggesting a keen interest in intraday trading among the younger generation.

From the back-testing of strategies, it was found that technical indicators will not work every time.

- All traders have different points of view and trading strategies
- Position sizing and risk management are critical aspects of risk management.
- Technical analysis does not always guarantee a successful trade.
- These findings provide a comprehensive overview of the study, covering gender, age, education, occupations, belief in technical analysis, psychological factors, risk management, news impact, future research needs, limitations, participant engagement, market bias, trader support, and the importance of the risk.

The recommendations are:

- Enhance Gender Diversity: Encourage more female participation to ensure a balanced understanding of technical analysis on intraday trading.
- Develop personalized educational programs on technical analysis, including risk management and psychological aspects for traders of different backgrounds.
- Practice Simulated trading: So that traders can learn trading without playing with cash, thereby they can develop confidence.
- Explore Combined Approach: Investigate the combined use of technical and fundamental analysis for more informed trading decisions and a diversified trading approach.
- The main tools used by the above-mentioned 11% are Trade Ideas, MetaTrader 5, and QuantConnect.

## 6 Conclusion

In summary, this study investigated how well technical analysis worked for intraday trading as well as trader behavior. The survey results depicted that most of the Odessa professional development of adult educators thought of positive technical analysis. The study illustrated that traders need to be taught how to use technical analysis and qualitative analysis tools in detail so that, even in fast-paced intraday trading, they would be able to use each of the indicators and techniques proficiently. The study further opined that there was a need to have management strategies that would foresee the gains and the losses that come with most intraday trading activities. It was also striking that the study appreciated the triangles that various emotions bring to the trading processes and therefore emphasized the essence of emotional intelligence and strategy that integrates technical and fundamental analysis and new decision-making and risk-taking for wider and profitable ones.

## References


1. Tissaoui, K.: The intraday pattern of trading activity, return volatility, and liquidity: evidence from the emerging tunisian stock exchange. Int. J. Econ. Financ. **4**(5), (2012). https://doi.org/10.5539/ijef.v4n5p156
2. Tissaoui, K.: The intraday pattern of trading activity, return volatility, and liquidity: evidence from the emerging tunisian stock exchange. Int. J. Econ. Financ., **4**(5), (2012b). https://doi.org/10.5539/ijef.v4n5p156
3. Schwager, J.D.: Market wizards: interviews with top traders. (1988). https://ci.nii.ac.jp/ncid/BB12043588
4. Barber, B.M., Odean, T.: All that glitters: the effect of attention and news on the buying behavior of individual and institutional investors. Rev. Financ. Stud. **21**(2), 785–818 (2007). https://doi.org/10.1093/rfs/hhm079
5. Baird, A.J.: Electronic trading masters: secrets from the pros! (2001). https://ci.nii.ac.jp/ncid/BA52573342
6. Sukumar, S.K.S., Mokhade, A.S.: Stochastic modeling using an ensemble of machines and deep learning for intraday stock trading. SSRN Electron. J. (2022). https://doi.org/10.2139/ssrn.4137577
7. Bartram, S.M., Branke, J., Motahari, M.: Artificial intelligence in asset management. SSRN Electron. J. (2020). https://doi.org/10.2139/ssrn.3692805
8. Alvior, E.B., Cabaluna, A.Y.: The handbook of technical analysis + test bank: the practitioner's comprehensive guide to technical analysis: book review. J. Econ. Financ. Account. Stud., **4**(2), 01–03, (2022). https://doi.org/10.32996/jefas.2022.4.2.31
9. Cox, J.C., Ross, S.A., Rubinstein, M.: Option pricing: a simplified approach. J. Financ. Econ. **7**(3), 229–263 (1979). https://doi.org/10.1016/0304-405x(79)90015-1
10. Hsu, P., Taylor, M.P., Wang, Z.: Technical trading: Is it still beating the foreign exchange market? J. Int. Econ. **102**, 188–208 (2016). https://doi.org/10.1016/j.jinteco.2016.03.012
11. Kondratieva, T., Razveeva, I., G.S Onore, Muntyan, E.R.: Implementation of an effective method of technical analysis for determining the market trend on the metatrader 5 platform. *Izvestiâ ÛFU. Tehničeskie Nauki*, **5**, 63–75. (2022). https://doi.org/10.18522/2311-3103-2022-5-63-75
12. Richard, E.M., Polla, D.L., Assir, M.Z., Contreras, M., Shahzad, M., Khan, A.A., Razzaq, A., Akram, J., Tarar, M.N., Blanpied, T.A., Ahmed, Z.M., Jamra, R.A., Wieczorek, D., Van

Bokhoven, H., Riazuddin, S., Riazuddin, S.: Bi-allelic variants in METTL5 cause Autosomal-Recessive intellectual disability and microcephaly. Am. J. Hum. Genet. **105**(4), 869–878 (2019). https://doi.org/10.1016/j.ajhg.2019.09.007

13. Spörer, J.: Backtesting of algorithmic cryptocurrency trading strategies. SSRN Electron. J. (2020). https://doi.org/10.2139/ssrn.3620154
14. Bhatt, K.A.: Investment and trading pattern of individuals dealing in the stock market. SIJ Trans. Ind. Financ. Busi-Ness Manag. **07**(01), 01–08 (2019). https://doi.org/10.9756/sijifbm/v7i1/0102580102
15. Kourtidis, D., Šević, Ž, Chatzoglou, P.: Investors' trading activity: a behavioral perspective and empirical results. J. Socio-Econ. **40**(5), 548–557 (2011). https://doi.org/10.1016/j.socec.2011.04.008
16. Shefrin, H., Statman, M.: The contributions of Daniel Kahneman and Amos Tversky. J. Behav. Financ. **4**(2), 54–58 (2003). https://doi.org/10.1207/s15427579jpfm0402_01
17. Dulock, H.L.: Research design: descriptive research. J. Pediatr. Oncol. Nurs. **10**(4), 154–157 (1993). https://doi.org/10.1177/104345429301000406
18. Murphy, J.J.: Study guide for technical analysis of the financial markets: a comprehensive guide to trading methods and applications. In *New York Institute of Finance eBooks*. (1999). http://ci.nii.ac.jp/ncid/BA58903111
19. Edelman, D., Davy, P.: Adaptive technical analysis in the financial markets using machine learning: a statistical view. In *Studies in fuzziness and soft computing* pp. 1–16. (2004). https://doi.org/10.1007/978-3-540-39972-8_1
20. Baralis, E., Cagliero, L., Cerquitelli, T., Garza, P., Pulvirenti, F.: Discovering profitable stocks for intraday trading. Inf. Sci. **405**, 91–106 (2017). https://doi.org/10.1016/j.ins.2017.04.013
21. Кусаев, З: KEY PATTERNS OF TECHNICAL ANALYSIS FROM THE POINT OF VIEW OF THE ELLIOTT WAVE THEORY. Прогрессивная Экономика, *4*, 83–92. (2024). https://doi.org/10.54861/27131211_2024_4_83
22. Agarwal, S.: A study on participation of women in Indian stock market as traders. Ijcrt— Int. J. Creat. Res. Thoughts (IJCRT), **9**(2), 1316–1319. (2021). https://www.ijcrt.org/download1.php?file=IJCRT2102163.pdf
23. Lukeman, J.: The market maker's edge: day trading tactics from a Wall Street insider. In *McGraw-Hill eBooks*. (2000). http://ci.nii.ac.jp/ncid/BA47607489

# Barriers to Sustainable Practices in the Construction Industry—A Bibliometric Analysis and Thematic Classification

**Angela Benham and Lijeesh Pullot**

**Abstract** Several studies explored the challenges involved in adopting sustainable practices in the construction industry from the perspective of different stakeholders, mainly developers, architects, consultants, and contractors. These challenges include financial constraints, lack of awareness, human resource issues, government policies, and market dynamics. However, a comprehensive bibliometric analysis and thematic classification of these barriers covering different aspects of sustainable construction is scarce owing to the fragmented nature of the literature. The objectives of the study are—(1). To examine the literature on barriers to incorporating sustainable practices in building construction, operation and demolition (2). To provide a thematic classification of the barriers (3). To identify research gaps and suggest avenues for further research. The articles were retrieved from the Scopus database and refined using PRISMA guidelines. 221 studies were included in the bibliometric analysis. Biblioshiny was used to identify the publication trends, most relevant countries, authors, publications, and highly cited articles. Furthermore, 30 empirical studies were analysed using NVivo 12 software to classify the barriers into the following themes: finance-related, attitude and behaviour-related, knowledge and awareness-related, government-related and market-related. The sub-themes cover various issues related to the development of sustainable buildings: prefabrication, Building Information Modelling, Cloud Computing, procurement of sustainable materials, energy management, managing sustainable projects and construction and demolition waste management. These comprehensive insights could help practitioners and policymakers develop strategies to drive the construction industry towards achieving its sustainability goals.

**Keywords** Bibliometric analysis · Sustainable construction barriers · Thematic classification

A. Benham (✉) · L. Pullot
Christ University, Bangalore, India
e-mail: angela.benham@res.christuniversity.in



M. Al Mubarak (ed.), *Sustainable Digital Technology and Ethics in an Ever-Changing Environment*, Studies in Systems, Decision and Control 237,
https://doi.org/10.1007/978-3-031-86708-8_4

# 1 Introduction

With the increasing emphasis on environmental protection, it is significant to focus on incorporating sustainability into business operations and processes. As defined by the United Nations Brundtland Commission in 1987, sustainable development is "the development that meets the needs of the present without compromising the ability of future generations to meet their own needs." During the last decade, global efforts to reduce $CO_2$ emissions to tackle global warming have increased tremendously. The construction sector, a key player in this global movement, provides the necessary infrastructure for business activities, generates employment opportunities, and boosts the economic growth of nations. Nevertheless, the industry contributes significantly to environmental degradation. Thus, the construction industry worldwide has been adopting sustainable practices such as energy management practices, water conservation, and effective construction and demolition waste handling to reduce the negative effects of its activities on the natural environment and society. However, there are barriers to implementing sustainable practices, a challenge that is not unique to any region but is a global concern.

This study aims

(1) To examine the literature on barriers to incorporating sustainable practices in the construction, operation and demolition of buildings;
(2) To provide a thematic classification of barriers and
(3) To identify the research gaps in extant literature and provide directions for further research.

# 2 Literature Review

Research on sustainability issues in the construction industry has gained momentum in the last decade. Several researchers have recently contributed to the literature on challenges faced by construction professionals in making building operations sustainable Durdyev et al. [10] Fitriani and Ajayi, [13]; Omopariola et al [24], Pradhananga et al. [27]. Oyewobi and Jimoh [25] stated that the country lacked a framework for sustainable procurement regulations, and the government failed to promote the procurement of sustainable building materials. In a survey conducted in three countries, namely, the USA, India and Afghanistan, it was found that the major impediments to developing Sustainable Projects were fewer possibilities for staff training to enhance their knowledge of sustainable concepts, minimal client awareness of the importance and benefits of sustainability, insufficient backing from the government, inadequate policies and regulations, high cost of sustainable materials, and difficulties in funding sustainable projects Shaker et al [28]. A study in China revealed that the main obstacle to prefabricated construction was the high cost compared to traditional buildings Hong et al. [14]. Furthermore, a recent study in China revealed that the main challenges to Prefabricated Construction were insufficient industry norms and standards, inadequate management experience and insufficient professional expertise Wang et al. [30]. Oke et al. [21] stated that network issues and restricted access to data were Nigeria's main obstacles to adopting cloud computing.

Despite several researchers' significant contributions, research efforts in the domain are limited. Table 1 shows the obstacles to Sustainable Practices in the Construction Industry.

**Table 1** Barriers to sustainable practices in the construction industry

| Barriers | Reference |
|---|---|
| Lack of financing schemes for EMT | (Iqbal et al. [17]) |
| Lack of funding for C&DWM | (Al-Otaibi et al. [5], Negash et al. [20]) |
| High cost of implementing BIM | (Alghuried [4]; Huang et al. [15]) |
| Perceived additional cost of SM | (Akadiri [2]) |
| High transportation cost of PM | (Li et al. [19]) |
| Industry's resistance to change | (Al-Yami and Sanni-Anibire [6]; Olawumi et al. [22]) |
| Lack of trust in Cloud Computing | (Omer et al. [23]) |
| Lack of top management support | (Fathalizadeh et al. [11]; Karji et al. [18]) |
| Difficulties in gaining expertise | (Ahmed et al. [1]) |
| Insufficient knowledge and awareness of Sustainable Construction | (Akindele et al. [3]; Durdyev et al. [10]; Fitriani and Ajayi [13]; Hwang et al. [16]) |
| Lack of understanding of the benefits of SP | (Fathalizadeh et al. [11, 12]; Serik et al., 2020) |
| Lack of information on SM | (Akadiri [2]) |
| Lack of exemplary SP | (Omopariola et al. [24]) |
| Lack of comprehension of the processes required for BIM | (Al-Yami and Sanni-Anibire, [6]; Olawumi et al. [22]) |
| Inadequate Sustainable Building codes and regulations | (Ahmed et al. [1]) |
| Lack of legal enforcement of C&DWM | (Negash et al. [20]) |
| Lack of incentives from the government | (Hai et al., 2020; Serik et al., 2020; Zulu et al. [31]) |
| Low demand for SP | (Dalirazar and Sabzi, [8]; Djokoto et al. [9]) |

* EMT—Energy Management Technology, C&DWM—Construction and Demolition Waste Management, BIM—Building Information Modelling, SM—Sustainable Material, SP—Sustainable Project, PM—Prefabricated Module

TITLE-ABS-KEY ( barrier AND construction AND sustain* ) AND PUBYEAR > 2013 AND PUBYEAR < 2025 AND ( LIMIT-TO ( DOCTYPE , "ar" ) ) AND ( LIMIT-TO ( PUBSTAGE , "final" ) ) AND ( LIMIT-TO ( SRCTYPE , "j" ) ) AND ( LIMIT-TO ( LANGUAGE , "English" ) )

**Fig. 1** Search string. *Source* Scopus database

# 3 Research Methodology

The studies selected for Bibliometric analysis and thematic classification were retrieved from the Scopus Database. The initial search for articles was performed in the title/abstract/keyword domain between 2014 and 2024. The search string for this purpose is shown in Fig. 1. The last search was performed on 7 September 2024. The results were imported to Microsoft Excel for further refining using the PRISMA guidelines. In the first step, the titles of the papers were manually screened to select the studies that contained the keywords "barrier", "challenge", or "obstacle". The process yielded 318 results. In the second step, the abstracts of the papers were subject to manual screening. This aimed to eliminate papers that were not focused on the challenges faced by the construction industry in adopting sustainable practices. This process resulted in the final selection of 221 articles. In the last step, the full papers were scrutinised to select the most relevant articles for thematic analysis. This process yielded 30 articles, as shown in Fig. 2.

# 4 Bibliometric Results

## *4.1 Trends in Publication*

The number of articles published annually shows an increasing trend from 2019 onward. 61 out of 221 articles chosen for review were published in 2023, followed by 33 and 29 published in 2022 and 2021, respectively, as shown in Fig. 3. Although, all of the articles published in 2024 could not be included in the review since the study was conducted in the middle of 2024. Nevertheless, the results indicate that the interest among the researchers in the field has been growing significantly over the years, which is most likely the result of increasing awareness among the global community about the adverse effects of construction on the environment and society.

## *4.2 Top Most Cited Countries*

As shown in Table 2, China obtained 1009 total citations for articles. This is followed by Hong Kong and Australia receiving 881 and 413 total citations for articles, respectively. Hong Kong obtained the highest average article citations, 80.10, followed by

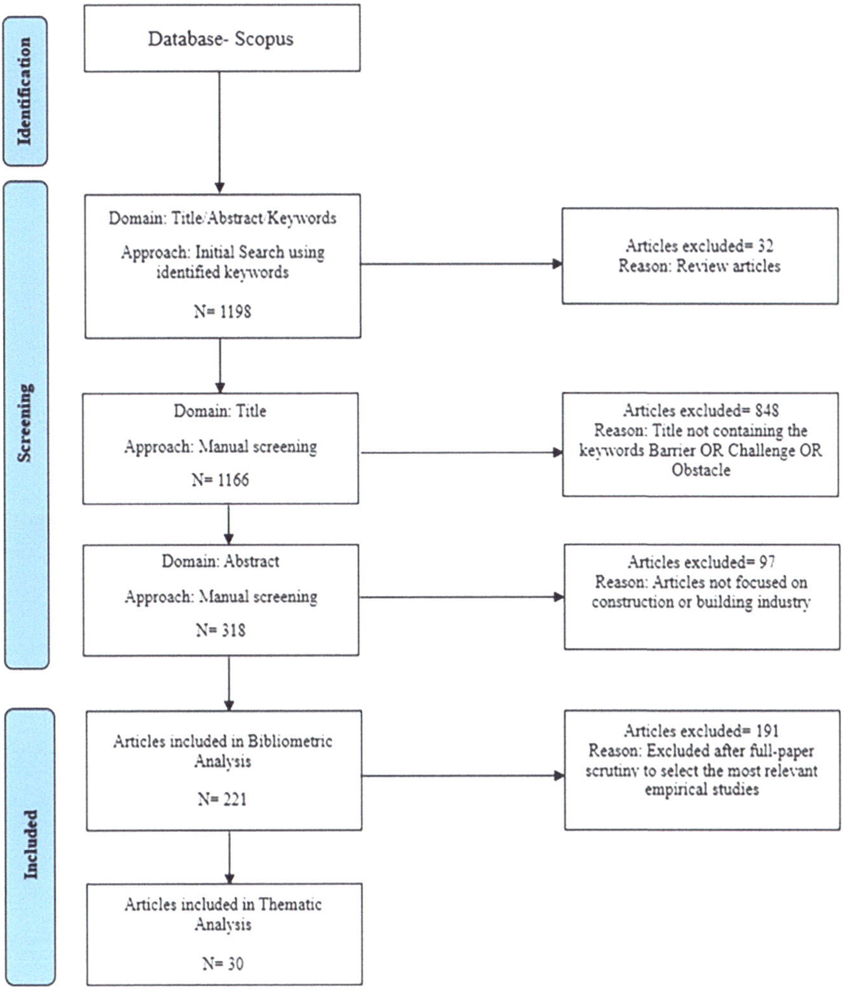


**Fig. 2** PRISMA flowchart. *Source* Authors' creation

New Zealand and Italy, scoring 68 and 67, respectively. This indicates that these countries focus on producing articles of high quality that have been published in reputed journals with a high impact factor, like the Journal of Cleaner Production.

## 4.3 Most Relevant Authors

Table 3 displays the top ten authors who contributed to the research on barriers to Sustainable Construction. These top ten authors produced 57 articles, accounting for

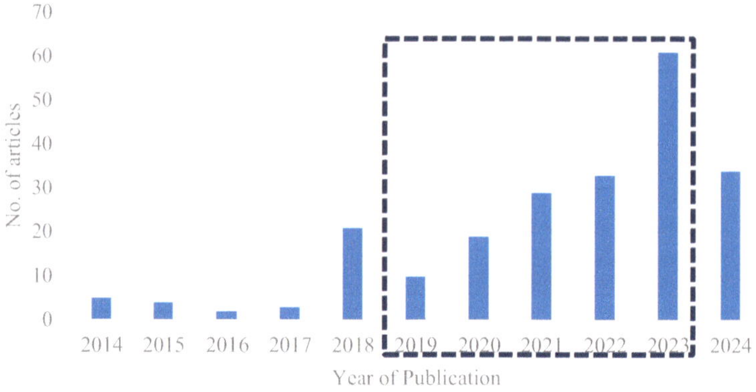


**Fig. 3** Publication trend. *Source* Authors' creation

**Table 2** Top most cited countries

| Country | Total citations | Average article citations |
|---|---|---|
| China | 1009 | 45.90 |
| Hong Kong | 881 | 80.10 |
| Australia | 413 | 22.90 |
| United Kingdom | 340 | 17.00 |
| New Zealand | 272 | 68.00 |
| USA | 204 | 22.70 |
| Malaysia | 157 | 13.10 |
| Italy | 134 | 67.00 |
| Saudi Arabia | 134 | 10.30 |
| Canada | 128 | 32.00 |

*Source* Biblioshiny

26% of the literature reviewed. Kineber AF and Oke AE produced 8 articles each, while Chan APC, Darko A and Durdyev S produced 6.

## 4.4 Most Relevant Sources

The journals with the highest number of articles on obstacles to adopting Sustainable Practices in the Construction Industry are shown in Table 4. Sustainability (Switzerland) published the most articles, 24 out of 221. This is followed by Buildings and the Journal of Cleaner Production, which published 17 articles each. The top ten journals published 102 out of 221 articles, accounting for 46% of the articles reviewed.

**Table 3** Most relevant authors

| Authors | Articles |
|---|---|
| Kineber AF | 8 |
| Oke AE | 8 |
| Chan APC | 6 |
| Darko A | 6 |
| Durdyev S | 6 |
| Hosseini MR | 6 |
| Edwards DJ | 5 |
| Martek I | 4 |
| Othman I | 4 |
| Singh AK | 4 |

*Source* Biblioshiny

**Table 4** Most relevant sources

| Sources | Articles |
|---|---|
| Sustainability (switzerland) | 24 |
| Buildings | 17 |
| Journal of cleaner production | 17 |
| Engineering, construction and architectural management | 13 |
| International journal of construction management | 9 |
| Smart and sustainable built environment | 6 |
| Built environment project and asset management | 5 |
| Journal of building engineering | 4 |
| Journal of engineering, design and technology | 4 |
| Ain shams engineering journal | 3 |

*Source* Biblioshiny

## 4.5 Most Highly Cited Articles

Table 5 shows the top ten highly cited articles on impediments to Sustainable Construction. With 349 citations, "Barriers to promoting prefabricated construction in China: A cost–benefit analysis", published in the "Journal of Cleaner Production" in 2018, is the most cited article. "Crital barriers to green building technologies adoption in developing countries: The case of Ghana", published in the "Journal of Cleaner Production" in 2018, is the second most cited article with 342 citations. "Prioritising barriers to adopt circular economy in construction and demolition waste Management, published in "Resources, Conservation and Recycling" in 2018, is the third most cited article with 296 citations.

Table 5 Most highly cited articles

| Paper | DOI | Total citations |
|---|---|---|
| Hong j, 2018, j clean prod | https://doi.org/10.1016/j.jclepro.2017.10.171 | 349 |
| Chan apc, 2018, j clean prod | https://doi.org/10.1016/j.jclepro.2017.10.235 | 342 |
| Mahpour a, 2018, resour conserv recycl | https://doi.org/10.1016/j.resconrec.2018.01.026 | 296 |
| Darko a, 2018, j clean prod | https://doi.org/10.1016/j.jclepro.2018.07.318 | 175 |
| Olawumi to, 2018, j build eng | https://doi.org/10.1016/j.jobe.2018.06.017 | 154 |
| Djokoto sd, 2014, j sustain dev | https://doi.org/10.5539/jsd.v7n1p134 | 148 |
| Durdyev s, 2018, sustainability | https://doi.org/10.3390/su10020392 | 125 |
| Hussain k, 2019, build environ | https://doi.org/10.1016/j.buildenv.2019.106225 | 124 |
| Durdyev s, 2018, j clean prod | https://doi.org/10.1016/j.jclepro.2018.08.304 | 123 |
| Giorgi s, 2022, j clean prod | https://doi.org/10.1016/j.jclepro.2022.130395 | 112 |

*Source* Biblioshiny

# 5 Thematic Classification of Barriers

The barriers were subject to thematic analysis to identify the main themes. Thematic analysis is used to summarise data, identify patterns within the dataset, and analyse and interpret it systematically to generate codes and themes guided by the research questions. It emphasises producing high-quality results through rigorous analysis Clarke and Braun [7].

## *5.1 Finance-Related Barriers*

These include an absence of funding plans to implement energy management technologies, the high cost of implementing Building Information Modelling (BIM), perceived additional costs of sustainable materials, the high transportation cost of prefabricated modules, and inadequate funding for construction and demolition waste management.

## 5.2 Attitude and Behaviour-Related Barriers

These include the absence of collaboration among project stakeholders to implement energy management practices, insufficient attention paid to waste management of construction and demolition, trust issues in adopting cloud computing and other sustainable digital technologies, reluctance to adopt new and sustainable technologies, insufficient backing from top management and absence of skilled professionals to implement sustainable practices.

## 5.3 Knowledge and Awareness-Related Barriers

These include limited knowledge of stakeholders of the benefits of sustainable practices, inadequate knowledge of sustainable purchasing, insufficient information on sustainable materials, absence of exemplary sustainable projects, lack of technical skills to implement cloud computing, lack of comprehension of the processes needed for Building Information Modelling, the low awareness level of waste management of construction and demolition processes and lack of data sharing between construction firms and suppliers.

## 5.4 Government-Related Barriers

These include inadequate codes and regulations for sustainable building, absence of legislation on sustainable procurement, lack of legal enforcement of construction and demolition waste management and lack of government incentives to promote sustainable construction practices.

## 5.5 Market-Related Barriers

These barriers include the restricted supply of sustainable materials, lack of access to sustainable technologies in the local market and the lack of client interest in sustainable buildings.

## 6 Research Gaps and Directions for Further Research

A descriptive Analysis of the bibliometric parameters shows a surge of interest in the topic among practitioners, researchers, and academicians worldwide. This is evident from the increasing publication trend since 2019, attributed to the growing emphasis on incorporating sustainability into construction practices globally. Although developing countries contribute more to the existing literature than developed countries, the contributions of the countries, particularly China, the UK and Australia, are significant. The most relevant authors are Kineber AF, Oke AE and Chan APC. The most highly cited sources are "Sustainability" (Switzerland), "Buildings", and "Journal of Cleaner Production". The most highly cited articles are "Barriers to Promoting Prefabricated Construction in China: A Cost–benefit Analysis" and "Critical Barriers to Green Building Technologies Adoption in Developing Countries: The Case of Ghana", the second most cited article.

The thematic analysis reveals that the main barriers to sustainable practices are finance-related, attitude and behaviour-related, knowledge and awareness-related, government-related and market-related. The sub-themes cover various issues related to different aspects of the development of sustainable buildings: prefabrication, Building Information Modelling, Cloud Computing, purchase of sustainable materials, energy management, managing sustainable projects and effective handling of construction and demolition waste. Future researchers can examine the organisational-level and project-level challenges involved in integrating sustainability into the building and demolition operations of the construction industry.

## 7 Conclusion

The study contributes to a better understanding of the barriers to adopting Sustainable Practices in the construction sector. It performs a bibliometric analysis of the extant literature to understand the current trends in publication, as well as the most relevant authors, sources, and articles in the field. It provides a theme-wise classification of the barriers, enhancing the current knowledge of the challenges stakeholders face in adopting sustainable practices and enabling stakeholders to develop strategies to overcome them. It provides opportunities for future researchers to gain deeper insights into the barriers at the organisational and project levels. Furthermore, it enables collaboration between practitioners, research organisations, and policymakers to address the barriers and drive the construction industry towards creating a sustainable built environment.

## References


1. Ahmed, A.M., Sayed, W., Asran, A., Nosier, I.: Identifying barriers to the implementation and development of sustainable construction. Int. J. Constr. Manag. **23**(8), 1277–1288 (2021). https://doi.org/10.1080/15623599.2021.1967577
2. Akadiri, P.O.: Understanding barriers affecting the selection of sustainable materials in building projects. J. Build. Eng. **4**, 86–93 (2015). https://doi.org/10.1016/j.jobe.2015.08.006
3. Akindele, O.E., Ajayi, S., Toriola-Coker, L., Oyegoke, A.S., Alaka, H., Zulu, S.L.: Sustainable construction practice in Nigeria: barriers and strategies for improvement. Built Environ. Proj. Asset Manag. **13**(4), 590–609 (2023). https://doi.org/10.1108/bepam-06-2022-0085
4. Alghuried, A.: Measuring the benefits and barriers of the implementation of BIM in sustainable practice in the construction industry of Saudi Arabia. Sustainability **15**(19), 14323 (2023). https://doi.org/10.3390/su151914323
5. Al-Otaibi, A., Bowan, P.A., Daiem, M.M.A., Said, N., Ebohon, J.O., Alabdullatief, A., Al-Enazi, E., Watts, G.: Identifying the barriers to sustainable management of construction and demolition waste in developed and developing countries. Sustainability **14**(13), 7532 (2022). https://doi.org/10.3390/su14137532
6. Al-Yami, A., Sanni-Anibire, M.O.: BIM in the Saudi Arabian construction industry: state of the art, benefit and barriers. Int.J. Build. Pathol. Adapt. **39**(1), 33–47 (2019). https://doi.org/10.1108/ijbpa-08-2018-0065
7. Clarke, V., Braun, V.: Thematic analysis. J. Posit. Psychol. **12**(3), 297–298 (2016). https://doi.org/10.1080/17439760.2016.1262613
8. Dalirazar, S., Sabzi, Z.: Strategic analysis of barriers and solutions to development of sustainable buildings using PESTLE technique. Int. J. Constr. Manag. **23**(1), 167–181 (2020). https://doi.org/10.1080/15623599.2020.1854931
9. Djokoto, S.D., Dadzie, J., Ohemeng-Ababio, E.: Barriers to sustainable construction in the Ghanaian construction industry: consultants perspectives. J. Sustain. Dev., **7**(1). (2014). https://doi.org/10.5539/jsd.v7n1p134
10. Durdyev, S., Zavadskas, E., Thurnell, D., Banaitis, A., Ihtiyar, A.: Sustainable construction industry in cambodia: awareness. Driv. Barriers. Sustain. **10**(2), 392 (2018). https://doi.org/10.3390/su10020392
11. Fathalizadeh, A., Hosseini, M.R., Silvius, A.G., Rahimian, A., Martek, I., Edwards, D.J.: Barriers impeding sustainable project management: a Social Network Analysis of the Iranian construction sector. J. Clean. Prod. **318**, 128405 (2021). https://doi.org/10.1016/j.jclepro.2021.128405
12. Fathalizadeh, A., Hosseini, M.R., Vaezzadeh, S.S., Edwards, D.J., Martek, I., Shooshtarian, S.: Barriers to sustainable construction project management: the case of Iran. Smart Sustain. Built Environ. **11**(3), 717–739 (2021). https://doi.org/10.1108/sasbe-09-2020-0132
13. Fitriani, H., Ajayi, S.: Barriers to sustainable practices in the Indonesian construction industry. J. Environ. Planning Manage. **66**(10), 2028–2050 (2022). https://doi.org/10.1080/09640568.2022.2057281
14. Hong, J., Shen, G.Q., Li, Z., Zhang, B., Zhang, W.: Barriers to promoting prefabricated construction in China: a cost–benefit analysis. J. Clean. Prod. **172**, 649–660 (2018). https://doi.org/10.1016/j.jclepro.2017.10.171
15. Huang, B., Lei, J., Ren, F., Chen, Y., Zhao, Q., Li, S., Lin, Y.: Contribution and obstacle analysis of applying BIM in promoting green buildings. J. Clean. Prod. **278**, 123946 (2021). https://doi.org/10.1016/j.jclepro.2020.123946
16. Hwang, B., Shan, M., Lye, J.: Adoption of sustainable construction for small contractors: major barriers and best solutions. Clean Technol. Environ. Policy **20**(10), 2223–2237 (2018). https://doi.org/10.1007/s10098-018-1598-z
17. Iqbal, M., Ma, J., Ahmad, N., Hussain, K., Usmani, M.S., Ahmad, M.: Sustainable construction through energy management practices in developing economies: an analysis of barriers in the construction sector. Environ. Sci. Pollut. Res. **28**(26), 34793–34823 (2021). https://doi.org/10.1007/s11356-021-12917-7

18. Karji, A., Namian, M., Tafazzoli, M.: Identifying the key barriers to promote sustainable construction in the United States: a principal component analysis. Sustainability **12**(12), 5088 (2020). https://doi.org/10.3390/su12125088
19. Li, Z., Zhang, S., Meng, Q., Hu, X.: Barriers to the development of prefabricated buildings in China: a news coverage analysis. Eng. Constr. Archit. Manag. **28**(10), 2884–2903 (2020). https://doi.org/10.1108/ecam-03-2020-0195
20. Negash, Y.T., Hassan, A.M., Tseng, M., Wu, K., Ali, M.H.: Sustainable construction and demolition waste management in Somaliland: regulatory barriers lead to technical and environmental barriers. J. Clean. Prod. **297**, 126717 (2021). https://doi.org/10.1016/j.jclepro.2021.126717
21. Oke, A.E., Kineber, A.F., Abdel-Tawab, M., Abubakar, A.S., Albukhari, I., Kingsley, C.: Barriers to the implementation of cloud computing for sustainable construction in a developing economy. Int. J. Build. Pathol. Adapt. **41**(5), 988–1013 (2021). https://doi.org/10.1108/ijbpa-07-2021-0098
22. Olawumi, T.O., Chan, D.W., Wong, J.K., Chan, A.P.: Barriers to the integration of BIM and sustainability practices in construction projects: a Delphi survey of international experts. J. Build. Eng. **20**, 60–71 (2018). https://doi.org/10.1016/j.jobe.2018.06.017
23. Omer, M.M., Kineber, A.F., Oke, A.E., Kingsley, C., Alyanbaawi, A., Rached, E.F., Elmansoury, A.: Barriers to using cloud computing in sustainable construction in Nigeria: a fuzzy synthetic evaluation. Mathematics **11**(4), 1037 (2023). https://doi.org/10.3390/math11041037
24. Omopariola, E.D., Olanrewaju, O.I., Albert, I., Oke, A.E., Ibiyemi, S.B.: Sustainable construction in the Nigerian construction industry: unsustainable practices, barriers and strategies. J. Eng. Des. Technol. (2022). https://doi.org/10.1108/jedt-11-2021-0639
25. Oyewobi, L.O., Jimoh, R.A.: Barriers to adoption of sustainable procurement in the nigerian public construction sector. Sustainability **14**(22), 14832 (2022). https://doi.org/10.3390/su142214832
26. Pham, H., Kim, S., Luu, T.: Managerial perceptions on barriers to sustainable construction in developing countries: Vietnam case. Environ. Dev. Sustain. **22**(4), 2979–3003 (2019). https://doi.org/10.1007/s10668-019-00331-6
27. Pradhananga, P., Elzomor, M., Kasabdji, G.S.: Barriers and drivers to the adoption of sustainable construction practices in developing countries: a case study of Venezuela. J. Arch. Eng. **27**(3). (2021). https://doi.org/10.1061/(asce)ae.1943-5568.0000483
28. Shaker, M.R., Eustace, B.S., Erukala, H.K.G., Patel, R.G., Mohammed, M.B., Jabri, M.A., Desai, K., Goyal, R., Chang, B.: Analysis of survey on barriers to the implementation of sustainable projects. Sustainability **14**(24), 16830 (2022). https://doi.org/10.3390/su142416830
29. Tokbolat, S., Karaca, F., Durdyev, S., Calay, R.K.: Construction professionals' perspectives on drivers and barriers of sustainable construction. Environ. Dev. Sustain. **22**(5), 4361–4378 (2019). https://doi.org/10.1007/s10668-019-00388-3
30. Wang, Q., Shen, C., Guo, Z., Zhu, K., Zhang, J., Huang, M.: Research on the barriers and strategies to promote prefabricated buildings in china. Buildings **13**(5), 1200 (2023). https://doi.org/10.3390/buildings13051200
31. Zulu, S.L., Zulu, E., Chabala, M., Chunda, N.: Drivers and barriers to sustainability practices in the Zambian Construction Industry. Int. J. Constr. Manag. **23**(12), 2116–2125 (2022). https://doi.org/10.1080/15623599.2022.2045425

# Students Perception on Corporate Social Responsibility in Coimbatore District

**S. Hemaprasanna, J. Karthikeyani, K. Suresh Babu, R. Muthukumar, and M. Janani**

**Abstract** Corporate Social Responsibility (CSR) is the emerging topic today, though is dated back to 1953. As corporate utilise all the resources from the society, they are more responsible to the society, it is the responsibility of the corporates to safeguard the society. Corporate Social Responsibility is closely related to employees, customers, general public and society, to the company itself and so on. This study focuses on the level of perception of college students on Corporate Social Responsibility (CSR) in Coimbatore district. College students were selected as respondents of this study since these students are the entrepreneurs of tomorrow, future managers, future executives and so on. Convenience sampling is the sampling method used for study. The findings show that of seven variables tested, gender at ten per cent level parents occupation at five per cent level, knowledge on CSR at five per cent level are significantly associated with corporate social responsibility.

**Keywords** Corporate social responsibility · College students · Awareness · Perception

S. Hemaprasanna · J. Karthikeyani · K. Suresh Babu · R. Muthukumar
Department of Commerce, PSG College of Arts and Science, Coimbatore, India
e-mail: hemaprasanna@psgcas.ac.in

J. Karthikeyani
e-mail: karthikeyani@psgcas.ac.in

K. Suresh Babu
e-mail: sureshbabu@psgcas.ac.in

M. Janani (✉)
Department of Professional Accounting and Finance, Kristu Jayanti College, Bengaluru, India
e-mail: janani.m@krsitujayanti.com



M. Al Mubarak (ed.), *Sustainable Digital Technology and Ethics in an Ever-Changing Environment*, Studies in Systems, Decision and Control 237,
https://doi.org/10.1007/978-3-031-86708-8_5

# 1 Introduction

Corporate Social Responsibility (CSR) is the emerging topic today, though is dated back to 1953. Internationally, Norway is ranked first in adopting the CSR, it is followed by France and Italy. The USA has the lowest score. In India CSR become very effective as it is made compulsory as per the Section 135 of the Companies Act, 2013 which came into force from April 2014. Corporates are the integral part of the society. Their activities definitely have impact on the society. American economist Howard Bowen framed the word Corporate Social Responsibility in the year 1953. Corporate Social Responsibility defines the responsibility of the corporates on the society. Though its origin has dated back to 1950s, it is hot topic even today. Corporates take everything from the society for their survival and existence, what the corporates have to do is something in return which benefits the society. Corporate Social Responsibility is closely related to employees, customers, general public and society, to the company itself and so on. Corporate Social Responsibility attracts employees, motivates them and retain them. To the customers, it attracts customer, enhances customer loyalty, gives quality products at nominal price. To the company it increases the reputation of the company and the company can have easy access to the capital through Corporate Social Responsibility. To the general public and society, by way of utilising the resources in such a way that they are not depleted and pave the way for sustainable development.

Corporate social responsibility (CSR) describes dedication of a company to do business as per ethical terms. It has social, economic and environmental impact. All the public limited companies are socially accountable to itself, its community and its stakeholders. Sustainability is the need of the hour. Sustainability refers to the utilisation of the resources in such a way that it is available to the future generation. Today in India corporates are growing tremendously, which makes the responsibility of the corporates more crucial. Corporate Social Responsibility is wider concept which covers production of eco-friendly products, disposing waste without spoiling the environment, effective utilisation of resources, good working environment and fair pay to employees, doing business following ethical practices, fulfilling the expectations of the stake holders, earning normal profits and so on.

# 2 Literature Review

De-los-Angeles et al. [8]—analysed the influence of corporate social responsibility upon business profitability. The objective of the author was to show there is neutral relationship between CSR and income of the business. Stefanie Hiss [14]—focused on the changing responsibility of the corporates in Germany. The study portrays the change in the structure of the corporates and change in the functions of the corporates in Germany. Chen and Bouvain [15]—examined top performing companies in four developed economies namely US, UK, Australia and Germany. Khurshid et al.

[10]—explored the perception of students doing post graduate in management studies on CSR. The respondents of the study were selected from Saudi university. Pal and Sarkar [16] examined knowledge of CSR among the people living in Bangladesh. The results of the study showed that most of the people have good awareness level in CSR. Zhang and Ahmad [11]—exhibits that CSR is advantageous to the society. Conclusion of the study portrays that CSR improves image of the corporates among the customers. KO, kim, Choi [17] examined anti-consumer awareness and CSR. The results of the study showed there exists negative relationship between CSR perception and anti-consumer awareness. Mario Rosario Gonzalez-Rodriguez, Maria Carmen Diaz-Fernandez—investigated perception and awareness level of customers of hotels on CSR. The study explained the role and importance of CSR in building the image of the hotels. Nitamoah et al. [3]—addressed several questions relating to Corporate Social Responsibility. The study was conducted by collecting data from the customers of UT bank Ghana and Barclays bank Ghana Ltd. Conclusion of the study states that the interrelationship with awareness level of CSR, commitment of the firm and performance of the organization was good. Agripina, Sagabho, Baleche Asha [1]—assessed the activities of the companies towards CSR. The results of the study exhibited that companies in Mwanza concentrated on health, education and entertainment of the people as a part of implementing Corporate Social Responsibility.

## 3 Objectives of the Study

1. To examine the perception of the college students regarding Corporate Social Responsibility
2. To find out factors associated with perception on Corporate Social Responsibility

## 4 Hypotheses

$H_{01}$: Gender is not associated with the perception on corporate social responsibility.

$H_{02}$: Degree pursing and perception of students on CSR are not associated the each other.

$H_{03}$: There is no significant relationship between Family income and perception on CSR.

$H_{04}$: Parent education is not associated with the perception on corporate social responsibility.

$H_{05}$: There is no significant relationship between Occupation of parents and perception on CSR.

$H_{06}$: Knowledge on CSR is not associated with perception on corporate social responsibility.

## 5 Research Methodology

Both primary and secondary data are used in the study. Data have been collected with a sample of college students as respondents. The data have been collected from one hundred and six students in Coimbatore district by administering a structured questionnaire. Convenience sampling method has been applied in order to obtain their views on corporate social responsibility. Secondary data is also applied in the study for theory and review purposes. Percentage analysis, mean, standard deviation and Chi-square test were the tools applied for the study.

### *5.1 Students Perception on Corporate Social Responsibility*

**I Students' Socio-Economic Profile**

The socio-economic profile of students broadly includes gender, degree pursuing, type of family, family income, parent education, parent occupation and knowledge on CSR, which are explained in the following paragraphs. Table 1 has been prepared in order to analyse the student respondents' socio-economic profile.

Out of the total 106 respondents who are college students, 57.5% are female students and 42.9% are male students. 94.33% of the respondents are UG and the 5.67% are PG students. Majority of the students live in nuclear family compared with Joint family. 44.3% Family income of the students are below 50,000. 38.7% students' parents are degree holders. 54.7% students are self-employed. 57.5% of the students are aware of CSR. Magazines and newspapers are the main source of knowledge about CSR. Most of the students stated that CSR should be included as part of their curriculum.

**II Variables Associated with Perception on CSR**

Rosak-Szyrocka et.al (2022) stated that the companies are doing CSR programs will improve the opinion and reputation among the stakeholders. A 'Perception Index' has been framed in order to analyse the variables that are associated with perception of students on Corporate Social Responsibility. There are four options given under each questions. The scores are assigned to each question which are relating to opinion of students on corporate social responsibility. Five point Likert's scale has been used to assign the scores. Chi-square test has been used to test the variables associated with Perception Index. The level of confidence chosen are ten, five and one per cent. The following table represents the variables associated with perception of students on corporate social responsibility.

**Table 1** Socio-economic profile of students

| Profile | No. of respondents | Percentage to total (%) |
|---|---|---|
| 1. Gender<br>Male<br>Female<br>Transgender | 45<br>61<br>– | 42.9<br>57.5<br>-- |
| 2. Degree Pursuing<br>UG<br>PG | 100<br>6 | 94.33<br>5.67 |
| 3. Type of family<br>Joint<br>Nuclear | 24<br>82 | 22.6<br>77.4 |
| 4. Family income<br>Below 50,000<br>50,001–75,000<br>Above 75,000 | 47<br>27<br>32 | 44.3<br>25.7<br>30.2 |
| 5. Parents education<br>Higher secondary<br>Degree<br>Post graduate<br>Professional | 38<br>41<br>11<br>16 | 35.8<br>38.7<br>10.4<br>15.1 |
| 6. Occupation of parents<br>Self-employed<br>Agriculturist<br>Employed in public companies<br>Employed in private companies<br>Professionals | 58<br>8<br>9<br>24<br>7 | 54.7<br>7.5<br>8.5<br>22.6<br>6.6 |
| 7. Knowledge about CSR<br>Yes<br>No | 61<br>45 | 57.5<br>42.5 |
| 8. Sources of getting knowledge on CSR<br>Through friends and relative<br>Through parents<br>From newspapers and magazines<br>CSR activities done in our village<br>Others (specify) | 14<br>9<br>29<br>1<br>53 | 13.2<br>8.5<br>27.4<br>0.9<br>50 |
| 9. CSR as a part of curriculum<br>Yes<br>No | 51<br>55 | 48.1<br>59.9 |
| 10. CSR in curriculum<br>As a course<br>As a module in course | 18<br>88 | 17<br>83 |
| 11. Opinion on CSR in curriculum<br>Should it to be included<br>Not to be included | 89<br>17 | 84.8<br>15.2 |

# 6 Gender and Level of Perception

*$H_{01}$: There is no Significant Relationship Between Gender and Perception on Corporate Social Responsibility*

Larrán et al. (2018) found that the variable gender is associated with perception on CSR. From Table 2 it can be inferred that male students have high level of as compared to female students. This may because; the male students perceive society-related information positively. The calculated $x^2$ value is greater than the table value at ten per cent level, it is inferred that gender influences the perception level on CSR. Therefore, the null hypothesis is rejected. Similar result has been observed in the study of Eweje and Brunton (2010) in New Zealand that females are more ethically disposed than males. But, Felix Okechukwu Ugwuozor (2020) observed that gender does not influence students' perception of CSR.

# 7 Degree Pursuing and Level of Perception

*$H_{02}$: Degree Pursing and Perception of Students on CSR Are not Associated the Each Other.*

Ugwuozor (2020) found that the opinion of students differs according to their different levels of study. The course or the contents are included in their curriculum then the students may have knowledge on CSR. "Academic degree positively influences the students' perceptions of CSR" is accepted" is observed by Premendra Kumar Singh et al. (2014). From the above table, the percentage is high with the PG students who have high level of perception. It can be said that PG students have high perception level as compared to UG students. This may be because, the PG students may have knowledge on corporate social responsibility. However, as the calculated $x^2$ value is less than the table value at ten per cent level, hence, degree does not influence the perception level on CSR. Therefore, the null hypothesis is accepted.

# 8 Family Income and Level of Perception

*$H_{03}$: Family Income Does not Influence Perception on Corporate Social Responsibility*

There are also corporate social responsibilities in which case consumers' reaction in terms of loyalty is positively influenced regardless of their income level (Ovidiu-Ioan MOISESCU, 2015). It can be seen from the table that the percentage of students who have low perception level is high with those whose monthly family income is above Rs. 75,000. The percentage of students who have high perception level is high with the same group of students. On comparison of the percentages, it can be

**Table 2** Variables associated with perception of students on CSR

| Gender and level of perception | | | | | | |
|---|---|---|---|---|---|---|
| Gender | Level of perception | | | Total (106) | $x^2$ Value | Table Value |
| | Low (22) | Moderate (67) | High (17) | | | |
| Male | 10 (22.22%) | 24 (53.3%) | 11 (24.44%) | 45 (100.00%) | d.f.: 2 $x^2$ = 4.733 | 1% = 9.210 5% = 5.991 10% = 4.605 |
| Female | 12 (19.67%) | 43 (70.49%) | 6 (9.83%) | 61 (100.00%) | | |
| *Degree pursuing and level of perception* | | | | | | |
| Degree | Level of perception | | | Total (106) | $x^2$ Value | Table Value |
| | Low (22) | Moderate (67) | High (17) | | | |
| UG | 20 (20.00%) | 65 (65.00%) | 15 (15.00%) | 100 (100.00%) | d.f.: 2 $x^2$ = 2.569 | 1% = 9.210 5% = 5.991 10% = 4.605 |
| PG | 2 (33.33%) | 2 (33.33%) | 2 (33.34%) | 6 (100.00%) | | |
| *Family income and level of perception* | | | | | | |
| Family Income | Level of perception | | | Total (106) | $x^2$ Value | Table Value |
| | Low (22) | Moderate (67) | High (17) | | | |
| Up to Rs. 50,000 | 9 (19.2%) | 34 (72.3%) | 4 (8.5%) | 47 (100.0%) | d.f.: 6 $x^2$ = 9.381 | 1% = 16.812 5% = 12.592 10% = 10.645 |
| Rs. 50,001 – Rs. 75,000 | 5 (18.5%) | 17 (63.0%) | 5 (18.5%) | 27 (100.0%) | | |
| Above Rs. 75,000 | 8 (25.0%) | 16 (50.0%) | 8 (25.0%) | 32 (100.0%) | | |
| *Parent education and level of perception* | | | | | | |
| Parent education | Level of perception | | | Total (106) | $x^2$ Value | Table Value |
| | Low (22) | Moderate (67) | High (17) | | | |
| Higher secondary | 9 (23.7%) | 26 (68.4%) | 3 (7.9%) | 38 (100.0%) | d.f.: 6 $x^2$ = 4.663 | 1% = 16.812 5% = 12.592 10% = 10.645 |
| Degree | 8 (19.5%) | 23 (56.1%) | 10 (24.4%) | 41 (100.0%) | | |
| Post graduate | 2 (18.2%) | 8 (72.7%) | 1 (9.1%) | 11 (100.0%) | | |

(continued)

**Table 2** (continued)

| Gender and level of perception | | | | | | |
|---|---|---|---|---|---|---|
| Gender | Level of perception | | | Total (106) | $x^2$ Value | Table Value |
| | Low (22) | Moderate (67) | High (17) | | | |
| Professional | 3 (18.8%) | 10 (62.5%) | 3 (18.8%) | 16 (100.0%) | | |
| *Parent occupation and level of perception* | | | | | | |
| Parent occupation | Level of Perception | | | Total (106) | $x^2$ Value | Table Value |
| | Low (22) | Moderate (67) | High (17) | | | |
| Agriculturist | 3 (37.5%) | 5 (62.5%) | - | 8 (100.0%) | d.f.: 8 $x^2$ = 16.392 | 1% = 20.090 5% = 15.507 10% = 13.362 |
| Employed in private companies | – | 17 (70.8%) | 7 (29.2%) | 24 (100.0%) | | |
| Employed in public companies | 1 (11.1%) | 8 (88.9%) | - | 9 (100.0%) | | |
| Self-employed | 17 (29.3%) | 33 (56.9%) | 8 (13.8%) | 58 (100.0%) | | |
| Professionals | 1 (14.3%) | 4 (57.1%) | 2 (28.6%) | 7 (100.0%) | | |
| *Knowledge on CSR and level of perception* | | | | | | |
| Knowledge on CSR | Level of Perception | | | Total (106) | $x^2$ Value | Table Value |
| | Low (22) | Moderate (67) | High (17) | | | |
| Yes | 7 (11.5%) | 42 (68.9%) | 12 (19.7%) | 61 (100.0%) | d.f.: 2 $x^2$ = 7.869 | 1% = 9.210 5% = 5.991 10% = 4.605 |
| No | 15 (33.3%) | 25 (55.6%) | 5 (11.1%) | 45 (100.0%) | | |

said that the students whose family income is above Rs. 75,000 have high level of perception. However, as the calculated $x^2$value is less than the table value at ten per cent level, there is no significant association between family income and perception level. Therefore, the null hypothesis is accepted that Family income does not influence perception on corporate social responsibility.

## 9 Parent Education and Level of Perception

*$H_{04}$: Parent Education is not Associated with the Perception on Corporate Social Responsibility*

The perception of students regarding corporate social responsibility may depend on the parent's education. The percentage of students with low level of perception is high with those whose parent's educational qualification is higher secondary. The percentage of students who have high level of perception is high with those whose parent's educational qualification is degree. Hence, it can be said that the educational qualification of parents influence the perception level on CSR. However, as the calculated $x^2$ value is lesser than the table value at ten per cent level, it is inferred that educational qualification of parents does not influence the perception of students. Therefore, the null hypothesis is accepted that Parent education is not associated with the perception on corporate social responsibility.

## 10 Parent Occupation and Level of Perception

*$H_{05}$: Parent Occupation is not Associated with Perception on Corporate Social Responsibility*

The percentage of students who have low perception level is high with those parents are agriculturist. The percentage of students who have high level of perception is high with those whose parents are employed at private companies. Hence, it can be said that occupation of parents influence the perception of students. As the calculated $x^2$ value is greater than the table value at five per cent level, it can be said that occupation of parents influence the perception of students. Therefore, the null hypothesis is rejected.

## 11 Knowledge on CSR and Level of Perception

*$H_{06}$: Knowledge on CSR is not Associated with Perception on Corporate Social Responsibility*

The percentage of students who have high level of perception is high with those who have knowledge on CSR. Therefore, it is observed that the students who have knowledge on CSR have high perception level on CSR. "Professional experience positively influence the students' perception of CSR" found out by Premendra Kumar Singh et al (2014). As the calculated $x^2$ value is greater than the table value at five per cent level, there is a significant relationship between knowledge on CSR and level of perception on CSR. Therefore, the null hypothesis is rejected.

## 11.1 Implications of Findings

Out of the total 106 respondents who are college students, 57.5% are female, 94.33% of the respondents are UG. Majority of the students live in nuclear family compared with Joint family. 44.3% Family income of the students are below 50,000. 38.7% students' parents are degree holders. 54.7% students are self-employed. 57.5% of the students are aware of CSR. Magazines and newspapers are the main source of knowledge about CSR. Most of the students stated that CSR should be included as part of their curriculum. Chi square test has been used in order to analyse the variables associated with perception of students on corporate social responsibility. Out of seven variables tested, gender at ten per cent level parents occupation at five per cent level, knowledge on CSR at five per cent level are significantly associated with corporate social responsibility.

The study is confined only to the Coimbatore district and hence the results cannot be generalized. Due to time constraint only, data has been collected from the college students. Corporates utilises all the resources from the society. Hence, corporates have to pay back to the society in some way or other. Corporates are responsible to the society in different forms. Responsible to the consumers of the product to give quality product at minimal cost. Actually, the foremost responsibility of the corporates is to sell quality products to the consumers at minimal cost that is corporates should earn only normal profit. Earning normal profit is the utmost responsibility that the corporates should practice. The companies are responsible to the employees in the form of giving fair wages, good working environment, fair holidays and so on. On the whole, corporates should see that their acts should not harm the society in any way. They should use the resources in such a way that the resources are not depleted. College students are the future entrepreneurs, future managers, future executives and so on. Hence, creating awareness among the college students on Corporate Social Responsibility is of utmost important and it paves the way for sustainable development. Further, the study may be enhanced by analysing the opinion of the college students on regulation regarding CSR and may identify the factors associated with the perception on Corporate Social Reporting.

## References

1. Agripina, S., Asha, B.: Awareness on corporate social responsibility among selected companies in Mwanza Tanzania. Int J Eng Bus Manag **6**(2), 125–134 (2022)
2. Mahmood A, Bashir J (2020) How does corporate social responsbility transform brand reputation and brand equity? Economic and Non-economic perspectives of CSR
3. Nitamoah EB, Egyiri PO, kwamega M (2014) Corporate social responsibility awareness, firm commitment and organizational performance. J Hum Resour Sustaninability Stud, 2(2):12
4. Eweje, G., Brunton, M.: Business ethics: ethical perceptions of business students in a New Zealand university: do gender, age and work experience matter? A Eur Rev **19**(1), 95–111 (2010)

5. Ugwuozor FO (2020) Students' perception of corporate social responsibility: Analyzing the influence of gender, academic status, and exposure to business ethics education. Bus Ethics A Eur Rev, 1–11
6. Rosak-Szyrocka J , Żywiołek1 J, Shengelia N, Stverkova H, Espírito Santo P, Ladislav Pilař (2022) Employee perception of CSR and its effects on the company's image production engineering archives. 28(3):210–216
7. Larrán, M., Andradcs, J., Herrera, J.: An examination of attitudes and perceptions of Spanish business and accounting students toward corporate social responsibility and sustainability themes. Revista de Contabilidad **21**(2), 196–205 (2018)
8. de-los-Angeles Gil-Estallo M, Giner-de-la-Fuente F, Griful-Miquela C (2009) Benchmarking corporate social responsibility within Spanish companies. Int Adv Econ Res, 15(2):207–225
9. Gonzalez-Rodriguez, M.R., Diaz-Fernandez, M.C.: Customers' corporate social responsibility awareness as antecedent of repeat behaviour intention. Corp Soc Responsib Environ Manag Responsib Environ Manag **27**(3), 1294–1306 (2020)
10. Khurshid MA, Aali AA, Soliman AA, malik J, khan T (2013) Awareness of corporate social responsibility in an emerging economy. Life Sci J, 10(4):2229–2240
11. Zhang Q, Ahmad S (2021) Analysis of corporate social responsibility execution effects on purchase intention with the moderating role of customer awareness. Sustainability, 13(8):4548
12. Moisescu OI (2015) Demographics-based differences in the relationship between perceived CSR and customer loyalty in the dairy products market, Vol. 10, No. 2,Spring, pp. 118–131
13. Singh PK, Sunder RG, Mahamuni PN et al (2023) Corporate social responsibility & students' perceptions: evidence from Indian higher education institutions, F1000Research 2023, 1–14
14. Hiss, S.: From implicit to explicit corporate social responsibility: institutional as a fight for myths. Bus Ethics **19**(3), 433–451 (2009)
15. Chen S, Bouvain P (2009) Is corporate responsbility converging? a comparison of corporate responsibility reporting in the USA, UK, Australia and Germany. J Bus Ethics, 87(1):299–317. Springer
16. Pal SK, Sarkar PC (2022) Promoting education to expand awareness and perception on Corporate Social Responsibility. J Soc Sci, 71(1–3):25-34
17. KO SH, kim JY, Choi Y (2023) consumers'corporate social resoponsibility percepetion and anti-consumer awareness: roles of compassion and corporate social responsibility authenticity in South Korea. Natl Libr Med, 13(8):62

# Masculinity, Emotional Regulation and Suicidal Ideation Among Young Adult Men

**Celeste Sam, Ekta Singh, Fathimath Sahala, Sana Hyder Ali, and Soumya Simon**

**Abstract** This study discusses the relationship between masculinity norms, emotional regulation and suicidal ideation among young adult males in India according to the gender role norms model and Self-Determination Theory (SDT) given by Mahalik. The study was conducted on 247 Indian males aged 18 to 25, to examine the societal expectations of masculinity, emotional regulation strategies, and suicidal ideation intersect within this demographic. The data was analyzed using regression analysis and correlation analysis using SPSS. The results showed that men exhibiting elevated levels of emotional suppression, dysregulation, and adherence to traditional masculinity norms were at heightened risk of suicidal ideation. Conformity to traditional masculinity norms showed no direct correlation with emotional suppression and dysregulation, emotional dysregulation emerged as a significant predictor of suicidal ideation. But emotional suppression did not exert a significant impact on suicidal ideation. This study draws attention to the importance of targeted interventions addressing emotional regulation challenges in this age group and recommend suicide prevention strategies that take in mind the relationship between masculinity and emotional regulation.

**Keywords** Masculinity · Norms · Emotional regulation · Suicidal ideation · Young adult males

## 1 Introduction

The complex relationship between adherence to masculinity norms, emotional regulation, and suicidal ideation is crucial for addressing the challenges that young adult males face when it comes to their mental health. In India where traditional gender roles play a significant role, expectations of masculinity can impact the well- being of individuals.

C. Sam · E. Singh (✉) · F. Sahala · S. H. Ali · S. Simon
Department of Psychology, Kristu Jayanti College Autonomous, Bengaluru, India
e-mail: soumya@kristujayanti.com



M. Al Mubarak (ed.), *Sustainable Digital Technology and Ethics in an Ever-Changing Environment*, Studies in Systems, Decision and Control 237,
https://doi.org/10.1007/978-3-031-86708-8_6

Conformity to masculine norms encompasses societal expectations of masculinity, nonconformity involves diverging from these expectations. Mahalik's model of gender role norms explains how sociocultural influences shape these expectations [1].

The Self-Determination Theory (SDT) is a psychological framework to understand human motivation and behavior. It describes three types of motivation: amotivation, controlled motivation, and autonomous motivation, which aligns with the socialization strategies: Parental Control Need Reinforcement (PCNR), Parental Control Persistence (PCPR), and autonomy support [2]. SDT also identifies three modes of emotional regulation: dysregulation, suppressive regulation, and emotional integration.

Suicidal ideation is often understood as thoughts or contemplations about suicide or self-harm. However, there is a lack of a universally agreed-upon definition which presents challenges for clinicians etc. [3].

**Structure of the Paper**

Suicide continues to be a major national public concern in India. Suicidal ideation relates to thoughts and contemplation about suicide or self-harm. This paper introduces the variables of the study followed by comprehensive review of literature. The methodology section indicates the aim, objectives and hypothesis of this study, research design employed, and information about the sample and the measures used. It includes the results and discussion of the study and finally the conclusion which highlights key findings, implications, and limitations.

## *1.1 Review of Literature*

**Masculinity and Emotion** Recent studies, on masculinity and emotions suggest a relationship between expectations and how men manage their feelings [4]. Roche et al. addresses how masculinity influences control and its effects on well-being, indicating that difficulties in handling emotions can make issues worse like body image concerns and the desire for a more muscular physique [5]. Vidal et al. and Benemann et al. further study the connection between masculinity and aggressive tendencies with Vidal focusing on aggression in relation to gender stereotypes while Benemann examines the concept of honour ideology [6, 7]. Vescio et al and Short et al study the impact of threats to masculinity including increased feelings of shame and guilt on health across different scenarios [8, 8]. In contrast Doorn et al. present opposing findings by suggesting that there's no significant association between masculine norms and depression exists [10]. Stanaland et al and Borduas et al study the changing dynamics of identities during events like the COVID 19 crisis and their consequences on individuals well-being [11, 12]. Lastly Wagner and Reifegerstes work sheds light on obstacles hindering discussions around struggles among men highlighting issues such, as loneliness and societal expectations regarding masculinity [13].

**Masculinity and Suicide** The literature of masculinity and suicide has emphasized the substantial effect of male norms on mental well-being and suicidal behavior. King et al. and Griffin et al. argue that conforming to traditional masculine values like independence is correlated with higher rates of suicide ideation, which necessitates a rethinking of these attitudes in terms of preventative measures [14, 15]. The increased vulnerability of men caused by toxic masculinity and societal forces such as Covid-19 pandemic or remote areas is stressed [16]. Emotional regulation and specialized interventions are important for avoiding major psychiatric outcomes as well as suicides [17]

**Emotion and Suicide** Studies on the impact of emotion regulation on suicide ideation indicate that the emotional processes behind suicide ideation are critical to understanding and preventing suicidal thoughts. Specifically, Swee et al. show that low belongingness and interference of emotion regulation skills increase suicide ideation, suggesting the possible protective role of emotion regulation through adaptive coping skills [18]. Turton et al. & Turton et al. attest to the complicated relationships of emotion dysregulation, attachment insecurity, and suicidal behaviours, suggesting the importance of attachment compared to emotion regulation [19, 20]. Reflecting the importance of attachment to emotional processes, Turton et al. & Turton et al. suggest therapeutic work on attachment and emotion regulation processes concurrently [19, 20]. Laghaei et al. invite us to consider the role of emotion regulation difficulties, finding evidence of mediation between childhood traumatic events, depression symptoms, and suicidal ideation [21]. Each of these studies highlight the importance of thinking comprehensively about suicide prevention and the emotional and relational processes that affect suicidal behaviour.

## 2 Research Methodology

The aim of the study is to find out the impact of Emotional Regulation and Adherence to Masculinity norms on Suicidal Ideation among young adult males in India. There are four main objectives of the study: (1) to investigate the relationship between masculinity and, emotional suppression and emotional dysregulation; (2) to investigate the relationship between, emotional suppression and emotional dysregulation and suicidal ideation; (3) to assess the relationship between masculinity and suicidal ideation; (4) to assess the impact of adherence to masculinity norms, emotional suppression and emotional dysregulation on suicidal ideation in men.

There are four hypotheses from these objectives: (1) Higher adherence to traditional masculinity norms will be positively associated with increased emotional suppression and emotional dysregulation among men; (2) Increased emotional suppression and emotional dysregulation will be positively associated with higher levels of suicidal ideation among men; (3) Higher adherence to traditional masculinity norms will be positively associated with higher levels of suicidal ideation among men; (4) Men with higher levels of emotional suppression and emotional dysregulation and traditional masculinity norms are at a higher risk of suicidal ideation.

The study employed a cross- sectional research design, which is quantitative research methodology. Statistical tools used for the analysis of data were, spearman's correlation and multiple regression analysis.

The data in this study was collected through online using three questionnaires and a convenient sampling technique was used to collect data among Indian young adult males. The sample includes 247 Indian young adult males ranging from 18 to 25 years. The inclusion criteria were young adults who identify themselves as a male, aged between 18–25 years old who resided in India.

## 2.1 Measures

The Conformity to Masculine Norms Inventory—22 (CMNI -22), Developed by Mahalik et al. in was used to assesses conformity to masculinity norms using 22 items [1]. It has a 4-point Likert scale and it covers 11 factor-validated masculinity norms, including winning, emotional control, risk-taking, pursuit of status, primacy of work, violence, dominance, power over women, 'playboy,' self-reliance, and homophobia. Emotion Regulation Inventory Scale (ERI) Developed by Roth et al. in 2009, consists of 18 items measuring three styles of emotion regulation: integrative, suppressive, and dysregulated [22]. Suicidal Ideation Attributes Scale (SIDAS): Developed by Van Spijker et al., it evaluates the severity of suicidal thoughts across five attributes - frequency, controllability, closeness to attempt, level of distress, and impact on daily functioning [23].

## 2.2 Results and Discussions

The study aims to examine the interplay among masculinity norms, emotional regulation and suicidal ideation and to find out the impact of Emotional Regulation and Adherence to Masculinity norms on suicidal ideation among young adult males in India. The data was gathered from 247 Indian young male adults between 18 and 25 years old and analyzed using Microsoft Excel and SPSS 25.

Table 1 suggests that the variables significantly deviate from a normal distribution as indicated by the Kolmogorov–Smirnov and Shapiro–Wilk tests. Thus, the data of these variables are not normally distributed.

Table 2 shows a weak positive correlation (rho $= 0.240$, $p < 0.01$) between adherence to masculinity norms and suppressive emotional regulation. However, adherence to masculinity norms does not significantly impact emotional dysregulation. Emotional dysregulation and emotional suppression are significantly correlated (rho $= 0.230$, $p < 0.01$), indicating a weak positive relationship between the two variables.

**Table 1** Tests of normality

| | Kolmogorov-smirnov[a] | | | Shapiro–wilk | | |
|---|---|---|---|---|---|---|
| | Statistic | df | Sig | Statistic | df | Sig |
| Total M | 0.063 | 247 | 0.018 | 0.987 | 247 | 0.030 |
| Integrative ER | 0.082 | 247 | 0.000 | 0.971 | 247 | 0.000 |
| Suppressive ER | 0.078 | 247 | 0.001 | 0.986 | 247 | 0.013 |
| Dysregulated ER | 0.096 | 247 | 0.000 | 0.983 | 247 | 0.004 |

a. Lilliefors Significance Correction.

**Table 2** Spearman's rank correlation

| Variables | n | M | SD | 1 | 2 | 3 | 4 |
|---|---|---|---|---|---|---|---|
| 1. Suppressive ER | 247 | 20.08 | 4.862 | – | 0.230** | 0.081 | 0.240** |
| 2. Dysregulated ER | 247 | 18.18 | 4.825 | 0.230** | – | 0.263** | 0.000 |
| 3. Total S | 247 | 17.70 | 8.456 | 0.081 | 0.263** | – | 0.058 |
| 4. Total M | 247 | 31.95 | 5.522 | 0.240** | 0.000 | 0.058 | – |

**. Correlation is significant at the 0.01 level (2-tailed)

A study by Cunningham et al. suggests that rigid adherence to masculinity norms may result in emotional dysregulation, affecting mental health outcomes [4]. Additionally, Roche et al. showed that greater adherence to traditional masculinity norms among adolescent boys correlated with fewer emotion regulation difficulties [5].

Suppressive emotional regulation (ER) does not significantly correlate with suicidal ideation (S) among men (rho = 0.081, p = 0.203), while dysregulated ER shows a significant positive correlation with S (rho = 0.263, $p < 0.01$). This indicates that emotional dysregulation, but not suppression, is linked to higher suicidal ideation. Swee et al. and Turton et al. found associations between emotion dysregulation and suicidal ideation [18, 19]. Additionally, Laghaei et al. identified emotional regulation deficits as a significant mediator between childhood trauma and suicidal ideation, underlining the importance of addressing these deficits in prevention efforts [21]. However, findings suggest that suppressive emotional regulation may not be a significant factor contributing to suicidal ideation among men, contrasting with existing literature and contradicting the hypothesis of a positive relationship between the two variables. Additionally, masculinity norms and suicidal ideation do not have statistically significant correlation.

Table 3 shows that the regression model with Dysregulated ER, Total Masculinity, and Suppressive ER as predictors significantly predicts suicidal ideation (Total S) with a significance level of 0.000.

In Table 4, The positive coefficient for Dysregulated ER (B = 0.525, $p < 0.01$) indicates that higher levels of dysregulated emotional regulation are associated with higher levels of suicidal ideation. Total Masculinity and Suppressive ER do not have statistically significant effects on suicidal ideation in this model. The $R^2$ value is 0.134

**Table 3** Anova

| Model | | Sum of squares | df | Mean square | F | Sig |
|---|---|---|---|---|---|---|
| 1 | Regression | 2351.108 | 3 | 783.703 | 12.498 | 0.000[b] |
| | Residual | 15,237.119 | 243 | 62.704 | | |
| | Total | 17,588.227 | 246 | | | |

a. Dependent Variable: Total S
b. Predictors: (Constant), Dysregulated ER, Total M, Suppressive ER

**Table 4** Regression analysis

| Variables | | | | β | t | Sig | 95.0% CI | |
|---|---|---|---|---|---|---|---|---|
| | | B | SE | | | | LL | UL |
| | Total M | 0.174 | 0.095 | 0.114 | 1.834 | 0.068 | −0.013 | 0.361 |
| | Suppressive ER | 0.161 | 0.112 | 0.093 | 1.433 | 0.153 | −0.060 | 0.382 |
| | Dysregulated ER | 0.525 | 0.109 | 0.300 | 4.803 | 0.000 | 0.310 | 0.741 |
| | $R^2$ | 0.134 | | | | | | |

a. Dependent Variable: Total S

(13.4%), suggesting that approximately 13.4% of the variance in suicidal ideation can be explained by the predictors in the model. However, there are other factors influencing suicidal ideation not accounted for by these predictors.

The results partially support Hypothesis 4 by confirming the role of Dysregulated ER but do not support associations with adherence to traditional masculinity norms or emotional suppression. Further research is needed to understand the nuanced relationship among emotional regulation, masculinity norms, and suicidal ideation among men, as no studies have explored these three variables together.

## 3 Conclusion

Contrary to the first hypothesis, the link between adherence to traditional masculinity norms and increased emotional suppression and dysregulation was not strongly supported. When it comes to the second hypothesis emotional dysregulation did predict suicidal ideation significantly but emotional suppression did not have significant effect. The third hypothesis was rejected stating that higher adherence to traditional masculinity norms did not directly correlate with increased suicidal ideation. However, the results did showcase that males exhibiting elevated levels of emotional suppression, dysregulation, and adherence to traditional masculinity norms were at risk of suicidal ideation.

The results emphasize critical need to address emotional dysregulation in suicide prevention strategies tailored for young adult males in India. Incorporating emotional regulation and adherence to traditional masculinity norms into suicide risk assessment protocols is crucial for informing targeted interventions. Further research is needed to understand the complex relationship between masculinity and suicidal ideation in males. This study was only limited to young adults aged 18-25, further research across age groups is essential to understand age- related differences. The study employed a convenience sampling technique, which may limit the generalizability of the findings to the broader population.

## References


1. Mahalik, J.R., et al.: Development of the conformity to masculine norms inventory. Psychol. Men Masculinity **4**(1), 3–25 (2003). https://doi.org/10.1037/1524-9220.4.1.3
2. Ryan, R.M., Deci, E.L.: Self-determination theory and the facilitation of intrinsic motivation, social development, and well-being. Am. Psychol. **55**(1), 68–78 (2000). https://doi.org/10.1037/0003-066X.55.1.68
3. Harmer, B., Lee, S., Rizvi, A., Pal, S., Jackson, D., Saadabadi A: Suicidal Ideation, in *StatPearls*, Treasure Island (FL): StatPearls Publishing, 2024. Accessed: 14 Sep 2024. [Online]. Available: http://www.ncbi.nlm.nih.gov/books/NBK565877/
4. Cunningham, M.L., et al.: 'Big boys don't cry': examining the indirect pathway of masculinity discrepancy stress and muscle dysmorphia symptomatology through dimensions of emotion dysregulation. Body Image **34**, 209–214 (2020). https://doi.org/10.1016/j.bodyim.2020.05.014
5. Roche, C., et al.: Gender compatibility and drive for muscularity among adolescent boys: examining the indirect effect of emotion regulation difficulties. Body Image **39**, 213–220 (2021). https://doi.org/10.1016/j.bodyim.2021.08.009
6. Malonda-Vidal, E., Samper-García, P., Llorca-Mestre, A., Muñoz-Navarro, R., Mestre-Escrivá, V.: Traditional Masculinity and Aggression in Adolescence: its Relationship with Emotional Processes. Int. J. Environ. Res. Public Health **18**(18), 9802 (2021). https://doi.org/10.3390/ijerph18189802
7. Benemann, H., McCartin, H., Russell, T., Cash, D., King, A.: Sadistic masculinity: masculine honor ideology mediates sadism and aggression. Personal. Individ. Differ. **206**, 112118 (2023). https://doi.org/10.1016/j.paid.2023.112118
8. Vescio, T.K., Schermerhorn, N.E.C., Gallegos, J.M., Laubach, M.L.: The affective consequences of threats to masculinity. J. Exp. Soc. Psychol. **97**, 104195 (2021). https://doi.org/10.1016/j.jesp.2021.104195
9. Short, S., Davis P.E., Gheyoh Ndzi, E.: An exploration of masculinity, social support and depression in new and experienced fathers. Midwifery, 123:103715, (2023). https://doi.org/10.1016/j.midw.2023.103715
10. Van Doorn, G., Teese, R., Gill, P.R.: Prospective associations between hegemonic masculinity and incident depression/depressive symptoms: results from a national sample of Australian emerging adult men. Personal. Individ. Differ. **179**, 110899 (2021). https://doi.org/10.1016/j.paid.2021.110899
11. Stanaland, A., Gaither, S., Gassman-Pines, A.: When is masculinity 'fragile'? an expectancy-discrepancy-threat model of masculine identity. Personal. Soc. Psychol. Rev. **27**(4), 359–377 (2023). https://doi.org/10.1177/10888683221141176
12. Borduas, C., Kehler, M., Knott-Fayle, G.: School spaces, school places: shifting masculinities during the COVID-19 pandemic. Int. J. Educ. Res. **120**, 102211 (2023). https://doi.org/10.1016/j.ijer.2023.102211

13. Wagner, A.J.M., Reifegerste, D.: Real men don't talk? relationships among depressiveness, loneliness, conformity to masculine norms, and male non-disclosure of mental distress. SSM - Ment. Health **5**, 100296 (2024). https://doi.org/10.1016/j.ssmmh.2024.100296
14. King, T.L., et al.: Expressions of masculinity and associations with suicidal ideation among young males. BMC Psychiatry **20**(1), 228 (2020). https://doi.org/10.1186/s12888-020-2475-y
15. Griffin, L., Hosking, W., Gill, P.R., Shearson, K., Ivey, G., Sharples, J.: The gender paradox: understanding the role of masculinity in suicidal ideation. Am. J. Mens Health **16**(5), 155798832211238 (2022). https://doi.org/10.1177/15579883221123853
16. Khan, A.R., Ratele, K., Arendse, N.: Men, suicide, and covid-19: critical masculinity analyses and interventions. Postdigital Sci. Educ. **2**(3), 651–656 (2020). https://doi.org/10.1007/s42438-020-00152-1
17. Oliffe, J.L., Kelly, M.T., Montaner, G.G., Seidler, Z.E., Ogrodniczuk, J.S., Rice, S.M.: Masculinity and mental illness in and after men's intimate partner relationships. SSM—Qual. Res. Health **2**, 100039 (2022). https://doi.org/10.1016/j.ssmqr.2022.100039
18. Swee, G., Shochet, I., Cockshaw, W., Hides, L.: Emotion regulation as a risk factor for suicide ideation among adolescents and young adults: the mediating role of belongingness. J. Youth Adolesc. **49**(11), 2265–2274 (2020). https://doi.org/10.1007/s10964-020-01301-2
19. Turton, H., Berry, K., Danquah, A., Pratt, D.: The relationship between emotion dysregulation and suicide ideation and behaviour: a systematic review. J. Affect. Disord. Rep. **5**, 100136 (2021). https://doi.org/10.1016/j.jadr.2021.100136
20. Turton, H., Berry, K., Danquah, A., Green, J., Pratt, D.: An investigation of whether emotion regulation mediates the relationship between attachment insecurity and suicidal ideation and behaviour. Clin. Psychol. Psychother. **29**(5), 1587–1598 (2022). https://doi.org/10.1002/cpp.2735
21. Laghaei, M., Mehrabizadeh Honarmand, M., Jobson, L., Abdollahpour Ranjbar, H., Habibi Asgarabad M.: Pathways from childhood trauma to suicidal ideation: mediating through difficulties in emotion regulation and depressive symptoms. BMC Psychiatry, 23(1):295, 2023. https://doi.org/10.1186/s12888-023-04699-8
22. Roth, G., Assor, A., Niemiec, C.P., Ryan, R.M., Deci, E.L.: The emotional and academic consequences of parental conditional regard: comparing conditional positive regard, conditional negative regard, and autonomy support as parenting practices. Dev. Psychol. **45**(4), 1119–1142 (2009). https://doi.org/10.1037/a0015272
23. Van Spijker, B.A.J., et al.: The suicidal ideation attributes scale (SIDAS): community-based validation study of a new scale for the measurement of suicidal ideation. Suicide Life Threat. Behav. **44**(4), 408–419 (2014). https://doi.org/10.1111/sltb.12084

# Ethical Challenges in Technology Development

**Noof Alsaud, Sara Saleh, and Muneer Al Mubarak**

**Abstract** This paper identifies and discusses the ethical dilemmas and considerations that arise during the process of creating and implementing new technologies. It aims to explore the ethical implications and potential risks associated with technological advancements and provide insights into how these challenges can be addressed. The paper is based on literature reviews of previous studies, different ethical challenges in the development of technology are discussed. The relationship between ethics and technology is ever-changing. As technology advances, new ethical challenges emerge, requiring ongoing reflection and proactive measures. Cultural differences exist, with some cultures having strict standards against dishonesty. In the information technology (IT) field, transparency and honesty are highly valued, as dishonesty is seen as unacceptable. By embracing ethical considerations, technology developers can navigate challenges more effectively. This leads to responsible and trustworthy technologies that benefit society while minimizing harm. The emergence of ethical challenges in technology development is deeply rooted in our core values and continues to evolve alongside developments. Effectively addressing these challenges by applying ethical principles is essential for the responsible development of technology that ultimately benefits society.

**Keywords** Technology · Ethical challenges · Artificial Intelligence · Ethics

## 1 Introduction

In the modern era, technology has become deeply ingrained in our daily lives, revolutionizing how we conduct work, communicate with others, and navigate our surroundings. Our lives are increasingly being lived in a world of information, a unique world that perfectly blends digital and analog, offline and online. In other

N. Alsaud · S. Saleh · M. Al Mubarak (✉)
Ahlia University, Manama, Bahrain
e-mail: malmubarak@ahlia.edu.bh



M. Al Mubarak (ed.), *Sustainable Digital Technology and Ethics in an Ever-Changing Environment*, Studies in Systems, Decision and Control 237,
https://doi.org/10.1007/978-3-031-86708-8_7

words, we no longer live online or offline but rather online [7]. From artificial intelligence and robotics to biotechnology and virtual reality, technological advancements have revolutionized industries and opened new possibilities. However, alongside the benefits of these innovations, there are also ethical challenges that arise in the realm of technological development. The ongoing surge in artificial intelligence (AI) is accompanied by persistent demands for the application of ethical principles. These calls aim to effectively channel the subversive capabilities of emerging AI technologies [12]. The growing advancements in Artificial Intelligence (AI) have captured significant attention from the media and the public. As AI systems, including robots, chatbots, avatars, and other intelligent agents, transition from being viewed as mere tools to autonomous entities and collaborators, there is a heightened emphasis on exploring the ethical implications of these systems. Understanding the ethical impact of AI is crucial as it shapes societal perceptions, determines trust levels, and ultimately influences the overall impact and acceptance of AI in our society. The ability of our society and systems to address these ethical questions will play a significant role in shaping the future of AI and its existence in our daily lives [5].

This research purpose to explore the ethical dilemmas that emerge in the process of developing and implementing new technologies. The central focus of the study is to identify and analyze the ethical challenges that arise during various stages of technology development. As technology continues to advance at an unprecedented pace, ethical considerations often lag behind, resulting in potential harms and unintended consequences. Issues such as privacy, data security, algorithmic biases, job displacement, and the impact on marginalized communities have become increasingly prevalent, raising questions about the moral implications and responsibilities of technology developers.

The need for this research stems from the critical importance of addressing these ethical challenges proactively and integrating responsible practices into the development and deployment of new technologies. The findings of this study will benefit various stakeholders. Firstly, technology developers and innovators will gain valuable insights into the ethical considerations that should be integrated into the early stages of technology development. These insights will help them navigate potential pitfalls and design technologies that align with ethical principles. Additionally, policymakers and regulators will benefit from a better understanding of the ethical challenges in technology development, enabling them to craft appropriate regulations and frameworks that promote responsible innovation. Furthermore, society will benefit from a more informed and engaged public discourse on the ethical implications of technology, fostering a culture of responsible technology use and development. To ensure a common understanding, it is important to define the main terms used in this research. "Ethics" refers to the study of moral principles and values that guide human behavior and decision-making. In the context of technology development, ethics involves considering the potential impacts of technology on individuals, communities, and society at large. It encompasses concepts such as transparency, fairness, privacy, accountability, and the equitable distribution of benefits and burdens. Technology development refers to the process of creating, designing, and implementing new technologies or improving existing ones. It encompasses various stages, including research, design, testing, and deployment.

In the context of the research problem, it is essential to consider the broader context and factors that contribute to the ethical challenges in technology development. The pace of technological advancement often outstrips the ability of society to fully comprehend and anticipate the ethical implications of new technologies. This creates a gap between innovation and responsible development, leading to potential harm and unintended consequences. Additionally, the profit-driven nature of technology development can sometimes prioritize commercial interests over ethical considerations, exacerbating the challenges and risks associated with emerging technologies. Moreover, the complexity and interconnectedness of technology systems introduce additional ethical challenges. Technologies often have far-reaching impacts, affecting multiple stakeholders and ecosystems. For example, the development of autonomous vehicles raises questions about safety, liability, and the potential displacement of human drivers. Similarly, the deployment of facial recognition technologies raises concerns about privacy, surveillance, and potential biases. Furthermore, the increasing reliance on artificial intelligence and algorithmic decision-making introduces ethical challenges related to transparency, fairness, and accountability. As AI systems become more autonomous and capable of making decisions that affect individuals' lives, questions arise about the biases embedded in algorithms, the lack of explainability, and the potential for discrimination. AI is not merely a utility that requires regulation once it reaches maturity. It is a potent and transformative force, representing a novel type of intelligent agency that is already redefining our daily lives, the way we interact with others, and even the environments we inhabit [8].

## *1.1 Scope of Study*

The scope of studying ethical challenges in technology development is broad and covers various aspects of the ethical implications and considerations associated with the creation, implementation, and use of technology. Here are some key areas within the scope of studying ethical challenges in technology development:

### 1.1.1 Facial Recognition Technology

Facial recognition technology has become increasingly prevalent in various applications, including surveillance, authentication, and social media. Ethical concerns arise due to the potential invasion of privacy, the collection and storage of facial data without consent, and the risk of biases and discriminatory practices in facial recognition algorithms. There is a need to examine the ethical implications of facial recognition technology and establish guidelines for its responsible use [9].

### 1.1.2 Artificial Intelligence in Hiring and Recruitment

The use of Artificial Intelligence algorithms in hiring processes has the possibilities to streamline and automate recruitment. However, ethical questions arise regarding the fairness and transparency of these algorithms. Biases in training data can result in discriminatory hiring practices, disadvantaging certain groups. Ensuring fairness, accountability, and transparency in AI-based hiring systems is crucial to mitigate bias and promote equal opportunities.

### 1.1.3 Algorithmic Bias and Discrimination

Algorithmic Bias and Discrimination refers to unfair outcomes generated by AI algorithms based on factors like race, gender, and socioeconomic status. It can perpetuate inequalities and impact areas like college admissions and job opportunities. Causes include biased training data, prejudiced programming, incomplete data, feedback loops, and lack of diversity in development teams. Detecting and mitigating bias involves methods like data auditing, fairness metrics, transparency, counterfactual testing, and bias reporting. Challenges include complex algorithms, ethical dilemmas, data availability, bias amplification, and a lack of standardization. To reduce bias, encourage diversity, follow ethical guidelines, use fairness tools, conduct regular audits, and promote transparency and accountability [19].

### 1.1.4 Ethics of Autonomous Systems

Ethical considerations can arise when developing and deploying autonomous systems, such as self-driving cars and drones. Safety is a primary concern, as accidents involving autonomous systems can have severe consequences. Additionally, questions arise regarding the allocation of responsibility between humans and autonomous systems, especially in situations where decisions must be made in complex and uncertain environments. Ethical frameworks and guidelines are needed to govern the behavior and decision-making processes of autonomous systems.

## 2 Literature Review

The rapid advancement of technology has undoubtedly brought numerous benefits and opportunities to society. However, these advancements have also given rise to various ethical challenges in the field of technology development. This literature review aims to provide an overview of the theoretical foundations that underpin these ethical challenges. In the realm of technological development, ethical dilemmas often arise due to the potential impact of technology on individuals, communities, and society as a whole. Key concepts and theories are examined to shed light on these dilemmas and considerations.

## 2.1 *Ethical Theory*

Ethical theories provide frameworks for understanding and evaluating ethical challenges in technology development. Utilitarianism, for example, emphasizes maximizing overall societal welfare, often measured in terms of utility or happiness. When applied to technology development, utilitarianism prompts the examination of the impact of technologies on individuals and society, weighing potential benefits against potential harms. Deontological theories, such as Kantian ethics, focus on the inherent moral principles and duties that guide human actions. In the context of technology development, deontological theories emphasize principles such as respect for autonomy, human dignity, and the protection of privacy and confidentiality.

## 2.2 *Value Sensitive Design*

Value Sensitive Design (VSD) is an approach that integrates ethical considerations into the design and development of technologies. VSD emphasizes the importance of understanding and incorporating human values throughout the entire technological development process. By involving stakeholders and considering their values, VSD aims to create technologies that align with ethical principles and meet societal needs. This approach recognizes that technology is not value-neutral and seeks to minimize potential negative impacts while promoting positive social values [21].

## 2.3 *Responsible Innovation*

Responsible innovation is a framework that encourages technology developers to proactively consider and address the ethical, social, and environmental impacts of their innovations. It emphasizes the integration of ethical considerations into the entire innovation process, from the early stages of design and development to the deployment and use of technologies. Responsible innovation involves engaging stakeholders, anticipating potential risks and unintended consequences, and adopting a precautionary approach to minimize harm. By promoting responsible practices, this framework aims to ensure that technological advancements align with societal values and contribute to sustainable and equitable development [16].

## 2.4 Ethical Dimensions of Technology

Technology development raises various ethical dimensions and challenges. Privacy and data security, for instance, are critical concerns in the era of big data and pervasive surveillance. The collection, storage, and analysis of personal data raise questions about consent, transparency, and the protection of individuals' privacy rights. Algorithmic biases and discrimination also pose ethical challenges. As technologies increasingly rely on algorithms and artificial intelligence, biases embedded in data and algorithms can perpetuate social inequalities and unjust outcomes. Additionally, the impact of technology on employment and job displacement raises ethical questions related to fairness, equality, and the well-being of individuals and communities.

## 2.5 Ethical Governance and Regulation

Effective governance and regulation play a crucial role in addressing ethics challenges in technology development. Regulatory frameworks need to keep pace with technological advancements and provide clear guidelines for responsible technology development and deployment. Ethical governance involves setting standards, Creating mechanisms for accountability, and promoting transparency in decision-making. Additionally, interdisciplinary collaborations between technologists, ethicists, policymakers, and other stakeholders are essential for developing comprehensive and context-specific approaches to ethical challenges [16].

This literature review has provided an overview of the theoretical foundations that underpin the ethical challenges in technology development. Ethical theories, such as utilitarianism and deontological ethics, provide frameworks for evaluating the moral implications of technological innovations. Value Sensitive Design and responsible innovation offer approaches to integrating ethical considerations into the design and development of technologies. The ethical dimensions of technology, including privacy, algorithmic biases, and job displacement, highlight the complex ethical challenges that emerge in the context of technological advancements. Finally, ethical governance and regulation play a crucial role in addressing these challenges and ensuring responsible technology development. By understanding these theoretical foundations, researchers, policymakers, and technology developers can gain valuable insights into the ethical considerations that arise during technology development. This comprehension can guide the development of responsible practices, regulatory frameworks, and decision-making processes that balance innovation with ethical responsibility. Ultimately, by addressing ethical challenges in technology development, we can strive for advancements that align with societal values, promote well-being, and contribute to a more equitable and sustainable future.

## 2.6 *Ethical Challenges in Technology*

Ethical challenges in technology development have become increasingly prominent as technological advancements continue to reshape various aspects of our lives. This literature review aims to provide an overview of the key ideas and insights presented in the literature regarding the ethical challenges that arise in the field of technology development. The articles reviewed are grouped based on common themes and ideas, offering a comprehensive understanding of the current discourse surrounding ethics in technology development. Ethical challenges in technology refer to the complex moral dilemmas and considerations that arise because of the development, deployment, and use of technology in various domains. These challenges arise due to the potential impact of technology on individuals, society, and the environment.

### 2.6.1 Privacy and Data Security

Protecting data privacy is crucial in today's technology-driven world. It is necessary to address risks such as identity theft and cybercrime. Data privacy protection is complex due to social and technical risks and ethical considerations. However, implementing a comprehensive approach and utilizing frameworks can help mitigate these risks and ensure effective data privacy (Lee et al., [16].

### 2.6.2 Social Implications and Equity

The social implications and equity considerations of technological advancements are crucial areas of study. This literature review focuses on articles that explore the social implications and equity concerns arising from the use of technology. By examining key research in this field, we aim to gain insights into the potential impacts of technology on society and the importance of promoting equity in technological development.

### 2.6.3 Ethical Considerations in AI and Automation

The integration of AI and automation in various domains has led to significant transformations in sectors such as healthcare, finance, and transportation. These advancements are driven by machine learning and deep learning techniques, which heavily rely on data for training and testing AI models. By analyzing large datasets, AI models can uncover complex patterns and trends that traditional data-analysis methods may struggle to detect. This enables them to learn from the data they are trained on and make accurate predictions. However, alongside these advancements, ethical considerations have become increasingly important. The ethical implications of AI and automation encompass the potential impact on individuals, society, and the overall ethical framework. It is crucial to address these considerations to ensure that AI technologies are developed and deployed in a responsible and ethical manner [1].

### 2.6.4 Facial Recognition Technology

Facial recognition technology is a form of biometric technology that examines and verifies individuals by analyzing their unique facial characteristics. It uses algorithms and machine learning techniques to detect and match unique patterns and characteristics of a person's face from images or video footage [11].

### 2.6.5 Environmental Sustainability

The rapid advancement of technology has led to concerns about its environmental impact. Ethical challenges include minimizing electronic waste, reducing energy consumption, and developing environmentally sustainable practices in manufacturing and disposal. Adopting eco-friendly design principles, promoting recycling and responsible consumption, and incorporating environmental considerations into technology development are essential for sustainability [4].

### 2.6.6 Digital Divide and Access

The digital divide is related to the disparity in technology access and availability of digital resources among various groups or communities. Ethical dilemmas emerge in guaranteeing equitable access to technology, promoting digital literacy, and providing equal opportunities for online resources. Bridging the digital divide requires efforts to provide affordable and accessible technology infrastructure, digital skills training, and inclusive policies to prevent further marginalization of underprivileged communities [23].

Jobin et al. [13] specifically explore the ethical considerations and principles underlying AI development and deployment. They analyze the existing AI ethics guidelines and discuss the convergence and divergence of ethical principles in the field of AI. They address the need for ethical guidelines in the development and implementation of AI technologies. It emphasizes the importance of transparency, justice, fairness, non-maleficence, responsibility, and privacy as key ethical principles in AI. The authors also highlight the significance of integrating ethical analysis with guideline development and implementation strategies. Therefore, it explores the ethical dimensions and principles that should guide the responsible development and use of AI technologies. They address the ethical considerations and challenges associated with the development and deployment of artificial intelligence (AI) technologies.

Floridi et al. [8] propose an ethical framework for a socially beneficial AI society. They discuss the opportunities and risks associated with AI and present a set of ethical principles and recommendations to guide the responsible development and use of AI. They emphasize the importance of transparency, fairness, accountability, and the protection of human values and rights in the design and implementation of AI systems. Overall, the authors provide a comprehensive examination of the ethical dimensions of AI and offers a framework for fostering a "good AI society" that aligns with ethical principles and values.

Munoko et al. [20] explore the ethical considerations associated with the use of artificial intelligence (AI) in the field of auditing. In recent years, there has been an increasing adoption of AI technologies in various industries, including auditing. AI offers the potential to enhance efficiency, accuracy, and effectiveness in auditing processes. However, its implementation raises important ethical concerns that need to be addressed. AI systems can be influenced by biases present in the data used for training. The authors may discuss the ethical considerations related to ensuring fairness in auditing practices, addressing potential biases, and avoiding discriminatory outcomes. The use of AI in auditing may raise questions about accountability and transparency. They may explore the ethical requirements for auditors to understand and explain the decisions made by AI systems, as well as the importance of transparency in auditing processes. Auditing often involves handling sensitive financial and personal data. They may discuss the ethical implications of using AI in auditing and the need to ensure robust privacy and data protection measures are in place. The introduction of AI in auditing may impact the role of auditors and their professional judgment. The authors may examine the ethical considerations associated with striking a balance between human judgment and the use of AI tools, ensuring human oversight, and maintaining the auditor's ethical responsibilities. They may also explore the ethical decision-making processes involved in using AI in auditing. It may discuss the importance of ethical frameworks and guidelines for auditors to navigate the ethical dilemmas that may arise when utilizing AI technologies. Overall, the authors likely provide a comprehensive analysis of the ethical implications of AI in auditing, highlighting the need for ethical guidelines, standards, and practices to guide the responsible and ethical use of AI in this field. The authors offer recommendations on how auditors and organizations can address these ethical challenges while leveraging the benefits of AI technology in auditing processes.

Gerke et al. [10] examine the ethical implications of AI-driven healthcare, the chapter likely addresses issues such as privacy, transparency, fairness, and patient autonomy. It may explore the ethical considerations surrounding the use of AI algorithms, data protection, and the impact of AI on the doctor-patient relationship. The authors may also discuss the need for ethical guidelines and policies to ensure responsible and ethical implementation of AI technologies in healthcare. As they said to successfully establish an AI-driven healthcare system under the motto "Health AIs for All of Us," several crucial factors must be considered and addressed which include ensuring high levels of data protection and privacy, maintaining cyber resilience and cybersecurity, promoting algorithmic fairness, implementing transparency and regulatory oversight, upholding safety and effectiveness standards, and establishing an appropriate liability framework for AI systems. It is essential to not only reevaluate existing regulatory frameworks and update them to accommodate technological advancements but also engage in public and political discussions regarding the ethical implications of AI-driven healthcare, including its impact on the workforce and society at large. While AI holds significant potential for improving healthcare, it is essential to proactively tackle the ethical and legal challenges that lie ahead to fully harness this potential.

Kuyumdzhieva [14] investigates the ethical challenges that emerge in the context of the digital era, particularly within the field of medical research. With the increasing integration of digital technologies in healthcare and research practices, new ethical considerations and dilemmas arise. The author likely explores these challenges and provides insights into how they affect medical research. The collection, storage, and sharing of sensitive patient data in the digital era raises concerns about privacy and security. The author may explore the ethical implications of safeguarding patient information and ensuring data protection in medical research. Informed consent is a crucial ethical requirement in medical research. The author may discuss how digital technologies impact the process of obtaining informed consent, such as remote consent procedures, online consent forms, and the need to ensure participants fully understand the implications of their participation. The use of digital tools and technologies in medical research can impact transparency and accountability. The author may examine the ethical considerations surrounding the transparent reporting of research methods, data collection, and analysis to ensure the reproducibility and reliability of results. The author may explore how the digital era affects the autonomy of research participants, considering issues such as the use of wearable devices, monitoring technologies, and the potential for excessive data collection that may infringe on participant autonomy. Responsible Use of AI and Machine Learning: As artificial intelligence (AI) and machine learning algorithms become more prevalent in medical research, ethical questions arise regarding their development, deployment, and potential biases. The author may discuss the need to ensure the responsible and fair use of these technologies in medical research. Overall, the author likely provides a comprehensive examination of the ethical challenges specific to the digital era in the context of medical research. It may offer recommendations and insights into how researchers, institutions, and policymakers can navigate these challenges while upholding ethical principles and ensuring the well-being and rights of research participants.

Stahl [22] discusses the shift from computer ethics and the ethics of AI to a broader consideration of the ethics of digital ecosystems. It explores the ethical challenges and considerations that arise when examining the interconnected networks of technologies, data, and users within digital ecosystems. The author advocates for a holistic and systemic approach to ethics in the digital landscape, addressing issues such as privacy, data governance, algorithmic decision-making, cybersecurity, and societal impact. By expanding the ethical scope, the author aims to provide insights into the complex ethical dimensions of digital ecosystems and advocate for a more comprehensive ethical framework. Maharani et al. [17] discusses the relationship between technological progress, the development of artificial intelligence (AI), and ethical considerations. It recognizes the growing importance of ethical paradigms in the context of advancing technology. The authors likely discuss the ethical implications associated with AI development and deployment. Topics such as algorithmic bias, privacy concerns, transparency, accountability, and the societal impact of AI are likely to be explored. The authors may examine different ethical frameworks that can guide responsible AI development and usage. The focus of the authors is likely to emphasize the need for integrating ethical considerations into the design, implementation, and governance of AI systems. It underscores the significance of ensuring that AI technologies are developed and utilized in a manner that aligns with ethical principles and societal values.

Moor [18] provides an overview of computer ethics and its relationship to information technologies. The author explores the ethical implications of computer technology, addressing issues such as privacy, security, social impact, and moral responsibility. The author likely delves into the unique challenges posed by computers and offers ethical frameworks and guidelines for responsible decision-making in the use of technology and development. Hagendorff [12] examines the field of AI ethics and evaluates existing guidelines in this area. As the author explores the ethical concerns linked to artificial intelligence (AI) and assesses the effectiveness of guidelines in addressing these concerns. The author likely discusses topics such as algorithmic bias, privacy, transparency, and accountability in relation to AI technology. However, the author offers critiques of current guidelines and makes recommendations for improving them, highlighting the need for comprehensive and practical ethical frameworks in the use of AI systems and development.

Etzioni and Etzioni [7] discuss the importance of considering ethics in the development and use of artificial intelligence (AI) systems. The authors highlight various ethical challenges associated with AI, such as privacy, algorithmic bias, accountability, and social inequality. They argue for the integration of ethical principles into AI design and propose strategies such as establishing guidelines, providing ethics training, and fostering interdisciplinary collaborations. They emphasize the need to proactively address ethical considerations to ensure the responsible and ethical use of AI. Dignum [5] provides an overview of the ethical implications of artificial intelligence (AI). The author emphasizes the need to address the ethical dimensions of AI due to its potential impact on society, including privacy, fairness, accountability, and human autonomy. The author calls for interdisciplinary collaboration and presents the special issue as a platform for discussing specific ethical considerations in AI. Overall, they highlight the significance of ethics in the context of AI and sets the stage for further exploration of ethical issues in subsequent articles.

Bostrom and Yudkowsky [2] discuss issues such as value alignment, decision-making, and the potential societal impact of AI systems. The authors emphasize the need to align AI with human values and goals, and they address risks such as unintended harm and loss of control. They also touch on philosophical questions pertaining to intelligence, consciousness, and moral agency. Overall, it confirms the significance of incorporating ethics into AI development to assure responsible outcomes. Wies et al. [24] discuss issues such as privacy, informed consent, confidentiality, and algorithmic bias. It highlights the potential benefits of digital mental health interventions, such as increased accessibility and personalized care. However, it also addresses ethical challenges, including data protection, the need for informed consent, and the responsible use of AI algorithms. Overall, they emphasize the importance of considering and addressing ethical concerns to ensure the effective and ethical use of digital technologies in mental health support for young individuals.

Chaet et al. [3] specifically discuss ethical considerations and guidelines for practicing telehealth and telemedicine. As telehealth and telemedicine involve the use of technology to provide healthcare services remotely, there are several ethical implications that need to be addressed. They likely explore topics such as patient confidentiality, informed consent, privacy and security of patient information, appropriate use of technology, maintaining the doctor-patient relationship, and ensuring equitable access to care. The authors may discuss the ethical challenges that arise in the context of telehealth, provide recommendations for ethical practice, and highlight potential issues that healthcare providers should be aware of when delivering care through telehealth platforms.

## 3 Conclusion

The rapid advancement of technology has introduced various ethical challenges. These challenges include concerns regarding privacy and surveillance in technologies like facial recognition, the decision-making process of autonomous vehicles and the ethical implications of those decisions, the impact of social media on mental health and user well-being, biases and discrimination in AI-based hiring processes, data privacy and the responsible use of big data analytics, and the ethical considerations surrounding genetic engineering technologies. These ethical challenges underscore the importance of core values such as individual privacy, fairness, transparency, accountability, and the responsible use of technology. Understanding and addressing these challenges is essential to ensure that technology is developed and implemented in a way that respects human rights, fosters equality, protects privacy, and safeguards the overall well-being of society.

By openly acknowledging and actively confronting these ethical dilemmas, developers, policymakers, and society can collaboratively shape technology to align with these values. This entails taking deliberate actions to address privacy concerns, ensuring that data collection and usage prioritize individual rights and consent. It also involves fostering fairness by mitigating biases and discrimination in algorithmic decision-making processes, fostering transparency in the design and implementation of AI systems. Moreover, accountability is a critical aspect that needs to be upheld in technological development. This involves establishing clear lines of responsibility for the actions and consequences of technology, ensuring that developers and organizations are held accountable for any harm or negative impacts that may arise. To achieve these goals, it is essential to cultivate a culture of responsible and ethical technology development. This requires a comprehensive approach that involves education, training, and awareness-building among developers, policymakers, and society at large. By instilling ethical principles and values, we can collectively work towards a more ethically sound and responsible approach to technology. Ultimately, by embracing ethical considerations and actively addressing these challenges, we can unleash its maximum potential of technology in a way that benefits society while minimizing potential risks and harms. It is through a steadfast commitment to ethical principles that we can ensure a future where technology promotes human well-being, respects individual rights, and contributes to a fair and equitable society.

## 3.1 *Theoretical Implications*

- Global and Intercultural Ethics: Ethical challenges in technology development have theoretical implications for global and intercultural ethics. They highlight the need to consider diverse cultural, social, and ethical perspectives in technology development to avoid ethnocentrism and cultural bias. Theoretical considerations involve exploring the ethical dimensions of technology in different cultural contexts, the challenges of reconciling divergent ethical values, and the potential for developing global ethical frameworks that account for cultural diversity.
- Ethical Innovation and Future Directions: Ethical challenges in technology development prompt theoretical considerations about ethical innovation and the ethical implications of emerging technologies. They raise questions about the responsibility of technology developers to anticipate and address ethical concerns in technological advancements, and the potential for ethical innovation to shape future directions of technology. Theoretical implications involve exploring the ethical dimensions of emerging technologies, considering the long-term societal impact of technological innovation, and envisioning ethical frameworks that guide the development of future technologies.

## 3.2 *Practical Implications*

- Ethical Guidelines and Policies: Practical implications of ethical challenges in technology development involve the development and implementation of clear ethical guidelines and policies. These guidelines provide practical frameworks that steer decision-making processes, promote ethical awareness, and ensure responsible technology development. They outline ethical considerations, principles, and best practices that guide developers in addressing ethical challenges effectively.
- Ethical Impact Assessments: Conducting ethical impact assessments is a practical implication of addressing ethical challenges in technology development. These assessments involve systematically evaluating the potential ethical implications and risks associated with the development and deployment of technology. They help identify and mitigate potential harms, ensure compliance with ethical standards, and inform decision-making regarding technology design, implementation, and use.

## 3.3 Limitations

When conducting a literature review on ethical challenges in technology development, it is important to consider the limitations of the existing studies.

- Limited scope: Many studies may focus on specific aspects of technological development or ethical challenges within a narrow context. Which will be limited scope as the paper may not provide a comprehensive understanding of the broader ethical challenges in technology development.
- Lack of empirical evidence: The paper depends only on literature reviews, which can make it difficult to draw concrete conclusions or recommendations.

Exploring the ethical challenges in technology development is an important area for future research. Here are some potential areas for further investigation:

- Ethical complexity: Future research can contribute to a more thorough understanding of ethical challenges and inform the development of responsible and ethical practices in the field.
- Rapidly evolving landscape: Future research can contribute to the timely understanding and resolution of ethical challenges, ensuring that ethical considerations keep pace with technological advancements. This will help guide the responsible development and deployment of technologies and promote ethical practices in the field.
- Bias and subjectivity: Future research can enhance the credibility and objectivity of research findings on ethical considerations in technology development. This will contribute to a more robust and reliable understanding of ethical challenges, helping guide decision-making and policy development in the field.

## References


1. Aldoseri, A., Al-Khalifa, K.N., Hamouda, A.M.: Re-thinking data strategy and integration for artificial intelligence: concepts, opportunities, and challenges. Appl. Sci. **13**(12), 7082 (2023)
2. Bostrom, N., Yudkowsky, E.: The ethics of artificial intelligence. In Artificial intelligence safety and security, pp. 57–69 (2018)
3. Chaet, D., Clearfield, R., Sabin, J.E., Skimming, K.: Council on ethical and judicial affairs american medical association. Ethical practice in telehealth and telemedicine. J. Gen. Intern. Med., 32:1136-1140, (2017)
4. Chen, X., Despeisse, M., Johansson, B.: Environmental sustainability of digitalization in manufacturing: a review. Sustainability **12**(24), 10298 (2020)
5. Dignum, V.: Ethics in artificial intelligence: introduction to the special issue. Ethics Inf. Technol. **20**(1), 1–3 (2018)
6. Etzioni, A., Etzioni, O.: Incorporating ethics into artificial intelligence. J. Ethics **21**, 403–418 (2017)
7. Floridi, L.: Soft ethics, the governance of the digital and the general data protection regulation. Philos. Trans. R. Soc. A Math. Phys. Eng. Sci. **376**(2133), 20180081 (2018)

8. Floridi, L., Cowls, J., Beltrametti, M., Chatila, R., Chazerand, P., Dignum, V., Luetge, C., Madelin, R., Pagallo, U., Rossi, F., Schafer, B., Valcke, P., Vayena, E.: AI4People-an ethical framework for a good ai society: opportunities, risks, principles, and recommendations. Mind. Mach. **28**(4), 689–707 (2018)
9. Gangarapu, K.R.: Ethics of facial recognition: Key issues and solutions, (2022)
10. Gerke, S., Minssen, T., Cohen, G.: Ethical and legal challenges of artificial intelligence-driven healthcare. In Artificial intelligence in healthcare, pp. 295–336 (2020)
11. Gillis, A. S. (2022). What is facial recognition?—Definition from WhatIs.com. TechTarget
12. Hagendorff, T.: The ethics of AI ethics: an evaluation of guidelines. Mind. Mach. **30**(1), 99–120 (2020)
13. Jobin, A., Ienca, M., Vayena, E.: The global landscape of AI ethics guidelines. Nat. Mach. Intell. **1**, 389–399 (2019)
14. Kuyumdzhieva, A.: Ethics challenges in the digital era: focus on medical research. In Ethics and Integrity in Health and Life Sciences Research (pp. 45–62) (2018)
15. Lee, W.W., Zankl, W., Chang, H.: An ethical approach to data privacy protection. (2016)
16. Li, W., Tan Yigitcanlar, T., Browne, W., Nili, A.: The making of responsible innovation and technology: an overview and framework. Smart Cities **6**(4), 1996–2034 (2023)
17. Maharani, S.D., Khairul, A.M.I.N., Taufiqoh, A.F.: Technological progress, artificial intelligence development and ethical paradigms. Wisdom **22**, 104–117 (2022)
18. Moor, J.H.: What is computer ethics? In The Ethics of Information Technologies (pp. 15–24) (2020)
19. Mukherjee, S. (2023). Beneath the Surface: Exploring Algorithm Bias in AI. Goglides Dev.
20. Munoko, I., Brown-Liburd, H.L., Vasarhelyi, M.: The ethical implications of using artificial intelligence in auditing. J. Bus. Ethics **167**, 209–234 (2020)
21. Salehi, N.: The glossary of human-centered AI. Niloufar's Substack. (2023)
22. Stahl, B.C.: From computer ethics and the ethics of AI towards an ethics of digital ecosystems. AI Ethics **2**(1), 65–77 (2022)
23. Taylor, K.: The Digital Divide: What it is, and what's being done to close it. (2023)
24. Wies, B., Landers, C., Ienca, M.: Digital mental health for young people: a scoping review of ethical promises and challenges. Front. Digit. Health **3**, 697072 (2021)

# Ethical Ramifications and Remedial Approaches on Bias in Artificial Intelligence

**Shaima Isa Ali, Shahd Faisal Albadoo, and Muneer Al Mubarak**

**Abstract** This paper explores the ethical implications of bias in AI systems and proposes remedial approaches to address this issue. It adopts a literature review approach to examine existing research, case studies, and ethical frameworks related to bias in AI. The analysis reveals different types of biases that can arise in AI systems, including algorithmic bias, data bias, and representation bias. The findings highlight the pervasive nature of bias in AI systems and its potential consequences, such as perpetuating discrimination, reinforcing stereotypes, and exacerbating social inequalities. One limitation of this paper is its reliance on existing literature and case studies, which may not encompass all the nuances and developments in the field. Further empirical research and real-world implementation of proposed remedial approaches are necessary to validate their effectiveness. Nonetheless, the paper provides practical implications for AI developers, policymakers, and organizations. It emphasizes the importance of incorporating ethical considerations into the design, development, and deployment of AI systems. Addressing bias in AI systems is crucial as unchecked bias can preserve systemic discrimination, reinforce existing biases, and marginalize certain groups. By tackling bias in AI, organizations and policymakers can contribute to a more inclusive and equitable society. This paper adds value by synthesizing existing research and frameworks on bias in AI and providing a comprehensive overview of its ethical ramifications and remedial approaches. Its originality lies in its focus on the ethical dimensions of bias in AI and its practical recommendations for addressing this issue.

**Keywords** Artificial Intelligence (AI) · Algorithmic bias · Systemic bias · AI decision-making · Socioeconomic bias

S. I. Ali · S. F. Albadoo · M. Al Mubarak (✉)
Ahlia University, Manama, Bahrain
e-mail: malmubarak@ahlia.edu.bh



M. Al Mubarak (ed.), *Sustainable Digital Technology and Ethics in an Ever-Changing Environment*, Studies in Systems, Decision and Control 237,
https://doi.org/10.1007/978-3-031-86708-8_8

# 1 Introduction

(AI) has the potential to transform our lives, revolutionizing industries such as healthcare, finance, and transportation. However, along with the benefits, AI systems can also perpetuate bias and discrimination, leading to ethical concerns [20]. Bias in AI refers to the partial treatment or favoritism shown towards positive individuals or groups based on characteristics such as contest, gender, or socioeconomic status. This research focuses on exploring the ethical ramifications of bias in AI and proposing remedial approaches to address this critical issue. The ethical suggestions of bias in AI are important. Biased AI systems can spread and amplify societal biases and discrimination, reinforcing existing inequalities and marginalizing vulnerable populations [10]. For example, biased algorithms used in hiring processes can perpetuate gender or racial biases, leading to unfair outcomes and hindering diversity and inclusion. Such biases erode public trust in AI systems and can have far-reaching social consequences [10, 20].

To address these ethical concerns, remedial approaches are crucial. This research aims to identify and evaluate various strategies to mitigate bias in AI. These approaches may include improving data collection and preprocessing techniques to ensure representative and unbiased datasets. Enhancing algorithmic fairness through rigorous testing and evaluation can help classify and correct biases in AI models. Promoting variety and inclusivity in AI development teams can bring diverse perspectives and reduce bias in decision-making processes. Additionally, implementing robust ethical guidelines and regulations can provide a framework for developers and organizations to ensure transparency, accountability, and fairness in AI systems [7]. The proposed remedial approaches are aimed at promoting fairness, transparency, and accountability in AI systems. By addressing bias in AI, we can work towards building trustworthy and socially responsible technology. This research contributes to the ongoing discourse on responsible AI development by identifying the ethical ramifications of bias and providing practical recommendations for developers, policymakers, and users Kamaruddin, [12].

The research problem addressed in this paper focuses on the ethical implications of bias in AI. Beyond technical intricacies, it is imperative to comprehend how biased Artificial intelligence (AI) systems can result in discrimination, unfair treatment, and the propagation of misinformation. This research aspires to contribute to the growing body of knowledge surrounding bias in Artificial intelligence (AI), shedding light on the ethical challenges posed by biased systems and delineating strategies to address them effectively.

The findings of this research will help raise awareness about the potential harms caused by bias in AI and emphasize the importance of positive measures to ease these issues. By promoting ethical practices and remedial approaches, we can ensure that AI systems are fair, unbiased, and aligned with societal values. Ultimately, addressing bias in AI is crucial for building a more equitable and inclusive future where AI benefits all sections of society Parikh, [16]. The need for research on the ethical ramifications and remedial approaches for bias in artificial intelligence stems from the increasing prevalence and influence of AI systems in various domains of society. Bias in AI can have far-reaching consequences, perpetuating inequality, discrimination, and unfair treatment. It is imperative to address this issue to ensure that AI systems are developed and organized responsibly, with fairness and societal well-being in mind Roselli, [17].

While the field of bias in (AI) research is still in its nascent stages, its importance cannot be overstated, especially as AI systems continue to permeate diverse sectors. The beneficiaries of this research span a wide spectrum of stakeholders:

**Individuals**: Those disproportionately affected by bias in Artificial intelligence (AI), often vulnerable populations, stand to benefit from research that identifies and mitigates such biases. Inaccurate or discriminatory Artificial intelligence (AI) outcomes can have real-world consequences, and understanding the ethical implications is crucial for ensuring equitable treatment.

**Organizations**: Entities developing or utilizing Artificial intelligence (AI) systems need insights into ethical guidelines for responsible AI integration. As Artificial intelligence (AI) becomes integral to decision-making processes within organizations, it is imperative to navigate the ethical nuances to avoid perpetuating biases that may exist within the data or system design.

**Policymakers**: The regulatory landscape surrounding AI is evolving, and policymakers play a pivotal role in shaping ethical frameworks and guidelines. Research on bias in AI aids policymakers in crafting informed policies and regulations that mitigate biases and ensure responsible AI deployment.

The research on ethical ramifications and remedial approaches for bias in AI will benefit multiple stakeholders. First, the public and individuals who interact with AI systems will benefit from improved fairness and reduced discrimination. AI systems that are free from bias will lead to more equitable outcomes in parts such as employment, finance, criminal justice, healthcare, and education. Individuals from marginalized or underrepresented groups, who are often disproportionately affected by bias, stand to gain the most from these advancements. AI developers and researchers will also benefit from this research. It will provide them with a deeper understanding of the ethical challenges related to bias in AI & equip them with tools and approaches to develop more responsible and unbiased AI systems. By incorporating ethical considerations early in the development process, developers can avoid unintended biases and ensure the overall integrity of their AI models Dwork, [6].

Bias in Artificial intelligence (AI) is an intricate and multifaceted challenge that lacks a one-size-fits-all solution. Recognizing this complexity is paramount as we navigate through the ethical implications of biased AI systems. However, understanding the ethical ramifications and diverse ways in which bias manifests in Artificial intelligence (AI) provides a foundation for developing effective strategies to mitigate its impact Veale and Binns, [22]. One crucial approach to bias mitigation is training Artificial intelligence (AI) systems on representative data. This entails ensuring that the datasets used for training are reflective of the broader population, minimizing the risk of perpetuating existing biases. Additionally, transparency and accountability in the design of AI are pivotal. Systems should be constructed to enable the identification and rectification of biases, fostering a culture of responsible AI development [7]. Beyond technical measures, policy and regulation play a dynamic role in mitigating bias in AI. Governments and regulatory bodies can implement assessments to evaluate the prospective for bias in AI systems, ensuring that organizations developing or utilizing AI adhere to ethical standards. This multifaceted strategy, encompassing technical, organizational, and regulatory aspects, aims to cultivate a fair and ethical landscape for AI deployment Dwork, [6].

## 2 Main Terms and Definitions

### *2.1 Artificial Intelligence*

- Artificial Intelligence (AI) refers to the imitation of human intelligence in technologies that are programmed to think and learn like humans. It involves the development of procedures and models that allow computers to do tasks such as speech recognition, problem-solving, decision-making, and pattern recognition Parikh, [16].
- Artificial Intelligence is a branch of computer science that focuses on the creation and development of intelligent machines capable of performing tasks that typically require human intelligence. These tasks may include natural language processing, visual perception, and complex problem-solving Roselli, [17].
- Artificial Intelligence is the field of study and engineering that aims to make intelligent machines capable of replicating or surpassing human intelligence in performing specific tasks. It includes various subfields such as machine learning, natural language processing, and robotics [14].

### *2.2 Bias in Artificial Intelligence*

Bias in Artificial Intelligence (AI) refers to the presence of systematic errors or prejudices in AI systems, leading to unfair outcomes. This bias can stem from biased training data, flawed algorithms, or the amplification of existing societal biases. When the training data is incomplete, unrepresentative, or reflects historical discrimination, AI models can inherit and perpetuate those biases. Additionally, flawed algorithms can introduce biases during data processing and analysis. It is crucial to address and mitigate bias in AI to ensure fair and equitable outcomes for all individuals and to promote the responsible and ethical use of AI technologies. It occurs when the data used to train AI models is biased or when the algorithms themselves inadvertently reflect and perpetuate existing biases present in society. Bias in AI occurs when machine learning systems produce results that are skewed or discriminatory towards certain individuals or groups based on reasons such as contest, gender, or socioeconomic status. These biases can lead to unequal treatment, and limited opportunities, and reinforce societal inequalities. Bias in Artificial Intelligence is the presence of partiality or unfairness in the decision-making of AI systems. It can arise from biased training data Kamaruddin, [12].

## 2.3 Ethical Implications

1. Ethical implications refer to the potential consequences or impacts that arise from the application of certain actions, decisions, or technologies, which raise moral questions and considerations regarding their rightness or wrongness. It involves examining the ethical dimensions and evaluating the potential effects on individuals, society, and the environment Gichoya, [8].
2. Ethical implications are the moral considerations and dilemmas that emerge when making choices or implementing practices that have the potential to affect individuals or society. These implications often involve questions of fairness, justice, privacy, transparency, accountability, and the overall impact on human well-being and values Melsión, [14].
3. Ethical implications involve the examination and assessment of the ethical consequences and responsibilities associated with particular actions, policies, or technologies. It encompasses the evaluation of potential benefits, risks, unintended consequences, and conflicts of interest, as well as the need to uphold fundamental principles such as autonomy, dignity, equality, and respect for human rights (Haussler, [10].

Research problems on the ethical implications and remedial approaches to bias in artificial intelligence is an important and contemporary topic in the field of (AI). This topic concerns the ethical impacts that AI-based decisions and actions can have on society and humanity, and how to deal with bias that may arise in these decisions and actions Seyyed, [19]. Bias in AI refers to the presence of unwanted biases in the data on which AI models are trained, and this may lead to unfair decisions. For example, if an AI model is qualified on data that contains biases toward a population, that bias may transfer to the decisions the model makes. Bias in AI refers to biases or preferences arising from the data or algorithms used in AI systems. Bias can be caused by many factors, including the method of data collection, the design of algorithms, and choices about criteria and variables of interest. Bias in AI can lead to unfair or unequal outcomes for different individuals or groups DeCamp and Lindvall, [5].

There are several common forms of bias in AI that can impact the fairness and accuracy of its outcomes. One form is data bias, which arises when the data used to train AI systems is unbalanced or limited. This imbalance can lead to biased results, as the data may favor a specific group or fail to adequately represent diverse conditions.

Another type of bias is algorithmic bias, which occurs when the algorithms employed in AI systems exhibit biased behavior or favor certain classes or groups. This bias can perpetuate discriminatory outcomes and unequal treatment. Lastly, systemic bias refers to bias that is inherent in the overall design of the AI system. This bias can manifest through the selection of certain performance metrics or through prioritizing specific outcomes over others, leading to unequal or unfair results. Addressing and mitigating these forms of bias is crucial for ensuring the ethical and equitable use of AI technology.

The task of reducing bias in AI requires ongoing efforts to improve data quality, design unbiased algorithms, and ensure transparency and accountability in the decision-making process. Recognizing and addressing biases is an essential part of developing ethical and fair AI systems. Achieving balance and justice in AI is an important ethical challenge. Research should address this issue by exploring methods and techniques that can be used to deal with biases in artificial intelligence. Remedial approaches to bias in AI can include Kiyasseh, [13].

1. Improving data quality: This can be achieved through methodologies such as data balancing, source diversification, and quality checking to ensure that the data on which models are trained reflects the diversity of societies and does not contain unwanted biases.
2. Transparent disclosure: AI developers must be transparent regarding the training process, algorithms used, and data used. Users and the public can better understand how decisions are made and can check for biases.
3. Ethical Audit: External Audit Bodies Can Evaluate and Scrutinize Potential Biases in AI Models

This study will focus on the ethical implications of bias in AI. We will examine the different ways in which bias can occur in AI systems and the potential consequences of such bias. We will also discuss several strategies for mitigating bias in AI systems.

## 3 Literature Review

Artificial intelligence (AI) stands at the forefront of technological advancement, permeating various facets of our daily lives and shaping the way we interact with information, services, and even each other Burrell, [4]. From the algorithms governing our social media feeds to the autonomous decision-making capabilities of self-driving cars, AI has become an important part of contemporary society. However, this rapid integration brings forth a consequential challenge: the pervasive issue of bias within AI systems Burrell, [4]. As AI systems evolve in sophistication and complexity, they become increasingly susceptible to bias, reflecting the inherent biases present in the data upon which they are trained Buolamwini and Gebru, [3]. The consequences of biased AI are not merely technical glitches but extend into the realms of ethics, fairness, and societal impact Burrell, [4].

Burrell [4] embarks on a comprehensive exploration of the multifaceted dimensions of bias in AI, aiming to scrutinize its ethical implications and propose effective strategies for mitigation. As AI systems become omnipresent, it is crucial to acknowledge the diverse contexts in which bias can emerge. Bias in AI is not a monolithic phenomenon but manifests in various forms, posing challenges that necessitate a nuanced understanding. For instance, biased data, where AI systems are trained on datasets skewed against certain demographic groups, can lead to perpetuating societal inequalities and reinforcing discriminatory practices. Similarly, design bias arises when the creators' perspectives and biases inadvertently influence the architecture and decision-making of AI systems. Algorithmic bias, on the other hand, is rooted in the very algorithms that govern AI, reflecting favoritism or discrimination toward specific groups.

To ensure clarity and coherence throughout the paper, key terms are precisely defined. AI encompasses the field of computer science dedicated to generating intelligent agents capable of reasoning, learning, and autonomous action. "Bias in AI" refers to the inclination of AI systems to produce results skewed in favor of or against specific groups. "Ethical implications" denote the potential negative consequences of bias in AI, encompassing discrimination, unfair treatment, and the dissemination of misinformation. Bias in artificial intelligence (AI) is a complex and challenging issue. AI systems are trained on data, and if the data is biased, the AI system will be biased. Bias in AI can have a number of negative consequences, such as discrimination, unfair treatment, and the spread of misinformation. There are several theoretical frameworks that can be used to understand bias in AI. One common framework is the concept of implicit bias [8]. Implicit bias refers to the unconscious biases that we all have, which can affect decision-making, even when we are not aware of them. Another common framework is the concept of algorithmic bias [1]. Algorithmic bias refers to the bias that can be introduced into AI systems by the algorithms that are trained on biased data. For example, if an algorithm is trained on data that is biased against certain groups of people, the algorithm will be more likely to produce biased results against those groups.

The literature on bias in AI is growing rapidly. However, research in this area is still in its early stages. One of the challenges of researching bias in AI is the difficulty in identifying and measuring bias. Additionally, research on bias in AI is often fragmented, with different researchers focusing on different aspects of the problem.

Despite these challenges, there has been a significant amount of progress made in researching bias in AI in recent years. Some of the key findings from this research include:

- Bias in AI can occur in a variety of ways, including data bias, design bias, and algorithmic bias.
- Bias in AI can have several negative consequences, such as discrimination, unfair treatment, and the spread of misinformation.
- There are a number of strategies that can be used to mitigate bias in AI, such as training AI systems on representative data, designing AI systems to be transparent

and accountable, and developing policies and regulations that help to mitigate bias in AI Burrell, [4].

"Managing bias in AI" by Roselli et al. [16] published in the Companion Proceedings of the 2019 World Wide Web Conference, and "Addressing bias in AI in health care" by Parikh, Teeple, and Navathe, published in JAMA in 2019. These papers tackle the important issue of bias in artificial intelligence (AI) and discuss strategies to mitigate bias in different contexts. Roselli et al. [16] focused on the all-purpose problem of bias in AI systems. It explores the potential biases that can arise in AI models and the impact of biased AI systems on various aspects of society. The authors emphasize the importance of addressing bias in AI to ensure fairness, transparency, and ethical use of these technologies. They discuss various sources of bias, including biased training data, biased algorithms, and biased decision-making processes. The paper also presents a framework for managing bias in AI, which involves identifying and measuring bias, understanding its causes, and taking appropriate actions to mitigate bias.

Parikh, Teeple, and Navathe [15] conducted research that specifically focuses on the application of AI in healthcare and the potential biases that can arise in this domain. In their paper titled "Addressing bias in artificial intelligence in health care," they delve into the challenges of bias in AI systems used in healthcare. The authors highlight the importance of addressing data bias, algorithmic bias, and systemic bias to ensure fairness and equity in healthcare AI applications [15]. Their work sheds light on the need for mitigating bias in AI systems to promote ethical and unbiased healthcare practices. The authors highlight how biases in AI systems used in healthcare can have significant consequences for patient outcomes and healthcare disparities. They discuss different types of biases that can occur, such as racial or ethnic bias, gender bias, and socioeconomic bias. The paper also explores the challenges of bias in AI systems, including issues related to data quality, algorithmic bias, and interpretability of AI models. The authors propose several strategies to address bias in AI healthcare applications, such as using diverse and representative training data, continuously monitoring, and evaluating AI systems for bias, and involving stakeholders in the development and validation of AI models. Both studies emphasize the critical want to address bias in AI systems to ensure fairness, equity, and ethical use of these technologies. They provide valuable insights into the sources of bias, the probable penalties of biased AI, and strategies to mitigate bias. The papers contribute to the ongoing dialogue on the ethical and responsible development and deployment of AI systems in various domains, including healthcare and broader societal contexts.

Nelson [14] discusses the issue of bias in artificial intelligence in their paper titled "Bias in artificial intelligence" published in the North Carolina medical journal. The author explores the presence of bias in AI systems and its implications. The paper emphasizes the importance of understanding and addressing bias in AI to ensure equitable outcomes [14], p. 220–222).These papers delve into the topic of bias in artificial intelligence (AI) and provide insights into the challenges associated with bias and strategies to mitigate it. In "Bias in artificial intelligence" by

Nelson, the author highlights the issue of bias in AI systems and its implications. The paper discusses how bias can emerge in AI algorithms due to various factors, including biased training data, algorithm design, and decision-making processes. Nelson emphasizes that biases in AI can preserve and increase existing societal biases, important to unfair outcomes and exacerbating disparities. The paper also explores the ethical considerations of bias in AI and calls for increased awareness and efforts to mitigate bias in AI systems to ensure fairness and equity.

Gichoya et al [8] conducted a study on mitigating bias in AI titled "AI pitfalls and what not to do: mitigating bias in AI." The authors specifically focus on the context of AI, with a particular emphasis on the field of radiology. They discuss common pitfalls and challenges associated with bias in AI systems and provide recommendations to address them. The study highlights the importance of diverse and representative training data, rigorous evaluation of AI models for bias, and the need for transparency and interpretability in AI algorithms. The research by Gichoya and colleagues (2023) emphasizes the active engagement of stakeholders, including clinicians and researchers, in mitigating bias in AI through the development, validation, and deployment processes.

Both papers underwrite the growing body of research on bias in AI and provide valued understandings into its causes, implications, and potential solutions. They emphasize the ethical considerations surrounding bias in AI and call for the development and adoption of responsible practices in AI development and deployment. By shedding light on the challenges and offering practical recommendations, these papers aim to promote awareness and guide researchers, practitioners, and policymakers in mitigating bias in AI systems to ensure fair and equitable outcomes. The two papers "Using Explain ability to Help Children Understand Gender Bias in AI" by Melsión, Torre, Vidal, and Leite, published in the Communication Design and Children consultation in 2021, and "Quantifying inductive bias: AI learning algorithms and Valiant's learning framework" by Haussler, published in Artificial Intelligence in 1988. These papers explore different aspects of AI, focusing on the topics of explain ability and inductive bias.

The paper "Using Explain Ability to Help Children Understand Gender Bias in AI" by Melsión et al. addresses the prominent issue of gender bias in AI and proposes a novel approach to help children understand and recognize this bias. The authors recognize that AI systems can exhibit gender bias due to biased training data or algorithmic design choices. They argue that it is crucial to educate children about these biases from an early age to foster a critical understanding of AI technologies. The paper presents an interactive system that uses explain-ability techniques to provide children with insights into how AI algorithms can perpetuate gender biases. The system allows children to explore and question the underlying mechanisms of AI, making them more aware of the potential biases in these systems. On the other hand, the paper "Quantifying inductive bias: AI learning algorithms and Valiant's learning framework" by Haussler focuses on the concept of inductive bias in AI learning algorithms. Inductive bias refers to the assumptions or preferences that an AI algorithm incorporates during the learning process. The paper explores Valiant's learning framework, which provides a theoretical basis for understanding inductive bias in AI algorithms. Haussler discusses various types of inductive biases, such as language bias, preference bias, and search bias, and how they impact the learning process and the generalization capabilities of AI systems. The paper also presents mathematical formalizations to quantify and compare the inductive biases of different learning algorithms.

Both papers contribute to the understanding of key aspects of AI. "Using Explain ability to Help Children Understand Gender Bias in AI" highlights the importance of explain ability in AI systems, particularly in the context of gender bias, and proposes a method to educate children about these biases. "Quantifying inductive bias: AI learning algorithms and Valiant's learning framework" explores the concept of inductive bias in AI algorithms, providing insights into how different biases can influence the learning process and the behavior of AI systems. Together, these papers shed light on important considerations in the development and use of AI technologies, with a focus on fairness, transparency, and understanding of AI algorithms.

"An Overview of Artificial Intelligence Ethics: Issues and Solution for Challenges in Different Fields" by Santhoshkumar et al. [17].

This article likely provides an overview of the ethical considerations surrounding artificial intelligence (AI) and explores the challenges and issues that arise in different fields. It may discuss topics such as algorithmic bias, privacy concerns, transparency, accountability, and the social impact of AI. The authors might propose solutions or strategies to address these challenges and promote ethical practices in the development and deployment of AI technologies.

"Might artificial intelligence become part of the person, and what are the key ethical and legal implications?" by Bublitz [2]:

This article appears to explore the concept of AI becoming integrated into human beings and the ethical and legal implications that arise from such integration. It may discuss topics like AI implants, brain-computer interfaces, or the potential blending of human and AI capabilities. The author might examine issues related to personal identity, autonomy, privacy, consent, and the legal frameworks that would need to be developed to regulate such scenarios.

"Compliance to GDPR Data Protection and Privacy in Artificial Intelligence Technology: Legal and Ethical Ramifications in Malaysia" by Kamaruddin et al. [11].

Likely focuses on the compliance of artificial intelligence technology with the General Data Protection Regulation in the situation of data protection and privacy in Malaysia. It might discuss the ethical and legal considerations related to the collection, use, and processing of personal data in AI systems. The authors may examine the challenges and ramifications of GDPR compliance for AI applications in Malaysia and propose strategies to ensure adherence to the legal and ethical requirements.

## 3.1 Literature Review Findings

1. The study titled "Addressing Bias in Artificial Intelligence in Health Care" by Parikh, Teeple, and Navathe [15] specifically focuses on the issue of bias in artificial intelligence within the healthcare domain. The authors examine the challenges and risks associated with bias in AI models and algorithms used in healthcare. They also provide suggestions and recommendations for addressing and mitigating bias to promote fairness and equity in healthcare AI applications [15].
2. The study titled "Managing Bias in AI" by Roselli, Matthews, and Talagala [16] focuses on the management of bias in AI. The study reviews various bias management tools and practices used to address bias in AI models. It also provides tips and strategies for reducing bias and enhancing fairness in the outcomes of AI systems. The research was published in the Companion Proceedings of the 2019 World Wide Web Conference [16].
3. The study titled "Bias in Artificial Intelligence" by Nelson [14] sheds light on the presence of bias in artificial intelligence systems in general. The study illustrates that AI models can demonstrate bias due to factors such as limited training data or algorithm design. Additionally, the paper provides suggestions and recommendations for mitigating bias in order to reduce its impact. The research was published in the North Carolina Medical Journal [14].
4. The study titled "AI Pitfalls and What Not to Do: Mitigating Bias in AI" by Gichoya et al. [8] addresses potential pitfalls and strategies for mitigating bias in AI. The study emphasizes the importance of designing and implementing AI models in a manner that ensures fairness and avoids bias. The research was published in The British Journal of Radiology [8].
5. The study titled "Using Explainability to Help Children Understand Gender Bias in AI" by Melsión, Torre, Vidal, and Leite [13] focuses on helping children understand gender bias in AI. The study introduces an interactive system that utilizes illustration techniques to demonstrate how AI algorithms can reinforce gender bias. The research was published in the proceedings of the Interaction Design and Children conference [13].
6. The study titled "Quantifying Inductive Bias: AI Learning Algorithms and Valiant's Learning Framework" by Haussler [9] explores the concept of inductive bias in AI learning algorithms. The research specifically focuses on Valiant's learning framework and examines how different learning algorithms exhibit varying levels of inductive bias. The study delves into the quantification of inductive bias and its impact on the learning process. The paper was published in the journal Artificial Intelligence [9].

## 4 Conclusion

Bias in AI has significant ethical ramifications that can result in unfair and discriminatory outcomes. It is crucial to address this issue and develop remedial approaches to mitigate bias in AI systems. The paper has outlined various strategies that can be employed to tackle bias in AI, including careful data collection and preprocessing, algorithmic transparency and explain ability, regular auditing and evaluation, diverse development teams, ethical guidelines and regulations, user feedback and recourse mechanisms, and bias mitigation techniques.

One of the primary remedial approaches is ensuring that the data used to train AI models is varied, characteristic, and free from bias. This involves implementing robust data collection and preprocessing techniques that identify and mitigate potential biases. By carefully curating training data, we can reduce the likelihood of biased outcomes in AI systems. Algorithmic transparency and explain ability are also crucial in addressing bias. AI systems should be designed to be transparent, allowing users to understand how decisions are made. This transparency facilitates the identification and mitigation of biases in the underlying algorithms. Techniques such as interpretable machine learning, and model-agnostic approaches can be employed to achieve this goal. Regular auditing and evaluation of AI systems are essential to detect and address biases that may arise over time. By endlessly monitoring the system's performance and analyzing its outcomes, potential biases can be identified, and corrective actions can be taken. Building diverse development teams is another important strategy for mitigating bias in AI systems. By including individuals from different backgrounds and perspectives, diverse teams are more likely to identify and address biases during the development process. This leads to the creation of more inclusive and fair AI systems.

Establishing ethical guidelines and regulations is crucial to safeguard responsible and accountable development and deployment of AI systems. Governments, organizations, and industry bodies should collaborate to establish clear guidelines that address bias and discrimination concerns. By adhering to these guidelines, developers can build AI technologies that are mindful of ethical considerations. User feedback and recourse mechanisms are essential in addressing bias in real-world scenarios. AI systems should provide users with mechanisms to provide feedback and seek recourse if they believe they have been subjected to biased outcomes. This not only helps in identifying and rectifying biases but also fosters trust and accountability in AI systems. Researchers and developers should continue to explore and develop bias mitigation techniques. Techniques such as debiasing algorithms, fairness-aware learning, and counterfactual fairness aim to reduce bias and ensure fair treatment across different groups. Ongoing research in this field is crucial to stay at the forefront of bias mitigation efforts.

In conclusion, addressing the ethical ramifications of bias in AI requires a multifaceted approach. By implementing careful data handling, algorithmic transparency, continuous evaluation, diverse teams, ethical guidelines, user feedback mechanisms, and ongoing research in bias mitigation techniques, we can work towards building

AI systems that are fair, inclusive, and accountable. It is essential for stakeholders in the AI community to collaborate and prioritize the development of unbiased AI systems that benefit society.

## 4.1 Theoretical Implications

This paper contributes to the theoretical literature on bias in AI by providing a comprehensive overview of the different ways in which bias can occur in AI systems and the potential consequences of such bias. Additionally, this paper proposes a number of hypotheses about the factors that contribute to bias in AI systems. These hypotheses can be tested in future research to develop a better understanding of bias in AI.

## 4.2 Practical Implications

The findings of this paper have a number of practical implications for organizations that are developing or using AI systems. First, organizations need to be aware of the possible bias in AI systems. Second, organizations need to take steps to alleviate bias in AI systems, such as training AI systems on representative data, designing AI systems to be transparent and accountable, and developing policies and regulations that help to alleviate bias in AI.

## 4.3 Study Limitations and Future Research

There are several limitations to this study. First, this study is a conceptual paper. Additional research is needed to test the hypotheses that are proposed in this paper and to develop effective strategies for justifying bias in AI systems. Second, this study motivations on the ethical implications of bias in AI. However, there are also several other important implications of bias in AI, such as legal implications and social implications. Future research should explore these other implications of bias in AI. The final limitation of this paper is its reliance on existing literature and case studies, which may not encompass all the nuances and developments in the field. Further empirical research and real-world implementation of proposed remedial approaches are necessary to validate their effectiveness.

## References


1. Barocas, S., Selbst, A.D.: Big data's disparate impact. Calif. Law Rev. **104**(3), 671–732 (2016)
2. Bublitz, J.C.: Might artificial intelligence become part of the person, and what are the key ethical and legal implications? AI Soc. **37**(1), 1–12 (2022)
3. Buolamwini, J., Gebru, T.: Gender shades: intersectional accuracy disparities in commercial gender classification. In Conference on fairness, accountability and transparency, pp. 77–91. PMLR. (2018)
4. Burrell, J., Mulligan, D.K., Kluttz, D.N.: Report from the first AFOG summer workshop. (n.d.)
5. DeCamp, M., Lindvall, C.: Mitigating bias in AI at the point of care. Science **381**(6654), 150–152 (2023)
6. Dwork, C., Hardt, M., Pitassi, T., Reingold, O., Zemel, R.: Fairness through awareness. In Proceedings of the 3rd conference on innovations in theoretical computer science, pp. 226–237, ACM. (2012)
7. Ghillyer, A. W.: Business ethics now (6th ed.). McGraw-Hill Education. (2021)
8. Gichoya, J.W., Thomas, K., Celi, L.A., Safdar, N., Banerjee, I., Banja, J.D., Laleh, S.-K., Trivedi, H. Purkayastha, S.: AI pitfalls and what not to do: mitigating bias in AI. Br. J. Radiol. 96(1150), 20230023, (2023)
9. Greenwald, A.G., Krieger, L.H.: Implicit bias: scientific foundations. Calif. Law Rev. **94**(4), 945–967 (2006)
10. Haussler, D.: Quantifying inductive bias: AI learning algorithms and valiant's learning framework. Artif. Intell. **36**(2), 177–221 (1988)
11. Jones, A.: Ethical considerations in AI bias mitigation. J. Ethics Technol. **5**(2), 45–62 (2020)
12. Kamaruddin, S., Mohammad, A.M., Saufi, N.N.M., Rosli, W.R.W., Othman, M.B., Hamin, Z.: Compliance to GDPR data protection and privacy in artificial intelligence technology: legal and ethical ramifications in Malaysia. In 2023 International conference on disruptive technologies (ICDT), pp. 284–288. IEEE. (2023)
13. Kiyasseh, D., Laca, J., Haque, T.F., Otiato, M., Miles, B.J., Wagner, C., Hung, A.J.: Human visual explanations mitigate bias in AI-based assessment of surgeon skills. Npj Digit. Med., 6(1), 54 (2023)
14. Melsión, G.I., Torre, I., Vidal, E., Leite, I.: Using explainability to help children understand gender bias in AI. In interaction design and children, pp. 87–99 (2021)
15. Nelson, G.S.: Bias in artificial intelligence. N. C. Med. J. **80**(4), 220–222 (2019)
16. Parikh, R.B., Teeple, S., Navathe, A.S.: Addressing bias in artificial intelligence in health care. JAMA **322**(24), 2377–2378 (2019)
17. Roselli, D., Matthews, J., Talagala, N.: Managing bias in AI. In companion proceedings of the 2019 world wide web conference, pp. 539–544, (2019)
18. Santhoshkumar, S.P., Susithra, K., Prasath, T.K.: An overview of artificial intelligence ethics: issues and solution for challenges in different Fields. J. Artif. Intell. Capsul. Netw. **5**(1), 69–86 (2023)
19. Seyyed-Kalantari, L., Zhang, H., McDermott, M.B., Chen, I.Y., Ghassemi, M.: Underdiagnosis bias of artificial intelligence algorithms applied to chest radiographs in under-served patient populations. Nat. Med. **27**(12), 2176–2182 (2021)
20. Sharma, S., & Graydon, M. S. (2021). Social bias in AI and its implications.
21. Smith, B.: The impact of bias in AI systems. J. Artif. Intell. Res. **12**(3), 187–203 (2019)
22. Veale, M., Binns, R.: Fairer machine learning: techniques for mitigating bias. arXiv preprint arXiv:1611.09350. (2017)

# Sustainable Employee Retention Strategies in Dynamic Environment of Higher Educational Institution

**K. Rajeswari, Niranchana Shri Viswanathan, Katya, and S. Jayashree**

**Abstract** Employee retention plays an important role in the higher educational institutions. Educational institutions follow some important strategies to retention the employees. The main intension of planning the strategies towards employee retention is to minimize the employee turnover rate for sustainable growth of the educational institutions in long run. Educational institutions are struggling to implement strategies to retain the capable manpower in workplace. Increased employee turnover will consequently have an impact on efficiency and sustainability of the educational institutions. Employee retention is essential to an organization to accomplish its objectives and to enhance the financial steadiness. Higher educational institutions have to follow the effective retention strategies to encourage the employee to remain in their workplace for maximum period of time. Whenever the employees get dissatisfaction in their job they are switch over to other educational institution within short span of time. The higher educational institutions have to motivate the employees by providing incentives, rewards and recognition which will increase the job satisfaction level of employees. The policies and practices regarding work culture have to be conversed to employees clearly when they step into the educational institutions which will be helpful to reduce the turnover ratio. At present the stress level of the employees in higher educational institutions have been increased due to poor work culture which directly increases the turnover ratio of employees. The current study is focused towards the educational institutions interventional strategies to look after the employees retention such as Training and development, flexible work schedule, promote a work-life balance, Work Recognition, compensation, emphasis on team work, on boarding Programs, Mentorship, workers welfare measures and opportunity for growth. The objective of the study is to ascertain that how the work culture

K. Rajeswari (✉) · Katya
REVA University, Bengaluru, India
e-mail: rajashiwarikarupp.14@gmail.com

N. S. Viswanathan
Sapthagiri NPS University, Bengaluru, India

S. Jayashree
Kristu Jayanti College (Autonomous), Bengaluru, India



M. Al Mubarak (ed.), *Sustainable Digital Technology and Ethics in an Ever-Changing Environment*, Studies in Systems, Decision and Control 237,
https://doi.org/10.1007/978-3-031-86708-8_9

and HRM Practice may affect the employee retention in Educational Institution. A study collected data through structured questionnaire by choosing 250 samples from higher educational institutions located in and throughout the city of Bangalore. The result of this study revealed that employee retention strategies significantly reduce the employees' turnover rate.

**Keywords** Strategies · Employee retention · Higher educational institutions · Sustainability

## 1 Introduction

Employee retention strategies will help the educational institutions to prevent the employees from huge turnover. The retention of employees will create more opportunities to the educational institutions for their growth and development. Educational institution has to frame the policies and practices to retain the quality employees. Employees those who know about the vision and mission of the educational institutions in a better way definitely they will work towards the improvement of the institutions and they put their hard work towards accomplishing the objectives of the educational institutions. When the educational institutions satisfy the employees towards its retention strategies it will helpful to them to survive for the long period of time in the same institution. Employees who have more experience in higher educational institutions will have more knowledge and skills which will be helpful to them to provide hundred percent perfection towards the completion of their tasks. Employers in educational institution have to create development opportunities by offering free workshops, seminars and training when the management is not providing such opportunities then the employees might feel the only way to progress in their careers and skills is possible towards changing the job. Employees will like to stay in the educational institutions when they get attractive benefits from management. Educational institutions have to promote healthy work environment through career focused training which provides a perfect way for employees to progress in the institution and encourages them to stay back in their job. Thus, the research is done to ascertain the different employee strategies that prevent employees' turnover and thus it is necessary to examine how higher educational institutions factors of retention will impact employees satisfaction in educational establishments.

## 2 Statement of the Problem

Retaining employees is the exertion taken by the educational institutions to hold them as a most valuable asset. It is essential to manage the employee turnover rate if the organization has to survive in long run. Retention of employee is the way to reduce the cost involved in recruiting the employees. In Educational institutions retaining the experienced employee will helpful to increase the work standard and also helpful to enhance the quality of work life. High employee turnover rates can destruct the work of the employees.

The positive work environment will increase the peace in the mind of the employee towards job and it will also create healthy relationship between employer and employee. The employees in educational institutions are service motive and they will give more importance to their personal growth and development so the management has to create more opportunities for the career growth of workforces.

As a result, there are several variables involved in Employee Retention strategies which have to be followed by higher educational institution. The purpose of the study is to investigate and evaluate the variables that impact employee retention in higher education establishments in the Bangalore city region.

## 3 Objectives of the Study

The primary goal of the study is to concentrate on institutes of higher education.

- To identify the elements that influence higher education institutions employee retention rates.
- To examine the variables affecting higher education institutions employee performance.
- To analyze the connection between higher education institutions Employee performance and employee retention rates.

To provide a recommendation based on the findings of research.

## 4 Literature Review

Erasmus et al. [1], As stated in their article titled "Employee Retention in a Higher Education Institution: An Organisational Development Perspective". The study primary goals of the research were to find out the turnover trends based on demographic and performance related variables. They investigated 4651 employees regarding retention and voluntary turnover decisions. A close-ended questionnaire was used to gather data. The cross-sectional study was conducted with structures questionnaire and the data were analysed using both descriptive and inferential statistics for each demographic category. They found that the institutions turnover rate was acceptable and not operating normally or properly and its marginal turnover as a result of underperforming personnel standard rating shows that employee resigned willingly. They concluded that there is a positive correlation and significant beta values that has been found between organization citizenship, Leadership, remuneration and the employee's decision to remain with or leave the organization.

Silbert Jose [2], carried out research on "The Impact of Human Resource Practices on Employee Retention in the Education Sector" While researching, he attempted to analyse the HR procedures regarding staff retention in Indian education sector. Researcher collected secondary data for analysis purpose and he found that the existence of a substantial connection between variables impacting staff retention activities of institution associated with Indian education. Hence the research paper suggested that retain strategies of employees are family benefits, incentives for seminars and conferences and increase in remuneration and perks have to be regulated by human resource managers. It has been concluded that most of the private universities has found the way to retain the employees by having separate departments for performing this function instead of leaving it to academicians and administrators who are not well trained to perform the role effectively and it have to be practiced by Indian state and central universities.

Jam and Jamal [3], undertook a study entitled "Impact of Green Human Resources Management Practices on Organizational Sustainability and Employee Retention: An Empirical study related to Educational Institutions". The study objective is to investigate the influence of green human recourse management practice on organizational longevity and staff retention in educational establishments in south Punjab, Pakistan. They randomly selected 150 samples from educational institutions of south Punjab to analysed the influence of green human resource management practice on staff retention. The result of the study also demonstrated that of green human resource management practice has a major impact on staff retention and organizational sustainability of institutions of higher learning in South Punjab. They concluded that educational institutions must follow the right strategies and tactics to retain the employee which will also add value to the institutions.

## 5 Research Methodology

Descriptive research methodologies are used in this study to determine the elements that influence employee retention strategies towards dynamic environment of higher educational institution. The research conducted by selecting 250 samples from higher education institutions in Bangalore city region. Data collecting methods employ random sampling techniques. Multiple regressions and one-way ANOVA were used to tabulated and analyse the data.

## 6 Findings of the Study

It is analyzed from the above Table 1, that there is significant relationship between factors of impact employee's retention in higher educational institutions. It is also found that employee retention factors such as Training and development shows significant group difference are evident because the p-value = 0.007 is below0.05, flexible work schedule have a significant difference which indicated by the p-value = 0.005 which is less than 0.05, promote a work-life balance has a significant variation were p-value = 0.003 is less than 0.05,Work Recognition has low p-value = 0.004 which shows that there is substantial difference and it is less than 0.05, compensation variable indicates that there is significantly varied between groups were p-value = 0.001 and shows that it is less than 0.05, emphasis on team work specifies that there is a significant difference in the p-value 0.008 which is less than 0.05, on boarding Programs in which it is identified that there is a notable changes shows the p-value = 0.005 that is smaller than 0.05, Mentorship indicates the substantial connection between variables that is p-value = 0.004 in the meanwhile it is less than 0.05,Workers welfare measures has a significant p-value = 0.006 which indicates that it is less than 0.05, opportunity for growth has large p-value = 0.007 which indicates that there is a significant variation in groups view and it is less than 0.05. Overall Variables shows that there is significant relationship between impacts of employee retention strategies in dynamic environment.

It is clear from the above Table 2, regression model which generated a statistical significance outcome that there is an influence on Employee retention strategies in dynamic environment of higher educational institutions ($F = 82.620$, $p < 0.05$ and $R^2 = 0.662$). The analysis further shows that 'Training and development' is the most reliable indication of deviation in Employee retention (βeta = 1.055, t-value = 23.524 and p< = 0.05), Flexible work schedule (βeta = 0.157, t-value = 3.361 and p< = 0.05), Work Recognition (βeta = 0.108, t-value 2.310 and p< = 0.05), compensation (βeta = 0.134, t-value = 2.845 and p< = 0.05), on boarding Programs (βeta = 0.158, t-value = 5.236 and p< = 0.05),workers welfare measures (βeta = 0.275, tt-value = 1.345 and p< = 0.05) and opportunity for growth (βeta = 0.152, t-value = 2.471 and p< = 0.05) have a major influence on the employee retention towards higher educational institutions in the meanwhile Promote a work-life balance, Emphasis on team work and Mentoring does have any impact towards employee retention strategies.

## 7 Suggestions

- Enrich the employee skills through training and development opportunities it will create a new path for their career development.
- The higher educational institutions have to encourage the creativity of employees and allow them to put forward new ideas or creativities without anxiety to execute the plan in successful manner.
- Employees performance based incentives has to be provided which will take them forward for improve their performance towards work.
- Higher educational institutions have to organize recognition programs through which employees achievements and contributions towards work have to be recognized.

**Table 1** Variables that impact employee retention strategies in dynamic environment

| Training and development | | | | | |
|---|---|---|---|---|---|
| Variables | Sum of squares | Degree of freedom | Mean square | Frequency | Significance |
| Between group(combined) | 1.361 | 2 | 1.640 | 7.390 | 0.007 |
| Within group(combined) | 202.507 | 247 | 1.912 | | |
| | 203.868 | 249 | | | |
| *Flexible work schedule* | | | | | |
| Variables | Sum of Squares | Degree of Freedom | Mean square | Frequency | Significance |
| Between group(combined) | 2.048 | 2 | 1.018 | 5.043 | 0.005 |
| Within group(combined) | 63.754 | 247 | 0.553 | | |
| | 65.802 | 249 | | | |
| *Promote a work-life balance* | | | | | |
| Variables | Sum of squares | Degree of freedom | Mean square | Frequency | Significance |
| Between group(combined) | 0.484 | 2 | 0.242 | 4.183 | 0.003 |
| Within group(combined) | 194.967 | 247 | 1.705 | | |
| | 195.451 | 249 | | | |
| *Work recognition* | | | | | |
| Variables | Sum of squares | Degree of freedom | Mean square | Frequency | Significance |
| Between group(combined) | 6.559 | 2 | 3.279 | 9.055 | 0.004 |
| Within group(combined) | 132.233 | 247 | 1.558 | | |

(continued)

**Table 1** (continued)

| Training and development | | | | | |
|---|---|---|---|---|---|
| | 138.792 | 249 | | | |
| *Compensation* | | | | | |
| Variables | Sum of squares | Degree of freedom | Mean square | Frequency | Significance |
| Between group(combined) | 0.028 | 2 | 1.014 | 11.352 | 0.001 |
| Within group(combined) | 84.564 | 247 | 0.723 | | |
| | 84.592 | 249 | | | |
| *Emphasis on team work* | | | | | |
| Variables | Sum of squares | Degree of freedom | Mean square | Frequency | Significance |
| Between group(combined) | 0.319 | 2 | 0.614 | 7.077 | 0.008 |
| Within group(combined) | 82.456 | 247 | 0.454 | | |
| | 82.775 | 249 | | | |
| *On boarding programs* | | | | | |
| Variables | Sum of squares | Degree of freedom | Mean square | Frequency | Significance |
| Between group(combined) | 0.014 | 2 | 0.520 | 9.035 | 0.005 |
| Within group(combined) | 52.121 | 247 | 1.424 | | |
| | 52.135 | 249 | | | |
| *Mentorship* | | | | | |
| Variables | Sum of squares | Degree of freedom | Mean square | Frequency | Significance |
| Between group(combined) | 0.231 | 2 | 1.254 | 9.043 | 0.004 |

(continued)

**Table 1** (continued)

| Training and development | | | | | |
|---|---|---|---|---|---|
| Within group(combined) | 65.176 | 247 | 0.421 | | |
| | 65.407 | 249 | | | |
| *Workers welfare measures* | | | | | |
| Variables | Sum of squares | Degree of freedom | Mean square | Frequency | Significance |
| Between group(combined) | 0.201 | 2 | 1.287 | 5.053 | 0.006 |
| Within group(combined) | 84.245 | 247 | 1.428 | | |
| | 84.446 | 249 | | | |
| *Opportunity for growth* | | | | | |
| Variables | Sum of squares | Degree of freedom | Mean square | Frequency | Significance |
| Between group(combined) | 0.214 | 2 | 0.245 | 9.053 | 0.007 |
| Within group(combined) | 89.742 | 247 | 0.489 | | |
| | 89.956 | 249 | | | |

**Table 2** Relationship between employee performance and employee retention in higher education institutions

| Independent variables | Co-efficient | t-value | Significant *p*Value | Collinearity statistics | |
|---|---|---|---|---|---|
| | Beta | | | Tolerance | VIF |
| (Constant) | 2.751 | 58.826 | 0.000 | | |
| Training and development | 1.055 | 23.524 | 0.000* | 1.000 | 1.000 |
| Flexible work schedule | 0.157 | 3.361 | 0.001* | 1.000 | 1.000 |
| Promote a work-life balance | − 0.029 | − 0.621 | 0.535** | 1.000 | 1.000 |
| Work recognition | 0.108 | 2.310 | 0.021* | 1.000 | 1.000 |
| Compensation | 0.134 | 2.845 | 0.006* | 1.000 | 1.000 |
| Emphasis on team work | − 0.084 | − 1.798 | 0.073** | 1.000 | 1.000 |
| On boarding programs | 0.158 | 5.236 | 0.000* | 1.000 | 1.000 |
| Mentorship | − 0.034 | − 1.257 | 0.542** | 1.000 | 1.000 |
| Workers welfare measures | 0.275 | 1.345 | 0.005* | 1.000 | 1.000 |
| Opportunity for growth | 0.152 | 2.471 | 0.006* | 1.000 | 1.000 |
| R Square = 0.662, R Square Adjusted = 0.612 | | | | | |
| F Value = 82.620, Significance = 0.000 | | | | | |

- Government has to formulate initial strategies like offering training and development opportunities to retain the employees in the higher educational institutions.

## 8 Conclusion

In the present scenario, higher educational institutions play a crucial role in retaining the employee for effective functioning of institutions. Employee retention in higher educational institutions has to focus on professional development and employee engagement initiatives and they have to create a supportive work environment. Retention of talented employee's is essential for achieving the institutional goals. Thus, the study found out that substantial correlation were identified between training and development, flexible work schedule, work recognition, on boarding programs, workers welfare measures and opportunity for growth in the mean while promote a work life balance, Emphasis on team work and Mentoring does have any impact towards employee retention strategies. Therefore, it is concluded that employee retention strategies has to be framed by higher educational institutions in order to reduce the factors which impact employee retention.

## References


1. Erasmus et al: Employee retention in a higher education institution: an organizational development perspective. : Research Gate Publication, Vol **27**, Issue no 2, Pg no 33–63. (2015). ISSN:0256-8853
2. Silbert Jose, S.V.: The Impact of Human Resource Practices on Employee Retention in the Educational Sector. A J. Compos. Theory, **VII**(VII), 87–93. (2019). ISSN: 0731-6755
3. Jam, M., Jamal, W.N.: Impact of green human resources management practices on organizational sustainability and employee retention: an empirical study related to educational institutions. Int. Res. Assoc. Sustain. Dev. **2**(1), 38–48. (2020)
4. Vu, H.M., Ekonomika, C.: Strategic human resource management, organizational culture and academic staff retention: literature review. Manag. Inovace. **12**(2), 5–13. (2020). ISSN: 1804-1299
5. Sala, O. et al: F1000 Research Article Determinants of retention strategies and sustainable performance of academic staff of government-owned universities in Nigeria. **9**(902), 1–19. (2021)
6. Goyal, C., Jambur, A.: employee retention for sustainable business and development: strategies and best practices. Empyreal Publishing House, vol **1**, 76–83, ISBN:978-93-93810-98-4
7. Gobal, B.B.: Factors that determine academic staff retention and commitment in private tertiary institutions in Botswana: empirical review. Adv. Res. J. Manag. Bus. Stud. **1**(9), 278–299. (2012). ISSN: 2315-5086
8. Parhyar et al: The impact of employees retention strategies and sustainable organizational performance: a case study of hyderabad business schools. Int. Res. J. Manag. Soc. Sci. **2**(3), 122–133. (2021)
9. Verma, S., Kaur, G.: Faculty retention dynamics: investigating the role of HR climate, trust, and organizational commitment in higher education context. SAGE Open, 1–14, (2024) https://doi.org/10.1177/21582440241233372
10. Molotsi et al: Enhancing Academic Staff retention in an open distance E-Learning Higher Education Institution in South Africa. Sabinet Afr. J.S, Indep. J. Teach. Learn. **18**(1), 39–58 (2023)

# To Study the Impact of Government Subsidies, Financial Literacy, and Policy Awareness in Consumer Adoption of Green Finance with the Mediating Effect of Perceived Financial Benefits—SEM Analysis

**Neerupa Chauhan, Reenu Mohan, Panduranganagouda Honnali, S. Gokilavani, and Mallikarjun Biradar**

**Abstract** This study analyzes the influence of government subsidies (GS), government policy awareness (GPA), financial literacy (FL) and perceived financial benefits (PFB) on consumer adoption of green finance products (CA). As the demand for sustainable financial solutions is increasing globally, It is import to understand various factors that are driving consumer behavior toward green finance adoption. To analyze these relationships, a quantitative approach has been adopted, where with the help of structural equation modeling (SEM) the data collected for 392 respondants has been analyzed for diverse customers. The results of the study shows that perceived financial benefits has an significant influence on promoting consumer adoption of green finance products. Whereas Government subsidies, policy awareness and Financial Literacy indirectly impact adoption, which are mediated by perceived financial benefits. Although government subsidies alone were not found to significantly influence adoption behavior directly, their effect is enhanced when combined with financial literacy and policy awareness. These findings suggest that financial literacy programs and increased policy awareness can increase the effectiveness of government incentives in fostering green finance adoption. The study concludes with recommendations for policymakers to develop multi-faceted strategies that include financial education, targeted subsidies, and clear policy communication to promote sustainable financial behaviors.

**Keywords** Green finance · Government subsidies · Financial literacy · Consumer adoption · Sustainable finance

---

N. Chauhan (✉) · R. Mohan · P. Honnali · S. Gokilavani
Kristu Jayanti College (Autonomous), Bengaluru, Karnataka, India
e-mail: neerupa.c@kristujayanti.com

M. Biradar
Koppal University, Koppal, Karnataka, India

 
M. Al Mubarak (ed.), *Sustainable Digital Technology and Ethics in an Ever-Changing Environment*, Studies in Systems, Decision and Control 237,
https://doi.org/10.1007/978-3-031-86708-8_10

# 1 Introduction

The global move towards environmental sustainability has further highlighted green finance, referring to the financing of financial products and services intended to support sustainability projects (mainly for reduction in carbon emission), eco-friendly development or preservation & conservation. Governments around the world have introduced policies and subsidies to promote these products among consumers as they recognize that there is a necessity for addressing climate change factors, and attaining sustainable economic development Taghizadeh-Hesary and Yoshino, [19]. Nonetheless these initiatives, consumer demand for green finance products has remained low which served to ask what determines this choice of behavior Baker et al. [4]. Subsidies also have a much needed role to play in introducing green finance products because they help to reduce the financial barriers for consumers Mazzucato and Semieniuk [15]. The goal is for these incentives to lower prices and ultimately help make sustainable investing more mainstream. These subsidies are said to be the most effective way of encouraging consumption by consumers Olson, [16] but it is unknown what impact this has on consumer adoption because some literature argues that other factors including governmental policy awareness and financial literacy can mediate or moderate such relationship Carrington et al. [7]. The Perceived Financial Benefit (PFB) is one of the key mediating variables in this context. Consumers are more willing to accept the green finance product, when realize that long-term financial benefits of course includes potential savings or gains Chen and Chang, [8]. Financial literacy is also a key factor, as people who are more financially literate in general tend to realize that green investments have value Atkinson and Messy, [3]. In addition, this knowledge of government policy related to green finance expand the consumers' understanding about benefits and incentives that provided which lead them towards adoption Choi and Kim, [9].

The rapidly growing demand of green financial products facilitates to conduct this study, that attempts to consider how government subsidies, financial literacy, and government policy awareness leads to consumer adoption of green finance products, with Perceived Financial Benefit (PFB) acting as a mediating factor. This paper is expected to provide a detailed analysis of these links on what makes potential customers purchase green finance products.

# 2 Review of Literature with Theoretical Background

Green finance is a pragmatic approach of mitigating environmental problems by providing financial products and services that enable sustainable development such as the green bonds, renewable energy financing, eco-friendly investment Baker et al. [4]. Mazzucato and Semieniuk [15] states that government subsidies can effectively lower the entry barrier for green investments by making these more affordable. Among the types of products so subsidized are solar energy systems and electric vehicles, with an overall goal to induce consumer involvement in a green economy Berke et al. [6]. Theophore and Mushiri [21] stated that governmental finance support acts as a stimulant to facilitate the adoption of sustainable practices in the short term, but its long-term implications for sustainability have been questioned by others Carrington et al. [7]. Consumer trust on government initiatives and perceived subsidy effectiveness plays crucial role shaping adoption behaviours Chen and Chang, [8]. The perceived financial benefit (PFB) With respect to the consumer behaviour in green finance, is a core determinant for Financial returns or savings derived from adopting green financial products Chen and Chang, [8]. The Theory of Planned Behavior (TPB) Ajzen, [1] represents a theory that is very useful in understanding how perceived financial benefits can affect individuals intentions to engage with green finance. According to the TPB, perceived financial benefits, as a form of behavioral belief, can positively influence attitudes toward green products, thereby increasing the likelihood of adoption. Several empirical studies have established the relationship between perceived financial benefits and consumer adoption. For instance, Kaufmann et al. [12] found that individuals who perceive higher financial returns from renewable energy investments are more likely to adopt such products. Similarly, studies on eco-friendly investments and green bonds highlight the importance of financial benefits as a critical factor for encouraging consumer participation Tang and Zhang, [20].

As stated by Lusardi and Mitchell [14] "financial literacy as the ability of individuals to use knowledge and skills in making informed investment, savings decisions". Financially literate consumers are able to appreciate the long term financial advantages of green investments and government subsidies, which can positively enhance their perceived ease Atkinson and Messy. [3]. Additionally, higher financial literacy also results in common eco-good behaviors Xie et al. [22]. According to the Behavioral Finance Theory, those who are more financially literate might be able to cope better with financial uncertainties and feel lower risk in considering a new type of green products Lo, [13]. Therefore, financial literacy helps to modify the relationship between perceived monetary benefits and consumers´ adoption. Compared to financial literacy, another determinant of consumer adoption behavior is government policy awareness—a key factor by which consumers can be informed about policies and regulations associated with green finance Choi and Kim, [9]. Consumer awareness to government subsidy, tax rebates & environmental regulation help the consumer in making a decision for such green financial product Berke et al. [6]. The Diffusion of Innovations Theory (DOI) (Rogers, [18]) states the framework that helps to understand the role of policy awareness inadoption of green finance. As per this theory, awareness of an innovation a new product is the first step in the decision-making process, followed by interest, evaluation, trial, and adoption. Lo [13] statest "individuals who are more aware of green finance policies are more likely to adopt green financial products". Choi and Kim [9] further added that policy awareness has a positive influence on consumer behavior by reducing informational barriers, making it easier for consumers to understand and evaluate the benefits of green finance. This underscores the importance of governmental efforts to disseminate information and raise awareness about available green financial products and incentives. The premise of this study is based on theoretical construction adapted

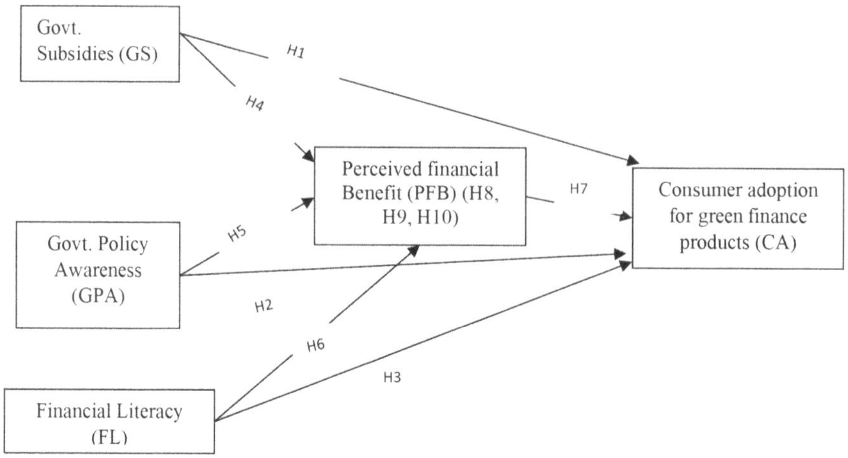


**Fig. 1** Theoretical model

from Theory of Planned Behavior (TPB) Ajzen, [1]; Rogers, [18]). Therefore, TPB is the model of personal behavior in order to explain how perceived financial benefits influence consumer adoption intention Ajzen, [1]. On the contrary, DOI can clarify how government policy consciousness and subsidies as external drivers may accelerate green finance products take-off. Moreover the study applies Behavioral Finance Theory using financial literacy as a moderating factor underlines that consumer may have better assessment of their own liable risk and to engage with green financial products. Based on the premises thoratical model has been framed (Fig. 1).

# 3 Objective of the Study

The objective of this study is to investigate the impact of government subsidies (GS), financial literacy (FL), perceived financial benefits (PFB), and government policy awareness (GPA) on the adoption of green finance products among consumers (CA).

## *3.1 Hypothesis*

Based on review of literature and theoretical background, following null hypothesis has been framed.

- $H_{01}$: Government subsidies (GS) doesnot have an influence on consumer adoption of green finance products (CA).

- $H_{02}$: Government policy awareness (GPA) doesnot influences consumer adoption of green finance products (CA).
- $H_{03}$: Financial literacy (FL) doesnot influences consumer adoption of green finance products (CA).
- $H_{04}$: Government subsidies (GS) doesnot influence perceived financial benefit (PFB).
- $H_{05}$: Government policy awareness (GPA) doesnot influences perceived financial benefit (PFB).
- $H_{06}$: Financial literacy (FL) doesnot influences perceived financial benefit (PFB).
- $H_{07}$: Perceived financial benefit (PFB) doesnot has a effect on consumer adoption of green finance products (CA).
- $H_{08}$: Perceived financial benefit (PFB) doesnot mediates the relationship between government subsidies (GS) and consumer adoption of green finance products (CA).
- $H_{09}$: Perceived financial benefit (PFB) doesnot mediates the relationship between government policy awareness (GPA) and consumer adoption of green finance products (CA).
- $H_{010}$: Perceived financial benefit (PFB) doesnot mediates the relationship between financial literacy (FL) and consumer adoption of green finance products (CA).

## 4 Research Methodology

To conduct this study quantitative approach has been used, where data has been collected through Likert scale structured questionnaire for 392 respondants to measure consumer adoption of green finance products. To analyze the collected data, Exploratory Factor Analysis using the Maximum likelihood and Varimax rotation through SPSS has been done. For validation of the factor structure Confirmatory factor Analysis (CFA) (Fig. 2) has been conducted through AMOS. Further with Structure equation modelling (SEM) using AMOS (Fig. 3) research hypotheses has been assessed.

The demographic analysis (Table 1) shows that in collected data, representation of males is 74% and females representing 26%. In education, 55.4% holds bachelor's degree and 44.6% have a master's degree or higher. In terms of Income distribution, it has been seen that 19.6%, earns less than ₹3,00,000, 36.2% earns between ₹5,00,001 and ₹10,00,000, 28.3% respondants income range from ₹3,00,000 to ₹5,00,000 where as 15.8% respondants earn more than ₹10,00,000. For product preferences, it has been seen 40.3% holds sustainable credit cards, 21.09% green bonds, 16.6% savings accounts with high-interest rates, 11.2% with eco-friendly insurance policies and 9.9% are having renewable energy investments.

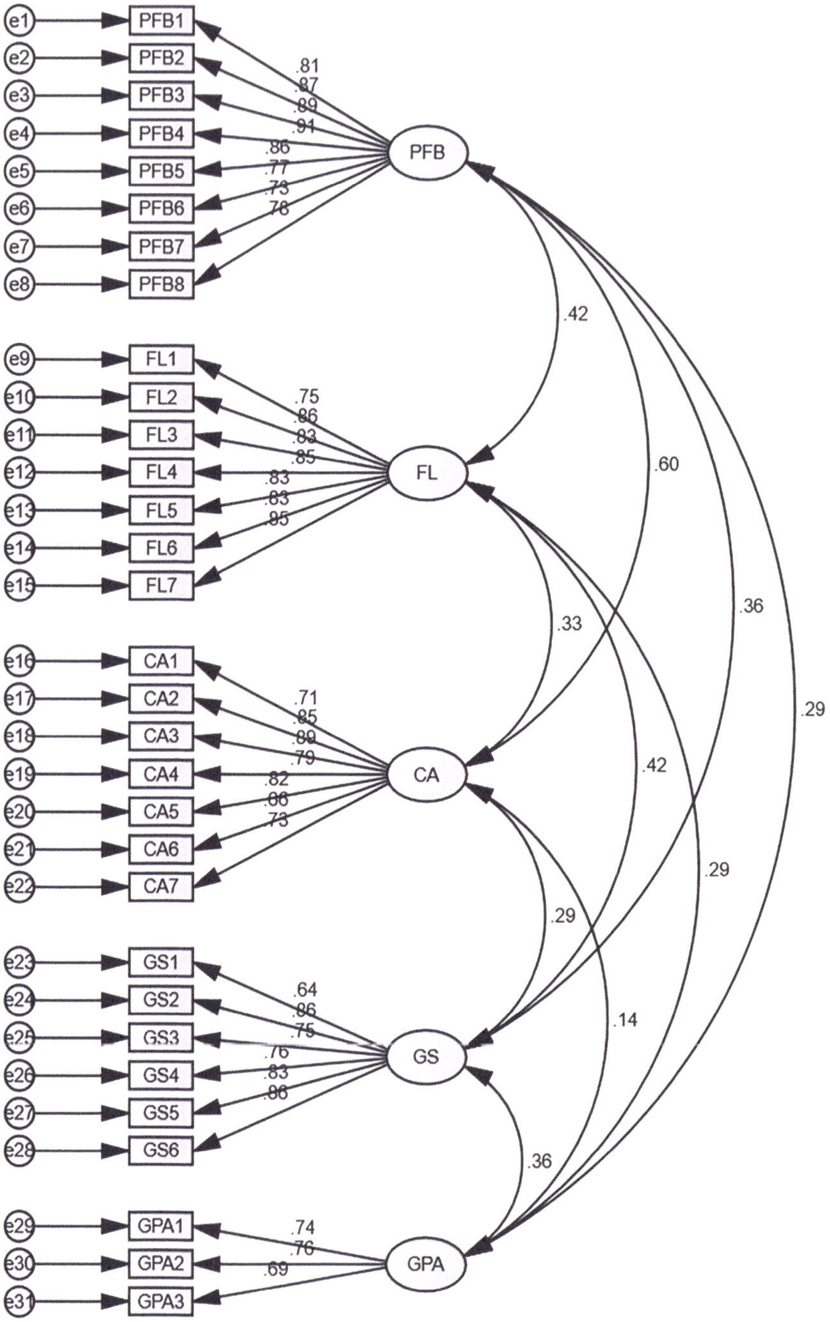


**Fig. 2** CFA model

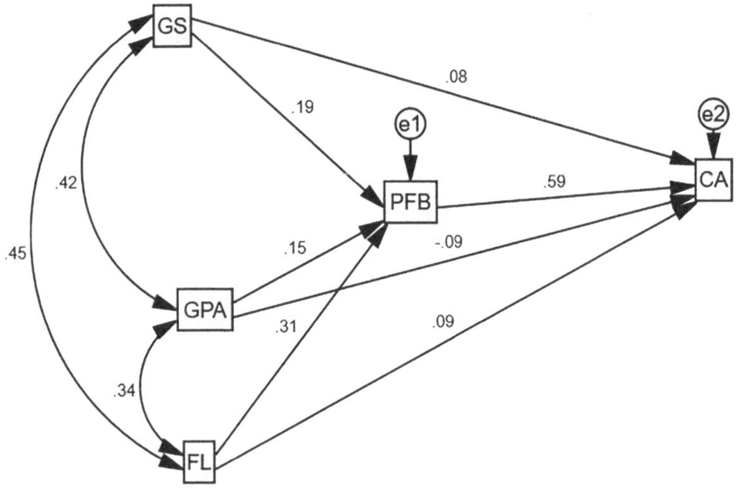


**Fig. 3** SEM model-results

**Table 1** Demographic analysis

| | | Count | Column N % |
|---|---|---|---|
| Gender | Male | 290 | 74.0% |
| | Female | 102 | 26.0% |
| Education | Bachelor's degree | 217 | 55.4% |
| | Master's degree or higher | 175 | 44.6% |
| | Less than ₹3,00,000 | 77 | 19.6% |
| Income_Level | ₹3,00,000–₹5,00,000 | 111 | 28.3% |
| | ₹5,00,001–₹10,00,000 | 142 | 36.2% |
| | More than ₹10,00,000 | 62 | 15.8% |
| | Savings account with high-interest rates | 65 | 16.6% |
| Product_Type | Green bonds | 86 | 21.9% |
| | Sustainable credit cards | 158 | 40.3% |
| | Eco-friendly insurance policies | 44 | 11.2% |
| | Renewable energy investments | 39 | 9.9% |

## 4.1 Exploratory Factor Analysis

Exploratory Factor analysis using the maximum likelihood method with Varimax rotation is used for analyzing the factor structure and correlation between items included in the scale. The results of rotated factor matrix are provided in the following tables.

**Table 2** KMO and bartlett's test

| | | |
|---|---|---|
| Kaiser–Meyer–Olkin measure of sampling adequacy | | 0.919 |
| Bartlet's test of sphericity | Approx. chi-square | 9603.769 |
| | df | 465 |
| | Sig | 0.000 |

The KMO value is above 0.50 (Table 2) which indicates that criteria of sampling adequacy has been met. The Bartlett test of sphericity is statistically significant ($P < 0.05$).

Table 3 shows the results of exploratory factor analysis (EFA) using Rotated Factor Matrix for identification of underlying constructs in the data (Hair et al, [11]). Factor 1 corresponds to Perceived Financial Benefit (PFB) (0.667–0.837), Factor 2 for Financial Literacy (FL) (0.672–0.836). Factor 3 for Consumer Adoption (CA) (0.647–0.873) Factor 4 Government Subsidies (GS) (0.68–0.834) and Factor 5 Government Policy Awareness (GPA) (0.668–0.719).

## 4.2 *Confirmatory Factor Analysis*

To perform Confirmatory Factor Analysis [2] AMOS version 18 has been used. The model is assessedto check reliability, convergent validity, and discriminant validity. The results for the same has been shown in Fig. 2 and calculations of the model in Table 4.

Table 4 represents the results of reliability and convergent validity. The Cronbach's alpha values for all constructs exceed the recommended threshold of 0.70 Nunnally, [16], with the Perceived Financial Benefit (PFB) construct having the highest alpha value (0.828) and Government Policy Awareness (GPA) the lowest (0.732). Composite reliability (CR) values for all constructs are above 0.90, indicating a high level of reliability Hair et al, [11]. The Average Variance Extracted (AVE) for each construct is greater than 0.50, satisfying the requirement for convergent validity Fornell and Larcker, [10].

In Table 5, shows the discriminant validity, Fornell and Larcker, [10] "square root of the AVE for each construct is higher than its correlations with other constructs". The results confirm that each construct is distinct from others in the model.

## 4.3 *Hypotheses Testing (Structural Model)*

Figure 3 shows the The structural path diagram exhibiting the relationships between Government Subsidies (GS), Government Policy Awareness (GPA), Financial Literacy (FL), Perceived Financial Benefit (PFB), and Consumer Adoption (CA).

**Table 3** Rotated factor matrix

| | Factor | | | | |
|---|---|---|---|---|---|
| | 1 | 2 | 3 | 4 | 5 |
| GS1 | | | | 0.608 | |
| GS2 | | | | 0.824 | |
| GS3 | | | | 0.716 | |
| GS4 | | | | 0.695 | |
| GS5 | | | | 0.798 | |
| GS6 | | | | 0.834 | |
| GPA1 | | | | | 0.719 |
| GPA2 | | | | | 0.710 |
| GPA3 | | | | | 0.668 |
| PFB1 | 0.756 | | | | |
| PFB2 | 0.771 | | | | |
| PFB3 | 0.811 | | | | |
| PFB4 | 0.826 | | | | |
| PFB5 | 0.837 | | | | |
| PFB6 | 0.714 | | | | |
| PFB7 | 0.667 | | | | |
| PFB8 | 0.740 | | | | |
| CA1 | | | 0.659 | | |
| CA2 | | | 0.812 | | |
| CA3 | | | 0.873 | | |
| CA4 | | | 0.745 | | |
| CA5 | | | 0.748 | | |
| CA6 | | | 0.533 | | |
| CA7 | | | 0.647 | | |
| FL1 | | 0.672 | | | |
| FL2 | | 0.823 | | | |
| FL3 | | 0.774 | | | |
| FL4 | | 0.806 | | | |
| FL5 | | 0.798 | | | |
| FL6 | | 0.799 | | | |
| FL7 | | 0.836 | | | |

Extraction Method: Maximum Likelihood
Rotation Method: Varimax with Kaiser Normalization[a]
a. Rotation converged in 6 iterations

**Table 4** Reliability and convergent validity

| Variables | Items | Standardized factor loadings | Cronbach alpha (CA) | Composite reliability (CR) | Average variance extracted (AVE) | Maximum shared variance (MSV) |
|---|---|---|---|---|---|---|
| Perceived financial benefits | PFB1 | 0.81 | 0.828 | 0.946 | 0.689 | 0.358 |
| | PFB2 | 0.875 | | | | |
| | PFB3 | 0.886 | | | | |
| | PFB4 | 0.906 | | | | |
| | PFB5 | 0.863 | | | | |
| | PFB6 | 0.773 | | | | |
| | PFB7 | 0.727 | | | | |
| | PFB8 | 0.784 | | | | |
| Financial literacy | FL1 | 0.749 | 0.828 | 0.939 | 0.687 | 0.177 |
| | FL2 | 0.859 | | | | |
| | FL3 | 0.827 | | | | |
| | FL4 | 0.853 | | | | |
| | FL5 | 0.825 | | | | |
| | FL6 | 0.833 | | | | |
| | FL7 | 0.851 | | | | |
| Consumer adoption of green finance products | CA1 | 0.714 | 0.777 | 0.916 | 0.61 | 0.358 |
| | CA2 | 0.846 | | | | |
| | CA3 | 0.888 | | | | |
| | CA4 | 0.786 | | | | |
| | CA5 | 0.822 | | | | |
| | CA6 | 0.658 | | | | |
| | CA7 | 0.728 | | | | |
| Govt. subsidies | GS1 | 0.642 | 0.784 | 0.907 | 0.622 | 0.175 |
| | GS2 | 0.859 | | | | |
| | GS3 | 0.754 | | | | |
| | GS4 | 0.762 | | | | |
| | GS5 | 0.83 | | | | |
| | GS6 | 0.861 | | | | |
| Govt. policy awareness | GPA1 | 0.742 | 0.732 | 0.776 | 0.536 | 0.131 |
| | GPA2 | 0.764 | | | | |
| | GPA3 | 0.689 | | | | |

**Model Fitness**: X2 = 1399.010, df = 424, X2/df = 3.3, RMSEA = 0.077, TLI = 0.887, PCFI = 0.817, CFI = 0.897.

**Table 5** Discriminant validity

| | PFB | FL | CA | GS | GPA |
|---|---|---|---|---|---|
| PFB | **0.83** | | | | 0.291*** |
| FL | 0.421*** | **0.829** | | | 0.293*** |
| CA | 0.598*** | 0.333*** | **0.781** | | 0.141* |
| GS | 0.363*** | 0.418*** | 0.287*** | **0.788** | 0.362*** |
| **GPA** | | | | | **0.732** |

* < 0.050, ** < 0.010, *** < 0.001

There is a direct effect of Government Subsidies on Consumer Adoption (β = 0.08). Where as, Perceived Financial Benefit significantly influences Consumer Adoption (β = 0.59). Government Policy Awareness (GPA) also shows a positive effect on Perceived Financial Benefit (β = 0.15) and Consumer Adoption (β = 0.31), suggesting that awareness of policies has a considerate impact on adoption behavior. It has been seen that, Financial Literacy (FL) impacts both Perceived Financial Benefit (PFB) (β = 0.34) and Consumer Adoption (CA) (β = 0.09), emphasizing the role of financial understanding in shaping adoption decisions.

The findings (Table 6) shows that Government Subsidies (GS) have a positive impact on Perceived Financial Benefit (PFB) (β = 0.204, $p < 0.001$), supporting the hypothesis that subsidies enhance consumers' perception of financial benefits associated with green finance products. Likewise, Government Policy Awareness (GPA) (β = 0.131, $p = 0.002$) and Financial Literacy (FL) (β = 0.260, $p < 0.001$) are positively associated with PFB, indicating that awareness of relevant policies and financial understanding contribute to higher perceived benefits. It has been seen that Perceived Financial Benefit (PFB) is the predictor for Consumer Adoption (CA) (β = 0.566, $p < 0.001$), confirming that perceived financial advantages are important in adoption of green finance products. However, the direct impact of Government Subsidies (GS) on CA is not statistically significant ($p = 0.085$), and Financial Literacy (FL) also shows an insignificant direct effect ($p = 0.058$), suggesting that these variables influence adoption indirectly through PFB. Interestingly, Government Policy Awareness (GPA) has a negative direct effect on CA (β = -0.078, $p = 0.033$).

## 4.4 Mediation Testing

Mediation analysis (Table 7) has been conducted based on the analysis of indirect effects [5]. To perform the analysis bootstrap procedures (200 samples) and bias-corrected bootstrap confidence interval (90%) has been used for direct and indirect effects.

Results of the analysis (Table 7) reveals that Perceived Financial Benefit (PFB) plays a significant mediating role between Government Subsidies (GS), Government

**Table 6** Regression weights

| H. No | Paths | Estimate | S.E | C.R | P | Remarks |
|---|---|---|---|---|---|---|
| H1 | PFB < GS | 0.204 | 0.057 | 3.608 | *** | Supported |
| H2 | PFB < GPA | 0.131 | 0.042 | 3.085 | 0.002** | Supported |
| H3 | PFB < FL | 0.260 | 0.042 | 6.249 | *** | Supported |
| H4 | CA < PFB | 0.566 | 0.043 | 13.187 | *** | Supported |
| H5 | CA < GS | 0.084 | 0.049 | 1.724 | 0.085 | Rejected |
| H6 | CA < GPA | −0.078 | 0.036 | −2.126 | 0.033* | Supported |
| H7 | CA < FL | 0.070 | 0.037 | 1.899 | 0.058 | Rejected |

*** < 0.001, ** < 0.01, * < 0.05

**Table 7** Mediation analysis

| H. No | Path | Total Effects | Direct effects | Indirect effects | Remarks |
|---|---|---|---|---|---|
| H8 | GS > PFB > CA | 0.199 | 0.084 | 0.115 | Hypothesis supported as indirect effects are statistically significant |
| H8 | GPA > PFB > CA | −0.003 | -0.078 | 0.074 | Hypothesis supported since indirect effects are statistically significant |
| H10 | FL > PFB > CA | 0.217 | 0.070 | 0.147 | Hypothesis supported since indirect effects are statistically significant |

* < 0.05, ** < 0.01, *** < 0.001

Policy Awareness (GPA), Financial Literacy (FL), and Consumer Adoption (CA), as all indirect effects are significant ($p < 0.05$).

## 5 Conclusion

This study has examined how different factors such as government subsidy, financial literacy and perceived financial benefits can encourage potential consumers to adopt green finance products. Consumer Adoption is mostly affected by perceived financial benefits (PFB), and thus shows that the presence of economic incentives can drive sustainable finance behaviour in a significant way Chen and Chang, [8]. Government subsidies and policy awareness were ancillary factors, which served to modulate consumer behaviour in each market; although their effect on adoption as a direct factor was more limited. Financial literacy was also identified as a moderator, suggesting that consumers with higher levels of financial education understand better the risks and benefits associated to green products Lusardi and Mitchell, [14] thus leading them to increase its adoption.

While the government is creating an ecosystem for green finance investments, consumer awareness and trust in policies upholding were found preponderantly new approaches deployed by institutions Choi and Kim, [9]. Policymakers and financial institutions need these insights to drive the transition toward a green economy. This study has widened the understanding of what drives green finance products adoption, but there are still several other areas to explore in order extend this research. One way to start would be more qualitative work on consumer behavior in different cultural and economic contexts, which could shed light into the global diffusion of green finance. For instance, this may help to identify some differences in consumer attitudes towards green finance between developing and developed countries that could be used as a basis for designing policy measures Baker et al, [4]. In addition, future research may investigate the capacity of digital finance platforms and fintech innovations for supporting green financial enablement since consumers behavior as well as their payment methods have experienced transformation by digital technologies Taghizadeh-Hesary and Yoshino, [19]. In addition, use of behavioral insights in design of policies could also be considered to ensure alignment between financial incentives and consumer motivations. This highlights the need for a comprehensive policy that mixes financial literacy, public policy and incentives to increase sustainable economic activity.

## References

1. Ajzen, I.: The theory of planned behavior. Organ. Behav. Hum. Decis. Process. **50**(2), 179–211 (1991). https://doi.org/10.1016/0749-5978(91)90020-T
2. Arbuckle, J.: AMOS 18 user's guide. Crawfordville: AMOS Development Corporation, (2009)

3. Atkinson, A., Messy, F.-A.: Measuring financial literacy: Results of the OECD / International Network on Financial Education (INFE) pilot study. OECD Working Papers on Finance, Insurance and Private Pensions, No. 15. (2012). https://doi.org/10.1787/5k9csfs90fr4-en
4. Baker, M., Bergstresser, D., Serafeim, G., Wurgler, J.: Financing the response to climate change: The pricing and ownership of U.S. green bonds. National Bureau of Economic Research Working Paper No. 25194. (2018). https://doi.org/10.3386/w25194
5. Baron, R. M., Kenny, D. A.: The moderator–mediator variable distinction in social psychological research: conceptual, strategic, and statistical considerations. J. Pers. Soc. Psychol. **51**(6), 1173–1182 (1986). https://doi.org/10.1037/0022-3514.51.6.1173
6. Berke, J., Stahel, A., Fawcett, W., Dovey, K.: Environmental policy awareness and consumer behavior: evidence from the solar energy industry. Energy Policy **37**(12), 4857–4861 (2015). https://doi.org/10.1016/j.enpol.2009.06.024
7. Carrington, M.J., Neville, B.A., Whitwell, G.J.: Lost in translation: exploring the ethical consumer intention–behavior gap. J. Bus. Res. **67**(1), 2759–2767 (2014). https://doi.org/10.1016/j.jbusres.2012.09.022
8. Chen, C.C., Chang, Y.H.: Green perceived value: the antecedents and consequences of green purchasing behavior. J. Bus. Res. **66**(8), 101–107 (2013). https://doi.org/10.1016/j.jbusres.2012.12.014
9. Choi, J., Kim, D.H.: The impact of green policy awareness on green product adoption. J. Consum. Policy **28**(4), 53–64 (2005). https://doi.org/10.1007/s10603-005-3987-6
10. Fornell, C., Larcker, D.F.: Evaluating structural equation models with unobservable variables and measurement error. J. Mark. Res. **18**(1), 39–50 (1981). https://doi.org/10.1177/002224378101800104
11. Hair, J.F., Black, W.C., Babin, B.J., Anderson, R.E.: Multivariate data analysis (7th ed.). Pearson, (2010)
12. Kaufmann, S., Lehner, O.M., Sieber, C.: The role of financial returns in renewable energy investment decisions. Renew. Sustain. Energy Rev. **17**, 330–344 (2013). https://doi.org/10.1016/j.rser.2012.09.017
13. Lo, A.W.: Adaptive markets: Financial evolution at the speed of thought. Princeton University Press, (2016)
14. Lusardi, A., Mitchell, O.S.: The economic importance of financial literacy: theory and evidence. J. Econ. Lit. **52**(1), 5–44 (2014). https://doi.org/10.1257/jel.52.1.5
15. Mazzucato, M., Semieniuk, G.: Financing renewable energy: who is financing what and why it matters. Technol. Forecast. Soc. Chang. **127**, 8–22 (2018). https://doi.org/10.1016/j.techfore.2017.05.021
16. Nunnally, J.C.: Psychometric theory (2nd ed.). McGraw-Hill, (1978)
17. Olson, M.: The logic of collective action: Public goods and the theory of groups. Harvard University Press, (2014)
18. Rogers, E.M.: Diffusion of innovations (5th ed.). Free Press, (2003)
19. Taghizadeh-Hesary, F., Yoshino, N.: Sustainable solutions for green financing. Asia-Pacific Finan. Markets. **27**(2), 151–169 (2020). https://doi.org/10.1007/s10690-020-09300-z
20. Tang, Y., Zhang, X.: Consumer attitudes toward green finance: evidence from the green bond market. J. Clean. Prod. **250**, 119–202 (2020). https://doi.org/10.1016/j.jclepro.2019.119202
21. Theophore, R., Mushiri, S.: Government subsidy and the diffusion of renewable energy technologies. Renew. Energy J. **39**(3), 254–263 (2014). https://doi.org/10.1016/j.renene.2014.04.004
22. Xie, B., Jia, Y., Wang, H.: Financial literacy and green behavior. J. Clean. Prod. **250**, 119490 (2020). https://doi.org/10.1016/j.jclepro.2019.119490

# Nutrient-Rich Elixir the Health and Wellness Benefits of Tropical Fruit Cocos Nucifera

**Tinto Tom, E. Eswara Reddy, Jerrin Jose, S. J. G. Preethi, Uma Maheswary, and Akhila Variyar**

**Abstract** This review examines the diverse advantages of food and beverages derived from coconut, with a specific emphasis on virgin coconut oil (VCO), coconut water, and beverages derived from coconut. Coconut water, widely recognized as a "Nature's Health Elixir," is acclaimed for its isotonic mineral composition, enabling it to function as an all-natural hydrator and confer many health benefits. Fermented coconut water, Neera, and other coconut-based beverages, which possess inherent nutritional value and potential medicinal properties, are linked to many health benefits. These organic and wholesome food items, abundant in vital nutrients such as vitamins, minerals, and antioxidants, promote holistic well-being and bolster multiple facets of health, such as cardiovascular health, nutrient utilization, and metabolic rate. By including products derived from coconut in one's diet, one can augment the gustatory experience, inspire innovative cooking methods, and avail oneself of a wide range of nutritional advantages.

**Keywords** Coconut · Virgin coconut oil · Coconut water · Health · Coconut beverages

T. Tom (✉)
CMR University, Bengaluru, India
e-mail: tinto.tom@cmr.edu.in

E. Eswara Reddy
School of Economics and Commerce, CMR University, Bengaluru, India

J. Jose · U. Maheswary · A. Variyar
Department of English, Kristu Jayanti College, Bengaluru, India

S. J. G. Preethi
Department of Economics, Kristu Jayanti College, Bengaluru, India

 
M. Al Mubarak (ed.), *Sustainable Digital Technology and Ethics in an Ever-Changing Environment*, Studies in Systems, Decision and Control 237,
https://doi.org/10.1007/978-3-031-86708-8_11

## 1 Introduction

Food is essential to our lives, providing the necessary nutrients to sustain our bodies and minds. Maintaining good health and preventing chronic diseases hinge on the importance of a well-balanced diet. Consuming various foods from different groups is essential to get all the necessary nutrients. Vegetables and fruits provide essential vitamins, minerals, and fiber to sustain good health. Whole grains like brown rice, quinoa, and oats provide a steady energy source and are rich in fiber, which helps keep us full for longer. In the body, lean proteins such as chicken, fish, and legumes play a crucial role in tissue building and repair. Additionally, reducing the consumption of processed foods, often high in unhealthy fats, sugar, and salt, is essential. Instead, prioritize nutrient-rich whole foods that offer the body the necessary fuel for optimal functioning.

Natural and healthy foods nourish our bodies and promote overall well-being. These foods, such as fruits, vegetables, whole grains, lean proteins, and nuts, are minimally processed and rich in essential nutrients like vitamins, minerals, fiber, and antioxidants. By incorporating natural and healthy foods into our diets, we can support our immune system, improve digestion, maintain a healthy weight, and reduce the risk of chronic diseases. Choosing whole, unprocessed foods over highly processed options helps fuel our bodies with the nutrients they need to thrive. Embracing a diet rich in natural and healthy foods is a cornerstone of a balanced and sustainable approach to nutrition and wellness.

Coconut, a versatile fruit known for its numerous culinary and health benefits, offers many food items that can enhance taste and well-being. Each product brings unique advantages, from coconut water to coconut oil, milk, and shredded coconut. Coconut water, a natural electrolyte-rich beverage, replenishes fluids and minerals post-exercise or during dehydration. With its high smoke point and healthy fats, coconut oil is ideal for cooking and baking. It may also promote heart health and aid in weight management. Coconut milk adds a creamy texture to dishes while providing essential nutrients like vitamins C and E. Shredded coconut is a flavorful addition to desserts and baked goods, offering fiber for digestive health. Incorporating coconut-based food into one's diet can contribute to hydration, heart health, weight management, and culinary creativity. Embracing the benefits of coconut in various forms can elevate dishes' flavour profile and the nutritional value of meals.

This review paper explores various aspects of coconut products, beginning with "Coconut Water: Nature's Health Elixir," which delves into the natural benefits and widespread appeal of coconut water. The paper then shifts focus to "Emerging Trends in Virgin Coconut Oil," highlighting new developments and uses for this versatile oil. Finally, it examines the "Health Benefits of Coconut Beverages," discussing how different coconut-based drinks contribute to overall wellness.

## 2 Coconut Water: Nature's Health Elixir

Prades et al. [1] reviewed the uses and composition of coconut water. The study conveys that coconut water is considered an ancient tropical beverage. At present, the popularity of coconut water in the international market is improving. Even though it is regarded as a spot beverage, manufacturers pay more attention to coconut water as a natural, functional drink. It comprises sugar and minerals. Dwarf varieties of coconut trees are suitable for getting tasty products, especially water. Coconut water is not a typical fruit juice. It is an isotonic mineral containing well-balanced sugar and low acidity.

Naik et al. [2] researched recent advancements in the preservation and process of tender coconut water. The natural delicacy of coconut water, commonly available in the market with its outer husk intact, presents a challenge due to its short shelf life of around a week at room temperature. Meeting increasing customer demand requires preservation, processing, and marketing efforts. Key preservation challenges include addressing enzymatic activity, the Maillard reaction, microbiological growth, and product quality during processing. While traditional methods involve thermal treatment and chemical preservatives, these can compromise sensory and nutritional quality. Emerging preservation technologies such as ultrasonic, ultraviolet, high-pressure carbon dioxide, ohmic heating, and cold plasma offer promising alternatives with commercial potential.

Zulaikhah [3] conveyed tender coconut water and its health benefits. The research highlights tender coconut water as a nutritious, natural beverage from coconut trees prevalent in tropical regions, particularly abundant in Indonesia, the world's leading coconut-producing country. Tender coconut water offers various health benefits and is enriched with active compounds like methionine, L-arginine, cytokines, selenium, vitamin C, zinc, manganese, and copper. Its components, including L-arginine, play a role in reducing free radical production, thereby combating oxidative stress. Studies on animals and humans suggest that tender coconut water consumption may aid in antioxidative activity, lipid profile improvement, blood pressure regulation, cardio-protective effects, anti-inflammatory properties, diarrhea management, enhancement of hemoglobin levels, anti-diabetic effects, and antithrombotic properties.

Prado et al. [4] studied the potential health benefits of fermented coconut water. This study aimed to create an innovative fermented beverage that utilizes coconut water as the primary component, hence offering intrinsic health advantages without including dairy products. Out of the seven native lactic acid bacteria strains obtained from the fermentation of coconut water, Lactobacillus plantarum was chosen due to its capacity to hinder particular pathogens and its adaptability to different technical and gastrointestinal environments. The fermentation process was conducted for 8 h at 37 °C, employing yeast extract, soy protein hydrolysate, and sucrose. This discovery introduces coconut water as a viable substitute for dairy in producing nutritious cultured beverages, primarily targeting people who follow vegan, vegetarian, or lactose-intolerant diets.

Nasution et al. [5] researched thermal treatment to improve the aroma compound profile of mature coconut water from tall varieties. Mature coconut water is recognized for its less desirable flavor attributes than young coconut water. Thermal processing has the dual effect of extending the shelf life of coconut water and modifying its fragrance characteristics. Different thermal treatments were administered, namely 85 °C for 15 min, 90 °C for 10 min, and 95 °C for 10 min. The findings suggest that applying heat treatment impacted the levels of volatile aroma components present in fully mature coconut water. The study observed a notable rise of 2–20% in esters, predominantly fatty acid esters, and a corresponding decline of 2–21% in ketones, predominantly acetoin.

Fadlilah et al. [6] conveyed the benefits of tender coconut water, watermelon, and papaya for hypertension blood pressure. The study shows young coconut water, watermelon juice, and papaya juice significantly lower blood pressure. Specifically, papaya, watermelon, and young coconut water demonstrated statistical effectiveness in reducing systolic blood pressure. Moreover, papaya juice, young coconut water, and watermelon juice were identified as herbal interventions that were particularly effective in lowering diastolic blood pressure. The study suggests incorporating young coconut water, watermelon juice, or papaya juice into one's diet to maintain blood pressure within normal levels.

Prithviraj et al. [7] researched tender coconut water preservation with emerging non-thermal processing techniques. The nutritional and health-promoting characteristics of tender coconut water (TCW) are significant. The beverage's considerable nutritional value has led to a surge in global consumer demand. However, the challenges of processing and extending the shelf-life of TCW pose significant obstacles to fully harnessing its nutritional capabilities. Enzymatic browning and its accompanying metabolic events have a detrimental impact on the quality indices of TCW. Despite the widespread utilization of thermal processing procedures, it is imperative to explore alternate non-thermal processes due to their inherent limits and the resulting off-flavor and discoloration. The primary mechanisms underlying enzyme inactivation and antimicrobial action in non-thermal approaches are the conformational change of protein structure and cell destruction, respectively.

Rethinam and Krishnakumar [8] studied coconut water and its health benefits. The study asserts that coconut water stands out as a nutritional powerhouse, owing to its unique chemical composition comprising phytohormones, enzymes, antioxidants, vitamins, minerals, phenolic compounds, and other bioactive substances closely linked to its biological and pharmacological effects. With a range of therapeutic properties, coconut water exhibits antimicrobial, antibacterial, anti-inflammatory, antihypertensive, rejuvenating, hepatoprotective, hypolipidemic, and diuretic attributes. It demonstrates therapeutic efficacy in addressing gastrointestinal issues, diarrhea, infant malnutrition, and hypertension management. Supported by numerous medical investigations, coconut water enhances exercise performance, reduces inflammation, dissolves kidney stones, enhances kidney function, aids digestion, alleviates constipation, lowers the risk of heart disease, reduces hypertension, and improves cholesterol levels. Experimental findings in animals suggest that consuming tender

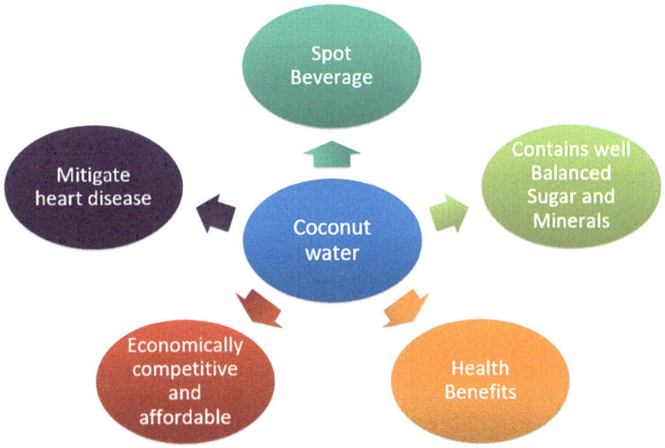


**Fig. 1** Benefits of coconut water

coconut water may mitigate the risk of heart disease, prevent Alzheimer's disease pathology, and deter osteoporosis.

Castro [9] studied about kamias jam with coconut water. Utilizing fruits for processing as a fundamental component is a viable approach to tackle the financial burden and challenge of achieving sufficient nutrition at a lower expense. This approach primarily focuses on maximizing the utilization of readily available and underutilized fruits as a source of sustenance and essential nutrients. The abundant nutritional composition of fruits offers customers a range of choices that can contribute to optimal health and overall well-being. Kamias emerged in coconut water jam as a viable gourmet substitute that exhibits potential for economic competitiveness within the market and accessibility to a wide range of individuals. This food item's unique and sometimes overlooked nutritional value positions it among the culinary elite. Figure 1 projects the health benefits of coconut water.

## 3 Emerging Trends in Virgin Coconut Oil

Marina et al. [10] researched virgin coconut oil as an emerging food. Since its debut, Virgin coconut oil (VCO) has become quite popular among the general public and the scientific community as a functional food oil. Clinical research indicates that virgin coconut oil (VCO) offers superior health benefits than refined coconut oil, mainly attributed to its enhanced antioxidant capacity. The primary distinction lies in the absence of the RBD (refining, bleaching, and deodorizing) process in VCO production, which preserves biologically active components such as phenolic compounds. Multiple studies have confirmed that VCO contains higher phenolic compounds than refined coconut oil, leading to more excellent antioxidant activity. Efforts have been directed towards developing methods for detecting adulteration of VCO. Advancements in understanding have facilitated the identification of new techniques to more accurately differentiate VCO from other vegetable oils, particularly refined coconut oil. Further research is necessary to establish practical purity standards for VCO.

Carandang [11] conveyed the health benefits of virgin coconut oil. It is processed differently from regular coconut oil. The latter is made from copra, or dried coconut meat, and goes through refinement to make edible oil. RBD (Refined, Bleached, and Deodorized) coconut oil, the refined oil produced, is primarily used as cooking oil. Since virgin coconut oil is already edible, refining the oil is not a step in production. The chemical qualities of RBD oil and VCO are essentially the same. Hence, the word "Virgin" only relates to the production process. They both contain the exact amounts of medium-chain fatty acids (MCFA), so their health effects would be the same.

Srivastava et al. [12] studied Virgin coconut oil (VCO), the fresh, mature kernel of the coconut that has not been heated or chemically refined. It can be obtained mechanically or naturally. The iodine value, saponification value, refractive index, fatty acid profile, specific gravity, and moisture content are marginally different from copra coconut oil (CCO). VCO offers various health advantages, including boosting the activity of antioxidant enzymes and avoiding the oxidation of low-density lipoprotein lipids. It was discovered that VCO blends could be stored for up to 12 months at various temperatures using both flexible and rigid packaging techniques. Virgin coconut meal (VCM), the byproduct of VCO extraction, is used to manufacture a variety of traditional Indian sweets, including ladoo and burfi, as well as baked items (biscuits and cake).

Yeap [13] says virgin coconut oil has antistress and antioxidant properties in living things. Virgin coconut oil (VCO) has been analyzed scientifically for some of its health-related advantages. Although it was discovered that medium-chain fatty acids may act as a functional dietary antidepressant, this effect was not tested in VCO, which is high in polyphenols and medium-chain fatty acids.

Neela Satheesh [14] reviewed the potential application of virgin oil. It claims that fats and oils are essential for human nutrition and health. Coconut oil serves primarily as a dietary resource and finds application in producing hair oil, soaps, cosmetics, and other commodities. A substantial quantity of coconut oil is made by over 93 countries globally, serving diverse purposes among individuals. Virgin Coconut Oil (VCO) is distinct from commercial coconut oil in its processing, as it is typically extracted from fresh coconut kernels without chemical treatments. It contains a high concentration of vitamins, minerals, and antioxidants, which makes it the most abundant oil. VCO is a coconut product of food quality and has been enhanced with value, making it suitable for many human uses. Volatile carbon monoxide (VCO) exhibits significant relevance across diverse food, health, cosmetics, and nanotechnology sectors.

Law et al. [15] researched the effects of virgin coconut oil as a supplement on breast cancer patients' quality of life. The research found that breast cancer patients who incorporated virgin coconut oil into their diet during six-cycle chemotherapy experienced enhanced functional and overall quality of life. There were notable differences in local and international quality of life measures between those who received the intervention and the control group. These findings indicate that virgin coconut oil may improve patients' functional and overall quality of life during chemotherapy. Moreover, the study suggests that consuming virgin coconut oil could alleviate symptoms and enhance functional well-being, such as body image, outlook on the future, and sexual function. Given these results, incorporating virgin coconut oil into the regimen of breast cancer patients shows promise in fostering a positive quality of life.

Josi et al. [16] researched coconut oil and immunity. Ayurveda medicine identified that coconut oil had healing power over 4000 years ago. Ancient texts convey that mother's milk has the same health benefits. The medium-chain fatty acids and monoglycerides found predominantly in coconut oil have incredible therapeutic properties, acting as natural antibiotics and modulating the immune system. This study reveals that when coconut oil is consumed or applied topically, it breaks down to release Lauric Acid and Monolaurin, which are proven antimicrobial agents. The literature is reviewed to assess coconut oil's antiviral, antibacterial, and antifungal properties. Not only do metabolites of coconut oil exhibit antibacterial activity but it has also been demonstrated that these extraordinary compounds do not promote the development of resistant organisms. In recent decades, the favourable health effects of coconut oil have never been more widely acknowledged. Never before in human history has it been so crucial to underline both the necessity and effectiveness of natural products renowned for their safety promise. Immunity is a hot topic in the contemporary climate, and the desire to modify immunity naturally is unprecedented and pervasive. Coconut oil and its derivatives can help create a robust and healthy future.

Pandiselvam et al. [17] studied cosmetics enhanced with organic coconut oil,for thousands of years, coconuts and their products have been essential foods. It's said that coconut oil is a superfood. Virgin coconut oil, or VCO, is trendy these days. VCO is more advantageous than copra oil because more bioactive ingredients, like vitamins and polyphenols, are preserved throughout the extraction process. Coconut often provides humans with valuable items like meat, milk, water, oil, and sugar. The manufacturing of cosmetics based on VCO is attracting a lot of interest from both academia and industry. VCO-containing beauty products are sold worldwide, including hair oil, massage oil, mouthwash, lip balm, shampoo, and moisturizers. Figure 2 indicates benefits of virgin coconut oil.

## 4 Health Benefits of Coconut Beverages

Prado et al. [4] studied the evaluation of a fermented coconut water beverage with possible health advantages. Coconut water is a beverage made from coconuts with nutritional value, practical health benefits, and natural moisturizing qualities. This study aimed to create a novel fermented beverage without dairy with intrinsic health benefits using coconut water as the main ingredient. Seven autochthonous lactic acid bacteria strains were identified through spontaneous coconut water fermentation. The

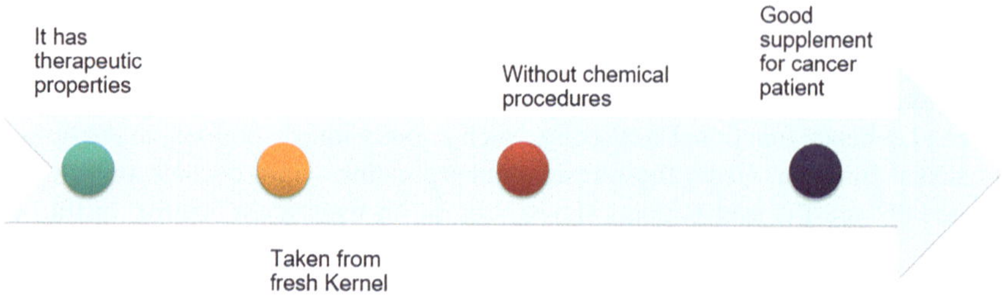


**Fig. 2** Benefits of VCO

selection of Lactobacillus plantarum was based on its inhibitory efficiency against particular pathogens and its capacity to flourish in specific technical and gastrointestinal environments. The soy protein hydrolysate, yeast extract, and sucrose were added to the fermentation, which was carried out for eight hours at 37 °C. A novel way for customers to receive healthy cultured beverages is coconut water, which is suitable for lactose intolerant, vegan, and vegetarian diets.

Chinnamma et al. [18] researched Neera as a healthy beverage from coconut palm. Nature has created nutritious substances in a way that cannot be replicated in laboratories or factories. The natural sap of fully-grown coconut trees is called coconut neera, and it contains all the vitamins, minerals, and nutrients needed for human health. Neera from the palm spadix may be harvested commercially without affecting the physiology of the tree. The fact that raw Neera has no alcohol shows that AFS is active. Neera's nutritional relevance as a health drink is underscored by the proximity of its essential elements and the existence of minerals, amino acids, and vitamins. The medicinal potential of Coconut Neera is shown by its high mineral content (15.2 mg and 100 mg of potassium and salt, respectively), improved levels of amino acids (7.11 mg, 14 mg, and 7 mg), and the presence of vitamin C (65 mg) and vitamin A (retinol, 4.25 IU) in the form of vitamin C. The statistics on oral toxicity and the glycemic index bolster the vitality of the substance.

Rethinam and Krishnakumar [19, 20] studied the value of adding coconut water. Coconut water is a highly favoured and well-acknowledged natural beverage for promoting good health and enhancing metabolism due to its inherent presence of bioactive enzymes. Fresh coconut water is most effectively used as a revitalizing carbonated beverage. Nevertheless, coconut water can also be employed to produce various premium items. Coconut water is regarded as an energy drink due to its ability to maintain an isotonic electrolyte balance. Moreover, it is marketed as a natural sports beverage that can be used prior to, during, or after physical exertion to provide optimal hydration and restoration appropriate for all intensity levels.

The potential or benefits of using coconut water in plant and animal biotechnology have been extensively studied. Mature coconut water is frequently utilized as a raw ingredient in creating vinegar and Nata de coco, which have numerous advantages.

Koner et al. [21] studied the pros and cons of natural and artificial beverages. The study conveys that a beverage is often a liquid for human consumption. For sating human thirst, there are thousands of beverages accessible today, including both soft drinks and strong drinks. In contrast to conventional drinks, natural beverages are unique since they focus entirely on human health, a pressing issue in the modern world. It is highly filled with nutrients, minerals, vitamins, and organic compounds that are helpful and safe for people to consume. Natural beverages are favoured by many people worldwide because of their enhanced nutritious qualities. It lowers the risk of chronic illnesses, heart disease, and type 2 diabetes.

Asha et al. [22] researched NEERA, a nonalcoholic drink. The study states coconut trees are the "tree of life" (Cocos nucifera). A natural health beverage popular in South Asian nations is a sap produced by tapping the unopened coconut palm inflorescence that is sweet, oyster white, and translucent. It has been determined that the sap from the inflorescence of the coconut is a nutritious bio-beverage full of carbs, minerals, vitamins, antioxidants, and amino acids. This health drink's most important feature is that it is diabetic-friendly due to its low glycemic index (GI 35), which results in very little sugar being absorbed into the blood. This non-alcoholic beverage is praised in conventional medical systems as extremely rejuvenating and is advised for the treatment/prevention of different diseases. They have a solid nephroprotective and hepatoprotective activity, which has been documented, increasing their market demand. It functions as a natural detoxifier,regular ingestion keeps the body relaxed and is regarded as a balanced food that enhances digestion. With its priceless offerings like coconut honey and sweeteners made from coconuts, this coconut nectar is gaining attention on a global scale.

Walter et al. [23] studied the safety of bottled coconut water. It examines the safety and sustainability aspects of fresh coconut water in Brazil, which has gained significant popularity due to its widespread availability in convenient plastic bottles. The researcher aims to demonstrate that providing safety measures and liability insurance for fresh coconut water represents both an opportunity and a responsibility for the Brazilian government, society, and other tropical nations. The safety of fresh coconut water is a matter of public health concern. It represents a wholesome and appealing product that contributes to the financial success of street vendors and micro-entrepreneurs.

Deepthy et al. [24] researched the standardization of coconut water-based beverages. Coconut water is widely regarded as a highly versatile natural product globally. This widely consumed beverage is recognized for its refreshing qualities and potential to contribute to overall health and nutrition. The environmental issue of coconut water wastage in oil mills is a matter of concern, particularly given the limited research on commercially produced beverages derived from mature coconut water.

Gordon and Jackson [25] researched appropriate technology for the quality and safety of coconut water. The study conveys that coconut water is a revitalizing liquid derived from the fruit of the coconut palm (Cocos nucifera L). Historically, coconut water was predominantly acquired as a freshly consumed beverage in regions where it was accessible, involving the act of opening a green coconut directly from the tree. However, in contemporary times, it has evolved into a thriving industry with a market value exceeding $300 million. This study delves into the scientific substantiation for the burgeoning sector while scrutinizing one case study conducted during the nascent stages of the industry's revival in the early 2000s. The content presented in the initial

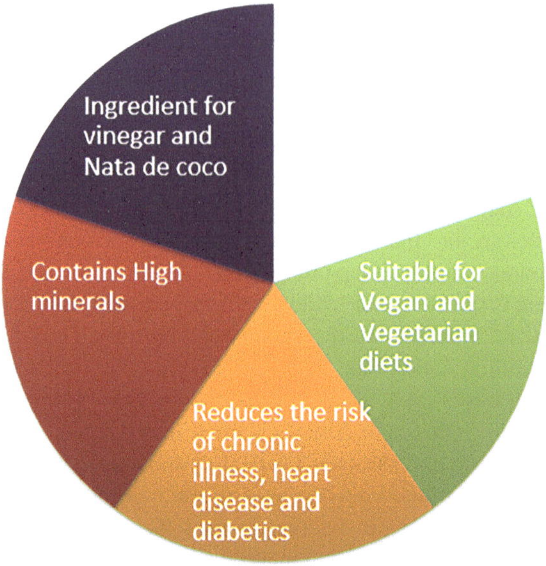


**Fig. 3** Coconut beverage: benefits

instance study has made notable advancements in comprehending and addressing the issues about the quality and longevity difficulties encountered by conventional bottlers of refrigerated coconut water. The study identified polyphenol oxidase (PPO), peroxidase (PO) enzymes, yeasts, and lactic acid bacteria as the primary factors responsible for coconut water's pink discoloration and spoilage. Additionally, the study suggested that pH could serve as an indicator for detecting spoilage. Figure 3 shows health benefits of coconut beverages.

## 5 Conclusion

Coconut products have gained popularity because of their supposed health benefits, especially for weight loss, skin health, and moisturizing. Expert opinions on their actual effects on health, however, differ widely. Due to its high saturated fat content, coconut oil has been endorsed for possible health advantages. However, organizations like the American Heart Association advise against it because it may harm cholesterol levels. On the other hand, coconut water is highly valued for its hydrating properties and nutritional value, making it a preferred choice for sportspeople and anybody looking for a natural alternative to sports drinks. However, there is little factual support for several inflated health claims associated with coconut products. Therefore, even if these items are made from coconuts and have some benefits, eating them in moderation and as part of a balanced diet is essential.

## References


1. Prades, A., Dornier, M., Diop, N., Pain, J.P.: Coconut water uses, composition and properties: a review. Fruits **67**(2), 87–107 (2012). https://doi.org/10.1051/fruits/2012002
2. Naik, M., Sunil, C.K., Rawson, A., Venkatachalapathy, N.: Tender coconut water: a review on recent advances in processing and preservation. Food Rev. Int. **38**(6), 1215–1236 (2022). https://doi.org/10.1080/87559129.2020.1785489
3. Zulaikhah, T.S.: Health benefits of tender coconut water (TCW). Int. J. Pharm. Sci. Res. **10**(2) (2019)
4. Camargo Prado, F., de Dea, L.J., Inaba, J., Thomaz-Soccol, V., Kaur Brar, S., Soccol, C.R.: Development and evaluation of a fermented coconut water beverage with potential health benefits. J. Funct. Foods. **12**, 489–497 (2015). https://doi.org/10.1016/j.jff.2014.12.020
5. Nasution, Z., Jirapakkul, W., Lorjaroenphon, Y.: Aroma compound profile of mature coconut water from tall variety through thermal treatment. Food Measure. **13**, 277–286 (2019). https://doi.org/10.1007/s11694-018-9942-x
6. Fadlilah, S., Sucipto, A., Judha, M.: Benefits of young coconut water, watermelon, and papaya for blood pressure among hypertension. Int. Med. J. (1994). https://www.researchgate.net/publication/357416483
7. Prithviraj, V., Pandiselvam, R., Babu, A.C., Kothakota, A., Manikantan, M.R., Ramesh, S.V., Beegum, P.P.S., Mathew, A.C., Hebbar, K.B.: Emerging non-thermal processing techniques for preservation of tender coconut water. LWT **149** (2021). https://doi.org/10.1016/j.lwt.2021.111850
8. Rethinam, P., Krishnakumar, V.: Health benefits of coconut water. In: Rethinam, P., Krishnakumar, V., (eds.) Coconut Water. Springer, Cham (2022). https://doi.org/10.1007/978-3-031-10713-9_9
9. Castro, C.J.: Development of Kamias (*Averrhoa bilimbi* L.) jam with coconut water. Int. J. Vocat. Educ. Train. Res. **7**(1), 1 (2021). https://doi.org/10.11648/j.ijvetr.20210701.11
10. Marina, A.M., Che Man, Y.B., Amin, I.: Virgin coconut oil: emerging functional food oil. Trends Food Sci. Technol. **20**(10), 481–487 (2009). https://doi.org/10.1016/j.tifs.2009.06.003
11. Carandang, E.V.: Health benefits of virgin coconut oil. Indian Coconut J.-Cochin-. **38**(9), 8 (2008)
12. Srivastava, Y., Semwal, A.D., Sharma, G.K.: Virgin coconut oil as functional oil. In: Therapeutic, Probiotic, and Unconventional Foods. Elsevier, pp. 291–301 (2018). https://doi.org/10.1016/B978-0-12-814625-5.00015-7
13. Yeap, S.K., Beh, B.K., Ali, N.M., Yusof, H.M., Ho, W.Y., Koh, S.P., Alitheen, N.B., Long, K.: Antistress and antioxidant effects of virgin coconut oil in vivo. Exp. Ther. Med. **9**(1), 39–42 (2015). https://doi.org/10.3892/etm.2014.2045
14. Satheesh, N.: Review on production and potential applications of virgin coconut oil. Annals. Food Sci. Technol. **115**(1) (2015). www.afst.valahia.ro
15. Law, K.S., Azman, N., Omar, E.A., Musa, M.Y., Yusoff, N.M., Sulaiman, S.A., Hussain, N.H.N.: The effects of virgin coconut oil (VCO) as supplementation on quality of life (QOL) among breast cancer patients. Lipids Health Dis. **13**(1) (2014). https://doi.org/10.1186/1476-511X-13-139
16. Joshi, S.: Coconut oil and immunity: what do we really know about it so far? J. Assoc. Physicians India (2020). https://www.researchgate.net/publication/342604136
17. Pandiselvam, R., Manikantan, M.R., Beegum, S., Mathew, A.C.: Virgin coconut oil infused healthy cosmetics (2019)
18. Chinnamma, M., Bhasker, S., Hari, M.B., Sreekumar, D., Madhav, H.: Coconut neera—a vital health beverage from coconut palms: harvesting, processing and quality analysis. Beverages. **5**(1) (2019). https://doi.org/10.3390/beverages5010022
19. Rethinam, P., Krishnakumar, V.: Global scenario of coconut and coconut water. In: Coconut Water. Springer, pp. 17–35 (2022). https://doi.org/10.1007/978-3-031-10713-9_2

20. Rethinam, P., Krishnakumar, V.: Value addition in coconut water. In: Rethinam, P., Krishnakumar, V. (eds.) Coconut Water. Springer, Cham. pp. 287–384 (2022). https://doi.org/10.1007/978-3-031-10713-9_8
21. Koner, S., Dash, P., Priya, V., Rajeswari, V.D.: Natural and artificial beverages: exploring the pros and cons. In: Koner, S., Dash, P., Priya, V., Rajeswari, V.D. (eds.) Natural Beverages: Volume 13: The Science of Beverages. Elsevier, pp. 427–445 (2019). https://doi.org/10.1016/B978-0-12-816689-5.00015-8
22. Asha, S., Ratheesh, M., Jose, S.P., Krishnakumar, I.M., Sandya, S.: NEERA: a nonalcoholic nutritious beverage from unopened inflorescence of coconut palm. In: Asha, S., Ratheesh, M., Jose, S.P., Krishnakumar, I.M., Sandya, S., (eds.) Natural Beverages: Volume 13: The Science of Beverages. Elsevier, pp. 339–360 (2019). https://doi.org/10.1016/B978-0-12-816689-5.00012-2
23. Walter, E.H.M., Kuaye, A.Y., Hoorfar, J.: Case study on the safety and sustainability of fresh bottled coconut water. In: Hoorfar, J., (ed.) Global Safety of Fresh Produce: A Handbook of Best Practice, Innovative Commercial Solutions and Case Studies. Elsevier, pp. 367–382 (2013). https://doi.org/10.1533/9781782420279.5.367
24. Deepthy, V., Divakar, S., Kumari, M., Anila, H.: Standardisation of coconut water based beverage
25. Gordon, A., Jackson, J.: Case study: application of appropriate technologies to improve the quality and safety of coconut water. In: Gordon, A., Jackson, J., (eds.) Food Safety and Quality Systems in Developing Countries. Elsevier. pp. 185–216 (2017). https://doi.org/10.1016/B978-0-12-801226-0.00007-4

# Green Versus Greenwashing: How Consumers Differentiate Authentic Green Marketing from Deceptive Practices

**Shilpa Chadichal, Prasanna Mohan Raj, and Sanjeev Padashetty**

**Abstract** *Purpose*: The study examines how effective green marketing is, as well as how greenwashing affects customer trust. Identifying the many aspects that influence consumer perception will provide insight into businesses and policymakers on how to promote true sustainability, prevent greenwashing, and allow consumers to make decisions that match their environmental beliefs. *Design/Methodology/Approach*: The method used is a quantitative research design, and the analysis is based on a survey. For this, the sample size is 254 consumers residing in Bengaluru, India. Regarding this, the stratified sampling method will be used to ensure that the selected sample is representative. This approach has been employed to gather significant information about consumer perception of green marketing and greenwashing. *Result*: It shows that perceived brand transparency has a positive impact on consumer trust in green marketing efforts. Furthermore, the more skeptical an individual is when it comes to advertising, the more capable they become at identifying actual greenwashing practices. *Conclusion*: The study also reveals that heightened awareness and knowledge of green marketing on the part of consumers enhance their ability to detect greenwashing. Greenwashing significantly erodes consumers' trust in products and ultimately affects purchasing behavior. Business recommendations include communicating real green practices to avoid being termed as greenwashing and to help regain consumer trust. *Originality/Value*: This provides new insights into how consumer awareness and brand transparency relate to the effectiveness and trust in green marketing. It gives practical suggestions to businesses regarding how they can communicate authentically and reduce greenwashing, hence contributing to the knowledge base in the area of sustainable consumerism.

S. Chadichal (✉) · S. Padashetty
Alliance School of Business, Alliance University, Bangalore, Karnataka, India
e-mail: shilpa.chadichal@alliance.edu.in

S. Padashetty
e-mail: sanjeev.padashetty@alliance.edu.in

P. M. Raj
Associate Professor, Department of Management Studies, Sri Sivasubramaniya Nadar College of Engineering, Chennai, India
e-mail: prasannamohanrajm@ssn.edu.in



M. Al Mubarak (ed.), *Sustainable Digital Technology and Ethics in an Ever-Changing Environment*, Studies in Systems, Decision and Control 237,
https://doi.org/10.1007/978-3-031-86708-8_12

**Keywords** Green · Greenwashing · Consumers · Quantitative research · Statistical analysis

## 1 Introduction

Over the years, green marketing has become a very potent tool for businesses in the face of growing environmental awareness [1, 10, 15]. Still, with this trend, greenwashing acts of deception through exaggeration or fabrication of environmental claims by businesses become common. It is also a fact that greater concern with greenwashing has produced more vigilant consumers who now face the challenge of distinguishing between genuine green marketing and misleading claims [8, 16, 19].

The study addresses the interlink between green marketing and greenwashing regarding how consumers identify and distinguish their perceptions of genuine environmental activism from those that have a hollow ring to them. The present research uses a quantitative survey to deeply understand consumer attitudes and behaviors. In this study, a survey has been used to collect data from a heterogeneous group of consumers to analyze awareness, skepticism, and trust in green marketing claims [4].

One of the strengths of this study is in its comprehensive approach wide range of consumers' perspectives are captured, which thus enables an integrated understanding of the problem being tackled [14]. The quantitative design, on its part, allows for both breadth and depth of data collection, thus making strong findings possible that are also statistically significant and rich in context. A partial limitation could be the reliance on self-reported data, which may be afflicted by social desirability response bias. Although the research is planned in such a manner to represent a wide variation of consumer responses, the results may not reflect opinions that are representative of niche markets or less informed segments of consumers.

This study aims the effectiveness of green marketing and how greenwashing affects consumer trust. This research would be able to compare those variables that affect the perceptions of consumers and help businesses that genuinely want to apply sustainability practices, as well as policymakers who need to regulate and minimize greenwashing. In conclusion, the study contributes to the broader debate on sustainable consumerism, aiming at empowering consumers with choices compatible with their environmental values. The main objectives of the study include.

1. To analyze consumer awareness and understanding of green marketing practices, including identifying greenwashing tactics.
2. To research the impact of greenwashing on consumer trust and purchase decisions associated with ecologically friendly products.
3. To make recommendations for organizations on how to successfully promote genuine green projects and avoid the hazards of greenwashing.

The remaining part of the research is organized into the following sections: The antecedent studies of the research are covered in the second section. In the third section, the methodology of the research is discussed in detail. In the fourth section, the empirical results of the data analysis will be presented. It analyzes the results of the questions about correlation, moderation, and mediation. It provides a succinct summary of the research findings, highlighting the most critical information and conclusions that have been drawn from the analysis of the data. Section 5 provides the Discussion of research. Section 6 gives the conclusion of the study. Section 7 provides the Practical implication. This looks at how the findings might impinge on industrial strategies, inform the value of policy decisions, and inform further studies.

## 2 Literature Review

### *2.1 Consumers' Environmental Knowledge (CEK) and Differentiation of Green Marketing*

Qayyum et al. [17] explored the adverse impacts of EPP, greenwashing, and green confusion on GBE and the moderating role of brand credibility in reducing the adverse impact of green marketing. Results indicate that excessive product packaging is positively predicting greenwashing and green confusion while greenwashing negatively influences GBE. However, brand credibility moderated the negative relation of greenwashing with GBE, hence reducing its negative effects. The study infuses the fact that marketing managers need to consider the consumers' concern for the environment and develop product and brand strategy that helps in environmental protection and sustainability.

Szabo and Webster [21] explored green marketing strategies that companies use to improve their environmental image and reach ecological consumers. It points out, however, that there is an ethical issue when the marketing might not truthfully reflect the way the company makes efforts toward being "green." They investigate the impact of perceived greenwashing on consumer perceptions and happiness when interacting with a company website. It also covers the relationship between website interactivity and greenwashing perceptions.

Sharma [20] aimed to narrow the gap in consumer attitude and actual purchasing behavior on green products by thematically analysing 232 studies that used a systematic review. The most relevant themes identified include personal factors influencing green purchases, green purchase behaviours, and the green marketing mix. Major influencers of green consumer behavior include eco-environmental concerns, eco-labelling, past experiences, and perceived usefulness. Barriers to green consumption also contribute to this attitude-behavior gap, such as lack of environmental knowledge, price, perceived risks, organizational image, and trust and willingness to pay. It has proposed a green purchase decision model and pointed out possible avenues for future research to better explain these dynamics to enhance the effectiveness of green marketing.

Tan et al. [22] explored the green marketing factors influencing purchase intention through the mediators of green brand trust and image. It also explores the associations between the constructs of green knowledge, attitude towards the environment, green image, green trust, and purchase intention. For this, it empirically surveyed 300 respondents from Pakistan. SPSS-AMOS was used in the study to evaluate the structural equation modelling to test the hypotheses cited above. Results show that practical green marketing strategies, such as green knowledge and environmental attitude, bear a significant positive relation with purchase intention related to green products.

Kaur et al. [11] assessed the impact of green marketing mix strategies on the green purchase intention of millennial consumers of personal care products in an emerging economy. The study investigates which of the green product attributes, green places, and green promotional strategies motivational drivers are leading to the development of purchase intention among millennials while considering the moderating effect of environmental attitudes and demographic variables. The study applies structural equation modelling and evaluates 405 responses from millennials by considering multi-group analysis about control variables. The regression results indicated that dimensions of green product attributes, places, and promotion strategies are significant drivers affecting green buying intention.

## *2.2 Perceived Brand Transparency (PBT) and Consumer Trust*

Rahman and Nguyen-Viet [19] analysed how greenwashing is perceived by consumers and gives the effect of greenwashing upon purchasing intentions. It searches for the effects of green advertising receptivity, non-deception, green brand image, and transparency on green brand trust, and how green brand trust itself affects purchasing intentions. It also investigates the mediating role of green brand trust and the moderating impact of non-deception upon green advertising receptivity.

Wiederstein [23] investigated how each of the dimensions of sustainable fashion marketing on social media contributes to the creation of an authentic, coherent, and trustworthy post. Particularly, it has looked upon the effects of green appeals-concrete versus abstract-and benefit associations-self-oriented versus other-oriented-on sustainable purchasing intention. This research also explores whether concrete green appeals, with appropriate benefit associative sets, result in the highest purchase intent, the greatest perceived authenticity, less confusion, and therefore greater perceived trustworthiness. It also examines whether perceived authenticity, confusion, and trustworthiness mediate the relationship between green appeals and purchase intentions.

Guerreiro and Pacheco [9] investigated the impact of customer perceptions on greenwashing, which in turn drives green purchase behavior, focusing attention on green trust, consumer brand engagement, and green word-of-mouth as mediators. They surveyed a sample of 302 participants who viewed both high and low-involvement green ads and analysed data through PLS-SEM. The findings indicate that while green-washing expectations did not affect the green purchase decisions directly, they have an impact on green trust and green word-of-mouth, which, in turn, mediate the relationship between perceptions of greenwashing and purchasing intentions.

Deng and Yang [6] examined the factors affecting green customer citizenship behavior within a restaurant context. The role of green attributes transparency was emphasized. A conceptual model was developed that indicated how transparency enhances green brand image and green trust, which in turn facilitates green customer citizenship behavior. Data from 312 consumers in Taiwan are analysed by using structural equation modelling. These results evidence that transparency of the green attribute is an influential factor that affects the green brand image and green trust together, which raises green customer citizenship behavior.

Bhaduri and Copeland [3] investigated the role of brand schema, information transparency, and climate skepticism in the acceptance of pro-environmental messages from both known and lesser-known brands. Two online experiments conducted for the purpose involved a combined total of 510 participants. The measurements were message and treatment varied to establish their validities. PROCESS-a regression-based path analytic approach used for data analysis. In this vein, the results show that matching consumer schema with pro-environmental messages increased message persuasiveness, brand perceptions, and behavioural intentions toward the brand. High information transparency also increases consumer evaluations.

### *2.3 Consumer Skepticism and Identification of Greenwashing*

Butt et al. [5] discussed the "greenwashing" phenomenon where businesses make deceitful claims about their products being "green" and, at the same time, considered eco-friendly. This study explains the pressures from external stakeholders like consumers and regulators that force firms to conduct the practice of greenwashing. It was deduced that consumer and regulatory pressures influenced the organizations studied in adopting greenwashing to a great extent, although this was not so with the competitive pressure. Consumers deduce that businesses are either greenwashing to please them or avoid the hassle of the law. The present study underlines the negative impact of external stakeholders on the firm's greenwashing behavior and gives recommendations for its monitoring and reduction.

Kwak et al. [12] investigated how the perception of consumers on green marketing, authenticity, and purchase intention of a fashion brand changes after the consumers have been exposed to information concerning greenwashing. Data were obtained from 211 respondents, and it was found that after being exposed, the perception of green marketing is lower than before, as the authenticity, and the purchase intention of a brand of fashion. It proves that authentic and transparent green marketing campaigns are significant for fashion brands to maintain and enhance brand authenticity.

Aparna and Murugan [2] discussed the issue of greenwashing, a form of deceptive marketing in which companies try to fool the public into believing their products or services are 'green.' They do this through advertisements and branding strategies that are misleading. The study also highlights how companies take advantage of consumers' increasing environmental and sustainability consciousness to falsely convince them. The results of greenwashing could be very harmful to the environment and to the consumers. The study further investigates the prevalence in different industries and the implications of greenwashing.

Esha [7] discussed the current trend in consumption, moving to green-friendly and highly aware consumers of environmental sustainability. It reflects upon marketing strategies adopted by companies concerning green labelling and greenwashing and discusses how they deceive consumers. The study investigates how ethical perceptions could affect consumers' behavior towards green-labelled products and awareness of greenwashing. To that effect, data is gathered from 213 respondents in Estonia through an online survey method and the results have been analysed on SPSS.

Fella and Bausa [8] examined consumers' ability to detect greenwashing in the marketplace, where lots of new and existing green products are being brought to the forefront. This research finds that prompting consumers to think about the concept of greenwashing, raises one's ability to differentiate between honest green products and greenwashed ones. In other words, it means that perceived greenwashing is better identified by the consumer when prompted about it. The results show that the activation of the concept of greenwashing helps consumers unmask greenwashing in purchase contexts and points to a potential public policy measure to support consumers in navigating green products.

Lopes et al. [13] addressed how greenwashing affects consumers' attitudes toward consumption based on circularity. The concept of greenwashing is that which describes the firm practice of misrepresenting their products as environmentally responsible. This may affect the perception and attitudes of the consumer. From a sample of 826 responses from Portuguese consumers, the results indicated that greenwashing was indeed affecting the environmental concerns of consumers and their search for sustainable information in terms of positive influences on their intentions toward consumption based on circularity.

## 3 Research Methodology

See Fig. 1.

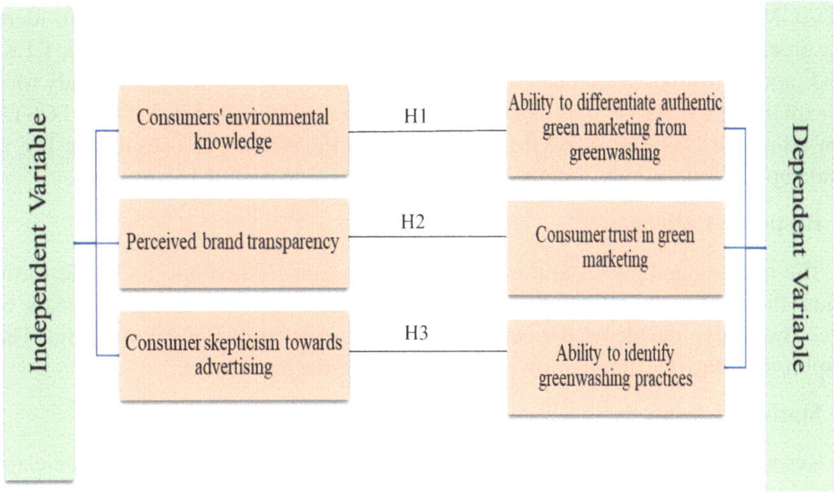


**Fig. 1** Conceptual framework

## *3.1 Hypotheses*

**Hypothesis 1 Null Hypothesis (H0):** Consumers' environmental knowledge does not influence their ability to differentiate authentic green marketing from greenwashing (ADAGMG) practices.

**Alternative Hypothesis (H1):** Consumers' environmental knowledge positively influences their ability to differentiate authentic green marketing from greenwashing practices.

**Hypothesis 2 Null Hypothesis (H0):** Perceived brand transparency does not impact consumer trust in authentic green marketing efforts.

**Alternative Hypothesis (H1):** Perceived brand transparency positively impacts consumer trust in authentic green marketing efforts.

**Hypothesis 3 Null Hypothesis (H0):** Consumers with higher levels of skepticism towards advertising are more likely to identify greenwashing practices.

**Alternative Hypothesis (H1):** Consumers with higher levels of skepticism towards advertising are not more likely to identify greenwashing practices.

## *3.2 Research Design*

- **Questionnaire Preparation**

The survey contains 30 questions assessing the dependency of two different kinds of variables: dependent and independent. Dependent variables (DV) will be the

ADAGMG, Consumer trust in green marketing (CTGM), and the ability to identify greenwashing practices (AIGP). Independent variables (IDV) will include CEK, PBT, and Consumer skepticism toward advertising (CSTA). The present study tries to explain how IDVs such as Consumer environmental knowledge, PBT, and CSTA can affect the DV: The ADAGMG, CTGM, and the AIGP. As a result, the DV is made up of 15 questions, and the IDV is likewise composed of 15 questions.

- **Response Collection**

The questionnaire used in this study has been transformed into a Google form. Thus, the consumers included in this research were between 25 and 44 years of age. The study considered the poll in the specific context of Bengaluru, India. The responses to the questionnaire were provided by citizens of India.

- **Statistical Analysis**

Responses gathered from the Consumer have been critically examined and elaborated upon by utilizing one of the renowned software SPSS for effective statistics analysis. Various features and techniques of the statistical method such as analysis regarding correlation, T-tests, regression, and descriptive statistics have been applied to analyze and interpret the data gathered during the study. The selected and implemented statistical technique was applied and is aptly used to carry out the testing of the hypothesis so that the aspect of the hypothesis is evaluated and tested for its reliability and validity. The use of these tests enabled us to systematize the relationships, trends, and correlations among relevant variables, which in turn gave us an enhanced understanding of the topic under research.

### *3.3 Online Survey*

The investigation in this research has involved a decent sample size of 254 participants who have voluntarily contributed their opinions and information. Both males and females were included in the research. To invite participants, several internet channels were exploited: social media, e-mail lists, and relevant forums. Data collection was carried out through an online survey using a structured questionnaire, which was specially designed to collect data on the ADAMGM, CTGM, AIGP, CEK, PBT, and CSTA, along with additional factors for research aims and hypotheses. The link to the survey has been set up on a secure online platform where data protection is guaranteed. The questionnaire consists of sections related to demographic data, ADAGMG, CTGM, AIGP, CEK, PBT, and CSTA. It conducted an online survey using a structured questionnaire, which was designed in such a way as to extract demographic information, perceptions of ADAGMG, CTGM, AIGP, CEK, PBT, and CSTA, among other factors, pertinent to the research objectives and hypotheses. Informed consent for every participant has been obtained even before the survey starts. Using the Random Sampling Approach, this research resulted in 254 valid samples and 0 invalid samples. The measurements are by using a 5-point Likert scale.

### 3.3.1 Design and Sample

The frequency distribution of demographic characteristics sheds light on the composition of the study's respondents. According to the age distribution, the majority of respondents are between the ages of 25 and 34 (39%), followed by those aged 35–44 (34.3%). This implies that the sample is primarily young to middle-aged, with fewer responses in the older age categories (45–54 at 10.6% and 55 + at 8.7%) and the youngest (18–24 at 7.5%). The gender distribution is relatively balanced, with a small female majority (51.6%) compared to males (48.4%). This near-even divide enables gender-based comparisons in the analysis of green marketing attitudes. In terms of education, the majority of respondents are highly educated, with 45.3% holding a Master's degree and 43.3% holding a Bachelor's. Only a small percentage have a high school diploma or less (1.2%), or some college experience (6.3%). This shows that the sample is primarily well-educated, which may affect their ability to distinguish between genuine green marketing and greenwashing. In terms of income, the largest group earns between $75,000 and $99,999 (46.1%), followed by $50,000–$74,999 (42.5%). This suggests a highly rich respondent base. Finally, the majority of respondents (45.7%) live in cities, with a sizable proportion in suburbs (41.3%) and a smaller number in rural areas (13%). This urban and suburban majority represents a demographic that is more likely to be exposed to green marketing initiatives.

The data provides insightful thoughts about respondents' capacity to distinguish between genuine eco-friendly statements and misleading marketing methods, popularly known as greenwashing. A sizable portion of respondents (45.3%) believe they can readily distinguish between legitimate and misleading marketing claims, with a further 28.7% strongly agreeing. This shows that the majority of people are confident in their ability to recognize genuine green marketing. However, a significant 23.2% disagree or strongly disagree, indicating a demographic segment that struggles with this distinction. When asked about the difficulty of identifying fraudulent envi ronmental claims, nearly half of respondents (48.4%) agreed, with 28.3% strongly agreeing, demonstrating that while many people are confident, a sizable proportion still find it difficult to distinguish greenwashing. This is reinforced by the confidence in recognizing true green marketing, with 50.4% agreeing and 26% strongly agreeing, but 20.9% still lacking trust. The cynicism toward environmental claims is clear, with 52.4% agreeing and 24.8% strongly agreeing that people frequently doubt brand statements, suspecting greenwashing. Finally, 47.2% agree and 30.3% strongly agree that they can tell the difference between genuine environmental initiatives and greenwashing, supporting the overall trend of cautious confidence among respondents. This research reveals a generally aware and suspicious customer base, with a mix of confidence and uncertainty about green marketing claims.

The data offers a mixed picture of customer trust and mistrust toward firms that promote their products as environmentally friendly. A large proportion of respondents

(48.8%) agree, with 29.1% strongly agreeing, that they trust brands that promote eco-friendly products, demonstrating a general willingness to believe in green marketing. However, a sizable minority (19.7%) disagree or strongly disagree, indicating underlying doubt. The replies to the question of whether they are dubious of green marketing claims show that 48.8% agree and 27.2% strongly agree. This implies that, while many consumers are predisposed to accept green marketing, a sizable proportion remain skeptical, doubting the veracity of these claims. When asked if they feel most corporations are honest in their environmental marketing, 49.6% agreed, with 24.4% strongly agreeing, indicating a majority belief in corporate honesty. However, the 21.3% who disagree or strongly disagree suggest that some customers continue to be skeptical. Confidence in purchasing from green marketing brands is relatively high, with 57.1% agreeing and 17.7% strongly agreeing, indicating that the majority of buyers feel secure in their purchases. Finally, 48.4% agree, with 26% strongly agreeing, that organizations employing green marketing techniques are genuinely devoted to sustainability, but the 22.1% who disagree or are undecided indicate that trust is not absolute. Overall, the data shows that while consumers are generally trusting of green marketing, they also maintain a healthy level of skepticism, ensuring that they remain cautious and analytical in their evaluation of such claims.

The data provides useful information about consumer knowledge and the capacity to recognize greenwashing tactics. The majority of respondents (51.2%) agree, with 24.8% strongly agreeing that people can tell when a company is overstating its environmental benefits. This indicates that most consumers are confident in their ability to discern exaggerated environmental claims, however, a significant 19.7% (disagree and strongly disagree) continue to struggle with this. When asked if they were aware of frequent greenwashing strategies, 56.3% agreed and 20.9% strongly agreed, showing a high degree of awareness among respondents. However, 18.9% disagree, indicating that not all consumers are aware of these fraudulent methods. Similarly, when it comes to recognizing deceptive environmental claims in commercials, 51.2% agree and 26.4% strongly agree, indicating that the majority find it rather simple to identify such claims, albeit 18.9% still find it difficult. Familiarity with the symptoms of greenwashing is also high, with 55.9% agreeing and 20.9% strongly agreeing, supporting the notion that most consumers are well-equipped to spot deceptive marketing. Finally, 52.8% agree, with 23.6% strongly agreeing, that they can tell when a brand isn't being completely honest about its environmental impact. This implies that a sizable proportion of consumers are cautious and critical when evaluating environmental claims, yet approximately 20% still express ambiguity or mistrust in their ability to detect dishonesty. Overall, the statistics show that, while the majority of consumers are aware of and confident in identifying greenwashing strategies, a sizable proportion may require further education or resources to better understand and recognize these deceptive activities.

The data includes information about respondents' self-assessed knowledge and awareness of environmental issues and sustainable practices. The majority of respondents (53.9%) agree, with 22% strongly agreeing that they are knowledgeable about environmental issues and sustainable practices. This suggests that individuals have a high level of awareness, yet 20.4% (disagree and strongly disagree) feel less knowledgeable, indicating that environmental education should be improved. When asked about the environmental impact of various items, 50.4% agreed and 23.6% strongly agreed, indicating that most respondents have a good grasp of these consequences. However, 21.7% either disagree or strongly disagree, implying that a sizable proportion may be unaware of how products impact the environment. The inclination to stay current on eco-friendly products and practices is clear, with 49.2% agreeing and 26% strongly agreeing that they keep up with the most recent information. This implies that most respondents are actively interested in learning about sustainability, while 20.1% still do not prioritize remaining informed. In terms of comprehending what makes a product truly ecologically friendly, 47.6% agree and 26.4% strongly agree, indicating that most respondents believe they understand green product criteria. Nonetheless, 22.1% disagree or are ambivalent, implying that some respondents may benefit from clearer rules or instructions on choosing truly eco-friendly products. Finally, 52.4% agree, with 22.8% strongly agreeing, that they have a good awareness of how their purchasing decisions influence the environment, indicating a typically informed and conscious customer base. However, 20.4% disagree or strongly disagree, indicating that a sizable proportion of respondents may require additional information about the environmental impact of their consumption habits. Overall, while the majority of respondents are educated and actively involved in sustainability, a sizable minority require additional education to fully comprehend and participate in eco-friendly practices.

The results represent respondents' perceptions of company transparency and honesty in environmental activities. Most respondents (53.9%) agree, with 20.1% strongly agreeing, that most corporations are upfront about their environmental activities, demonstrating widespread support for corporate transparency. However, 22.5% of respondents (disagree or strongly disagree) express doubt, indicating a high level of skepticism. When asked whether firms explain the environmental impact of their products, 47.6% agreed and 24.4% strongly agreed, indicating that the majority of respondents believe companies communicate adequately. However, 23.3% (disagree and strongly disagree) believe communication might be improved, demonstrating that not all customers are convinced. Trust in brands' transparency about their sustainability initiatives is slightly lower, with 45.7% agreeing and 20.5% strongly agreeing. The 27.6% who disagree or strongly disagree demonstrates a significant section of the population that is skeptical of brand claims, indicating that there is still potential for development in terms of consumer trust. When it comes to giving sufficient information to evaluate green claims, 48% agree and 30.3% strongly agree, indicating that the majority of respondents believe they have access to enough information to evaluate these claims. However, 20.1% (disagree and strongly disagree) believe that companies do not provide enough information, showing a gap that companies must fill to ensure that all consumers are well-informed. Finally, there is a relatively positive assessment of brands' honesty regarding the limitations of their environmental measures, with 48% agreeing and 30.3% strongly agreeing. However, 19.7% (disagree and strongly disagree) continue to dispute this honesty, indicating that while many customers value transparency, there is still uncertainty about full disclosure of environmental limits. Overall, while there is a broad conviction in corporate transparency and honesty, a sizable minority of respondents remain doubtful, highlighting the need for corporations to communicate more clearly and consistently about their environmental practices.

The statistics show that respondents had a high level of mistrust toward environmental claims stated in commercials. A plurality (50.4%) agrees, with 25.6% strongly agreeing, that people generally distrust advertising claims, including those claiming environmental benefits. This suggests that more than three-quarters of respondents are skeptical about commercials, implying widespread uncertainty about the veracity of marketing messaging. Similarly, 44.9% agree, with 20.1% strongly agreeing that people frequently question the veracity of green marketing statements. With 27.9% (disagree and strongly disagree) still having some trust, this data reveals a large divide in consumer confidence in green marketing, with a sizable portion remaining skeptical. When asked about their skepticism regarding the legitimacy of environmental claims in advertising, 45.3% agreed and 23.6% strongly agreed, confirming that a sizable majority of respondents doubt the reality of such statements. This sentiment is shared by 41.7% who agree and 24.4% who strongly agree that most green marketing is overstated or misleading. The fact that only 29.5% (disagree and strongly disagree) believe otherwise demonstrates how widespread doubt is. Finally, 51.2% agree, with 28.7% strongly agreeing, that people tend to distrust the environmental promises made by firms in their marketing, contributing to the general trend of consumer doubt and skepticism. With only 18.1% (disagree and strongly disagree) believing these promises, the research demonstrates a clear need for businesses to improve the credibility and transparency of their environmental marketing to recover customer trust.

### 3.3.2 Measures

Table 1 shows the descriptive statistics for the study's demographic characteristics and attitudes of green marketing giving a complete picture of the respondents. The sample's mean age of 2.74 and standard deviation (SD) of 1.038 indicate that it is predominantly made up of younger to middle-aged adults, most likely between the ages of 25 and 34. The gender mean of 1.52 shows a small majority of female responses, indicating a balanced representation with a minor bias towards women. Regarding education, the mean score of 3.44 and an SD of 0.724 indicate that the majority of respondents have at least a bachelor's degree, with some pursuing higher study. The income level has a mean of 3.43 and an SD of 0.795, putting most respondents in the moderate to high-income range. The geographic location means of 1.67 indicates that respondents are slightly more likely to live in suburbia, but there is also a considerable urban representation. In terms of perceptions of green marketing, the variable ADAGMG has a mean of 3.8150, indicating that respondents are typically confident in their ability to differentiate between genuine and deceptive green marketing techniques. CTGM and AIGP both had mean values close to 3.8, indicating a strong belief in green marketing and awareness of green products. CEK,

**Table 1** Descriptive statistics

| | N | Mean | Std. deviation | Std. error mean |
|---|---|---|---|---|
| Age | 254 | 2.74 | 1.038 | 0.065 |
| Gender | 254 | 1.52 | 0.501 | 0.031 |
| Education | 254 | 3.44 | 0.724 | 0.045 |
| Income | 254 | 3.43 | 0.795 | 0.050 |
| Location | 254 | 1.67 | 0.694 | 0.044 |
| ADAGMG | 254 | 3.8150 | 0.92907 | 0.05830 |
| CTGM | 254 | 3.7693 | 0.91738 | 0.05756 |
| AIGP | 254 | 3.7882 | 0.59326 | 0.03722 |
| CEK | 254 | 3.7551 | 0.94063 | 0.05902 |
| PBT | 254 | 3.7362 | 0.86761 | 0.05444 |
| CSTA | 254 | 3.6638 | 0.79774 | 0.05005 |

with a mean of 3.7551, indicates that respondents believe they have a reasonable understanding of environmental issues. PBT and CSTA, with averages of 3.7362 and 3.6638, respectively, highlight respondents' broad willingness to trust and support real green marketing, despite some variation in their replies, as demonstrated by the SDs. These findings point to a well-educated, environmentally sensitive respondent group that is rather confident in discerning between genuine and greenwashed marketing techniques.

# 4 Result

## *4.1 Reliability Statistics*

Table 2 displays the data's reliability statistics, which include a Cronbach's Alpha of 0.951 and the same value (0.951) for standardized items across 30 items. This high alpha value implies that the study's measurement instrument had outstanding internal consistency and reliability. It shows that the survey items are highly correlated and consistently measure the underlying construct, giving confidence in the survey replies' reliability.

**Table 2** Reliability statistics

| Cronbach's alpha | Cronbach's alpha based on standardized items | N of items |
|---|---|---|
| 0.951 | 0.951 | 30 |

**Table 3** ANOVA

| | | | Sum of squares | df | Mean square | F | Sig |
|---|---|---|---|---|---|---|---|
| Between people | | | 3821.889 | 253 | 15.106 | | |
| Within people | Between items | | 66.176 | 29 | 2.282 | 3.095 | < 0.001 |
| | Residual | Non-additivity | 5.762[a] | 1 | 5.762 | 7.824 | 0.005 |
| | | Balance | 5403.262 | 7336 | 0.737 | | |
| | | Total | 5409.024 | 7337 | 0.737 | | |
| | Total | | 5475.200 | 7366 | 0.743 | | |
| Total | | | 9297.089 | 7619 | 1.220 | | |

The values which have been super scripted with the letter "a" means, they are statistically significantly different under certain conditions

## 4.2 ANOVA

Table 3 shows the ANOVA findings from the data analysis. The "Between People" total of squares is 3821.889, with 253 degrees of freedom (df) and a mean square (MS) of 15.106. The "Within People" variance is divided into "Between Items" with a sum of squares (SOS) of 66.176, 29 df, and an MS of 2.282, providing an F-value of 3.095 at a significance level less than 0.001, suggesting substantial variations between items. The "Residual Non-additivity" reveals a total of squares of 5.762 with 1 df and an MS of 5.762, yielding an F-value of 7.824, which is significant at 0.005. The "Balance" and "Total" sections elaborate on the variance distribution, with the total SOS values of 5403.262 and 5475.200, respectively, adding context to the analysis. Overall, significant F-values and p-values imply meaningful differences, and a robust model fit to the data.

## 4.3 Hotelling's T-Squared Test

Table 4 presents the findings of Hotelling's T-squared test. The test statistic is 62.450, with an F-value of 1.915, df 29, 225, and a p-value of 0.005. The low p-value (0.005) suggests that the test result is statistically significant, implying that there are substantial differences between the means of the groups under consideration. The multivariate differences are assessed using Hotelling's T-squared test, and the significant result indicates that the groups differ in terms of the variables examined.

**Table 4** Hotelling's T-squared test

| Hotelling's T-squared | F | df1 | df2 | Sig |
|---|---|---|---|---|
| 62.450 | 1.915 | 29 | 225 | 0.005 |

### 4.4 T-Test

The T-test results in Table 5 show significant differences across all variables tested, with p-values less than 0.001 for each. This implies that the mean values of these variables differ significantly from the test value of zero, emphasizing their importance in the context of the study. Age has a mean difference (MD) of 2.740 with a t-value of 42.052, indicating a significant departure from the test value and an age-related influence in the sample. Similarly, Gender has a significant MD of 1.516 and a t-value of 48.243, indicating the importance of gender in the analysis. Education and income show higher MDs of 3.445 and 3.425, respectively, with t-values of 75.800 and 68.627. These data indicate that educational attainment and income levels are particularly important determinants. Geographical location also has a significant MD of 1.673 with a t-value of 38.421, indicating that geography has an important effect on the outcomes analyzed. For more specific factors such as ADAGMG, CTGM, AIGP, CEK, PBT, and CSTA, the MDs range from 3.66378 to 3.81496, with corresponding t-values ranging from 63.624 to 101.767. These high values indicate significant correlations and consistent importance across these variables, implying that these aspects are extremely important in the context of the study. Overall, the T-test results show robust significance across all variables, underscoring their relevance in the investigation.

**Table 5** T-test

| | Test value = 0 | | | | | |
|---|---|---|---|---|---|---|
| | t | df | Sig. (2-tailed) | Mean Difference | 95% Confidence interval of the difference | |
| | | | | | Lower | Upper |
| Age | 42.052 | 253 | < 0.001 | 2.740 | 2.61 | 2.87 |
| Gender | 48.243 | 253 | < 0.001 | 1.516 | 1.45 | 1.58 |
| Education | 75.800 | 253 | < 0.001 | 3.445 | 3.36 | 3.53 |
| Income | 68.627 | 253 | < 0.001 | 3.425 | 3.33 | 3.52 |
| Location | 38.421 | 253 | < 0.001 | 1.673 | 1.59 | 1.76 |
| ADAGMG | 65.442 | 253 | < 0.001 | 3.81496 | 3.7002 | 3.9298 |
| CTGM | 65.483 | 253 | < 0.001 | 3.76929 | 3.6559 | 3.8827 |
| AIGP | 101.767 | 253 | < 0.001 | 3.78819 | 3.7149 | 3.8615 |
| CEK | 63.624 | 253 | < 0.001 | 3.75512 | 3.6389 | 3.8714 |
| PBT | 68.631 | 253 | < 0.001 | 3.73622 | 3.6290 | 3.8434 |
| CSTA | 73.195 | 253 | < 0.001 | 3.66378 | 3.5652 | 3.7624 |

**Table 6** Correlation analysis

| | ADAGMG | CTGM | AIGP | CEK | PBT | CSTA |
|---|---|---|---|---|---|---|
| ADAGMG | 1.000 | 0.842 | 0.198 | 0.857 | 0.849 | 0.746 |
| CTGM | 0.842 | 1.000 | 0.229 | 0.809 | 0.826 | 0.766 |
| AIGP | 0.198 | 0.229 | 1.000 | 0.180 | 0.184 | 0.190 |
| CEK | 0.857 | 0.809 | 0.180 | 1.000 | 0.806 | 0.750 |
| PBT | 0.849 | 0.826 | 0.184 | 0.806 | 1.000 | 0.778 |
| CSTA | 0.746 | 0.766 | 0.190 | 0.750 | 0.778 | 1.000 |

## 4.5 Correlation Analysis

Table 6 shows the correlations between the variables ADAGMG, CTGM, AIGP, CEK, PBT, and CSTA. ADAGMG has a substantial positive connection with CTGM ($r = 0.842$) and CEK ($r = 0.857$), which suggests that when ADAGMG scores rise, so do scores on these variables. It also shows a good association with PBT ($r = 0.849$) and a modest correlation with CSTA ($r = 0.746$). CTGM has a strong correlation with ADAGMG ($r = 0.842$) and CEK ($r = 0.809$), indicating that greater CTGM scores are connected with higher scores in these variables. It also has a significant connection with PBT ($r = 0.826$) and CSTA ($r = 0.766$). AIGP shows poor correlations with all other variables, with r values ranging from 0.180 to 0.229, indicating minor associations. CEK had substantial associations with ADAGMG, CTGM, and PBT ($r = 0.857$, 0.809, and 0.806, respectively), as well as a moderate correlation with CSTA ($r = 0.750$). PBT has substantial associations with ADAGMG ($r = 0.849$), CTGM ($r = 0.826$), and moderate correlations with CSTA ($r = 0.778$). CSTA correlates strongly with ADAGMG ($r = 0.746$), CTGM ($r = 0.766$), and PBT ($r = 0.778$), and somewhat with CEK ($r = 0.750$). Overall, the data suggest strong interrelationships between ADAGMG, CTGM, CEK, PBT, and CSTA, whereas AIGP has fewer associations with these variables.

## 4.6 Correlation Coefficient

Table 7 shows the Kendall's tau_b correlation matrix. It demonstrates considerable positive connections between variables. ADAGMG has a good correlation with CEK ($\tau = 0.571$) and CTGM ($\tau = 0.537$), suggesting a strong relationship between these parameters. CEK displays a substantial positive correlation with PBT ($\tau = 0.503$), indicating a relationship between the two variables. AIGP had significant but lesser correlations with the other components, including CTGM ($\tau = 0.156$) and CSTA ($\tau = 0.150$). These correlations imply that while all the variables are positively associated, the strength of these interactions varies, with some elements being more strongly linked than others, indicating potential shared underlying influences.

**Table 7** Kendall's tau_b

| | ADAGMG | CTGM | AIGP | CEK | PBT | CSTA |
|---|---|---|---|---|---|---|
| ADAGMG | 1.000 | $0.537^{**}$ | $0.137^{**}$ | $0.571^{**}$ | $0.476^{**}$ | $0.395^{**}$ |
| CTGM | $0.537^{**}$ | 1.000 | $0.156^{**}$ | $0.481^{**}$ | $0.473^{**}$ | $0.440^{**}$ |
| AIGP | $0.137^{**}$ | $0.156^{**}$ | 1.000 | $0.141^{**}$ | $0.131^{**}$ | $0.150^{**}$ |
| CEK | $0.571^{**}$ | $0.481^{**}$ | $0.141^{**}$ | 1.000 | $0.503^{**}$ | $0.426^{**}$ |
| PBT | $0.476^{**}$ | $0.473^{**}$ | $0.131^{**}$ | $0.503^{**}$ | 1.000 | $0.448^{**}$ |
| CSTA | $0.395^{**}$ | $0.440^{**}$ | $0.150^{**}$ | $0.426^{**}$ | $0.448^{**}$ | 1.000 |

Asterisk mark (*) in Kendaal's test, typically indicates the statistical significance of the correlation coefficient

**Table 8** Spearman's rho

| ADAGMG | CTGM | AIGP | CEK | PBT | CSTA | ADAGMG |
|---|---|---|---|---|---|---|
| ADAGMG | 1.000 | $0.677^{**}$ | $0.152^{*}$ | $0.711^{**}$ | $0.605^{**}$ | $0.524^{**}$ |
| CTGM | $0.677^{**}$ | 1.000 | $0.193^{**}$ | $0.626^{**}$ | $0.616^{**}$ | $0.584^{**}$ |
| AIGP | $0.152^{*}$ | $0.193^{**}$ | 1.000 | $0.181^{**}$ | $0.159^{*}$ | $0.188^{**}$ |
| CEK | $0.711^{**}$ | $0.626^{**}$ | $0.181^{**}$ | 1.000 | $0.636^{**}$ | $0.564^{**}$ |
| PBT | $0.605^{**}$ | $0.616^{**}$ | $0.159^{*}$ | $0.636^{**}$ | 1.000 | $0.577^{**}$ |
| CSTA | $0.524^{**}$ | $0.584^{**}$ | $0.188^{**}$ | $0.564^{**}$ | $0.577^{**}$ | 1.000 |

Asterisk mark (*) in Sperman test, typically indicates the statistical significance of the correlation coefficient

Table 8 shows the Spearman's rho correlation matrix. It shows strong, positive correlations between most variables. ADAGMG has a significant correlation with CEK ($\rho = 0.711$) and CTGM ($\rho = 0.677$), indicating a strong link among these parameters. CEK has a substantial connection with PBT ($\rho = 0.636$) and CSTA ($\rho = 0.564$), indicating their interconnectivity. CTGM has significant associations with PBT ($\rho = 0.616$) and CSTA ($\rho = 0.584$), suggesting a coherent link between these measures. While still significant, AIGP had lesser associations with the other components, especially with ADAGMG ($\rho = 0.152$) and CEK ($\rho = 0.181$). Overall, the matrix highlights positive linkages, with certain aspects being more tightly associated than others, indicating potential underlying connections.

## *4.7 Regression Analysis*

The regression analysis in Table 9 shows the strength of the correlations between the independent and DVs, as evaluated by the R, $R^2$, and F values. CEK (R-value = 0.857) has a very good association with the DV. The $R^2$ value of.735 shows that CEK can explain 73.5% of the DV's variance. The F value of 698.325 and a significance

**Table 9** Regression analysis

| | R | $R^2$ | Sum of squares | df | Mean square | F | Sig |
|---|---|---|---|---|---|---|---|
| CEK | 0.857[a] | 0.735 | 160.474 | 1 | 160.474 | 698.325 | < 0.001 |
| PBT | 0.826[a] | 0.682 | 145.289 | 1 | 145.289 | 541.359 | < 0.001 |
| CSTA | 0.190[a] | 0.036 | 3.226 | 1 | 3.226 | 9.474 | < 0.001 |

The values which have been super scripted with the letter "a" means, they are statistically significantly different under certain conditions

level of < 0.001 indicate a statistically significant link. PBT also has a substantial connection with the DV, as indicated by an R-value of.826. The $R^2$ value of.682 indicates that PBT accounts for 68.2% of the variation in the DV. The high F value of 541.359, with a significance level of < 0.001, emphasizes the importance of this link. In comparison, CSTA has a much lower correlation, with an R-value of.190 and a $R^2$ value of.036, indicating that it only explains 3.6% of the variation in the DV. The F value of 9.474 is significant at < 0.001, indicating a weaker association than CEK and PBT. Overall, the regression analysis demonstrates the strong influence of CEK and PBT, while CSTA has very little impact on the DV.

## 4.8 Coefficients

The coefficients Table 10 shows how each independent variable affects the dependent variable, as well as the statistical significance of these effects. In Model 1, the constant (intercept) is 0.636, which represents the predicted value of the dependent variable when CEK is zero. The unstandardized coefficient for CEK is 0.847, which means that for every one-unit increase in CEK, the dependent variable is anticipated to rise by 0.847 units. The standardized Beta value of 0.857 indicates that CEK has a considerable positive effect on the dependent variable. This effect is statistically significant, as confirmed by the t-value of 26.426 and p-value of < 0.001. In Model 2, the constant is 0.506, and the unstandardized coefficient for PBT is 0.873, implying that a one-unit increase in PBT results in an increase of 0.873 units in the dependent variable. The standardized Beta value of 0.826 indicates that PBT has a considerable positive effect. The t-value of 23.267 and significance level of < 0.001 confirm a statistically significant connection. Model 3 has a higher constant, 3.270, which reflects the dependent variable's baseline. The unstandardized coefficient for CSTA is 0.142, which means that a one-unit increase in CSTA causes a 0.142-unit rise in the dependent variable. The standardized Beta coefficient is 0.190, indicating a lesser effect than CEK and PBT. Despite this, the t-value of 3.078 and significance level of 0.002 shows that the connection remains statistically significant. Overall, CEK and PBT exhibit strong and significant positive impacts on the dependent variable, whereas CSTA's effect is weaker but still significant.

**Table 10** Coefficients

| Model | | Unstandardized coefficients | | Standardized coefficients | t | Sig |
|---|---|---|---|---|---|---|
| | | B | Std. error | Beta | | |
| 1 | (Constant) | 0.636 | 0.124 | | 5.125 | < 0.001 |
| | CEK | 0.847 | 0.032 | 0.857 | 26.426 | < 0.001 |
| 2 | (Constant) | 0.506 | 0.144 | | 3.514 | 0.001 |
| | PBT | 0.873 | 0.038 | 0.826 | 23.267 | < 0.001 |
| 3 | (Constant) | 3.270 | 0.172 | | 18.962 | < 0.001 |
| | CSTA | 0.142 | 0.046 | 0.190 | 3.078 | 0.002 |

**Table 11** Factor analysis

| | | |
|---|---|---|
| Kaiser–Meyer–Olkin measure of sampling adequacy | | 0.908 |
| Bartlett's test of sphericity | Approx. Chi-square | 1327.329 |
| | df | 15 |
| | Sig | < 0.001 |

## 4.9 *Factor Analysis*

Table 11 shows the factor analysis results. It shows a high level of sample adequacy, as shown by a Kaiser–Meyer–Olkin (KMO) value of 0.908, which is significantly higher than the permissible threshold of 0.6. This shows that the data are appropriate for factor analysis. Furthermore, Bartlett's Test of Sphericity yields a significant chi-square value of 1327.329 with 15 df and a p-value less than 0.001, showing that the correlations between items are substantial enough to support factor analysis. These findings confirm that the factor structure is suitable and that the variables are likely to have common components.

# 5 Discussion

The correlation analysis shows a strong positive relationship of CEK with other variables like ADAGMG and PBT, which indicates that a high level of environmental knowledge can be related to the better differentiation between green marketing and greenwashing. Regression analysis supports this further with a strong R-value of 0.857 and an $R^2$ of 0.735 for CEK, which is highly significant in predictive relationships. The coefficients reveal a high unstandardized coefficient, $B = 0.847$, and an equally high standardized Beta of 0.857. The t-value is highly significant at 26.426 with a p-value of 0.000. Therefore, the alternative hypothesis, H1, is accepted and demonstrates that environmental knowledge positively influences the consumers' ability to differentiate between authentic green marketing and greenwashing.

The regression analysis shows a very strong positive relation of PBT with the consumer's trust, having an R-value of 0.826 and $R^2$ of 0.682. The coefficients show an unstandardized coefficient, $B = 0.873$, and a standardized Beta of 0.826, both significant with a t-value of 23.267 ($p = 0.000$). These results support the fact that increased PBT enhances CTGM efforts. Consequently, the alternative hypothesis of the study is accepted as greater brand transparency exerts a positive effect on consumer trust in the authentic green marketing of the firm.

Regression analysis shows a weak positive association between consumer skepticism and the identification of greenwashing, with an R-value of 0.190 and an $R^2$ of 0.036. Coefficients showed a small unstandardized coefficient of $B = 0.142$, a standardized Beta of 0.190, with a significant t-value of 3.078, $p = 0.002$. The relationship is statistically significant; however, it is relatively weaker than other variables. As a result, the study's alternative hypothesis is accepted, as consumers who are more skeptical of advertising are not more likely to identify greenwashing methods.

## 6 Conclusion

Eventually, this research study contributes to the development of a profound understanding of how customers differentiate between real green marketing and misleading strategies for greenwashing. The analysis shows that consumer awareness and knowledge about green marketing are two critical determinants for recognizing greenwashing strategies. The more significant the environmental knowledge, the more capable the consumers in making a difference between factual and misleading claims of green. Greenwashing affects the trust of consumers to such an extent that once misled, they distrust their purchasing decision for any green product. The trust of the consumers can only be built and fostered by businesses if the communication of actual green efforts is perceived. This would diminish the risks associated with greenwashing to a minimum by offering a chance for businesses to market their products transparently and honestly along with scientifically verified environmental claims. This will not only add to brand credibility but also ensure consumer loyalty in the long run. Addressing these aspects, therefore, can help organizations make sure that their green marketing strategies are moulded to match genuine environmental concerns, hence steering away from greenwashing and contributing towards a more ecologically viable marketplace. The study emphasizes that continued vigilance and improvement are necessary for green marketing practices that meet consumer expectations of this industry and promote genuine environmental sustainability.

## 7 Implications

The study has once again put forth larger ramifications in the field of green marketing, as relates to businesses and consumers. For businesses, it means authenticity counts in environmental claims. Companies should ensure that their green marketing is aligned with sustainable practices, lest the pitfalls of greenwashing raise their head and consumer trust is eroded. Precisely, this is where clear communications with proper substantiation of claims go a long way in keeping up credibility and boosting brand loyalty. It therefore signals to consumers a vigilant approach in making choices, considering eco-friendly products. By developing appropriate knowledge of green marketing practices or possible greenwashing practices, consumers are well-informed in making choices and can support truly sustainable brands. Overall, this study calls for a collective approach by all stakeholders about genuine environmental responsibility and a reduction of deceptive practices in return for a better and more trustworthy market environment.

## References


1. Abbas, Q.: Sustainable marketing in the age of environmental consciousness. JETIR **11**, 317–326 (2024)
2. Aparna, P., Murugan, K. S.: Green washing: a deceptive marketing practice and its implications. In: 3rd International Conference on Reinventing Business Practices, Start-ups and Sustainability (ICRBSS 2023). Atlantis Press, pp. 122–132, February 2024
3. Bhaduri, G., Copeland, L.: Going green? How skepticism and information transparency influence consumers' brand evaluations for familiar and unfamiliar brands. J. Fash. Mark. Manag.: Int. J. **25**(1), 80–98 (2021)
4. Bosch, M., Obeso, E., Palao, A.: Greenwashing in the fast fashion industry: greenwashing in the fast fashion industry: the role of consumer education (2023)
5. Butt, A., Khan, S.N., Zakir, M.M., Rana, W., Laila, A.: Firms greenwashing practices and consumers' perception: role of marketing and non-marketing external stakeholders in firms greenwashing practices. Indian J. Econ. Bus. **20**(3), 689–706 (2021)
6. Deng, Y.Y., Yang, Y.C.: Exploring the role of green attributes transparency influencing green customer citizenship behavior. Br. Food J. **124**(5), 1473–1484 (2022)
7. Esha, A. S.: Ethical perception and consumer behavior regarding green label and greenwashing marketing (2023)
8. Fella, S., Bausa, E.: Green or greenwashed? Examining consumers' ability to identify greenwashing. J. Environ. Psychol. **95**, 102281 (2024)
9. Guerreiro, J., Pacheco, M.: How green trust, consumer brand engagement and green word-of-mouth mediate purchasing intentions. Sustainability **13**(14), 7877 (2021)
10. Jamal, F. N., Othman, N. A., Nizam, N. Z., Jelita, A., Rohmah, W., Dzakiyullah, N. R.: Green marketing: reviewing aspect of communication tools. Int. J. Sustain. Dev. Plan. **17**(4) (2022)
11. Kaur, B., Gangwar, V.P., Dash, G.: Green marketing strategies, environmental attitude, and green buying intention: a multi-group analysis in an emerging economy context. Sustainability **14**(10), 6107 (2022)
12. Kwak, H.S., Park, J.A., Lee, H.H.: How do consumers' perceptions of brands change?-Investigating a fashion brand's green marketing, authenticity, and purchase intention in the context of greenwashing. Res. J. Costume Cult. **30**(2), 189–207 (2022)

13. Lopes, J.M., Gomes, S., Trancoso, T.: The dark side of green marketing: how greenwashing affects circular consumption? Sustainability **15**(15), 11649 (2023)
14. Mariani, M.M., Perez-Vega, R., Wirtz, J.: AI in marketing, consumer research and psychology: a systematic literature review and research agenda. Psychol. Mark. **39**(4), 755–776 (2022)
15. Nath, P., Siepong, A.: Green marketing capability: a configuration approach towards sustainable development. J. Clean. Prod. **354**, 131727 (2022)
16. Peterson, V.J.: Gray areas in green claims: why greenwashing regulation needs an overhaul. Vill. Env't LJ **35**, 177 (2024)
17. Qayyum, A., Jamil, R.A., Sehar, A.: Impact of green marketing, greenwashing and green confusion on green brand equity. Span. J. Mark.-ESIC **27**(3), 286–305 (2023)
18. Rahman, S.U., Nguyen-Viet, B.: Towards sustainable development: coupling green marketing strategies and consumer perceptions in addressing greenwashing. Bus. Strateg. Environ. **32**(4), 2420–2433 (2023)
19. Shanor, A., Light, S.E.: Greenwashing and the first amendment. Columbia Law Rev. **122**(7), 2033–2118 (2022)
20. Sharma, A.P.: Consumers' purchase behaviour and green marketing: a synthesis, review and agenda. Int. J. Consum. Stud. **45**(6), 1217–1238 (2021)
21. Szabo, S., Webster, J.: Perceived greenwashing: the effects of green marketing on environmental and product perceptions. J. Bus. Ethics **171**, 719–739 (2021)
22. Tan, Z., Sadiq, B., Bashir, T., Mahmood, H., Rasool, Y.: Investigating the impact of green marketing components on purchase intention: the mediating role of brand image and brand trust. Sustainability **14**(10), 5939 (2022)
23. Wiederstein, C.: From greenwashing to green authenticity: overcoming greenwashing strategies in sustainable fashion marketing on Instagram. Master's thesis, University of Twente (2024)

# Gender-Inclusive Circular Economy Policies: Assessing the Effectiveness of Current Policies in Supporting Women Scrap Collectors in Karnataka Specifically Bangalore

**Madhu Druva Kumar, Cherian Thomas, M. Lokanadha Reddy, K. Lavanya, S. Varshini, and N. T. Jeevitha**

**Abstract** The focus areas of the circular economy include waste minimization and optimum utilization of available resources by way of more recycling, reuse, and sustainability. The informal sector in waste management is playing a very important role in this transition in India, wherein the women waste collectors comprise a significant part of the workforce. This paper examines the efficiency of inclusive circular economy policy in enabling and supporting the women waste collectors in Bangalore, Karnataka. Women waste collectors in the informal waste management sector face problems like resource gaps, unequal pay, and hazardous working conditions despite playing a critical role in this industry. This study will assess the effectiveness of existing policies in addressing the concerns and economically empowering women within this sector. It works on finding the gap and opportunities of policy implementation through secondary data analysis supported by primary fieldwork with the aim of making the transition to a circular economy equitable and inclusive for the concerned stakeholders.

**Keywords** Gender-inclusive policies · Circular economy · Women scrap collectors · Economic empowerment · Informal waste sector · Policy implementation

M. D. Kumar · C. Thomas (✉) · S. Varshini · N. T. Jeevitha
Department of Commerce PG, Kristu Jayanti College (Autonomous), Bengaluru, Karnataka, India
e-mail: cherian28@gmail.com

M. Lokanadha Reddy
REVA Business School, REVA University, Bengaluru, Karnataka, India

K. Lavanya
C B Bhandari Jain College, Bengaluru, Karnataka, India



M. Al Mubarak (ed.), *Sustainable Digital Technology and Ethics in an Ever-Changing Environment*, Studies in Systems, Decision and Control 237,
https://doi.org/10.1007/978-3-031-86708-8_13

# 1 Introduction

A circular economy is an economic model that is beneficial to businesses, society, and the environment by keeping resources in use for as long as possible. Unlike the more common linear economy where one produces, consumes, and disposes of goods in a linear fashion, a circular economy intends to eliminate waste by turning waste into resources, maintaining the ownership of products and materials, and enhancing ecosystems at the same time. It implies that at the end of the lifecycle of the product, there is a planned culling of the product and its components done according to a multiple use, restoration, and regeneration system. The need for this evolution is also motivated by the fact that a circular economy presents diverse prospects for sustainable growth potential due to the low waste produced and input reuse. Nevertheless, policies that promote the strategies and tactics to deal with the interests of all the stakeholders in such projects such as women scrap collectors are very important to the success of the transition. Women add to the informal waste management issue in the state of Karnataka in Bangalore. But women are restricted due to several factors including resources organization, pay ratios, and working environment related risks. So, there is a need to look at the existing circular economy policies on the issue of gender equality for women scrap collectors in Bangalore and appraise the effectiveness of the policies on women scrap collectors and figure out what recognizing areas that require more attention in order to facilitate balanced and sustainable development for everyone. This analysis will seek to understand the economic consequences and challenges that women scrap collectors in Bangalore aggravate, determine whether they'll be supported by more gender-inclusive policies of circular economy approaches, and evaluate how these policies are being practiced in reality to benefit those women.

**Structure of the paper:** Introduction, Literature review, Research Methodology, Results and Discussion, Conclusion.

# 2 Review of Literature

In a paper published by Estrada et al. [1] analyses how grassroot activism, NGO-municipality collaboration and women's empowerment led to realizing transformative potential for both solid waste management and social inclusion in the Indian city of Pune. The authors have analyzed the case of SWaCH Pune Seva Sahakari and Sanstha a predominantly-female social cooperative of waste pickers to show how supporting and enhancing informal labor can produce multiple positive outcomes. The study also further examines the role of women waste pickers in bringing about social inclusive changes.

In an article penned by Joshi [2], the author highlights the importance of integrating gender equality into the circular economy to ensure a sustainable and inclusive transition. The article underlines that women are often involved in lower-value, informal sectors like cycling and waste management and their contribution and knowledge in sustainable practices are ignored. Addressing these gender gaps is crucial for achieving a "just transition" and the circular economy can better align with social equity and environmental goals is the main contribution of this article.

Buch et al. [3] has told that the inclusive circular economy framework has aimed to empower waste pickers by building collaborative networks of stakeholders so that it can enable inclusion of waste pickers, then the second point to be noted was cooperative enterprise models can be established to integrate waste pickers into the formal economy also this paper has highlighted to enable waste pickers to manufacture the upcycled products access to technologies can be provided.

In an article written by Prakash et al. [4], the authors point out that despite the growing global attention to Circular Economy, such as the G20 Resource Efficiency Dialogue and the world Economic Forum's circularity gap report, social aspects, particularly gender inclusion, have been overlooked. The article further continues to highlight the undervalued contribution of women in traditional sustainable practices such as waste segregation and recycling of materials. The authors also feature the initiatives such as "Hasirudala" in Bengaluru, "Swach" in Pune, and "SEWA" in Gujarat and their potential to empower women in waste management through formalization and leadership training. The article concludes by enhancing the essence of unlocking circular economy's transformative potential by addressing gender hierarchies and recognizing women as key agents in sustainability efforts.

Bebasari [5] offers a thorough overview of the ways in which the circular economy approaches can contribute to women empowerment. The author contend that women are benefited with upcycling training as part of their engagement with the local governments in tackling waste issues. The author has pointed that the circular economy initiatives have given a space for liberal empowerment, but also mentions that additional skills for women to challenge the roots of inequality, the author has analyzed and linked that economic empowerment to a broader gender advocacy with the government and it has created an opportunity for women to participate in policy making. And these initiatives can enhance women empowerment by engaging with the local governments in future.

## 3 Research Methodology

Primary data toward fulfilling the first objective was collected by conducting fieldwork, in surveying and interviewing women scrap collectors in Bangalore. The reason behind this is to observe directly how changes in the broader economy-for example, demand for recyclable materials-affect their economic outcomes and working conditions. This will also enable a comparison of experiences by women and their male counterparts to find out if there are any gender-specific challenges. The primary data will then be analyzed using statistical methods, such as ANOVA, to deeply review the effectiveness and inclusiveness of the current policies in supporting women scrap collectors amid changing economic conditions.

The secondary data review was done for the second objective, which has to do with ascertaining how effective the current policies on circular economies are in Bangalore, and the effect these have on women waste collectors. Government reports, policy documents, academic research, and NGO publications will be reviewed in light of finding out existing gaps in policy enforcement and how well they address the needs of the women collecting scrap. The present analysis therefore focuses more on the inclusiveness of such policies and to what extent they have genuinely benefited women in terms of income, working conditions, and resource availability.

The study tested the following hypotheses:

Null Hypothesis: Women scrap collectors in Bangalore do not significantly benefit from the extensive implementation of circular economy policies, in terms of increased pay and improved working conditions.

Alternate Hypothesis: Women scrap collectors in Bangalore significantly benefit from the extensive implementation of circular economy policies, resulting in increased pay and improved working conditions.

## 4 Results and Discussion

The model summary mentioned in Table 1, reveals that the predictors—Demand for recycle products, Access to resources, Gender, Years of experience, and Physical Safety—explain 33.8% of the variability in the "Change of Income," as indicated by the $R^2$ value. However, after adjusting for the number of predictors and sample size, the Adjusted $R^2$ drops to 27.7%, suggesting that the model's explanatory power is somewhat diminished when considering the complexity introduced by the number of predictors. This indicates that while the model captures a moderate portion of the income change variability, there is still a substantial amount of variability unexplained.

The ANOVA table referred below (Table 2) indicates that the regression model, with 5 degrees of freedom for the predictors, is compared against 54 degrees of freedom for the residuals, totaling 59 degrees of freedom. This setup allows for assessing the overall significance of the model in explaining the variation in the "Change of Income." The degrees of freedom show that the model's predictors are

**Table 1** The model summary table

| Model summary[b] | | | | |
|---|---|---|---|---|
| Model | R | $R^2$ | Adjusted $R^2$ | Std. error of the estimate |
| 1 | 0.582[a] | 0.338 | 0.277 | 0.549 |

[a] Predictors: (Constant), Demand for recycle products, Access to resources, Gender, Years of experience, Physical Safety
[b] Dependent Variable: Change of Income
*Source* Author's Compilation

evaluated in relation to the residual variance, which helps determine if the included predictors collectively provide a statistically significant explanation for the changes in income. Here, the significance value is 0.000 which is less than 0.05 significant value, hence we reject the null hypothesis and accept the alternative hypothesis.

**$\beta$Gender $\neq 0$ or $\beta$Experience $\neq 0$ or $\beta$Resources $\neq 0$ or $\beta$Safety $\neq 0$ or $\beta$Demand $\neq 0$**

Fitted Line of a Regression Model:

$$Y = \beta 0 + \beta 1(X1) + \beta 2(X2) + \beta 3(X3) + \beta 4(X4)$$

$$\text{Change in Income} = 1.342 + 0.063\,(\text{Gender}) + 0.044(\text{Years of Experience}) + 0.196\,(\text{Access to resources}) - 0.023(\text{Physical Safety})$$

Table 3 of coefficients reveals that among the predictors, only "Demand for Recycle Products" significantly influences the "Change of Income," with a coefficient of 0.442 and a p-value of 0.000, indicating a strong positive impact. In contrast, "Gender," "Years of Experience," "Access to Resources," and "Physical Safety" do not significantly affect income changes, as their p-values are above the common significance threshold of 0.05. The collinearity statistics (Tolerance and VIF) are within acceptable ranges, suggesting that multicollinearity is not a concern in this model. Therefore, while "Demand for Recycle Products" is a key driver of income changes, the other predictors do not contribute significantly to explaining variations in income.

For the second objective, the following Government implemented policies have been analyzed and the effectiveness of the schemes are being assessed.

(a) **Swachh Bharat Mission (Urban)**

The Swachh Bharat Mission primarily focuses on creating solid waste infrastructure and sanitation services to reorganize the activities of waste pickers and improve their work life. On paper, the filling in of scrap collectors into the system should

**Table 2** The ANOVA table

ANOVA[a]

| Model | | Sum of squares | df | Mean square | F | Sig. |
|---|---|---|---|---|---|---|
| 1 | Regression | 8.319 | 5 | 1.664 | 5.525 | 0.000[b] |
| | Residual | 16.264 | 54 | 0.301 | | |
| | Total | 24.583 | 59 | | | |

[a] Dependent Variable: Change of Income
[b] Predictors: (Constant), Demand for recycle products, Access to resources, Gender, Years of experience, Physical Safety
*Source* Author's Compilation

**Table 3** The co-efficient table

| Coefficients[a] | | | | | | |
|---|---|---|---|---|---|---|
| Model | | Unstandardized coefficients | | Sig. | Collinearity statistics | |
| | | B | Std. error | | Tolerance | VIF |
| 1 | (Constant) | 1.342 | 0.518 | 0.012 | | |
| | Gender | 0.063 | 0.158 | 0.694 | 0.901 | 1.110 |
| | Years of experience | 0.044 | 0.072 | 0.549 | 0.940 | 1.063 |
| | Access to resources | 0.196 | 0.144 | 0.180 | 0.970 | 1.031 |
| | Physical safety | – 0.023 | 0.100 | 0.817 | 0.875 | 1.143 |
| | Demand for recycle products | 0.442 | 0.088 | 0.000 | 0.918 | 1.090 |

[a] Dependent Variable: Change of Income
*Source* Author's compilation

strengthen their economic position (by boosting job security and offering insurance along several lines such as health today; education for children in future and pensioning—theoretically). But the thing is that this formalization process has not been implemented evenly across the board. Only a small minority of such women scrap collectors have been issued the essential id-card needed to avail these benefits. It has made small gains around working conditions, although these are hardly where you would want the industry to be. Over 800 million women work in hazardous work environments underpaid, unprotected from harassment or discrimination. This means that the policy framework is conceptually correct, but its implementation does not have level of consistency and penetration required to produce significant gains [6].

(b) **Solid Waste Management Rules, 2016**

The SWM Rules 2016 aim at enhancement in segregation, collection, and processing of wastes, with recognition to the role of the informal sector of the waste workers, specifically scrap collectors. While such rules attempt to integrate the informal waste workers into the mainstream waste management system, this varies in implementation. Most such women scrap collectors continue to operate outside this ambit of formalization, and their work is precarious, significantly unrecognized, or supported. One main challenge is the lack of awareness among these women regarding their rights under these regulations. This knowledge gap seriously impairs their access to government schemes, subsidies, or training programs that could provide them with enhanced economic security. Although all this is there, the real benefits rarely trickle down to the intended population because little information has flowed and less outreach has been done.

(c) **Solid Waste Management Round Table (SWMRT)**

The Solid Waste Management Round Table, most women scrap collectors do not know their rights and thus have limited access to government initiatives and training programs which empower them to improve their livelihoods. This represents a major problem of lack of policy awareness that inherently reduces the impact of these policies. Because they do not understand the regulations, women scrap collectors are exposed to risk and often work under exploitative and hazardous conditions [7].

(d) **Policy Impact on Working Conditions**

NGOs like Hasiru Dala, working directly with the waste workers of Bangalore, report mixed results regarding the effects of policies on female scrap collectors. While Hasiru Dala, in fact, has helped several women to gain formal recognition and access to benefits, it has also been active, advocating and supporting these workers. The identification cards it has facilitated, the access to health insurance, and educational programs for scrap collectors' children. None of these benefits are universally provided, and many women have been struggling to obtain these rights. Apart from organizations like Hasiru Dala, women scrap collectors face all sorts of harassment, gross lack of safety measures, and low wages compared to men. All these issues show the failure of policy implementation and its proper enforcement. Some of the other pressing demands are to strengthen outreach programs targeting women waste pickers. Educating them on their rights and entitled benefits is planned. Collaboration with NGOs like Hasiru Dala, community leaders and local governments is very necessary for ensuring than. This ensures that the information reaches whom it is intended for. Partnership building with organizations like Hasiru Dala can help in enhancing the support given to women scrap collectors, as such organizations bridge the gap between policy and practice by helping women access entitled benefits. Authorities should then ensure that there are consistent policy implementations across all regions. Regular monitoring and evaluation are highly necessary to ensure that the women scrap collectors are accorded the due protections and benefits.

## 5 Conclusion

In order to boost earnings in the recycling industry and provide assistance to scrap collectors various strategies must be implemented. The main objective should be to increase the demand, for recycled goods as it's the factor affecting income fluctuations. This goal can be accomplished by launching awareness campaigns targeted at the public offering incentives to businesses that utilize recycled materials and backing innovations, in recycling technology. At the time it is important to enhance initiatives that reach out to women who collect scrap in order to educate them about their rights and the support available, to them effectively; this can be done through collaborations with organizations such as Hasiru Dala and local community leaders and government bodies to ensure that information is disseminated efficiently and inclusively across different areas. Furthermore, it is crucial for authorities to enforce policies, throughout regions and monitor them regularly to make sure that women are receiving the benefits and protections they deserve. Moreover, it is crucial to redirect resources, from sectors towards projects that have a direct impact on revenue generation income disparities should be investigated further with additional studies required for a deeper understanding of the fluctuating incomes by implementing these tactics such demand stimulation efficient allocation of resources and increased assistance, for female collectors the industry can attain a more stable and fair financial growth.

## References

1. Estrada, M., Galvin, M., Maassen, A., Hörschelmann, K.: Catalysing urban transformation through women's empowerment in cooperative waste management: the SWaCH initiative in Pune India. Local Environ. **28**(7), 852–866 (2023)
2. Joshi, A.: Towards a gender-inclusive circular economy. Retrieved 13 Sep 2024, from CIL website. https://www.circularinnovationlab.com/post/towards-a-gender-inclusive-circular-economy (2022)
3. Buch, R., Marseille, A., Williams, M., Aggarwal, R., Sharma, A.: From waste pickers to producers: an inclusive circular economy solution through development of cooperatives in waste management. Sustainability **13**(16), 8925 (2021)
4. Prakash, R., Paterok, K., Paterok, R.P.A.D.: Women & circular economy: mainstreaming gender equity for sustainability. Retrieved 13 Sep 2024, from Feminism in India website. https://feminisminindia.com/2021/09/09/women-circular-economy-mainstreaming-gender-equity-for-sustainability (2021)
5. Bebasari, P.: The role of women in upcycling initiatives in Jakarta, Indonesia: a case for the circular economy in a developing country. In: The Circular Economy and the Global Southm, pp. 75–92. Routledge (2019)
6. Prime Minister's Office, Swachh Bharat Abhiyan, Government of India website. https://www.pmindia.gov.in/en/major_initiatives/swachh-bharat-abhiyan/
7. Haryana State Pollution Control Board: Municipal Solid Waste Management Rules website. https://hspcb.org.in/uploads/laws/MSW_Rules.pdf (2016)

# AI, Business and Education

# The Ethics of ChatGPT in Academia

**Husain Mohamed Tarish and Muneer Al Mubarak**

**Abstract** Technological advancement has taken shape at a significant rate over the recent years, further cementing itself as a vital part of life, more than it already is. Artificial Intelligence is one technological feat that introduced a new era of a never-ending cycle of advancements with new features and updates emerging daily. ChatGPT which is a Large Language Model that generates answers, research papers, essays, and much more based on the trained data that was collected from the entirety of the public internet. It started as any other AI model tool but soon grew to become the most advanced AI chatbot open to public use. In line with the continuous advancements of ChatGPT, ethical concerns have grown proportionally due to its massive capabilities that users can exploit and commit ethical misconduct, especially in the academic scene. Such ethical concerns include but not limited to plagiarism, cheating, impact on critical thinking, and spreading misinformation. This research paper explores these concerns and the state of ChatGPT use which reflects on academia by analysing literature findings and presenting the results of a survey done by students to examine whether these ethical concerns are grounded by evidence. It was found that most students do in fact unethically use ChatGPT for different purposes. Thus, through the analysis and findings, it was concluded that ChatGPT poses a threat on academic integrity and would require the establishment of advanced AI detection tools and ethical teaching and awareness amongst students.

**Keywords** Ethics · ChatGPT · AI · Academia

H. M. Tarish
The George Washington University, Washington, DC, USA

M. Al Mubarak (✉)
Ahlia University, Manama, Bahrain
e-mail: malmubarak@ahlia.edu.bh



M. Al Mubarak (ed.), *Sustainable Digital Technology and Ethics in an Ever-Changing Environment*, Studies in Systems, Decision and Control 237,
https://doi.org/10.1007/978-3-031-86708-8_14

# 1 Introduction

Over the years through the twentieth century, technology have been in rapid development, especially in the military field. Countries such as the United States of America have revolutionized the uses of technology due to the global competition present at the time. Fast forward to the twenty-first century and technological feats have surpassed the need for humans to manually input commands to direct computers as artificial intelligence have evolved to become a tool to significantly assist humans in a vast of operations. Patterns and the structure of existing data on the internet have enabled generative AI to create new data based on said data [9]. Data which us humans can use for our intended purposes which can vary from ethical to malicious uses.

ChatGPT is a large language model which is trained to process data from the internet which involves all varieties of topics available online to which it could produce an answer with accurate, high-quality output [11]. The authors have described it as an educated, expert human who has knowledge on a large range of subjects. The trained ChatGPT therefore not only provides answers, but it imitates a conversational pattern which humanizes its responses to the user's question or inquiry [9]. Due to its effectiveness in providing reliable and accurate answers, it is evident that it is potentially replacing conventional search engines, especially with students [12]. Hence, search engine providers such as Google have soon followed suit in introducing a competitive product to ChatGPT (Cohen, [3], while Microsoft has utilized similar technologies to ChatGPT to upgrade its Bing search engine and Edge browser by integrating AI into it, granting it the accuracy and reliability in results that ChatGPT offers [9]. Consequently, it is understandable that such a tool would be highly utilized by students in higher education by conducting research and generating text, codes, or entire papers based on the existing data that ChatGPT was trained on. For example, students in science, technology, engineering, and mathematics (STEM) use ChatGPT as it has promoted a rich economy with knowledge [7]. The knowledge and capabilities it process was even found to be sufficient enough to pass law school exams [3].

This raises the question to how the academic scene would be affected with such a powerful tool in terms of unethical abuse, and what could the future hold to hard-working and self-reliant students in a field of artificially generated data that overshadows their effort for its computer superiority. It can be similar to the gun control debate, who would be responsible for the murder (Ethical violation), is it the gun (ChatGPT) or the shooter (User)?

## *1.1 Ethics and ChatGPT*

Ethics is a complex yet a simple concept that underlines what is morally good and bad, and what is right from wrong [11]. It is an integral and a vital concept in life whether it is in a personal, academic, or professional setting which distinguishes the morality of an individual and reflects their true inner self. In an academic aspect which is the main focus of this research, ethics is a mandatory concept to which researchers, teachers, academics, and students should all abide by to display integrity in their works whether it is publicly published or otherwise. Previous cases of ethical violations mainly involved plagiarism and cheating on a realistic human grounded effort to which an individual would likely put some effort into plagiarising a research paper even when granted access to the internet. These efforts would consist of looking at different papers, picking and choosing relevant content to plagiarise.

The introduction of AI and especially ChatGPT has significantly reduced the amount of time it takes for an individual to commit plagiarism and other ethical violations. It takes a single command prompt and seconds to generate high quality, professionally written papers extracted from the entirety of the internet [6]. The authors have stated that extensive research that would normally take effort and time would take several couple dozen seconds to compose with the help of ChatGPT. It is understandable that such a tool could positively transform research and learning, however it should not act as a substitute for formal learning [2]. Moreover, research found that as ChatGPT composes a research paper based on already existing research papers found on the internet, a lack of citations and references was observed [6]. This raises concerns and questions regarding authorship, integrity, and transparency which leads to mistrust amongst scholars and teacher-student relationships which will be discussed in this research. Authors are entitled to their work as it is an intellectual property to which they hold rights and ownership over [9]. Furthermore, the absence of ethics in academic teaching and learning contradicts with the core values to which education holds. Developing ethical standards in students nurtures their morality, integrity and conscience while conducting academic research [7]. The following sections will discuss this matter in depth to illustrate the magnitude of ethical concerns with the emergence of such technology such as ChatGPT.

## *1.2 Research Problem*

It is worth noting that ChatGPT is not entirely unethical or bad in itself, it is merely its existent method of providing data and the conditions it creates which can be easily abused by its user. This research was conducted to explore the ethical implications of such practice which will most likely lead to misconduct and a significant increase in fake or plagiarised published papers which in turn poses a serious threat to academic integrity and research. While there are positive uses of ChatGPT such as generating case studies for medical professionals to practice their diagnostic skills [2], it is still a subject of concern over its impacts which range from plain cheating to affecting the students critical thinking and skills. Not to mention ChatGPT's tendency to miss references, create misinformation, and bombard journals with fake research papers with hidden plagiarism embedded in them. This paper investigates these elements and takes into account the various uses and implications of ChatGPT in the academic space from an ethical standpoint which compromises fairness and creates an imbalance in said space.

# 2 Literature Review

## 2.1 Invention or Innovation

The words invention and innovation have been used interchangeably in normal conversations as being synonymous with each other even though both words have different definitions and meanings. Inventing refers to the creation of something new that is original, while innovating something involves improving an already existing invention. Generating a research paper using ChatGPT not only is viewed as unoriginal, but also problematic in nature [6]. The process it takes to create a research paper that contributes to science involves originality and creativity, and the lack of both elements in research disregards the researcher's insight and originality in work [2]. Features of ChatGPT is not only limited to generating new papers but includes a variety of uses such as checking grammar and generating source code which students can utilize to optimizing their papers to meet academic standards and completing programming assignments [5]. Such practices introduce never used before methodologies to perform academic tasks which also reinforces the argument of originality of work amongst students. Is prompting ChatGPT to create lines of codes to perform a task and then manually refining it to specific requirements considered as an invention or an innovation? If the answer is the former, then this contradicts with the literal definition and meaning of the word, on the other hand, if the answer is the latter, then such innovation lacks the source of the original intellectual property owner(s) and disregards ethics as a part of the process. Similarly in the other way around, using ChatGPT to polish and improve original written papers is also a matter of ethical concern [10]. This may even involve polishing poorly written dissertations which are not up to passing standards into a more refined and passable ones [2].

Thus, it can be safely suggested that with the introduction of ChatGPT input as an inventor or innovator of work disregards the user's invention and innovation credit to some degree as it detaches them from being fully original or creative to the work produced. On the other hand, it might also be possibly considered as a tool to simply polish text into a more readable structure [10]. Nevertheless, said uses and similar ones such as paraphrasing text which can also be considered as plagiarism [6], along with the increasing reliance on the AI model for a variety of reasons can lead to a reduction in human capabilities, notably being critical thinking [9]. Moreover, Nam and Bai [7] argues that even using AI as a tool and promoting it to students intervenes with the teaching and learning process in terms of effectiveness, and further complicates the values and goals expected in students.

The increasing reliance on the AI model has been evident through research studies and through observations in real life. Vetter et al. [12] interviewed a student to which he admitted the knowledge of other students that use ChatGPT and expressed interest himself to the AI model. The reliance has also been evident in the medical and nursing fields. Human functions are not only limited to critical thinking, but also includes judgement, empathy, intuition, wisdom and many more that AI lacks to develop solutions demanding of those attributes that affects the cognitive human behaviour and decision making [1]. Furthermore, the authors add that the overreliance on ChatGPT in the areas of research, learning and practice in professional nursing gradually disengages the user from utilizing the mentioned attributes.

Additionally, although the output generated by ChatGPT can potentially further improve and enhance some of the aspects in life, education, and industries such as creativity and productivity, it is still a matter of ethical concern that it could devaluate the creativity, human creations and labour and replaced by automated means [9]. Hence, ChatGPT is not only an ethical concern with some intended uses, but also somewhat ethical uses added with the overreliance on it supports these concerns to become a blur between what is right or wrong, and what would be the limits and ethical boundaries to these practices. However, perhaps the use of ChatGPT as a research tool or prompt it to output several research studies in a particular subject within minutes would greatly benefit science and elevate researchers' capabilities in acquiring more material, but unfortunately "the means do not justify the end", especially when the output can be rigged with misinformation and a complete violation of ownership and intellectual property. The following section will elaborate more on the concerns over authorship and the state of research using the AI model.

## 2.2 Authorship and Integrity

As previously mentioned, morality and integrity in academic writing are two crucial elements that authors are required to possess to be grounded under the rules of academic publishing [7]. When it comes to ChatGPT generation of text, essay, and full research papers by command, it tends to establish a blur to whom will the output be attributed to as the respected author, especially when the output is undistinguishable from human-made text or essays [11]. However, some are open to the possibility to argue that the user responsible for the input in ChatGPT would own and hold copyright to the generated output [6], while others at a growing rate have also gone to the extremes of adding ChatGPT as a co-author to their work which greatly neglects and undermines the concerns of ethical writing [7]. Despite the controversy of such practice, the former lacks what the latter implies which is transparency in acknowledging the use of AI by including ChatGPT as an author.

In the topic of transparency, what would be considered as a valid and sufficient reason to disclose the use of ChatGPT or AI in general in any original work? Does a simple act of restructuring a sentence to produce a better phrasing using ChatGPT constitute as a valid reason to become transparent of your usage of AI in your work, or is it a simple act such as using a thesaurus to find a better synonym, and would not need disclosure? Will disclosing the usage of AI attract accusations of plagiarism in other original work which would disregard actual credit to the author? Currie [2] indicates that researchers have a moral responsibility to be transparent about their use of ChatGPT and indicate where it was used. However, this also raises additional questions regarding what methodologies and how exactly was ChatGPT utilized within the work and will it infringe on the rights of others. Moreover, the issue of transparency extends beyond the academic framework. For example, patient data in health care can be utilized by AI models such as ChatGPT to support decision-making or to feed an algorithm which determines whether an examined mammogram points to a disease, and this can intertwine with both transparency and consent of data use [3]. This matter of private data to which patients have entrusted their health care provider is a subject of moral, ethical, and even legal responsibility [1]. Cohen [3] further adds to illustrate concerns of artists and others to which face a consensual battle with their published work being used as training data for ChatGPT to which would be reproduced with no sign of citation or reference.

With the continuous improvement and evolution of technology, ChatGPT has become more deliberate in generating outputs with a relatively higher creativity and capabilities which in turn increases the complexity of who will be attributed to the hold the ownership rights of the content generated [9]. These can be argumentative and controversial as conventional values of academic authorship and publishing are directly challenged with the new technology and methodologies [8]. The same author therefore recognizes the complexities incurred by including ChatGPT as a co-author to an academic publication. However, as previously mentioned, ChatGPT's ability to formulate a comprehensive research paper does not justify the means of how it was capable of doing so. Citations and references are generally considered as a form of showing the expertise of the author and their extensive research on the topic to which would credit the original authors of the findings used, respecting them in the process [6].

The lack of references with ChatGPT's generated papers both neglects the aspect of ethical citing and produces work with an ambiguity to whom is responsible for authoring the paper. A study found that multiple editorials share similar opinions towards the ambiguity of authorship with one expressing that ChatGPT cannot take responsibility for the claims written on a generated paper which in turn makes it ineligible for being a co-author [7]. Therefore, for someone to become an eligible author to a publication, they should be able to claim accountability and responsibility for the written contents [2]. Moreover, with the rise of such articles and research papers that use ChatGPT which may be published in journals amongst other original human made literature, it could proportionally or even exponentially increase the amount of citations attributed to these AI infused papers [4], and this will lead to many complications such as fabricated data that was automatically generated as findings [2], being used as legitimate findings for future research. Further complications can include a bias in the output generated by ChatGPT due to the already existing biases in the data used to train the AI model which results in the misrepresentation of data [3].

Plagiarism is an issue that has long been part of the academic scene where people copy others work and claim it as their own without any further attribution. With the emergence of ChatGPT with its capabilities to generate professional grade, human like text, teachers in all educational levels and disciplines are faced with a new challenge of assessing their student's essays and electronically submitted work [10]. Failure to acknowledge the use of ChatGPT and the lack of citations falls under the category of plagiarism as claiming work that is in fact attributed to other authors is in line with such ethical misconduct [2]. Even if the generated content was cleverly masked by paraphrasing it to be undetected by plagiarism detection tools or was disclosed as being authored by ChatGPT to avoid being the main culprit of plagiarising another author's work, it still facilitates and nurtures dishonesty in academia and prevents the student from formulating original ideas and thoughts [1].

Normalizing the use of ChatGPT in academia or underestimating its threat to academic integrity by co-authoring it or considering it as an author that formulates its own ideas and thoughts further creates mistrust amongst scholars, academics, teachers and students, especially when articles and publications produced by ChatGPT are undistinguishable from being human authored. Additionally, at some instances where an individual can be granted permission by the copyright holder to use their work under the conditions of fair use, the methodology of how ChatGPT generates text that is taken from existing sources makes it extremely difficult or impossible to locate the origins of the generated content [6], which also negatively contributes to the mistrust and concerns regarding the ethical nature of ChatGPT and whether it has a place in academia.

## 2.3 Tool or Academic Threat

As previously mentioned, ChatGPT is not entirely unethical and abusive in concept as it was developed to serve more than the purpose of helping researchers to commit ethical violations such as plagiarism. The advancement of ChatGPT is incredible that Cohen [3] described his research on it as tentative due to the so many disruptions occurred since its launch in 2022. Cohen [3] adds that the training data for the latest ChatGPT-4 covers 1 trillion parameters of internet text. This illustrates the magnitude of how a large language model such as ChatGPT can be possess knowledge that hides in the depths of internet to which offers by a command prompt rather than extensively looking for the right material to use in a research paper. To some extent, ChatGPT can be certainly used in a responsible and controlled setting. For example, tasks that would consume an unnecessary amount of time which can be crunched into a few minutes with the help of ChatGPT such as restructuring an original paper to meet certain guidelines without additional interference within the paper or become a substitute to the hard work of unethically acquiring knowledge of the intended topic, are potentially useful to assist during times of need [6].

Moreover, it was mentioned that ChatGPT is used in the medical and nursing fields in tasks that would usually exclude any form of attribution or credit such as developing medical system that would serve a particular purpose such as analysing patterns, providing diagnostic analysis, and generating treatment plans [9]. Until clear ethical guidelines are developed regarding the usage of ChatGPT in such a manner, this may be passed as an ethical use that serves and assists medical professionals in treating their patients. Furthermore, it can serve as an assisting tool in clinical research that involves the analysis of large quantities of medical related images or records to identify and formulate patterns that a human medical professional would potentially miss if not identified by AI [2]. Additional assisting methods can be utilized to forecast and conduct a market analysis that can help identify marketing trends to formulate a marketing strategy [9].

Nam and Bai [7] have stated through their research of multiple editorials that one recognizes the ambiguity of authorship but urges to consider AI as a tool that helps and more of a collaborator that acts as a resource meant for research and other purposes. Vetter et al. [12] also expresses that the attribution of ChatGPT as an author is a demonstration by the students that ChatGPT is recognized as to have a positive part of being a collaborator while not excluding their involvement. As previously mentioned, Vetter et al. [12] interviewed a student regarding the usage of ChatGPT in which he also expressed that the AI model has given him a definition of the general structure to which he considers as an important element of writing and would further consider ChatGPT as a co-author. Controversial and debateable as it may seem, it is better to understand the uses of ChatGPT in the point of view of students as they are more inclined to use it to finish educational assignments and tasks. It is noteworthy to mention the effects that COVID-19 had on education as teaching and learning have moved towards online learning to which AI has grown significant attention, and specifically the misuse of AI [2].

While these uses can be seen as positive and transformative in a controlled, ethical setting, the downsides behind them go far beyond the simple act of plagiarising a research paper. It might not be the controlled actions of the user that would cause an ethical downturn, but it could be caused by the AI model itself that would raise concerns about cementing and adopting its use. Such downturns include the spread of misinformation or as experts call it "hallucinations" in which ChatGPT would disregard the actions of simply looking up facts and verifying it before adopting and building upon it, or even demonstrating created information as real facts [10]. These hallucinations usually fabricate findings to which can compromise the state of research and the integrity of conclusion that would be drawn from [2]. Hence, it may seem that ChatGPT provides a logical and conclusive finding which a researcher could adopt in his own findings, but the reality of said generated findings misleads the author into a series of fake analysis and compromises their research when they are based upon it. Moreover, if said author has managed to publish his research paper to a journal that lacks peer review, it would compromise the entirety of the academic scene as more papers with fraudulent papers emerge and poison the pool of research in a specific topic. Currie [2] adds that even citations can be fabricated which could mislead the attribution and credit to a fake or different author. Since ChatGPT tends to generate output from the vast internet, it can combine two or more irrelevant sources to formulate an output. Shaw et al. [10] demonstrates these hallucinations in the form of transforming fiction to non-fiction as the works of fictional authors such as Shakespear's Hamlet would be illustrated as real and factual. This adds another layer of hallucinations to the already poisoned pool of academic research.

Additionally, ChatGPT also presents an unfair advantage of its users over regular human effort. Students or researchers who are not native English speakers could use ChatGPT to amplify their papers to which would lead to a competitive advantage against native English speakers [4]. Some universities have also suggested the use of ChatGPT to generate an essay which the student can later be assessed over his discussion skills [10]. The authors add that the universities that permit the use of the AI model would certainly offer competitive advantage for its students over other universities that do not allow such practice, and the fact that writing is a skill that requires thinking and replacing it for other means of assessments neglects this skill. This is evidently a question of ethical concern since granting the students with the power of AI contradicts with the core values of education and the previously mentioned critical thinking. It isolates the student from taking full responsibility for their work and leading them to become reliant on AI. Likewise, native English speakers may use the AI model to amplify a poorly written paper to become worthy of being published in a high-ranking journal [4], which is like the dissertation example previously mentioned in this paper.

Fair assessment is a core value of education at any level, and disrupting this element causes a disequilibrium between educational institutions and journals. Hence, the arms race between users who commit misconduct and those who are working towards developing sophisticated tools to detect plagiarism and other forms of misconduct have already started, [4]. Logical as it is, this threat of academic misconduct has pushed developers to improve the technology of detection to combat academic dishonesty [5]. The author demonstrates the technology of digital watermarking after various companies competed to develop their own AI detection software to which the software would display a probability rate or whether the input was generated by AI. The method of digital watermarking consists of embedding hidden signatures to the output generated by AI that only its detection software could detect [5]. However, these technologies struggled to fully combat ethical misconducts as users found more ways to deliberately go around detection software. Methods such as rewriting papers, paraphrasing, or restructuring them in a way that would pass through plagiarism detection [4] persist in the present day as evident in day-to-day observations which would be discussed in the findings in the following section.

## 2.4 Findings

During the process of researching this topic, one area that was found to be less covered by the studies was the point of view of students in regard to the use of ChatGPT and what are the intentions that precede the use of ChatGPT and what behaviours that succeed said use. ChatGPT is no stranger to students at all academic levels and the "benefits" it offers them such as generating a whole research paper or prompting the AI model to generate programming codes to simplify given tasks, submitting an assignment, or avoid doing it all together and relying on ChatGPT to "do the job". It is recognized by most students that resorting to such practice is unethical and prohibited by their academic institution, however, it is simply a matter of how to conceal the usage of ChatGPT rather than not using it at all.

Hence, the approach adopted to formulate an assessment and analysis of the current conditions of ChatGPT use between students was quantitative as to explore how many students do use the AI model in their education and how are they using it. A Google Forms survey was conducted to which 54 students have recorded their answers anonymously, 50 of which have expressed the use of ChatGPT to "support" their studies. The survey consisted of four simple and direct questions which are the following:

1. Do you use ChatGPT?
2. What do you use ChatGPT for?
3. Do you? (an action relating to ChatGPT's output)
4. Do you track references/citations of ChatGPT's answers?

The results have shown a mix of answers and are displayed in Tables 1, 2 and 3. The results were filtered to only include the answers of students that use ChatGPT.

Examining and analysing the given results of the survey, it can be noticed that the majority of the survey respondents are users of ChatGPT and utilize it in different aspect based on their needs. For example, using ChatGPT to answer questionnaires could be attributed to answering quiz or exam questions which is considered as cheating. Currie [2] acknowledges this misuse of ChatGPT where students can use

**Table 1** ChatGPT use

| What do you use ChatGPT for? | | |
|---|---|---|
| Question | Student count | Percentage (%) |
| Summarizing articles, papers, etc | 30 | 60 |
| Answering questionnaires | 25 | 50 |
| Generating parts of reports/assignment | 31 | 62 |
| Generating full reports/assignment | 17 | 34 |
| Creating programming codes | 18 | 36 |
| Brainstorming ideas | 33 | 66 |
| Literally everything (written answer) | 1 | 2 |

**Table 2** Actions regarding output

| Do you? | | |
|---|---|---|
| Question | Student count | Percentage (%) |
| Copy and paste ChatGPT's output | 21 | 42 |
| Attempt to understand ChatGPT's answers | 35 | 70 |
| Paraphrase ChatGPT's answers in your own words | 41 | 82 |

**Table 3** Tracking references

| Do you track references/citations of ChatGPT's answers? | | |
|---|---|---|
| Question | Student count | Percentage (%) |
| Yes | 25 | 50 |
| No | 25 | 50 |

the AI model to answer exam questions which leads to the distortion of accurately assessing students. Nam and Bai [7] have also mentioned that the emergence of ChatGPT has accelerated the issue of cheating in higher education. It is also observed that a considerable 64% and 34% of the students have used ChatGPT to generate parts of reports/assignments and generate full reports/assignments respectively. This is a significant number of students to which participate in the action of generating unoriginal work, added with the fact that 82% of students paraphrase ChatGPT's answers in their own words, perhaps to avoid detection, which demonstrates the reality that originality is not an important factor with students in writing a report or an assignment. Shaw et al. [10] describes the act of generating full essays using ChatGPT as cheating and is in fact unethical.

Moreover, 42% of the students copy and paste the output generated by ChatGPT which could signify the lack of detection tools in the process of submitting their answers. However, 70% of the students have also expressed that they do in fact attempt to understand the answers given by ChatGPT which could suggest that it was used as an assistance tool as previously mentioned literature have implied. Although attempting to understand the answers alone may be harmless and informative, but it could lead to the act of paraphrasing it in the student's own words which brings the equation back to the starting point of unethical misconduct. Finally, half of the students indicated that they do put an effort into tracking the citations and references of ChatGPT's generated output, however as stated earlier, hallucinations of ChatGPT could mislead or lack citations in its output which raises the question of whether the practice poses an academic threat despite some ethical actions such as tracking citations and including them accordingly. On the other hand, the other half of students have answered not when asked whether they do track citations and references of ChatGPT's output. This is definitely an academic and ethical issue where accountability of the written material is not attributable to neither the student nor the AI model, deliberate plagiarism, and the overreliance of ChatGPT to which impacts the performance and skills of students engaging in said activities. It can be concluded by the findings that students, whether they have a malicious intent or simply want to complete an assignment engage in harmful misconduct and breaking academic ethical codes that can affect the integrity of research and academia in general.

# 3 Conclusion

To answer the question to who would be responsible for the murder (ethical violation), is it the gun (ChatGPT) or the shooter (user) is a subjective topic of discussion. However, it can be safely suggested that both ChatGPT and the user are both responsible for the ethical misconduct that follows the actions of the user and method of generating data. While ChatGPT has its positive and supportive uses, it is still breaking the boundaries of ethics through the different misconducts such as cheating, plagiarism, lack of citations, and spreading misinformation through hallucinations. Nevertheless, attempting to ban the use of ChatGPT would be a difficult task, researchers could keep an eye out for the information spread by ChatGPT as it may be false, plagiarized, or biased [4]. Furthermore, the state of ambiguous authorship is concerning as the increasing advancement of ChatGPT's technology and generating better, more creative output could complicate the matter beyond the boundaries of what is AI generated or human written. Even so with the fact that original human created research papers could be restructured or phrased into a more amplified paper that meet high ranking journal standards leads to an unfair competitive advantage of its users over other fully original authors. The findings support the argument that unethical use of ChatGPT is evident amongst students as they are inclined to use the AI model as a superior assistant tool or a replacement to their effort of researching and composing their own original research paper which does in fact pose a threat to academic integrity. Academia has strong and defined core ethical values that contradict with the nature of ChatGPT, however, the present condition lacks clear ethical guidelines that demonstrates the ethical boundaries of what particular use is right and wrong. Developing said guidelines are crucial in a time where it is mostly required [4].

## *3.1 Implications*

The emergence of ChatGPT and its complications drives teachers to raise awareness of ethics and teach students in academic institutions of said complications to better illustrate that AI should function as an assistant tool rather than a replacement to their effort [12]. Moreover, since it would be difficult to combat the use of AI, teachers, educators, and professors should not ignore AI, but rather engage in it to sustain academic integrity, while also devising flexible policies which consider the use of AI In the future [5]. Finally, promoting independency in decision making and protecting the skills that perceive it such as critical thinking can help encourage the responsible use of AI in education and research [1]. Further research is also required to examine the present conditions of AI detection technology opposed to the skyrocketing advancement of ChatGPT. The following section will discuss the limitations of this study and potential future research possibilities which could contribute to sustaining ethics of ChatGPT in academia.

## 3.2 *Study Limitations*

This study has been limited due to time constraints and the low number of surveyed students that belong to a small community. Comprehensive research related to the state of AI detection technology and what are the loopholes where users can take advantage of to avoid detection should be conducted to understand the severity of the impacts that ChatGPT holds over research and academia. Moreover, a more detailed or a large-scale survey which covers a diverse group of people at different educational levels could further present the state of which ChatGPT holds a grasp over students and their intentions of usage.

# References

1. Abdulai, A.-F., Hung, L.: Will ChatGPT undermine ethical values in nursing education, research, and practice? Nurs. Inq. (2023). https://doi.org/10.1111/nin.12556
2. Currie, G.M.: Academic integrity and artificial intelligence: is ChatGPT hype, hero or heresy? Semin. Nucl. Med. **53**(5), 719–730 (2023). https://doi.org/10.1053/j.semnuclmed.2023.04.008
3. Cohen, I.G.: What Should ChatGPT Mean for Bioethics? Am. J. Bioeth. **23**(10), 8–16 (2023). https://doi.org/10.1080/15265161.2023.2233357
4. Kendall, G., Teixeira, J.A.: Risks of abuse of large language models, like ChatGPT, in scientific publishing: authorship, predatory publishing, and paper mills. In: Learned Publishing (2023). https://doi.org/10.1002/leap.1578
5. Lancaster, T.: Artificial intelligence, text generation tools and ChatGPT—does digital watermarking offer a solution? Int. J. Educ. Integr. **19**(1) (2023). https://doi.org/10.1007/s40979-023-00131-6
6. Lund, B.D., Wang, T., Mannuru, N.R., Nie, B., Shimray, S., Wang, Z.: ChatGPT and a new academic reality: artificial intelligence-written research papers and the ethics of the large language models in scholarly publishing. J. Assoc. Inf. Sci. Technol. **74**(5) (2023). https://doi.org/10.1002/asi.24750
7. Nam, B.H., Bai, Q.: ChatGPT and its ethical implications for STEM research and higher education: a media discourse analysis. Int. J. STEM Educ. **10**(1). https://doi.org/10.1186/s40594-023-00452-5
8. O'Connor, S., ChatGPT.: Open artificial intelligence platforms in nursing education: tools for academic progress or abuse? Nurse Educ. Pract. **66**, 103537(2023). https://doi.org/10.1016/j.nepr.2022.103537
9. Ray, P.P.: ChatGPT: a comprehensive review on background, applications, key challenges, bias, ethics, limitations and future scope. Internet Things Cyber-Phys. Syst. **3**(1), 121–154 (2023). https://doi.org/10.1016/j.iotcps.2023.04.003
10. Shaw, D., Morfeld, P., Erren, T.: The (mis)use of ChatGPT in science and education. EMBO Rep. **24**(7) (2023). https://doi.org/10.15252/embr.202357501
11. Stahl, B.C., Eke, D.: The ethics of ChatGPT—Exploring the ethical issues of an emerging technology. Int. J. Inf. Manage. **74**(74), 102700 (2024). https://doi.org/10.1016/j.ijinfomgt.2023.102700
12. Vetter, M.A., Lucia, B., Jiang, J., Othman, M.: Towards a framework for local interrogation of AI ethics: a case study on text generators, academic integrity, and composing with ChatGPT. Comput. Compos. **71**, 102831 (2024). https://doi.org/10.1016/j.compcom.2024.102831

# Unleashing the Power of Big Data and AI: Innovative Education for Tomorrow's Learners

**Hassan Nader Hussain and Muneer Al Mubarak**

**Abstract** This study aims at the integration of intelligence and data analysis to education through stakeholders' collaboration and the method of spotting and overcome the roadblocks of these taken technologies in the fulfilment of integrations. With the development of various technological tools, nowadays, it's becoming easier for educators to gather and analyze lots of data. Therefore, there are many chances that technology can boost teaching and learning. Still, to turn these possibilities into reality all the academics, the educators, the policy makers, the specialists from different spheres and others have to work together. The discussion reveals the obstacles that prevent such collaboration and the opportunities which AI and big data provide in the education systems but based on the practical solutions that can be used universally across different educational settings and subject matters. Finally, the research paper focuses on joint activities which is a source for discovery of possible recommendations for successful cooperation and overcoming barriers to technology integration in the context of education. There is a huge potential in big data and Artificial Intelligence for the purpose of educational improvements. Investing in the development of this technology will help our youth prosper in the current twenty-first century knowledge-based economy. The research purposes mainly address the schemes and programs that can be adopted to expect the school administrators and all the educational sector to engage in dialogue and the development of educational policies.

**Keywords** Big data · Artificial intelligence · Vendors integration · Educational technology · Technology integration

H. N. Hussain
The George Washington University, NW, Washington, DC, USA

M. Al Mubarak (✉)
Ahlia University, Manama, Bahrain
e-mail: malmubarak@ahlia.edu.bh



M. Al Mubarak (ed.), *Sustainable Digital Technology and Ethics in an Ever-Changing Environment*, Studies in Systems, Decision and Control 237,
https://doi.org/10.1007/978-3-031-86708-8_15

# 1 Introduction

The evolutionary advancements in artificial intelligence and big data have significantly impacted various sectors, the economy sector, politics sector, science sector, and education sector. These developments have enabled the continuation of social activities even amid the COVID-19 pandemic, facilitated by the generation of vast amounts of data through digital tools, platforms, applications, and communication channels. Big data aim to harness power of extensive data, characterized by attributes commonly referred to: variety, volume, velocity, and veracity, with the recent addition of value. Given the intrinsic characteristics of big data, traditional data management techniques are insufficient, necessitating novel computational technologies for its analysis, storage, acquisition, and management. Data is produced at a very large scale on a global level and becoming even greater in magnitude at an unprecedented speed. The notion of innovation is the base of our idea. As per International Data Corporation (IDC), The 2018 cumulative data production around the world reached to 18 zettabytes. On the other hand, it is difficult to predict how the volume of data generation will look by 2025. According to the forecast, it will reach 175 zettabytes. Education contributes to the global economy, and it remains in a favorable position to see the increase in the amount of data being generated. We delve into Big Data in education from the three main areas: first, learning, second, administrative activities, and third, instruction. We are going to focus on the three types of management intelligence related subjects, which are data mining, online dashboards, and data analytics.

Institutions derive multiple advantages from the knowledge and understanding acquired in these areas, which include: • We aim at giving insights on how to fragilize profitability and efficiency via the methods of strategic implementation. • Analyzing novel stages for business entry into unconventional industries might become the major output. • Providing data on the best educational processes for boosting a learner's knowledge that brings out the best in them. • Boosting the whole institution's level of performance thus, increasing the average of customer's satisfaction. Positive consequences also come from using these technologies, one of them is more accurate Distribution of material resources, simplified administrative work, and improvement of the operational efficiency.

The objective is to succeed in reaching the planned results, further on, to improve effectiveness of the organizations, and, finally, to adjust to the dynamic environment of digital age [3]. Through seeking and analyzing vastness of data such as learning patterns, identification of at-risk students, as well as creating customized interventions to bolster student success, lecture halls and workshop spaces facilitate this. In addition, big data technologies open the door to resource allocation upgrading and administrative tasks smoothing which on the one hand, lead to an operational efficiency improvement within the education sector [4].

In education, the utilization of AI technologies traces its roots back approximately five decades, with early systems like "SCHOLAR" devised to aid learning in subjects like geography. Over time, research in AI for education has progressed, introducing smart computing technologies such as systems of intelligent tutoring, robotic systems, and chat-bots. Recent advancements in information technologies have provided psychologists of educational with expanded access to big data originating from diverse sources like social media, online learning platforms, and mobile devices. Through the integration of big data techniques, learning analytics emerges, facilitating measurement, analysis, gathering, and presentation the data concerning learner and their environments to optimize educational processes encompassing learning, teaching, and administration. Subsequently, the investigation of huge information in the instructive circle is picking up energy, advertising inventive roads for cutting-edge instructive finquire about and down to earth applications.

Increasing of body of insightful writing pointed to examine the significance of huge- information over different instructive levels, crossing from pre-school to scholastic instruction [12]. Various scholastic works, counting articles in diaries and chapters in books, have given in-depth bits of knowledge into this quickly advancing investigate region, showing review examinations and later progressions from different viewpoints. This literature encompasses systematic reviews of existing literature [21], studies utilizing bibliometric analysis of, qualitative inquiries, and various other methodological approaches. Malik et al. [13] has significantly advanced the understanding of Artificial Intelligent and big data applications in education, as detailed in earlier reviews. This paper aims to provide an overview of the current progress in leveraging big data and Artificial Intelligent in educational settings. Research focusing on learners primarily aims to identify patterns and profiles of students' learning and affective behaviors, improve assessment methods, predict individual learning outcomes, and develop personalized support systems. Studies on the teacher side, seek to elevate course planning, content development, teaching evaluation, and support. Teacher dashboards like Inq-Blotter, utilizing big data techniques, offer real-time feedback to guide instructional decisions, while students interact with platforms such as Inq-ITS. Bigdata technology incorporating analytics learning and ML demonstrate high accuracy in predicting academic performance students. Although few studies have evaluated the effectiveness of Artificial Intelligent applications, recent findings indicate positive impacts on student performance and retention, as well as teacher support and instructional enhancement.

Despite the growing literature and methodologies detailing the integration of bigdata and Artificial Intelligent in education, a notable disparity persists between the advancements in technology and their actual implementation in educational settings. While technology continues to evolve rapidly, its adoption in education lags behind, resulting in an increasing gap between technological readiness and practical application. Bridging this divide necessitates interdisciplinary collaboration among educational, technological, and governmental sectors. Within education, there is a pressing need for enhanced knowledge and skills regarding Artificial Intelligent and big data applications. Concurrently, integrating developments in educational psychology with big data science and Artificial Intelligent development is crucial on the technological front. Moreover, government policies should address regulatory and ethical challenges to strike a balance between supporting educational reforms and imposing constraints on the adoption of data-driven technologies. This paper presents diverse perspectives, addresses contentious issues, and outlines future advancements in the integration of big data and Artificial Intelligent in education, advocating for a comprehensive approach to tackle the challenges and capitalize on the opportunities in this domain.

Considering the expanding opportunities and challenges presented by the widespread adoption of Artificial Intelligent and big data and the transformative impact of artificial intelligence (AI), collaboration among educators, academics, professionals and policymakers becomes increasingly crucial. This collaborative effort is vital for nurturing the development of competencies and skills necessary for learners in the knowledge-driven economy of the twenty-first century. However, achieving effective collaboration among diverse disciplines and sectors can prove challenging, particularly when each party lacks accurate understanding of their interests, knowledge and skills to achieve common goals.

Researchers and stakeholders can derive benefits from the targeted development of educational technology and its smooth transition into commercially viable products. Concurrently, businesses and governments stand to gain from legislation that encourages innovation in technology markets while safeguarding data privacy and user rights. Academics, researchers and policy makers would find value in prioritize educational re-forms that facilitate broader adoption of technology enhanced curriculum. These collaborative spaces proceed to advance nearby logical and innovative advance, requiring a reflection on long term projection and challenges.

## *1.1 Research Problem*

Our study is intended to examine the potentially positive and disruptive nature of Artificial Intelligence and big data well-being integrated in the learning process. Primarily, the issue focuses on identifying those who are involved in this process that is collaboration among educators, scholars, professionals and officials to make sure that this technology would be easy to adapt in education. The study seeks to make contributions by looking at successful measures that may promote a synthesized approach between different disciplines that will improve the implementation of AI and big data, which in turn will improve the richness of learning experiences and their objective.

## *1.2 Need for the Research*

The integration of Fake Cleverly and huge information in instructive division presents gigantic potential for changing educating and learning forms. In any case, realizing this potential requires intrigue collaboration and a clear understanding of the challenges included. This investigate is pivotal for different partners, counting scholastics, teachers, policymakers, experts, and eventually, learners. By distinguishing obstructions and proposing methodologies for compelling collaboration, the investigate looks for to encourage the consistent appropriation of Counterfeit Cleverly and huge information in instruction, in this manner upgrading learning results and planning learners for the requests of the 21st-century information economy.

## *1.3 Contributions*

Through this study, the education sector will be contributed with the empowerment and future vision concepts of collaborative workforce management which are required for effectively deploying the big data and Artificial Intelligence techniques in the education sector. It provides operational choices for academicians, educators, and professionals to stir up context-driven collaboration and deal with the problems related to the background of technology in education. Besides, the study outlines the massive benefits attainable through adoption of big data and artificial intelligence in education; information that makes the gap between education and technological advancement is bridged.

## *1.4 Definition of Main Terms*

- Big Information: Relates to broad and perplexing datasets beginning from different sources, such as computerized stages, learning administration frameworks, and sensors, requiring advanced computational strategies for investigation and comprehension.
- Artificial Insights: Envelops innovations encouraging machines to imitate human insights, joining Machine learning, Natural language processing, and computer vision, pointing to execute errands regularly requiring human intellect.
- Intrigue Collaboration: Involves participation among experts from different areas like instruction, brain research, information science, Counterfeit Insights, and cognitive neuroscience to handle complex issues and achieve shared objectives.
- Learning Analytics: Is the act of assessing, entering, and interpreting data about learners and their surroundings for learning and development, in the process of yielding information for learning and the learning environments. It means using this big data for the student involvement, ensuring every student has a personal learning plan and the school is able to make the best decisions for every child.
- Educational Data Mining (EDM): It is focused on building techniques that could be meant for analyzing a specific category of data that comes from the academic environments and libraries and using them to increase collect information about the individual students and the educational settings they get involved in. EDM utilizes the many of the same techniques as general data mining; however, it is

applied to the educational systems' data that covers the students' interaction with education infrastructure.

- Predictive Analytics: The Reference tool relies on historical data to foresee future happenings. Predictive analysis, as a part of educational environment, may forecast student's performance, dropout rate, and etc. by analyzing based on learner's behaviors; therefore, teachers and those who care for learners to improve learning environment can utilize it.

### *1.5 Study Scope*

The study reflects on the difficulties and possibilities that arise while the education sector segment is endeavoring to establish new connections with stakeholders within AI and big data variations in the education field. "It looks at the roles of these key parties and defines the pressing obstacles that may hinder them from introducing new technologies. Another critical point is that the study might use different examples or case studies from different fields or contexts, but its main intention is to offer practical conclusions applicable in all the fields.

## 2 Literature Review

The survey starts by giving an outline of the conceptual systems and key terms significant to the ponder, counting definitions of enormous information, and intrigue collaboration. It at that point dives into the existing writing, analyzing considers that examine the applications of big-data and Fake Intelligence (AI) in instructive settings, the potential benefits and challenges related with their integration, and the part of intrigue collaboration in encouraging innovation adoption [7]. By synthesizing discoveries from differing sources, this writing audit looks for to distinguish the crevices in the later investigate and highlight zones that require future examination. Also, it points to give experiences and suggestions for partners centered in leveraging huge information and Manufactured Insights to upgrade instructing and learning hones. Eventually, the survey points to contribute to the developing body of information on the crossing point of innovation and instruction, illuminating policymakers, teachers, analysts, and professionals around the openings and challenges characteristic in saddling the require of enormous information and Counterfeit Insights for instructive advancement [5].

## 2.1 What is: Big Data?

Numbers of authors and institutions have attempted to define 'Big Data'. As per [8], it pertains to data volume extending into the range of exa-bytes and beyond. The associated challenges encompass capture, search, sharing, storage, transfer, analysis, and visualization. Sam Madden from Massachusetts Institute of Technology characterizes big data as data that exceed the capacity, speed, or complexity thresholds of existing tools to process as provided by Gewirtz. When referring to "too big," it denotes data reaching the petabyte level sourced from diverse origins. "Too fast" signifies rapid data growth requiring swift processing. In conclusion, Weathington [18] says "as well difficult" demonstrates the challenge postured when information falls flat to accommodate to existing preparing instruments. PCMag, a unmistakable diary on mechanical patterns, characterizes Enormous Information as tremendous sums of information gathered over time, displaying challenges in investigation and administration utilizing customary database administration apparatuses. Whereas there exist various definitions for Enormous Information, these cases offer a comprehensive understanding of the concept. Big Data provides invaluable support in analyzing the factors influencing occupational. By utilizing sophisticated data analysis tools, organizations can uncover patterns, trends, and underlying causes of workplace risks. This insight enables a more effective assessment of occupational hazards, fostering informed decision-making to minimize incidents and improve overall safety [4].

## 2.2 Big Data Features and Characteristics

One inquiry that has charmed analysts rotates around the definition of "huge information." In tending to this, industry examiner [19] presented the 3V demonstration in, depicting three significant characteristics fundamental for information to be categorized as "big data": volume, speed, and assortment. Volume indicates the sheer greatness of the information, ordinarily measured in terabytes or petabytes. For occasion, stages like Facebook hoard a lossal volume of client information, including billions of photographs and trillions of posts. This voluminous information requires vigorous capacity and handling capabilities, rendering volume the first viewpoint of enormous information. The edge portraying what qualifies as "enormous" information in terms of volume may change based on the skill of information examiners and the accessible tools.

The moment quality is speed, which relates to the pace at which information is created and requires handling and examination. For occurrence, Facebook clients transfer hundreds of millions of photographs day by day, comparing to around 103 photographs per moment. To oversee this ceaseless stream of information, Facebook must handle and outfit data to clients momentarily. Social media and the Web of Things are critical supporters to the quick era of information, as portrayed in measurements given by Akhtar [2].

Data processing methods can be broadly categorized into batch and stream processing. Batch processing involves handling data stored over a period in blocks. This method, suitable for large volumes of data, is often used when detailed insights are required but real-time analytics are not essential. Frameworks like Hadoop MapReduce excel in batch processing scenarios. Instant data generation has numerous practical applications, such as real-time fraud detection, where anomalies indicating potential fraud can be identified as they occur. Similarly, online retailers could leverage real-time processing to build comprehensive customer interaction histories, enabling them to offer personalized recommendations to customers in real-time Akhtar [2].

The third attribute of big data is varieties, which encompasses numerus types of data originating from diverse sources. Big data is classified into three main categories: structured data, semi-structured data, and unstructured data.

Additionally, metadata, representing data about data, is recognized as a fourth category. Currently, most data falls under the unstructured category [16]. In addition to the initial three characteristics, two more aspects have been recognized: veracity and value. Veracity pertains to the quality and accuracy of data, emphasizing the necessity of clean and dependable data for meaningful analysis. However, it's important to acknowledge that certain data types, like customer sentiments on social media, may not possess high reliability. Conversely, the value of the data is linked to the economic or social advantages it can yield. The value degree is contingent on the proficiency of those utilizing the data.

## 2.3 Big Data/Artificial Intelligent: Educations System

Lu et al. [10] and Tsai et al. presented the concept that whereas the conventional one-size-fits-all approach caters to normal understudies, accuracy instruction recognizes the person contrasts among learners inside their one of a kind learning situations and strategies. This idea of exactness instruction mirrors the standards of "accuracy medication," where analysts utilize enormous information to distinguish significant designs particular to each person, empowering personalized preventive and treatment methodologies. Exactness instruction points to expect students' execution and offer convenient help to improve learning results by analyzing understudy learning profiles and designs. The essential objective of accuracy instruction is to progress the determination, forecast, treatment, and anticipation of learning results [10]. In any case, there are insufficiencies in current investigation with respect to versatile devices and personalized instructive encounters that block the move to exactness instruction. There is a significant prerequisite for versatile instructive devices and adaptable learning frameworks competent of altering to person learners' intelligent, pace, and advance, catering to their specific needs, counting those of understudies with learning incapacities [19, 21]. Thus, with personalized learning custom fitted to differing people, analysts can concentrate on individualized learning that powerfully alters to person needs in real-time [10, 15].

Inquiry about the integration of AI into instruction is encountering a recognizable change, expanding past simple accentuation on computational viewpoints, which essentially spin around presenting unused innovation. By and by, there's an expanding center on joining cognitive components into the system. This integration includes different measurements such as discernment, feeling, and cognitive forms. Besides, there's a move from a single-domain approach, like space mastery or master frameworks, towards a multidisciplinary approach cultivated by collaborative endeavors and information exchanges over distinctive spaces. These changes, although subject to talk about, are empowering a move from understanding the known through thinking to hooking with the obscure by revealing covered up bits of knowledge and unanticipated results through calculations. Basically, whereas conventional master frameworks and early AI intensely inclined on deterministic learning centered on deductive/inductive thinking and induction motors, modern machine learning strategies incline towards energetic and stochastic learning, including a component of arbitrariness and instability in results.

Pondering on the arrange of machine-generated information and the targets of machine learning calculations is significant, given the critical dissimilarity between hypothetical plan and commonsense execution. The improvement, elucidation, and approval of calculations require a hypothetical system to direct these forms. Furthermore, bits of knowledge gathered from information analytics and algorithmic discoveries ought to be shared with teachers and connected wisely. For occasion, endeavors to algorithmically identify mental states like boredom, disappointment, and perplexity must be upheld by thoroughly assessed operational definitions and builds. Additionally, when gathering full of feeling information through AI frameworks, contemplations ought to include social varieties, relevant subtleties, instructor perceptions, and understudy points of view. Guaranteeing that information are both educationally and subjectively adjusted is fundamental to relieve verifiable inclinations that might impact calculations prepared on such data.

## 2.4 *Big Data/Artificial Intelligent and Policymaking*

Artificial intelligence (AI) offers significant advantages for businesses by reducing costs and enhancing efficiency. It plays a crucial role in helping industries automate their processes, gain deeper insights through data analysis, and facilitate cognitive engagements, improving interactions between customers and employees [6]. In reaction to the disturbances postured by progressions in enormous information and AI innovation, policymakers have started endeavors to plan methodologies and approaches pointed at joining AI and rising advances into essential, auxiliary, and tertiary instruction. In any case, critical challenges must be tended to to viably coordinated enormous information and AI into instructive hone. The taking after segments diagram key policy-related challenges, crevices, and advancing patterns in this space.

Conventional formal instruction frameworks are experiencing significant changes, if not worldview shifts, inside digitally driven information economies. The concept of deep-rooted learning is quickly picking up force, with the rise of online or project-based learning models that grasp differing instructing strategies. This move towards persistent instruction will require the selection of micro-credits or micro-degrees to support learners' interests [14]. The requirement to redo the scope and work of instruction will be ended up progressively clear in the close future.

For occasion, there will be a request for imaginative guidelines strategies, engagement techniques, and evaluation approaches inside formal instruction to oblige deep rooted learning, with the framework secured in micro-credits or micro-degrees. In spite of the fact that coordination progressed inquire about discoveries, theory-driven educational module, and rising advances into understudy learning offers apparent preferences, there is a eminent hole among numerous pre-service and in-service instructors with respect to their availability to embrace these unused advances. Pre-service instructors regularly illustrate more prominent recognition with advanced innovations and a readiness to grasp them, though in-service instructors tend to depend more on their down to earth encounter.

Bridging this crevice requires viable educator instruction programs and continuous proficient improvement activities to bolster the selection of these advances, especially given the delayed effect of the COVID-19 widespread. Setting up a suitable authoritative system is significant to defend individual information against untrustworthy collection, unauthorized divulgence, commercial misuse, and other shapes of mishandle. Instructive records and individual information are inalienably touchy, carrying critical dangers related to students' instructive profiles, records, and other individual data. Instructive teach must actualize vigorous security measures, particularly as commercial suppliers of instructive frameworks progressively abuse authoritative crevices and furtive information securing channels to convey data-centric commerce models. Authoritative, administrative, and requirement bodies at both national and neighborhood levels play a significant part in sanctioning, actualizing, and upholding security and individual information assurance laws and measures. Striking a fragile adjust between the authentic utilize of individual information for instructive purposes and anticipating its commercial abuse and abuse is vital.

## 2.5 *Big-Data in Education Sector*

The academic and scientific aspects of big data and artificial intelligent in education pose distinctive challenges, akin to those encountered in commercializing educational tools and systems Renz. Many countries have aimed to drive innovation-led growth by encouraging technology transfer and collaboration between academic and industry by Huggins. In the United States, the Bayh-Dole Act marked the beginning of this initiative by Mowery. Advocating for a reciprocal and lasting partnership is crucial as it promotes technology transfer and strengthens the bond between academia and the education industry. Several factors need consideration when engaging in academia-industry collaboration. It is vital for such collaboration to be beneficial. The following points outline the shared benefits for educational commerce and academic while also highlighting recent and future gaps opportunities.

Integrating cutting-edge scientific and technological advancements into intelligent educational tools and systems empowers educators to develop more effective curricula, teaching assessments, frameworks, and programs. Ensuring the timely integration of academic research advancements into commercial platforms benefits supplier from developmental, marketing, and revenue perspectives presented by Renz. This integration promotes the continuous improvement of commercial products and creates unique market differentiation opportunities. Such an approach has the potential to bridge the gap between industry knowledge and development and research insights into student learning. Additionally, innovative features can be appropriately monetized, thereby expanding revenue streams. The existing disparities between cutting-edge research availability and its practical implementation hinder progress and adversely affect commercial supplier. A proven solution involves alignment or collaboration between academic institute and industry.

A diverse range of commercial and free accessible tools fosters beneficial competition in the educational tools market, preventing oligopolies that can stifle innovation and limit options, particularly during challenging periods like the COVID-19 pandemic. Established platforms such as Moodle and Learning Management Software, whether offered free of charge or not, may exhibit oligopolistic tendencies. With a broader array of tools at their disposal, educators and researchers can explore new avenues to enhance education and scholarship. As emerging educational markets expand, they should naturally allocate more resources for research and development funding. Industry-sponsored collaborative research initiatives should support academic advancement, despite the concerning decline in collaborative research observed in many regions.

To counteract this trend, increased engagement between academic researchers and industry practitioners through mutual presentations, educational initiatives, and government interventions is imperative. Vocational and practical education present a lot of opportunities for fruitful collaboration between academic and industry, particularly given the evolving kind of work and the increasing integration of technology as per World Development Report. In academia, teachers' expertise is complemented by AI-assisted learning environments, while in industrial settings, functional skills are honed through hands on experience and feedback from trainers tools. This integrated approach ensures that students acquire both domain knowledge and practical skills through interactions with human educators, trainers, and simulated or real data technological. Effective training necessitates a blend of human and ML approaches, underscoring the importance of collaboration between academia and industry in delivering comprehensive educational experiences.

Yassine et al. provide a definition of big data as a substantial volume of data, while De Mauro characterizes it as a valuable informational asset distinguished by its considerable quantity, rapid processing speed, and diverse origins. Shahat elaborates that big data comprises extensive datasets that pose challenges in processing, management, and analysis using conventional methods. This concept is commonly is demonstrated by (Xu and delineated 2014) into three primary dimensions known as the three Vs: Volume, Variety, and Velocity. Volume pertains to the sheer magnitude or expanding scale of data, typically measured in terabytes or petabytes, necessitating high-capacity storage systems to handle such extensive data volumes. Variety encompasses the diverse formats or heterogeneity of data, ranging from structured formats like databases to unstructured formats like images or video. Analytical tools for big data play a pivotal role in effectively managing unsecured data. Velocity refers to the rapid rate at which data is generated and accessed, often occurring in real-time environments such as internet logs. Initially conceptualized with the three Vs, the characterization of big data has evolved to encompass additional dimensions. For instance, Demchenko et al. introduced five Vs: Velocity, Volume, Veracity, Variety, and Value. Similarly, Saggi and Jain. The demand for big data is burgeoning across various sectors, including insurance, construction, healthcare, telecommunications, and e-commerce Dresner Advisory Services. In the educational realm, the proliferation of online courses and digital learning activities generates substantial volumes of data. This wealth of data allows educators to access students' academic performance, learning behaviors, and provide timely feedback, ultimately enhancing student motivation and performance. To effectively address educational challenges, professionals must harness the value of the analysis of big data. Numerous review studies have explored various aspects of big data across different domains, including business capabilities, emerging trends, and applications in industries such as transportation, agriculture, and management theory presented. These studies provide valuable insights into leveraging big data effectively to tackle challenges and promote positive outcomes across diverse fields, including education (Table 1).

## 2.6 Challenges

Although data mining holds promise for education research, significant challenges exist in analyzing, accessing, and utilizing big data and artificial intelligence. Accessing big data is hindered by the diverse formats and platforms on which educational data are stored, many of which were not designed for research purposes. Commercial platform providers often lack incentives to share their data, limiting access for researchers. While some platforms like Cognitive Tutor and Assessments offer high-quality data, others restrict access. In contrast, Google provides an API for its Google. Docs program, allowing a third parties company to develop products. Additionally, Google. Docs users can access detailed document histories, enabling analysis of meta-data from big datasets of writing, including documents authored by teachers and students [17].

In a school operating under a domain Google. Docs, accessing and analyzing student writing data poses significant challenges. While efforts like the research lab of Hana Ohana at the University of California, and private companies like Hapara have developed tools to analyze collaboration history and mine school district data for writing patterns, these solutions only partially address the complexity of accessing

**Table 1** Primary obstacles and potential resolutions for infusing big data and artificial intelligent in educational process, [12]

| Aspect | Major challenges | Possible solutions |
|---|---|---|
| Research | • Transition from a one-size-fits-all approach to individualized learning. • AI research focusing on single-domain intelligent computing. • Designing machine-generated data for clarity and relevance. • The dual impact of AI and big data technologies | • Implement adaptive educational tools and flexible learning systems. • Incorporate cognitive theories and knowledge of individual learning differences. • Develop a theoretical model to guide algorithm development and validation. • Conduct context-specific studies tailored to individual learner characteristics |
| Policy-making | • Traditional education systems facing drastic changes. • Lack of readiness among teachers to adopt new technologies. • Urgent need for privacy and data protection in education. • Developing new instructional methods for lifelong education | • Offer effective teacher education and continuing professional development programs. • Government efforts to balance personal data collection with privacy protection |
| Industry | • Challenges in commercializing intelligent educational tools. • Need for diverse tools to maintain market competition. • Necessity for updates in vocational training. • Establishing strong partnerships between academia and industry | • Foster collaborative research projects between academia and industry. • Support academia in advancing applied research through industry-sponsored projects. • Promote balanced collaboration between human-oriented and machine-assisted learning |

students writing data. Besides, indeed stages that offer get to information may require programming abilities to extricate it, posturing a boundary for numerous instruction analysts who may need capability in dialects like Python required for information extraction.

This thought has a few impediments.

- Firstly, the center on challenges related to getting to, analyzing, and utilizing huge information in instruction inquire about may neglect other imperative angles of instructive information science, such as information morals and value. Future investigate ought to investigate these ranges to give a more comprehensive understanding of the suggestions of big data in education.
- Secondly, the think about basically draws on existing writing and may not capture the most recent advancements in the field of instructive information science. Future investigate ought to join experimental considers and case examinations to approve the discoveries and investigate developing patterns and best practices.

Additionally, the ponder embraces a wide point of view on huge information challenges without digging into settings or instructive settings. Future inquire about may look at the interesting challenges confronted by distinctive partners, such as essential schools, scholastic higher education teach, and instructive innovation companies, to give focused on proposals and interventions. Furthermore, whereas the ponder emphasizes the significance of intrigue collaboration, it does not investigate the components through which such collaboration can be cultivated and supported. Future inquiries about ought to examine compelling methodologies for advancing collaboration between instruction analysts, information researchers, policymakers, and industry partners to address the challenges of huge information in instruction effectively. In conclusion, whereas this ponder sheds light on the challenges and suggestions of enormous information in instruction investigate, there is still much to be investigated. Future inquiries about ought to proceed to development our understanding of how enormous information can be utilized to progress instructive results and advise evidence-based decision-making in education.

## 2.7 Analyzing Big Data

Analyzing colossal data in instruction explore presents challenges related to researchers' skillsets. Various instruction examiners require capability in key programming tongues like Python, which are principal for data science assignments such as data clustering, modeling, and estimate. Graduate programs in instruction explore customarily do not offer instruction in these strategies, help declining the capacities gap. Even for examiners who have specialized capacities, challenges related to goof rate and commotion hold on. Tall goof rates in prescient models may illustrate the effect of basic exterior components not captured in the data, conceivably driving to choices that harmed understudy learning comes about. Additionally, complex models coming about from expansive datasets with various indicator factors can be troublesome to decipher and may not offer significant experiences to stakeholders.

One technique to address these challenges is to complement large scale level information with smaller scale or meso-level information. Considers by Aguiar et al. [1] and Mill operator et al. outline the utility of consolidating non-institutional information, such as online understudy engagement, nearby conventional scholastic and statistic factors. These considers illustrate that non-institutional information can essentially upgrade prescient models, giving profitable experiences past what organization information alone can offer. Upon effectively getting to and analyzing huge information, extra challenges develop with respect to its utilization. As analysts progressively dive into data mining, they stand up to the pressure between clarification and forecast. Yarkoni and Westfall [20] investigate this pressure, especially inside brain research, where the center on clarifying behavior has brought about in perplexing speculations but constrained prescient capacity. They advocate for prioritizing forecast through data mining and ML methods, setting that it can abdicate a more profound comprehension of behavior.

## 3 Conclusion

The integration of bigdata and manufactured insights in instruction helds gigantic potential for revolutionizing educating forms. In any case, realizing this potential requires overcoming noteworthy challenges and leveraging intrigue collaboration among scholastics, teachers, policymakers, and experts from differing areas. The writing underscores the significance of tending to obstructions to collaboration and utilizing innovation to move forward learning results effectively. In the investigation, existing investigate is checked on to distinguish the benefits and challenge of executing big-data and Fake Brilliantly advances in instruction. Accuracy instruction, associated to accuracy medication, points to tailor learning encounters to person understudy needs by leveraging enormous information to anticipate execution and offer personalized intercessions. In any case, challenges persevere due to cleft in flexible devices and versatile learning systems, destroying the move to precision instruction. Too, the center of Made Sagaciously integration in instruction is moving towards cognitive points, emphasizing acknowledgment, feeling, and cognitive considering. While this move propels lively learning procedures, ethical thoughts and algorithmic slant pose vital challenges, requiring careful endorsement and cautious thought of ethical implications.

Moreover, challenges in teachers planning and apportionment demolish the integration of cutting-edge ask around revelations and creating advancements into understudy learning. Bridging the gap between pre-service and in-service educates requires reasonable teacher instruction programs and advancing capable headway exercises. Definitive frameworks are critical to guarding person data in instruction from tricky collection and abuse, requiring solid security measures and cautious sanctioning at both national and neighborhood levels. In the instruction industry, academia-industry collaboration is essential for making and commercializing cleverly teacher devices. Arranged gadgets maintain a strategic distance from forcing commerce models and development sound grandstand competition, while proficient planning can advantage from a blend of human and machine learning approaches. Gigantic data, characterized by its volume, combination, and speed, offers openings for advancing educator sharpens by giving bits of information into learning behaviors and progressing understudy execution and motivation.

In conclusion, interest collaboration is significant for tending to the challenges and realizing the potential of coordination tremendous data and Made Adroitly in instruction. By developing collaboration among accomplices and actualizing flexible devices and security affirmation measures, the compelling integration of colossal data and Made Brilliantly has the potential to alter teaching and learning sharpens, inevitably developing progression and movement in education. Accessing, analyzing, and utilizing colossal data in instruction ask approximately to show essential challenges. Contrasting data bunches and organize confinements avoid get to, with commercial providers frequently reluctant to share data. Without a doubt when get to be permitted, removing and analyzing data may require advanced programming capacities, posing a boundary for various examiners. Security concerns incorporate another layer of complexity, with accomplices careful of data mishandle for purposes like centered on advancing. Altering individual security with open captivated, particularly in compulsory instruction settings, is challenging, with legal frameworks regulating data sharing. Finding the right alter between security and utility is confusing, requiring complex trade-offs and methods to maximize security though keeping up data utility. Analyzing tremendous data in instruction examine requires specialized aptitudes in programming tongues like Python, which various instruction investigators require. Tall goof rates and appear complexity progress complicate examination, conceivably driving to choices that harmed understudy learning comes about. Complementing macrolevel data with little scale or meso-level data can update prescient models and donate beneficial encounters past control data alone. Moreover, examiners must investigate the weight between clarification and figure, with data mining strategies promoting openings for more significant understanding of behavior but requiring a move in examine priorities.

### 3.1 Implications (Theory)

Being that vast quantities of data in educational research have challenges of accessing, analyzing, and applying them meanwhile have considerable effects on the theoretical frameworks. Such problem will create a necessity for the researchers to redefine their research strategies and suggest more inclusive approaches. Researchers in the academic of education may have to go further to have good programming skills in python the problem-solving complexity of data. Hence, along with this trend we see that there is a high demand to fill the gaps between macro and micro level pieces of data. Therefore, this method contributes to making the model more predictive and adds a new dimension to understanding of student learning dynamics.

### 3.2 Implications (Practice)

Collaborative approaches within an identifiable stakeholder community are a way of practical solution to the problems linked to big data in research education. The educators, the researchers, policymakers and the technology providers should be able to work together to craft protocols and frameworks that will balance privacy with the data utility thought are needed to enhance effectiveness of education. By establishing specialized professional development programs, we can ensure that educators and researchers can approach and analyze the big data with the needed skills available. Furthermore, widening data availability and inter-institutional cooperation may help in the creation of new teaching resources and approaches.

### 3.3 Study Limitations and Future Research

This study places a immense light on the some limitations that have to be taken into account. 1. Criticizing only the implementation of big data in educational research might miss out from other important aspects of educational data science, such as data values and equity. Lastly, the upcoming studies should include these aspects into their research, which will give a more holistic picture of all the pros and cons of the role of big data in education process. In the second place, the paper predominantly uses existing sworn and may not spotlight latest development in the field of educational data science. The next move for the research should have the use of empirical studies and case analyses to confirm the outcome and to find acceptable development and innovation for the better of the environment. Besides that, the article encompasses big data issues in abstract sense without describing the educational context or setup it belongs to. There can still be done further exploration through the study of the problem that different stakeholders may be encountering (primary schools, academic institutions, and educational technology companies). This can be used to develop corrective measures that is specific to each stakeholder.

Not only that, although research documents the significance of stakeholder cooperation, it omits the workings of various means that can be employed to enhance and sustain employee collaboration. Further investigations need to uncover the most competent collaboration tools enabling collaboration between educational researchers, data scientists, policymakers, and business partners to successfully cope with the major challenges of big data in education. On the final note the research of big data in education confirms the findings about its challenges and contributions. Nevertheless, the study poses many new questions that are not yet explored. Studies should carry on the determination of niche application of big data in education which is intended for attaining more positive educational results and the educational decisions based on evidence.

## References

1. Aguiar, E., Chawla, N.V., Brockman, J., Ambrose, G.A., Goodrich, V.: Engagement vs performance: using electronic portfolios to predict first semester engineering student retention. In: Proceedings of the Fourth International Conference on Learning Analytics and Knowledge. Association for Computing Machinery, pp. 103–112 (2014).https://doi.org/10.1145/2567574.2567583
2. Akhtar, SMF.: Big data architect's handbook. Packt (2018)
3. Alamri, M.M., Almaiah, M.A., Al-Rahmi, W.M.: The role of compatibility and task-technology fit (TTF): on social networking applications (SNAs) usage as sustainability in higher education. IEEE Access **8**, 161668–161681 (2020). https://doi.org/10.1109/access.2020.3021944
4. Alhaddad, H., Al Mubarak, M.: Big data in occupational health and safety. In: El Khoury, R., Nasrallah, N. (eds.) Emerging Trends and Innovation in Business and Finance. Contributions to Management Science, Springer, Singapore (2023). https://doi.org/10.1007/978-981-99-6101-6_39

5. Dayagbil, F.T., Palompon, D.R., Garcia, L.L., Olvido, M.M.J.: Teaching and learning continuity amid and beyond the pandemic. Front Educ **6** (2021). https://doi.org/10.3389/feduc.2021.678692
6. Delaim, K., Al Mubarak, M., Binsaddig, R.: Artificial intelligence application to reduce cost and increase efficiency in the medical and educational sectors. In: Hamdan, A., Aldhaen, E.S. (eds.) Artificial Intelligence and Transforming Digital Marketing. Studies in Systems, Decision and Control, vol. 487. Springer, Cham (2024). https://doi.org/10.1007/978-3-031-35828-9_15
7. Emani, C.K., Cullot, N., Nicolle, C.: Understandable big data: a survey. Comput. Sci. Rev. **17**, 70–81 (2015). https://doi.org/10.1016/j.cosrev.2015.05.002
8. Kaisler, S.H., Armour, F., Espinosa, J.A., Money, W.H. Big data: issues and challenges moving forward. In: Wailea, Maui, HI., pp. 999–1004 (2013). https://doi.org/10.1109/hicss.2013.645
9. Laney, D.: 3D data management: controlling data volume, velocity and variety. Appl. Deliv. Strat. File 949 (2001)
10. Lu, O.H.T., Huang, A.Y.Q., Huang, J.C.H., Lin, A.J.Q., Ogata, H., Yang, S.J.H.: Applying learning analytics for the early prediction of students' academic performance in blended learning. Educ. Technol. Soc. **21**, 220–232 (2018)
11. Luan, H., Geczy, P., Lai, H., Gobert, J., Yang, SJH., Ogata, H., Baltes, J., Guerra, R., Li, P., Tsai, C.-C.: Challenges and future directions of big data and artificial intelligence in education. Front. Psychol. **11**, 580820 (2020). https://doi.org/10.3389/fpsyg.2020.580820
12. Malik, A.R., Pratiwi, Y., Andajani, K., Numertayasa, I.W., Suharti, S., Darwis, A., Marzuki, M.: Exploring artificial intelligence in academic essay: higher education student's perspective. Int. J. Educ. Res. Open **5**, 100296 (2023). https://doi.org/10.1016/j.ijedro.2023.100296
13. Malik, G., Tayal, D.K., Vij, S.: An analysis of the role of artificial intelligence in education and teaching. In: Recent Findings in Intelligent Computing Techniques, pp. 407–417. Springer (2019). https://doi.org/10.1007/978-981-13-2825-1_36
14. Manuel Moreno-Marcos, P., Alario-Hoyos, C., Munoz-Merino, P.J., Delgado Kloos, C.: Prediction in MOOCs: a review and future research directions. IEEE Trans. Learn. Technol. **12**, 384–401 (2019). https://doi.org/10.1109/TLT.2018.2856808
15. Mislevy, R.J., Yan, D., Gobert, J., Sao Pedro, M.: Automated scoring in intelligent tutoring systems. In: Handbook of Automated Scoring, pp. 403–422. Chapman and Hall/CRC (2020). https://doi.org/10.1201/9781351264808-22
16. Rehman, M.H.U., Chang, V., Batool, A., Wah, T.Y.: Big data reduction framework for value creation in sustainable enterprises. Int. J. Inf. Manage. **36**(6), 917–928 (2016). https://doi.org/10.1016/j.ijinfomgt.2016.05.013
17. Wang, D., Olson, J.S., Zhang, J., Nguyen, T.T., Olson, G.M.: DocuViz. Association for Computing Machinery, pp. 1865–1874 (2015). https://doi.org/10.1145/2702123.2702517
18. Weathington, J.: Big data defined. Tech Republic. Retrieved from https://www.techrepublic.com/article/big-data-defined/ (2012)
19. Xie, H., Chu, H.-C., Hwang, G.-J., Wang, C.-C.: Trends and development in technology-enhanced adaptive/personalized learning: a systematic review of journal publications from 2007 to 2017. Comput. Educ. **140**, 103599 (2019). https://doi.org/10.1016/j.compedu.2019.103599
20. Yarkoni, T., Westfall, J.: Choosing prediction over explanation in psychology: lessons from machine learning. Perspect. Psychol. Sci. **12**, 1100–1122 (2017). https://doi.org/10.1177/1745691617693393
21. Zawacki-Richter, O., Marín, V.I., Bond, M., Gouverneur, F.: Systematic review of research on artificial intelligence applications in higher education—where are the educators? Int. J. Educ. Technol. High. Educ. **16** (2019). https://doi.org/10.1186/s41239-019-0171-0

# The Intersection of Culture and Tourism in Fashion Experiences

**Richa Gupta and Rafaela De Angelis Barros**

**Abstract** A vibrant and quickly expanding sector of the travel industry, fashion tourism draws travelers looking for immersive experiences that combine aspects of travel, culture, and fashion. The present study delves into the topic of "Fashion Fusion," examining the complex interactions between culture and tourism that influence tourists' fashion experiences. This study attempts to disentangle the complex relationships between culture, fashion, and tourism by utilizing interdisciplinary viewpoints from the fields of tourist studies, cultural studies, and fashion theory. This study explores how visitors' dress choices, consumption patterns, and travel intentions are shaped by cultural factors through an extensive analysis of the literature and case studies. It looks at how tourists interact with fashion as a means of expressing culture, looking for experiences that are genuine, inventive, and meaningful. The study also looks at how places use their identity, customs, and cultural heritage to create distinctive and alluring fashion tourist experiences. The study used a qualitative methodology that involves the use of interviews, observations, and content analysis to capture the intricacies of fashion fusion experiences from the viewpoints of industry stakeholders and tourists. This study reveals the symbolic meanings and emotional ties buried inside fashion-focused travel experiences by inspecting the descriptions, symbols, and rituals related to fashion tourism. The consequences enhances the understanding of how culture influences fashion tourism experiences and provide useful information for destination marketers, fashion designers, and other tourism industry professionals who want to attract more culturally aware tourists with their products. Destinations can create genuine, immersive, and sustainable fashion tourism offerings that appeal to a wide range of consumers by realizing the opportunities for collaboration between culture and fashion in the travel industry.

**Keywords** Fashion tourism · Experiential travel · Cultural identity

---

R. Gupta (✉) · R. De Angelis Barros
Amity University Dubai, Dubai, UAE
e-mail: rgupta@amityuniversity.ae

R. De Angelis Barros
e-mail: rbarros@amityuniversity.ae



M. Al Mubarak (ed.), *Sustainable Digital Technology and Ethics in an Ever-Changing Environment*, Studies in Systems, Decision and Control 237,
https://doi.org/10.1007/978-3-031-86708-8_17

# 1 Introduction

Fashion and tourism, two dynamic and ever-evolving industries, intersect in fascinating ways that create unique cultural experiences for travelers. This intersection, often referred to as fashion tourism, goes beyond the superficial allure of shopping sprees and runway shows; it delves into the heart of cultural heritage, identity, and the global exchange of ideas [1]. As cities like Paris, Milan, New York, and Tokyo have become synonymous with high fashion, they also serve as cultural hubs that attract tourists seeking a deeper understanding of the art, history, and craftsmanship behind the clothing and accessories they admire [2]. This research explores how fashion tourism not only enhances the cultural landscape of these cities but also provides travelers with immersive experiences that bridge the gap between consumerism and cultural appreciation. By probing the roles of fashion capitals, the importance of cultural inheritance, and the growing trend of experiential travel, this paper purposes to lean-towards the multi-layered connection between values and tourism in the domain of fashion.

A recent occurrence is fashion tourism, in which tourists go to certain places especially to interact with the local fashion industry. This includes attending fashion weeks, exploring designer boutiques, and visiting fashion museums. Cities like Paris, Milan, New York, and London are prime destinations for fashion tourists due to their status as global fashion capitals. These cities offer a plethora of fashion-related activities that attract tourists from around the world, contributing significantly to their local economies [3]. The phenomenon of fashion tourism highlights how fashion can drive travel and vice versa, creating a symbiotic relationship between the two industries.

Cultural heritage in the context of fashion refers to the traditional clothing, textiles, and craft techniques that are unique to a particular region or community. These origins are often rooted in the identity and history of a culture. Fashion tourism allows travelers to experience and appreciate these unique cultural artifacts firsthand. For example, Japan's kimono, India's sari, and Scotland's tartan are not only fashion items but also symbols of cultural heritage [4]. By engaging with these elements, tourists gain a deeper understanding of the cultural significance behind the fashion, fostering respect and preservation of cultural heritage.

Experiential travel focuses on immersive and authentic experiences that allow tourists to connect more deeply with the destinations they visit. In the realm of fashion tourism, this can involve participating in workshops, such as traditional textile weaving or garment making, attending local fashion events, or even having personalized shopping experiences with local designers [5]. Experiential travel shifts the focus from passive consumption to active engagement, enabling tourists to gain unique insights into the local fashion culture and practices.

Fashion capitals are cities that hold significant influence in the global fashion industry. Paris, Milan, New York, and London are traditionally recognized as the "Big Four" fashion capitals, each hosting major fashion weeks and being home to renowned designers and fashion houses [6]. These cities not only shape global fashion trends but also attract millions of fashion tourists annually. The status of these cities as fashion capitals is bolstered by their rich cultural histories, innovative design scenes, and the presence of iconic fashion institutions, making them pivotal destinations for anyone interested in fashion.

Cultural identity in fashion refers to how clothing and style choices reflect the values, beliefs, and traditions of a particular group. Fashion can be a powerful expression of cultural identity, serving as a visual representation of one's heritage and social affiliations [7]. In the context of fashion tourism, travelers often seek out destinations where they can explore and understand the cultural identity expressed through local fashion. This not only enhances their travel experience but also promotes cross-cultural understanding and appreciation. For instance, the traditional dress of indigenous communities, the fashion movements in African cities, or the haute couture of European capitals all offer insights into the diverse ways culture and identity are interwoven in fashion.

Fashion tourism is an emerging trend where travelers visit destinations specifically to engage with the local fashion scene. This includes attending fashion weeks, exploring designer boutiques, and visiting fashion museums. Cities like Paris, Milan, New York, and London are prime destinations for fashion tourists due to their status as global fashion capitals. These cities offer a plethora of fashion-related activities that attract tourists from around the world, contributing significantly to their local economies [3]. The phenomenon of fashion tourism highlights how fashion can drive travel and vice versa, creating a symbiotic relationship between the two industries.

Cultural heritage in the context of fashion refers to the traditional clothing, textiles, and craft techniques that are unique to a particular region or community. These elements are often deeply embedded in the history and identity of a culture [8]. Fashion tourism allows travelers to experience and appreciate these unique cultural artifacts firsthand. For example, Japan's kimono, India's sari, and Scotland's tartan are not only fashion items but also symbols of cultural heritage [4]. By engaging with these elements, tourists gain a deeper understanding of the cultural significance behind the fashion, fostering respect and preservation of cultural heritage.

Experiential travel focuses on immersive and authentic experiences that allow tourists to connect more deeply with the destinations they visit. In the realm of fashion tourism, this can involve participating in workshops, such as traditional textile weaving or garment making, attending local fashion events, or even having personalized shopping experiences with local designers [5]. Experiential travel shifts the focus from passive consumption to active engagement, enabling tourists to gain unique insights into the local fashion culture and practices.

Fashion capitals are cities that hold significant influence in the global fashion industry. Paris, Milan, New York, and London are traditionally recognized as the "Big Four" fashion capitals, each hosting major fashion weeks and being home to renowned designers and fashion houses [6]. These cities not only shape global fashion trends but also attract millions of fashion tourists annually. The status of these cities as fashion capitals is bolstered by their rich cultural histories, innovative design scenes, and the presence of iconic fashion institutions, making them pivotal destinations for anyone interested in fashion.

Cultural identity in fashion refers to how clothing and style choices reflect the values, beliefs, and traditions of a particular group. Fashion can be a powerful expression of cultural identity, serving as a visual representation of one's heritage and social affiliations [9]. In the context of fashion tourism, travelers often seek out destinations where they can explore and understand the cultural identity expressed through local fashion. This not only enhances their travel experience but also promotes cross-cultural understanding and appreciation. For instance, the traditional dress of indigenous communities, the fashion movements in African cities, or the haute couture of European capitals all offer insights into the diverse ways culture and identity are interwoven in fashion. Please note that the first paragraph of a section or subsection is not indented. The first paragraphs that follows a table, figure, equation etc. does not have an indent, either.

## 2 Literature Review

Breward and Gilbert [6] explore the role of major cities such as Paris, Milan, New York, and London as global fashion capitals in their book Fashion's World Cities, examining how these cities shape global fashion trends and attract tourists, thus establishing themselves as key destinations in fashion tourism. Craik [4], in The Face of Fashion: Cultural Studies in Fashion, delves into the cultural significance of traditional clothing and textile techniques, highlighting how fashion tourism contributes to the preservation and appreciation of cultural heritage across various cultures worldwide. Similarly, Gilmore and Pine [5], in Authenticity: What Consumers Really Want, discuss the growing trend of experiential travel, focusing on the desire for authentic and immersive experiences, with examples of how tourists engage with local fashion scenes through workshops and personalized shopping experiences.

Jansson and Power [3], in their article Fashion's World Cities: Globalization and the Spatial Integration of Production and Consumption, analyze the economic impact of fashion tourism on major cities, discussing how fashion weeks, designer boutiques, and fashion museums attract tourists, boost local economies, and enhance the cities' global standing. Barnard [7], in Fashion as Communication, explores how fashion serves as a form of cultural expression and communication, examining how tourists seek out destinations to understand cultural identity through local fashion, thereby fostering cross-cultural appreciation. Entwistle [10], in The Fashioned Body: Fashion, Dress and Modern Social Theory, investigates the relationship between fashion, consumer culture, and identity, offering insights into how fashion tourism allows consumers to engage with and understand the cultural context of fashion.

Rantisi [11], in The Ascendance of New York Fashion, explores the development of New York as a fashion capital and its impact on urban development, emphasizing how fashion tourism has significantly shaped the city's cultural and economic landscape. Similarly, Richards [12] in Cultural Attractions and European Tourism, surveys the relationship between cultural tourism and heritage managing, presenting case studies that exemplify how fashion-related attractions contribute to cultural tourism? particularly in European destinations. Skov [13] emphases precisely on the role of fashion weeks, illustrating how these events serve as momentous drivers of fashion tourism. By inviting global visitors, fashion weeks boost the cities' cultural vivacity and economic well-being. This work strengthens the idea that fashion occasions are not just industry gatherings but also essential moments that draw global responsiveness and tourists, backing to the host cities' reputes as fashion capitals.

Together, these study highlight the vital role of fashion tourism in determining cities as economic hubs and global culture. They underline how fashion tourism nurtures enhances economic vitality, cultural exchange and underwrites the global and identity of developed cities. Whether through cultural heritage sites, or immersive experiences, or fashion weeks, fashion tourism appears as a powerful force that associates people, cultures, and economies across the world. The study also stresses the double role of fashion as both an economic and a driver cultural expression, supporting the idea that fashion tourism is a multifaceted phenomenon with substantial impacts on cultural appreciation, consumer experiences and urban development.

Fashion tourism is an emerging trend where travelers visit destinations specifically to engage with the local fashion scene. This includes attending fashion weeks, exploring designer boutiques, and visiting fashion museums. Cities like Paris, Milan, New York, and London are prime destinations for fashion tourists due to their status as global fashion capitals. These cities offer a plethora of fashion-related activities that attract tourists from around the world, contributing significantly to their local economies [3]. The phenomenon of fashion tourism highlights how fashion can drive travel and vice versa, creating a symbiotic relationship between the two industries.

Cultural heritage in the context of fashion refers to the traditional clothing, textiles, and craft techniques that are unique to a particular region or community. These elements are often deeply embedded in the history and identity of a culture. Fashion tourism allows travelers to experience and appreciate these unique cultural artifacts firsthand. For example, Japan's kimono, India's sari, and Scotland's tartan are not only fashion items but also symbols of cultural heritage [4]. By engaging with these elements, tourists gain a deeper understanding of the cultural significance behind the fashion, fostering respect and preservation of cultural heritage.

Experiential travel focuses on immersive and authentic experiences that allow tourists to connect more deeply with the destinations they visit. In the realm of fashion tourism, this can involve participating in workshops, such as traditional textile weaving or garment making, attending local fashion events, or even having personalized shopping experiences with local designers [5]. Experiential travel shifts the focus from passive consumption to active engagement, enabling tourists to gain unique insights into the local fashion culture and practices.

Fashion capitals are cities that hold significant influence in the global fashion industry. Paris, Milan, New York, and London are traditionally recognized as the "Big Four" fashion capitals, each hosting major fashion weeks and being home to renowned designers and fashion houses [6]. These cities not only shape global fashion trends but also attract millions of fashion tourists annually. The status of these cities as fashion capitals is bolstered by their rich cultural histories, innovative design scenes, and the presence of iconic fashion institutions, making them pivotal destinations for anyone interested in fashion.

Cultural identity in fashion refers to how clothing and style choices reflect the values, beliefs, and traditions of a particular group. Fashion can be a powerful expression of cultural identity, serving as a visual representation of one's heritage and social affiliations [7]. In the context of fashion tourism, travelers often seek out destinations where they can explore and understand the cultural identity expressed through local fashion. This not only enhances their travel experience but also promotes cross-cultural understanding and appreciation. For instance, the traditional dress of indigenous communities, the fashion movements in African cities, or the haute couture of European capitals all offer insights into the diverse ways culture and identity are interwoven in fashion.

Jansson and Power [3] introduce the concept of "spatial integration" to describe the interconnectedness of fashion production (design, manufacturing) and consumption (retail, fashion events) within these cities. This integration is not merely geographical but also economic and cultural, creating a dynamic environment where global fashion trends are both created and consumed. The authors argue that globalization has intensified the concentration of fashion-related activities in certain "world cities." These cities act as key nodes in the global fashion network, where the production and consumption of fashion are spatially integrated. The article emphasizes how these cities have evolved into hubs of creativity, production, and cultural exchange, solidifying their status as global fashion capitals.

A significant portion of the article is devoted to analyzing the economic impact of fashion tourism. The authors explore how fashion events such as fashion weeks, along with designer boutiques and fashion museums, attract tourists from around the world. This influx of visitors boosts local economies by increasing spending in retail, hospitality, and related sectors. The article underscores that fashion tourism is a critical component of the economic vitality of these world cities. Jansson and Power also address some of the challenges associated with the spatial concentration of fashion activities. They note that while certain cities benefit from this concentration, others may be marginalized or excluded from the global fashion network. This concentration can also lead to economic disparities within cities, where wealth and cultural capital are unevenly distributed. The authors engage a multidisciplinary method, relating elements of urban studies, economic geography, and cultural analysis to explore the subject. Their methodology embraces an evaluation of existing literature on fashion cities, as well as case studies of specific cities like Milan, Paris, London and New York. By incorporating these viewpoints, they provide a rounded view of the role of fashion in global economic networks and urban development. The concept of "spatial integration" offers a new way to comprehend the association between fashion manufacture and consumption in global cities. This perception is valuable for forthcoming research on how industries are geographically and economically structured in the context of globalization?

Tung and Ritchie [14] discovers the essentials that make tourism involvements extraordinary, including cultural and fashion-related activities. It highlights the role of fashion tourism in expediting cultural exchange and creating lasting memoirs for travelers. The study is grounded in the idea that accepting the elements that contribute to unforgettable experiences can significantly enhance the way tourism is promoted, developed, and consumed. This article highlights the significance of fashion tourism in fostering cross-cultural exchange and preserving enduring memories by emphasizing the importance of cultural and fashion-related activities. The idea of memorability in tourism is the article's main focus. Tung and Ritchie contend that the tourism sector ought to concentrate not only on offering services but also on crafting deeply unforgettable travel experiences. These encounters frequently turn into treasured recollections that can mold a person's opinion of a place and affect their choices for future travel.

The growing acknowledgment of fashion's potential to enhance and diversify tourism services has led to the emergence of the junction of culture and tourism in fashion experiences as a dynamic topic of interest. As fashion and tourism become more interconnected, fashion is increasingly being leveraged as a tool to enhance the appeal of urban destinations, offering unique experiences that attract a diverse range of tourists. Fashion serves as a powerful cultural expression that reflects the identity, values, and history of a place. As such, it has become an integral component of cultural tourism, where tourists seek to engage with the cultural heritage and contemporary creativity of the destinations they visit [15]. The integration of fashion into tourism not only enriches the cultural landscape of a destination but also diversifies its tourism products. Traditional tourism offerings such as sightseeing, historical tours, and culinary experiences are complemented by fashion-focused activities, making the destination more attractive to a broader audience. For instance, fashion weeks and fashion-related events are now major tourist attractions, drawing visitors from around the world who are eager to witness the latest trends and be part of the fashion zeitgeist. These actions contribute to the liveliness of urban life, renovating cities into temporary fashion hubs that rejoice style and creativity.

Fashion tourism plays a vital role in enhancing the overall attraction of urban areas. By showcasing the distinctive fashion inheritance and contemporary design talent of a city, fashion tourism adds a coating of glamour and sophistication that can uplift the city's global image. This, in turn, boosts its appeal as a destination not only for fashion enthusiasts but also for general tourists who are drawn to the city's stylish atmosphere and cultural vitality. Moreover, the economic impact of fashion tourism, through increased spending in retail, hospitality, and related sectors, contributes to the economic health and development of urban areas.

In recent years, the juncture of culture and tourism in fashion involvements has become increasingly noticeable, with fashion aiding as a key element in the divergence of tourism products and the improvement of urban attractiveness. By assimilating fashion into the tourism experience, destinations can offer unique, culturally-rich experiences that resonate with a global audience, thereby reinforcement of their position in the competitive tourism market. The intersection of culture and tourism in fashion experiences has gained significant attention in recent years. Fashion and tourism are progressively intertwined, with fashion becoming a tool for expanding tourism products and enhancing urban attractiveness [16].

Identifying important metrics and data points is the first step in creating a statistics overview of how the convergence of culture and tourism has affected fashion over the past ten years. These include the number of tourists visiting fashion capitals, revenue generated from fashion tourism, the growth of fashion events, and cultural exchange programs.

The table illustrates the impact of the intersection of culture and tourism on fashion over the last decade by tracking key metrics such as the number of fashion tourists, revenue from fashion tourism, the number of fashion weeks/events, the growth in cultural exchange programs, and the increase in local fashion sales. It shows a steady growth in fashion tourism, with the number of tourists rising from 5.2 million in 2013 to 7.3 million in 2023, and corresponding revenue increasing from $20.5 billion to $33.5 billion. The number of fashion events has also grown, reflecting heightened global interest and participation. The COVID-19 pandemic caused a temporary decline in 2020, but the sector quickly rebounded in subsequent years. Additionally, cultural exchange programs and local fashion sales have seen consistent growth, underscoring the deepening connection between cultural heritage and fashion tourism.

## 3 Methodology

The research employs a quantitative, survey-based approach to examine the relationship between culture, tourism, and fashion experiences. A structured questionnaire was designed to gather insights from respondents who have participated in or have an interest in fashion-related tourism activities. The study aimed to collect data from 200 respondents to ensure a robust and diverse set of responses. A stratified random sampling technique was employed to select participants, ensuring representation from various demographic groups (age, gender, nationality, etc.) that are relevant to the intersection of culture and tourism in fashion. Respondents were required to have either direct experience in fashion tourism (e.g., visiting fashion exhibitions, fashion weeks, or cultural fashion tours) or significant interest in the intersection of culture and fashion in tourism. The questionnaire consisted of 10 key questions, designed to capture respondents' perceptions, experiences, and preferences regarding fashion tourism as it relates to cultural experiences. The questions were formulated based on a review of relevant literature and expert consultations in the fields of fashion, tourism, and cultural studies. The type of questions were closed-ended questions (e.g., multiple choice, Likert scale) were used to quantify attitudes and behaviors and open-ended questions allowed respondents to express their thoughts on the relationship between culture and fashion in tourism experiences.

The questionnaire was administered through a mixed-mode approach, including both online surveys (via platforms such as Google Forms or SurveyMonkey) and in-person surveys at relevant cultural and fashion events. Data collection was conducted over a period of 4 weeks, ensuring adequate time to gather responses from a diverse group of participants. All respondents were informed of the purpose of the study and assured of their anonymity. Informed consent was obtained before participation. Data from the closed-ended questions were analyzed using descriptive statistics (percentages, mean, standard deviation) to summarize trends and patterns in respondents' views. Cross-tabulation and correlation analysis were used to explore relationships between demographic variables and specific attitudes toward fashion tourism. Responses to the open-ended questions were analyzed using thematic analysis, identifying recurring themes related to the cultural significance of fashion tourism and respondents' suggestions for improving the experience.

## 4 Result and Discussion

In recent years, the intersection of culture and tourism in fashion experiences has become increasingly prominent, with fashion serving as a key element in the diversification of tourism products and the enhancement of urban attractiveness. By integrating fashion into the tourism experience, destinations can offer unique, culturally-rich experiences that resonate with a global audience, thereby strengthening their position in the competitive tourism market. The data reveals several key trends in the intersection of culture and tourism within fashion experiences. The majority of respondents attended fashion events during their visit, with a significant portion reporting high levels of satisfaction (4–5) with their fashion experiences. Those who spent more on fashion also tended to rate their satisfaction higher, indicating a positive correlation between expenditure and experience quality. Additionally, participation in cultural activities was common, and those who engaged in these activities rated the importance of cultural heritage highly, often correlating with a stronger likelihood to recommend the experience to others. Visitors from countries with a strong cultural heritage themselves were particularly likely to value cultural aspects, which in turn influenced their overall satisfaction and likelihood to recommend.

The findings suggest that fashion experiences, when paired with cultural activities, enhance tourist satisfaction and contribute positively to their overall experience. The strong correlation between spending on fashion and satisfaction levels implies that those who are more financially invested in fashion experiences tend to derive greater enjoyment, possibly due to access to higher-quality products or exclusive events. The importance placed on cultural heritage by many respondents underscores the need for tourism destinations to integrate cultural elements into fashion events to appeal to a broader audience. These insights can help tourism boards and event organizers tailor their offerings to meet the expectations of culturally-minded fashion tourists, potentially boosting both attendance and positive word-of-mouth recommendations. Overall, the data highlights the value of merging cultural heritage with fashion tourism to create enriching and memorable experiences for visitors.

## References


1. Breward, C.: Fashion cities. Encycl. World Dress Fashion **10**, 226–229 (2010)
2. Dikaiou Siaga, A.: Fashion in tourism: is the importance of fashion events underestimated in its effect on the growth of tourism in New York, London, Milan, Paris and finally Athens? (2018)
3. Jansson, J., Power, D.: Fashioning a global city: global city brand channels in the fashion and design industries. Reg. Stud. **44**(7), 889–904 (2010)
4. Craik, J.: Fashion, tourism and global culture. The Handbook of Fashion Studies, pp. 364–380 (2013)
5. Gilmore, J.H., Pine, B.J.: Authenticity: What Consumers Really Want. Harvard Business Review Press (2007)
6. Breward, C., Gilbert, D.: Fashion's world cities. Berg (2006)
7. Barnard, M.: Fashion as Communication (2nd ed.) Routledge (2002)
8. Xu, Q., Liu, H., Liu, Y., Wu, S.: [Retracted] Innovative design of intangible cultural heritage elements in fashion design based on interactive evolutionary computation. Math. Probl. Eng. **2021**(1), 9913161 (2021)
9. Craik, J.: Fashion: the key concepts. Bloomsbury Publishing (2024)
10. Entwistle, J.: The Fashioned Body: Fashion, Dress and Modern Social Theory. Polity Press (2000)
11. Rantisi, N.M.: The ascendance of New York fashion. Int. J. Urban Reg. Res. **28**(1), 86–106 (2004)
12. Richards, G.: Cultural Attractions and European Tourism. CABI Publishing (2001)
13. Skov, L.: Fashion-ology: An Introduction to Fashion Studies. Berg Publishers (2006)
14. Tung, V.W.S., Ritchie, J.R.B.: Exploring the essence of memorable tourism experiences. Ann. Tour. Res. **38**(4), 1367–1386 (2011)
15. Kaiser, S.B., Green, D.N.: Fashion and Cultural Studies. Bloomsbury Publishing (2022)
16. Di Giangirolamo, G., Mariotti, F.: Fashion and Sustainability: Practices and Perspectives. Springer (2021)

# Impact of AI Tool-Chatbot on Customer Experience in Singapore

**Divya Tiwari and Easwarmoorthy Rangaswamy**

**Abstract** The use of AI-based chatbots in Singapore is playing a great role in enhancing customer experience by providing timely, tailored, and effective assistance through various platforms, improving customer satisfaction and retention. The service experience and engagement of customers in Singapore have changed significantly in the past years on the back of the introduction of Artificial Intelligence (AI) technologies. In different AI-powered technologies, chatbots have emerged as a significant game changer in repositioning customer experience in the contemporary world. The influence of AI chatbots on customer experience is especially notable in Singapore, considering the country's tech-savvy population and role as a regional engine for innovation and digital transformation. One of the primary causes of the broad adoption of AI chatbots in Singapore is the country's constant pursuit of technical improvement and digitization. AI is already being used throughout society to address every aspect of business in Singapore with adequate infrastructure and a government willing to encourage industriousness. In this regard, we sought to examine the multi-dimensional effect of AI chatbots commercialization on the customer in Singapore specifically identifying the best ways to embrace such technology to ensure exceptional customer service and interaction as well as exploring the changing position of AI chatbots as enablers and deliverers of outstanding customer experience management in Singapore.

**Keywords** AI Tool · Chatbots · Customer experience · Singapore · Transformation

D. Tiwari (✉) · E. Rangaswamy
Amity Global Institute, Singapore, Singapore
e-mail: tdivya@singapore.amity.edu

E. Rangaswamy
e-mail: moorthy@singapore.amity.edu

 
M. Al Mubarak (ed.), *Sustainable Digital Technology and Ethics in an Ever-Changing Environment*, Studies in Systems, Decision and Control 237,
https://doi.org/10.1007/978-3-031-86708-8_18

# 1 Introduction

To reproduce human-like interactions with consumers through text or audio exchanges, chatbots are conversational devices that employ artificial intelligence. Digital assistants are specifically engineered to understand client requests, provide relevant information, and perform tasks autonomously or with minimal human intervention. Chatbots use the methodology of natural language processing (NLP) and machine learning algorithms to read, interpret, and provide a response to the customer. Chatbots have seen great technological development over time which has made its pathway for using it in various fields of operations for corporates and businesses like managing customer acquisition, and customer retention while providing good after-sales services with 24*7 support services. Singapore has witnessed a change in the way businesses interact with customers from putting deals in front of them to securing the deal and between this all AI has a major role to play. The impact that AI has created on customer experience is noteworthy, considering the highly technologically advanced population around the country making the impact even larger. Resulted Singapore marks itself as one of the top countries when it comes to innovation and Digital advancement [1].The extent of use of Chatbots in Singapore is due to the business ecosystem in the country, which is largely driven by innovation and technological advancement, which also contributes to faster economic growth in the country. The use of Chatbot technology as customer support helps to cater to different languages with 24*7 availability, making it extremely useful for a country like Singapore which has multi-linguistic customers. This aspect of the Chatbot acts as a basis for including all the customers from different languages and provides them accessibility for their concerns easily, which works best for Singapore because of its multi-linguistic customer base [2]. Furthermore, the business ecosystem in Singapore is structured around extreme competition and high commitment to providing high-quality customer services, and in between this competition, AI chatbots stand as a method for businesses to strengthen their competitive edge.

# 2 Significance of AI Tool-Chatbot on Customer Experience in Singapore

Implementing AI chatbots in Singapore can greatly improve customer experience by offering prompt, tailored, and effective assistance over many platforms, ultimately resulting in heightened customer contentment and allegiance [3]. The implementation of Al Tool chatbots in Singapore, as well as in other regions, has substantial ramifications for customer experience. The following are many salient points that underscore their importance:

1. AI offers 24/7 service aid to customers which helps them to get support at any time. This enhances the overall satisfaction of the customers in Singapore.

2. Technology delivers quick answers to the customers, usually in less than a minute, hence improving overall customer satisfaction.
3. Powerful AI technology enables the easy processing of the collection of customer information data based on which it can provide personalized solutions to each customer. These personalized solutions cultivate the customer's feeling of reliability and worthiness of using this technology.
4. AI chatbots can handle large amounts of data, queries, and concerns of the customers, which enables the technology to provide solutions to a range of problems, ranging from common repeated concerns to complex unique problems. Hence this creates an even higher level of satisfaction in the employees.
5. AI chatbots can handle many customers' queries at the same time, making the technology a high level of scalability to use this technology in businesses, especially in Singapore where businesses have seen Enormous an increase in customer inquiries very frequently.
6. AI chatbots turn out to be a cost-effective method for organizations as they reduce the need for human resources to attend to regular queries, in place use them to better places in the organization for higher productivity.
7. The cost savings ability of the technology allows the business to use the funds for better opportunities and plans which allows the business to attract better customer satisfaction and better returns to the business.
8. Out of the many features of AI technology, the chatbot also allows the business to collect and analyse customer data to list the various concerning points for the business such as what are customers' preferences, and which product or service creates dissatisfaction among the employees. These insights help the business to plan for better product or service delivery to achieve maximum customer satisfaction.
9. Singapore has been a cosmopolitical country, having people from different linguistic backgrounds. AI technology allows people to interact with the tool in various languages which promotes customers' inclusiveness in applying AI tools in business.
10. The AI tool allows the integration of the data from the business's existing CRM, e-commerce platforms, and other applications used by the firm [4]. This allows the business to have a better plan of services for the customer and better efficiency in the working of the organization.

## 3 Review Literature

A study conducted by Araújo and Casais [5] has shown the impact of AI as a shopping assistant on the shopping behaviour of customers. The study shows that customers use AI for various support in shopping. This study shows various valuable information about how the customer has used AI for a better shopping experience which leads to better customer engagement and satisfaction. A study conducted by Behera et al. [6]

shows that many businesses create individual chatbots for the tailored conversation between customers from one product to another.

The study also shows how AI chatbot has the feature of offering personalized support to customers, which creates overall better satisfaction in the customers regarding business. The study also looked at concepts, applications, and different possible usages of this impressive technology which creates an individual offering to each customer which indeed leads to a higher level of satisfaction among customers.

In this study Borsci et al. [7] the author used a detailed process for scale development and pilot testing to test the experience of customers when they're using the chatbot. Study Borsci et al. [7] also studies the importance of using the same in getting better output in different areas of work powered by AI, providing a good viewpoint for professionals and scholars working in the areas of human–computer interaction. Also, the study shows the potential AI must better overall customer service and their experience. The study Hoyer et al. [8] explained well the potential of including technological advancements in the business like AI to provide real-time and more enhanced services to customers by making their overall experience by making them more immersive, personalized, and engaging.

The research done in Joshi [9] investigates understanding the assessment and acceptance of customer service chatbots by the millennial population in India. In the case of, Joshi [9] it was rather more focused on how millennials regard and interact with the chatbot technologies in terms of being used for customer service.

The study enhances knowledge of the factors affecting the uses and attitudes of the millennials towards chatbots in India, with the factors being ease of use, efficiency, and contentment. These insights may be beneficial, especially to organizations willing to understand how to integrate chatbots into their service provision.

The research pursued by Ohio State University Glasgow Faculty proposed in Kasilingam [10] tried to investigate what people think about shopping using smartphone chatting bots and their purposes. The study highlights the likely consequences of smartphone chatbots on the future shopping experience for consumers by investigating several features which are usability and functionality. In their work, Pillai et al. [11] discusses the use of advanced technology in the form of analytics AI-driven chatbots to manage employee experience for example. In their practical research, Pillai et al. [11] looked at different factors affecting the implementation of Employee Experience chatbots. The research delves into how any of these technologies might be taken up and what one might expect as benefits and challenges, especially to increase the level of employee satisfaction and involvement. The purpose of the research conducted by Tan [12] is to find out how consumers of general insurance products in Singapore respond to artificial intelligence chatbots in gap management. Tan [12] examined the way consumers responded to the integration of artificial intelligence chatbots in the insurance market. This study included many such characteristics as trust, contentment, and perceived value attached by respondents to chatbots for insurance purposes. The results of this study contribute new information on the level and dynamics of AI usage for the improvement of insurance distribution processes and customer experience in a country like Singapore. Valentine [13] made a very meticulous search in this effort to assess the interventions by AI technology in the customer journey including the initial perception phase to the remedial phase after purchase. The study shows the different dynamics that companies manage for better customer attention and relations using AI technologies and its impact on customer satisfaction and loyalty for the longer term for the product.

## 4 Objective of the Study

The analytical objectives were pinpointed to investigate the Impact of AI Tool-chatbot on Customer experience in Singapore. In more specific terms:

- To study the Impact of AI Tool-chatbot on Customer experience in Singapore.

## 5 Research Methodology

Secondary data can be divided into two categories: internal and external. Internal data is generated within the organization and includes annual reports and other easily accessible records. External data is generated and distributed by a variety of sources, including private organizations, commercial publications, and researchers. Knowledge is derived from a thorough literature review, which is published by either an organization or an individual. Information gained for a specific purpose usually refers to data gathered by an individual in Singapore. Quantifiable research is a systematic and rigorous approach to discovery that uses bigger sample sizes to collect and analyse data, information, and facts. The quantitative study was approved for this research. The data was analysed using a variety of approaches, including baseline surveys using structured questionnaires. Nonetheless, the survey methodology was chosen for this investigation. The survey approach is based on a structured opinion survey, in which questions are presented in a preset order and distributed to representative samples of the community to collect responses. Customers have been given surveys that use a five-point scale. Responders were given a designated space to provide specific information about various linked topics. During the stated time window, 110 Singaporean individuals responded. The sample picked includes a complete list of clients from various industries who use AI tool chatbots.

**Table 1** Demographic profile of respondents

| | Demographic | (N = 110) | Percentage (%) |
|---|---|---|---|
| Gender wise | Male (M) | 73 | 66.37 |
| | Female (F) | 37 | 33.63 |
| Age groupwise | 20–25 | 31 | 28.18 |
| | 25–30 | 42 | 38.18 |
| | 30–35 | 20 | 18.18 |
| | 35–40 | 08 | 7.27 |
| | 40 and above | 09 | 8.18 |
| Marital Status wise | Married | 42 | 38.18 |
| | Unmarried | 54 | 49.09 |
| | Others (Single/Divorced/ Widow) | 14 | 12.72 |
| Qualification wise | Graduation | 36 | 32.72 |
| | PG | 28 | 25.45 |
| | Professional degrees and diploma courses | 41 | 37.27 |
| | Others skill Courses | 05 | 04.54 |
| Mode of using AI chatbots | In laptop | 68 | 61.81 |
| | In mobile | 42 | 38.18 |
| Customer experience (Sector wise) | Retail/Manufacturing sector | 14 | 12.72 |
| | Service provider | 43 | 39.09 |
| | E-commerce Business | 32 | 29.09 |
| | Foreign/Import/Export | 21 | 19.09 |

# 6 Data Analysis and Interpretation

## *6.1 First, Let's Discuss the Demographic Profile in Table 1*

## *6.2 Inferential Study on the Factors of AI Tool-Chatbot*

**Null Hypothesis (H-1)**: No significant variance in M & F experience using Al Tool-chatbot in Singapore.

**Interpretation**: The impact of AI tool-chatbots on customer experience in Singapore might be influenced by various impactful factors, and it's quite important to consider both female and male customers in such a scenario. While customer experience depends on several individuals as well as contextual-based elements, here are indicating some of the significant factors that may play a strong role in the impact of AI chatbots for both female and male customers. As of now seen in Table 2: T-test for significant variance in M & F experience using AI Tool-chatbot in Singapore, the P_Value is < 0.001, so the hypothesis is not at all accepted and therefore, "there is significant variance in M & F experience in Singapore of using AI Tool-chatbot". Hence, it's proved that there is significant variance in M & F experience in Singapore of using AI Tool-chatbot. Only one factor i.e., "brand reputation and trust (P_Value = 0.124*)" is greater than the 5% significance level.

**Null Hypothesis (H-2)**: No significant variance in married and unmarried experience in Singapore of using AI Tool-chatbot.

**Interpretation**: The impact of AI tool-chatbots on customer experience in Singapore might be influenced by various impactful factors, and it's quite important to consider both married and unmarried customers in such a scenario as shown in Table 3 T-test for significant variance in married and unmarried experience in Singapore using AI Tool-chatbots. While customer experience depends on several individuals as well as contextual-based elements, here are some of the significant factors that may play a strong role in the impact of AI chatbots for both married and unmarried customers. As of now, the P_Value is < 0.001, so the hypothesis is not at all accepted and therefore,

**Table 2** T-test for significant variance in M & F experience using AI Tool-chatbots in Singapore

| Factors | Gender wise | | | | (T Value) | (P Value) |
|---|---|---|---|---|---|---|
| | (M) | | (F) | | | |
| | Mean scores | Std._Dev | Mean scores | Std._Dev | | |
| Customer expectations and preferences | 21.14 | 3.432 | 19.87 | 2.174 | 2.887 | < 0.045* |
| Brand reputation and trust | 27.65 | 2.567 | 27.42 | 2.215 | 3.998 | 0.124* |
| Quality of natural language processing (NLP) | 21.78 | 1.289 | 29.54 | 2.216 | 3.768 | < 0.001** |
| Training and maintenance | 31.45 | 2.379 | 33.45 | 3.745 | 2.309 | < 0.001** |
| Integration with existing systems | 33.61 | 2.346 | 24.78 | 3.162 | 3.521 | < 0.001** |
| User adoption and acceptance | 25.89 | 2.789 | 29.36 | 2.678 | 3.765 | < 0.001** |
| Regulatory environment | 22.76 | 2.230 | 31.62 | 3.234 | 2.675 | < 0.001** |
| Technology infrastructure | 31.68 | 1.678 | 24.65 | 3.923 | 3.167 | < 0.001** |
| Overall contribution | **215.96** | **22.247** | **220.69** | **21.987** | **4.654** | **< 0.001**** |

** @ 1% level of significance; * @ 5% level of significance

"there is significant variance in married and unmarried experience in Singapore of using AI Tool-chatbot". Hence, it's proved that there is significant variance in married and unmarried experiences in Singapore of using AI Tool-chatbot. Only one factor i.e., "training and maintenance (P_Value = 0.167*)" is greater than the 5% significance level.

**Table 3** T-test for significant variance in married and unmarried experience in Singapore using AI Tool-chatbot

| Factors | Marital Status wise | | | | (T_Value) | (P_Value) |
|---|---|---|---|---|---|---|
| | (Married) | | (Unmarried) | | | |
| | Mean scores | Std._Dev | Mean scores | Std._Dev | | |
| Customer expectations and preferences | 21.23 | 3.432 | 25.85 | 4.456 | – 2.546 | < 0.023* |
| Brand reputation and trust | 22.67 | 4.245 | 28.82 | 2.467 | – 3.875 | < 0.001** |
| Quality of natural language processing (NLP) | 26.13 | 3.878 | 31.93 | 3.378 | 2.287 | < 0.001** |
| Training and maintenance | 28.08 | 4.323 | 33.84 | 3.846 | – 2.954 | 0.167* |
| Integration with existing systems | 21.97 | 5.687 | 22.47 | 2.919 | – 3.845 | < 0.001** |
| User adoption and acceptance | 31.19 | 3.799 | 26.87 | 3.748 | – 2.908 | < 0.001** |
| Regulatory environment | 26.85 | 3.261 | 27.23 | 4.384 | 3.769 | < 0.001** |
| Technology infrastructure | 32.23 | 2.645 | 26.89 | 2.763 | 2.345 | < 0.00** |
| Overall contribution | **210.35** | **25.678** | **223.90** | **19.843** | **3.998** | **< 0.00**** |

** @ 1% level of significance; * @ 5% level of significance

## 7 Finding of the Study

This study reveals that the efficacy of AI chatbots is heavily influenced by the accessibility and dependability of internet connectivity and digital infrastructure. The advanced technology infrastructure in Singapore often facilitates the smooth functioning of these systems, hence enhancing the overall consumer experience. A multinational society, Singapore exhibits a wide range of language preferences and cultural peculiarities. To make sure the chatbots provide a service that meets the requirements of different demographics, the AI chatbots must work on this functionality and should work on these differences for better service to be provided for customers. Meanwhile, keeping par with data security to make sure the customer data is secure, the AI chatbot must comply with required standards such as the Personal Data Protection Act (PDPA). This compliance not only helps to meet regulatory standards but also impacts positively the brand image of the AI Chatbot. Also, customer engagement plays an important role when it comes to creating an impact on AI chatbots. These customer engagements are impacted by various factors like how friendly the user interface is, how easy it is to use, and how aware customers are of its usage with the level of digital literacy. Continuous Updates, Requirements for keeping it up to date, and learning to use it perfectly to get the best results are some important factors that contribute to the smooth and effective working of AI Chatbots. Also, keeping an eye on the feedback of customers regarding what they face as difficulties while operating the chatbots and working on those difficulties enhances overall efficiency and helps to provide better customer service. The amount of good reputation and trust that the companies can create among the customers will have a whole and direct impact on the customer's experiences while they are using the AI Chatbots. Customers while using the AI chatbots of companies that have created a strong reputation in their eye, tend to show more trust and belief in the information shared by the chatbot. Furthermore, the study shows that understanding customer expectations and choices also plays an integral role in forming the chatbot features as they impact the overall satisfaction of the customers. The interface of the Chatbot is impacted by various factors such as the platform of usage, expected response time, and level of customization.

## 8 Conclusion

Chatbots are conversational computers that employ AI to simulate natural human interaction in textual or auditory forms of communication. Chatbots work like computers which provide information for a set of questions but with a real-time interaction feature that enables the customer to experience personalized services. The interpretation and analysis of the conversation between both parties are finding information for the customer and doing small activities or tasks for customers without any human involvement. Chatbots use Natural Language Processing (NLP), Pre-Fixed scripts, and Machine Learning (ML) to interpret the conversation and provide relevant information as solutions to the customers. Over the period chatbots improved immensely from the day they were created; they can now be used in many industries such as E-commerce, Virtual assistance, and Client service can be implemented by using different channels and platforms such as website chat, social networking, and voice compatibility. The usage of AI Chatbots in CRM also goes along with the requirement of the Singapore business ecosystem as the country has a diverse population from different native language backgrounds and these AI tools meet the requirement of the different native languages customers to find the solution in their language by using the feature of different dialects in the tool which is best for a country like Singapore. Also, Singapore has marked itself as a leading country when it comes to the usage of innovation and technology in society overall for having higher productivity.

In addition to this, Companies by using the AI tool Chatbot in this tough competition market can gain a competitive edge over other competitors by providing better customer services with the help of AI chatbot and mark their standing in the market, especially in such activities which help in customer acquisition and customer retention. However, companies cannot run away from the fact that there are some obstacles to work around for the successful implementation of AI tools. A few of the major concerns to look at will be finding the best timing to involve human interaction rather than automated one, securing the information of customers, and improving the AI tool while using it and learning to incorporate it better in business over time.

## References

1. Kumar, P., Mokha, A.K.: Electronic customer relationship management (E-CRM) and customer experience: evidence from the banking industry. Natl. J. Arts Commer. Sci. Res. Rev. **7**(1), 32–37 (2020). https://www.kaavpublications.org/abstracts/electronic-customer-relationship-management-e-crm-and-customer-experience-evidence-from-the-banking-industry
2. Kapooria, P.J.: Product innovation as a key to exceptional customer experience: case of pinnacle specialty vehicles. Kaav Int. J. Econ. Commer. Bus. Manage. **6**(1), 237–240 (2019)
3. Dembla, N.: Impact of artificial intelligence on e-commerce. Kaav Int. J. Econ. Commer. Bus. Manage. **6**(1), 56–63 (2019). https://www.kaavpublications.org/spiabstracts/impact-of-artificial-intelligence-on-e-commerce
4. Chaudhary, A.: Indian e-commerce market on the edge: understanding the impact of the internet. Kaav Int. J. Econ. Commer. Bus. Manage. **5**(1), 358–361 (2018)
5. Araújo, T., Casais, B.: Customer acceptance of shopping-assistant chatbots. In: Marketing and Smart Technologies: Proceedings of ICMarkTech 2019, pp. 278–287. Springer Singapore (2020)
6. Behera, R.K., Bala, P.K., Ray, A.: Cognitive Chatbot for Personalized (2021)
7. Borsci, S., Malizia, A., Schmettow, M., Van Der Velde, F., Tariverdiyeva, G., Balaji, D., Chamberlain, A.: The Chatbot usability scale: the design and pilot of a usability scale for interaction with AI-based conversational agents. Pers. Ubiquit. Comput. **26**, 95–119 (2022)
8. Hoyer, W.D., Kroschke, M., Schmitt, B., Kraume, K., Shankar, V.: Transforming the customer experience through new technologies. J. Interact. Mark. **51**(1), 57–71 (2020)
9. Joshi, H.: Perception and adoption of customer service chatbots among millennials: an empirical validation in the Indian context. In WEBIST, pp. 197–208 (2021)

10. Kasilingam, D.L.: Understanding the attitude and intention to use smartphone chatbots for shopping. Technol. Soc. **62**, 101280 (2020)
11. Pillai, R., Ghanghorkar, Y., Sivathanu, B., Algharabat, R., Rana, N.P.: Adoption of artificial intelligence (AI) based employee experience (EEX) chatbots. Inf. Technol. People **37**(1), 449–478 (2024)
12. Tan, L.H.: Consumer reaction to the use of artificial intelligence chatbot on the distribution of general insurance in Singapore (2023)
13. Valentine, D.: Artificial intelligence: implications in the customer journey artificial intelligence & chatbots: implications in the customer experience. Doctoral dissertation (2022)

# Resilience and Perceived Stress—A Comparative Study Among Performing Arts Versus Non-performing Art Students in India

**P. V. Ankita, Mehak Mehreen, U. P. Riddhi Raj, S. Subasree, and Soumya Simon**

**Abstract** This comparative study investigates resilience and perceived stress levels among performing arts students and non-performing art students in India. Recognising the potential stressors and resilience-building factors with-in the performing arts realm, the study aims to shed light on the psycho-logical well-being of individuals engaged in creative endeavors. Through a mixed-methods approach, data was collected to examine various dimensions of resilience and perceived stress, including coping mechanisms, social support, and personal experiences. The findings are anticipated to contribute to a deeper understanding of the intersection between artistic engagement and mental health, offering insights for educators, practitioners, and policymakers in promoting well-being within the performing arts community.

**Keywords** Resilience · Perceived stress · Coping mechanisms

## 1 Introduction

This study examines the role of resilience, perceived stress, and artist type (performing or non-performing) in the psychological well-being of individuals in the arts. Resilience helps artists navigate challenges, while perceived stress indicates the impact of these challenges. By comparing performing and non-performing artists, the study explores differences in their resilience and stress levels. The research includes both quantitative and qualitative methods. Quantitative analysis assesses differences in resilience and stress perception. Qualitative interviews delve into coping strategies. This mixed-methods approach provides a comprehensive understanding of how artists manage stress and build resilience in their professional lives.

P. V. Ankita (✉) · M. Mehreen · U. P. Riddhi Raj · S. Subasree · S. Simon
Kristu Jayanti College Autonomous, Bengaluru, India

S. Simon
e-mail: soumya@kristujayanti.com



M. Al Mubarak (ed.), *Sustainable Digital Technology and Ethics in an Ever-Changing Environment*, Studies in Systems, Decision and Control 237,
https://doi.org/10.1007/978-3-031-86708-8_19

### 1.1 Perception of Stress, Performing Arts and Resilience

*Perception of stress* is an individual's assessment of their stress levels influenced by factors like personality, coping methods, social support, and culture. Lazarus and Folkman (1984) proposed the Cognitive Evaluation Theory [1] suggesting people analyze stress through cognitive evaluation. Research on this theory explores how individuals' stress perceptions are affected by their appraisals of stressors and coping resources. Performing artists' stress perception is complex, influenced by career nature, individual differences, and external pressures in the unpredictable and competitive performing arts industry, leading to heightened stress levels.

*Performing arts* encompass a variety of expressive forms such as music, dance, and theater, showcasing human creativity across different cultures. They contain a wide range of artistic expressions that showcase human ingenuity and are present in many different sectors of intangible cultural heritage, albeit to varying degrees. It serves as a medium to depict religious rituals, social activities, and emotional experiences.

*Resilience* is the ability to positively adapt and recover from adversity or significant life events, impacting an individual's overall well-being. This is crucial for both performing and non-performing artists as it helps manage stress, setbacks, and maintain psychological health. It also aids in coping with rejection, criticism, and uncertainties in their careers. Resilience is essential in preventing long-term negative effects like anxiety, depression, and burnout.

### 1.2 Theoretical Framework

In 1973, "Buzz" Holling introduced resilience theory, differentiating between ecological resilience, the ability of a system to absorb disturbances while maintaining its core functions, and engineering stability, the system's capacity to return to equilibrium after disruption [2]. Over time, resilience theory evolved to include both adaptation to new conditions and recovery to previous states (Folke 2006) [1]. Sheldon Cohen expanded the understanding of stress through the Perceived Stress Scale (PSS) in 1983, which built on Lazarus and Folkman's concepts of stress perception and cognitive appraisal, providing a measurable approach to perceived stress in everyday life.

### 1.3 Need and Significance

Studying resilience and perceived stress among performing artists is crucial due to the unique challenges they face, such as pressure to perform and financial instability.

Understanding how resilience affects their ability to cope can lead to tailored interventions and support systems to improve their well-being. The mental health implications are significant, with high levels of stress, anxiety, and burnout in the performing arts sector. Exploring resilience and stress perception can provide insights into mental health hurdles and strategies for promoting psychological well-being. Resilience also contributes to sustaining artistic excellence and enriching communities and cultural heritage, making this research essential for advancing individual welfare and societal impact in the industry.

### *1.4 Research Gap*

There is multiple research which says that people who are good at managing emotions, people with high EQ, people who are mindful, etc. can perform better forms of art. Here we are trying to prove the other way round where we are assuming that the practice of arts makes the person better at all other aspects of life compared to those who don't have any expression of art in their life being practiced on a daily basis.

Following this introduction, Sect. 2 provides a review of the relevant literature. Section 3 describes the methodology used in this study, while Sect. 4 presents the results and discussion Sect. 5 discusses the summary and conclusion of the entire paper.

## 2 Review of Literature

A study by Mavely et al. (2023) explored dentists' stress and lack of resilience during COVID-19. Emotional, social, and financial well-being were significantly affected. Stressors included discipline, isolation, perfectionism, compromised therapy, and time constraints. The research underscores the importance of stress management and resilience for dental workers' mental health amidst the pandemic [3].

Banisudha's (2023) study analyzed the impact of performing arts in higher education. The findings revealed that engaging in the arts boosts cognitive skills, helps students navigate social situations, instills civic responsibility, and builds the confidence they need to step outside their comfort zones [4].

Kigozi conducted a study on citizenship in Uganda, exploring how various community groups use performing arts to express culture and promote collaboration. The study highlights community-based performing arts as a tool for strengthening citizenship and bridging divides between different groups within the community [5].

Nixon and Leon (2023) conducted a study on the factors influencing the adaptability of performing arts professionals during the COVID-19 pandemic. Participants were interviewed to collect data on their experience during the lockdown, revealing a reliance on maladaptive coping strategies due to lack of organizational support [6].

In their 2023 study, Nixon and Davis investigate the impact of the COVID-19 pandemic on the resilience of permanently employed workers in England's performing arts sector. Differentiating their research from existing literature that primarily focuses on freelance artists and industry-wide effects, they employ a qualitative methodology involving semi-structured interviews. The findings reveal that many participants displayed maladaptive coping behaviours instead of resilient strategies. Notably, the workers expressed discomfort with the expectation of personal resilience, highlighting a significant gap in organizational support during the industry's reopening. This research emphasizes the critical need for enhanced institutional frameworks to cultivate resilience among workers in the performing arts sector [7].

In a 2015 paper, Pasquinelli and Sjöholm examine the spatial practices of resilience among visual artists in London amidst the shifting dynamics of precarious work in post-Fordist production regimes. They employ a multi-scalar approach to demonstrate how artists adapt to temporary and flexible work arrangements by developing skills for strategic adaptability. The authors argue that a resilient artistic career is not only shaped by individual effort but is also a geographically situated process, often unfolding through alternative communication channels and spaces outside the mainstream art system. This highlights the importance of context and innovative networking in fostering resilience in the face of career challenges [8].

# 3 Method

## 3.1 *Research Question*

To measure the level of resilience and stress perception among students who are performing arts students and non-performing arts students.

## 3.2 *Objectives*

- To observe if there is any significant difference in resilience among performing arts students and non-performing arts students.
- To observe if there is any significant difference in perception of stress among performing arts students and non-performing arts students.
- To identify the coping strategies employed by performing arts students.

## 3.3 Sample

The sample involves people of both undergraduate and postgraduate students pursuing their course in performing arts as well as students who don't pursue performing arts. Students should fall under the age criteria of 18–25 which is young adulthood.

**Inclusion criteria**

Students pursuing performing art must be full time students in that course.

**Exclusion criteria**

Anyone who falls out of the age criteria even if they are pursuing the course right now or not.

## 3.4 Tool Description

***Tool 1—Perceived Stress scale by Sheldon Cohen***: the validity is high with Cronbach's alpha ranging from 0.71 to 0.91, with test–retest reliability over 0.70.

***Tool 2—Resilience questionnaire by Dr Nicholson***: The validity is high with an alpha value of 0.800.

## 3.5 Procedure

Participants were asked to fill out a Google form for the quantitative portion of the study, with data collection following. Resilience and stress levels were assessed using questionnaires among performing arts and non-performing arts students. Results were then analyzed using T-test and SPSS software. The qualitative aspect involved one-on-one interviews with participants, with thematic analysis applied to the responses. This approach aimed to gather data for further analysis.

## 3.6 Research Design

The research design is a mixed method. It uses both qualitative and quantitative.

### 3.7 Hypothesis

**H1:** There is no significant difference in resilience among performing arts students and non-performing art students.

**H2:** There is no significant difference in stress perception among performing arts students and non performing art students.

### 3.8 Ethics

The research has maintained research ethics. The confidentiality was maintained and also informed consent was obtained.

## 4 Results and Discussion

### 4.1 Quantitative Results T Test

The T Test is the statistical analysis performed for these independent variables of stress and resilience. Prior to that, normality tests were performed to ensure and to determine whether the dataset has a normal distribution. The normality test proved that the data has normal distribution. Interpreting a non-significant result (high p-value) in Levene's test suggests that there is no statistically significant difference in the variances between the groups. Therefore, it is assumed that the assumption of equal variances is met, which is one of the assumptions required for the standard t-test for independent samples. With Levene's test result of 0.805, we proceed with the standard independent samples t-test, assuming equal variances, for comparing the means of the groups. This implies that the variability of the total stress variable across the groups is not significantly different, so the assumptions for the t-test are met.

According to Table 1, we find that with a p-value of 0.289 greater than 0.05, the null hypothesis of data being normally distributed is not rejected, indicating insufficient evidence of significant deviation from normal distribution. Figures 1 and 2 show a straight diagonal line from the bottom-left to the top-right ($y = x$) indicating that the two are normal distributions.

The p-value for Levene's test is above 0.05, indicating that variances between groups are not significantly different. Therefore, assumption of equal variances is met, allowing standard independent samples t-test. The variances in stress variables across groups are not significantly different, meeting t-test assumptions. Levene's test result of 0.902 confirms equal variances, supporting use of t-test for comparing means of groups in resilience.

Table 1 Tests of normality

| | Kolmogrov-Smirnov[a] | | | Shapiro–Wilk | | |
|---|---|---|---|---|---|---|
| | Statistic | df | Sig. | Statistic | df | Sig. |
| Total stress | 0.071 | 106 | 0.200* | 0.985 | 106 | 0.265 |
| a. Lilliefors Significance Correction | 0.074 | 106 | | 0.985 | | |
| | 0.184 | | | 0.289 | | |

* This is a lower bound of the true significance
[a] Lilliefors Significance Corrections

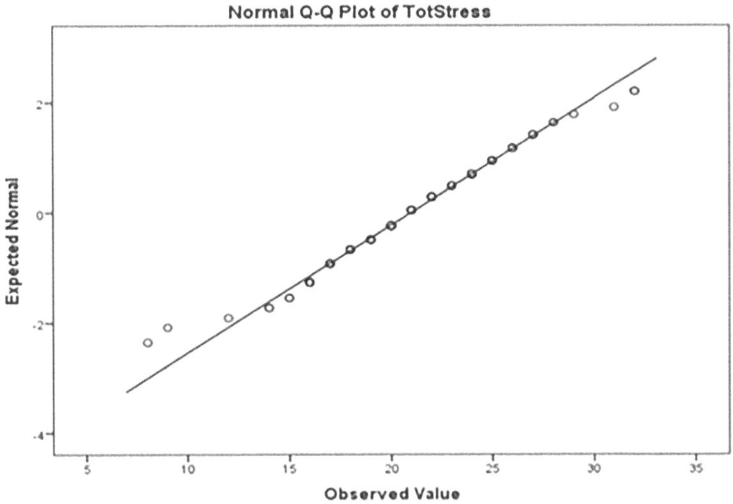


**Fig. 1** Normal Q-Q plot for TotStress

According to Table 2, the t-value compares mean differences to variability within groups. With t-values of 0.405 for total stress and 0.250 for total resilience, and df of 104, p-value and gt; 0.05 leads to not rejecting the null hypothesis, suggesting no significant difference between stress and resilience means.

## 4.2 *Qualitative Results (Interview)*

(1) Name: RB, Gender: Male, Date of interview: 08/04/2024

**Theme: Academic Stress and Coping Mechanisms**: Subtheme: Academic Challenges and Pressure, directionlessness, Emotional Response, Coping Strategies.

Verbatim: "I use dance as a means to express and process emotions, fostering catharsis and emotional well-being."

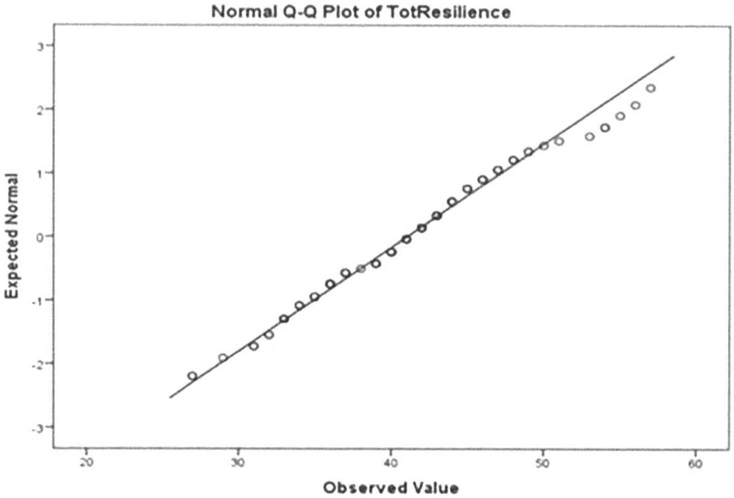


**Fig. 2** Normal Q-Q Plot TotResilience

**Table 2** T test for the independent samples

| Table for independent sample test for total perceived stress and total social desirability | | | | |
|---|---|---|---|---|
| | Performing artist | | Non-performing artist | |
| | M | SD | M | SD |
| Total stress (M + SD) | 21.16 | 4.434 | 20.78 | 4.244 |
| Total resilience (M + SD) | 41.18 | 6.221 | 40.88 | 6.070 |
| Sample size | | 56 | | 50 |
| T-value | | 0.450 | | 0.250 |
| P-value | | 0.653 | | 0.803 |

(2) Name: S.G, Gender: Female, Date of Interview: 08/04/2024

**Theme: Emotions and Coping**: Subtheme 1: Emotions, Challenging Emotions, Physical activities, Individual Coping, Internal Struggle, self-reliance and emotional stoicism. Realization. Verbatim: " Dancing isn't just a physical activity; it's how I release everything that's bottled up inside, especially when I'm overwhelmed."

(3) Name: SR, Gender: Female, Date of interview: 08/04/2024

**Theme**: Subtheme 1: Challenges: Adversity/Pressure, Difficult Feelings, Sustained Injury.

Subtheme 2: Maintaining Balance and Resilience-Consistent Practice Schedules: Setting Goals, Staying Discipline. Verbatim: " Dancing under pressure feels like I'm pushing through a storm—there's the constant adversity, the weight of difficult feelings, and sometimes the pain of a sustained injury, but I just keep going because stopping isn't an option as I have to stay focused and disciplined."

### 4.3 Discussion

Through the quantitative analysis it is found that both resilience and stress perception levels remained constant between both the groups—performing arts and non performing arts students remain the same. And through the qualitative analysis it is found that there is no significant difference between both the groups. While one of the limitations is the population size in itself from previous research, the relationship studied was the other way around. It said that people who have good resilience and stress perception. A better mental well -being as a whole are the ones who are good at activities like performing arts apart from just educational qualification but the aspects of creativity, expressions etc.

## 5 Summary

This study summarizes findings, infers from gathered data, and offers recommendations for future research in psychology and education. The study compared stress and resilience perceptions between performing and non-performing arts students in India using a mixed research approach with 104 participants. IBM SPSS program was used for data analysis.

T test significance level shows 0.65 for total stress and 0.63 for resilience. The significance level resilience and stress perception levels remained constant between both the groups.

### 5.1 Limitations

1. Sample size is too small.
2. The area is only limited to Bengaluru.
3. Cultural Context: The study may overlook the influence of cultural factors on resilience and perceived stress within the Indian context

### 5.2 Future Implications

A comparative study on resilience and perceived stress among performing arts students versus non-performing artists in India can have several future implications:

Tailored support programs can be developed for performing arts students and artists, mental health services improved, and resilience training programs implemented based on study findings to enhance coping skills in performing arts institutions and communities.

### *5.3 Conclusion*

There is no significant difference between performing and non-performing arts in terms of resilience and stress perception. In conclusion, while the study and literature review provide valuable insights into the complexities of resilience and perceived stress, and their implications for mental health among individuals who are performing artists and non-performing artists, there are notable limitations and opportunities for future research. The limitations highlighted, such as sample size restrictions, cultural specificity, and language barriers, point towards the need for more inclusive and comprehensive research methodologies. Future studies should strive to expand sample diversity, employ inclusive sampling methods, conduct cross-cultural comparisons, utilize multilingual measures, conduct longitudinal studies, and incorporate intersectional analyses to provide a more nuanced understanding of resilience and perceived stress across diverse populations. By addressing these suggestions, researchers can advance knowledge in this field and contribute to developing interventions to promote mental health and well-being among performing arts individuals. Overall, this research serves as a foundation for further exploration and underscores the importance of inclusive approaches to performing arts students.

## References

1. Folkman, S.: Stress: appraisal and coping. In: Gellman, M.D., Turner, J.R. (eds.) Encyclopedia of Behavioral Medicine. Springer, New York, NY (2013). https://doi.org/10.1007/978-1-4419-1005-9_215
2. Holling, C.S.: Resilience and stability of ecological systems. Annu. Rev. Ecol. Syst. **4**(1), 1–23 (1973). https://doi.org/10.1146/annurev.es.04.110173.00024
3. Mavely, A., Manohar, S., Basavaraju, S.: Perceived stress and resilience among dentists during the COVID-19 outbreak (2023)
4. Banisudha, B.: Analyzing the impact of performing arts in higher education. Int. J. Flexibility Mov. Res. (2023). https://www.ijfmr.com/research-paper.php?id=1549
5. Kigozi, B. Citizenship: the role of performing arts in Uganda. J. Visual Perform. Arts 1 (2023). https://doi.org/10.58721/jvpa.v1i1.245
6. Nixon, M., Davis, L.: Factors influencing the adaptability of performing arts professionals during the COVID-19 pandemic. J. Perform. Arts Res. **29**(4), 123–139 (2023)
7. UNESCO—Performing arts (such as traditional music, dance and theatre). Retrieved from 18 April 2024 (2022). From https://ich.unesco.org/en/performing-arts-00054
8. Nixon, M., Davis, L.: Understanding the factors that affected the resilience of performing arts workers during the COVID-19 pandemic. Cultural Trends (2023). https://doi.org/10.1080/09548963.2023.2240254

# Unraveling It Sophistication: Driving AIS Success in Medan's Food and Beverage MSMEs

**Risanty, Dede Ryansyah, Sambas Ade Kesuma, Adi Syah Putra, and Yuni Annastia Suherman**

**Abstract** This study examines the influence of IT sophistication on the success of Accounting Information Systems (AIS) in micro and small enterprises (MSEs) within the food and beverage sector in Medan City. A cross-sectional survey method was employed, collecting data from 108 cafés using questionnaires. The research focuses on four dimensions of IT sophistication: technological, informational, functional, and managerial. The findings reveal that technological, informational, and managerial sophistication positively affect AIS success, while functional sophistication shows no significant impact. This study underscores the importance of enhancing managerial capabilities and aligning technology with business needs to achieve optimal AIS outcomes, contributing to the understanding of digital transformation strategies in MSEs.

**Keywords** Technological sophistication · Information sophistication · Functional sophistication · Managerial sophistication · AIS success · Micro small enterprise

## 1 Introduction

Accounting Information System (AIS) is a procedure for collecting and recording data through identification and processing to produce useful information [17]. AIS benefits various business sectors, including Micro, Small, and Medium Enterprises (MSMEs), in mitigating and overcoming financial challenges [6]. However, the use of

Risanty (✉) · D. Ryansyah · S. A. Kesuma · A. S. Putra · Y. A. Suherman
Faculty of Economics and Business, University of North Sumatra, Medan, Indonesia
e-mail: risanty@usu.ac.id

S. A. Kesuma
e-mail: sambas@usu.ac.id

Y. A. Suherman
e-mail: yuni@usu.ac.id



M. Al Mubarak (ed.), *Sustainable Digital Technology and Ethics in an Ever-Changing Environment*, Studies in Systems, Decision and Control 237,
https://doi.org/10.1007/978-3-031-86708-8_20

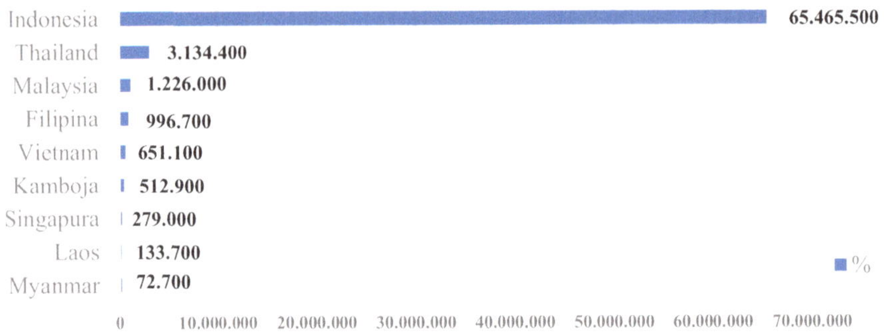


**Fig. 1** ASEAN countries' MSMEs by September 2022

AIS in several developing countries remains suboptimal, especially among MSMEs [3].

Indonesia has the highest number of MSMEs in the ASEAN region, totaling approximately 65.46 million units (Fig. 1). MSMEs play a vital role in Indonesia's economic growth, contributing 60.3% of GDP, employing 97% of the total workforce, and accounting for 14.4% of national exports [1]. Therefore, the Indonesian government needs to support MSME performance improvement through digitalization strategies and technological adoption [1].

MSMEs in the food and beverage sector hold the peak of technology adoption in the industrial sector in Indonesia (Fig. 2). Based on a survey by Boston Consulting Group and Telkom Indonesia [4], it shows that of the 3700 local MSMEs surveyed, 71% of MSMEs in the food and beverage sector have utilized technology in finding suppliers, and 69% of technology utilization in reaching customers. Furthermore, 26% of MSME respondents in the food and beverage sector have digitized in operating a business, be it sales, financial management, and other activities [4]. Of the total, there are also only 23% of MSME respondents in the food and beverage sector who use digitalization tools as a tool to analyze the market [4].

Information systems can be said to be successful if the company has done good planning in designing and implementing AIS in its company [10]. The utilization of AIS will not be optimal if the company does not have human resources and also financial capabilities ready to implement the system [20].

Contingency Theory in the Context of IT Sophistication in the research of Raymond and Paré [16], has made observations to see the influence of IT sophistication which was initially carried out in the context of MSMEs. As a development stage, recent studies try to apply the dimensions of IT sophistication in a broader context [2, 10, 14, 15]. The research explores four variables of IT sophistication: technological, information, functional, and managerial the main variables of IT sophistication contingent model are technological, information, functional, and managerial. These four variables will determine the success of AIS in MSMEs [15]. Therefore, this study uses the IT Sophistication Contingency Model to determine the success of AIS in Micro Small Enterprises in the Food and Beverage Sector in Medan City.

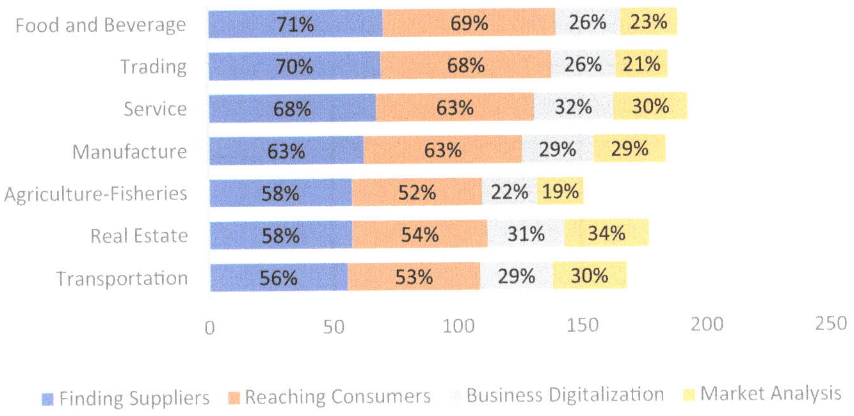


**Fig. 2** Proportion of MSMEs utilizing technology

## 2 Literature Review

### 2.1 *Accounting Information System (AIS)*

AIS is defined as the capability to access, manage, and use information effectively by applying information technology [12]. An AIS is an integrated system involving procedures, users, and technology designed to collect, process, and present accounting information [17].

### 2.2 *Success of AIS*

AIS success relates to the extent of perceived benefits that system users gain from using Accounting Information Systems [2]. AIS is considered successful when structured management through comprehensive planning of financial resources, human resources, information needs, implementation targets, and post-implementation strategies is carried out [10].

### 2.3 *Conceptual Framework*

The study's conceptual framework is based on contingency theory, which explains that there is no fixed factor in measuring organizational performance [19]. As in Fig. 3, in the context of IT sophistication, the four main variables that affect the performance

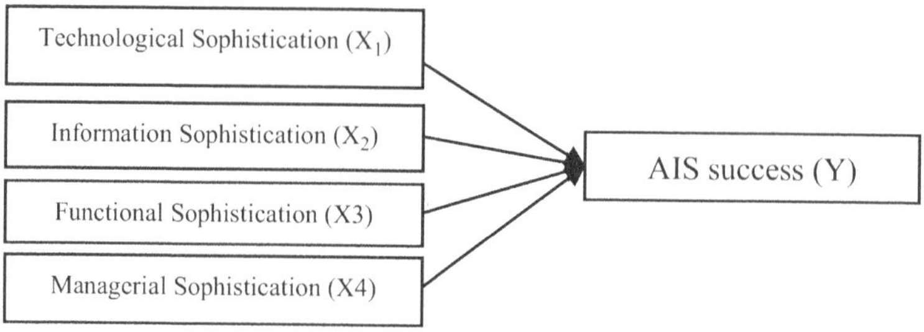


**Fig. 3** The study model

of information systems in a company include technological sophistication, information sophistication, functional sophistication, and managerial sophistication [2, 10, 16]. Technological Sophistication is closely related to the use of various types of hardware in supporting the performance of information systems [16]. Information sophistication is the use of integrated applications that have advanced features such as order entry, budget and production variance, budgeting, planning, and control [2]. The indicators of these variables are application portfolios and application integration [10, 16].

The role and participation of users in developing information systems is functional sophistication [11, 13, 5]. The indicators of these variables are IS user specialists [10]. Managerial Sophistication is related to top managers' participation in organizing IS implementation [16]. The indicators of these variables are the IT usage planning process [10].

## 3 Research Methods

This study utilized a survey method using questionnaires distributed directly to the target population. The population consisted of all cafés in Medan City as recorded by the Medan City Tourism Office in 2023, amounting to 156 cafés. Based on Cohen's sampling table, 108 respondents were selected as the sample size. Primary data were collected through surveys using a Likert scale of 1–5 points, administered via Google Forms.

## 4 Result

Several statistical tests were conducted, including validity, reliability, normality, multicollinearity, and heteroscedasticity tests, followed by multiple linear regression analysis, F-tests, and T-tests. These analyses were used to evaluate the impact of

**Table 1** Reliability test result

| Variabel | Nilai *Cronbach's Alpha* |
|---|---|
| AIS success (Y) | 0.612 |
| Technological sophistication ($X_1$) | 0.790 |
| Information sophistication ($X_2$) | 0.908 |
| Functional sophistication ($X_3$) | 0.802 |
| Managerial sophistication ($X_4$) | 0.649 |

independent variables on AIS success. The results showed that technological, informational, and managerial sophistication significantly contributes to AIS success, while functional sophistication does not.

## *4.1 Validity Test*

The validity test is used to see the validation of the questionnaire [18]. If the r value is positive and the number of r count > r table, then the variable is valid. Based on all the results, namely r count > r table, it means that all AIS Success statement items ($Y$), the four variables are valid.

## *4.2 Reliability Test*

Table 1 shows that all variable's values are greater than 0.6. It means that all items in the instrument are reliable.

## *4.3 Normality Test*

This test explains that the data is normally distributed if the significance > 0.05 [8]. Based on Table 2, the data result is normal.

## *4.4 Multicollinearity Test*

This test in Table 3 shows that the tolerance value > 0.10 and VIF value < 10.0. It is explained that each independent variable is declared free of indications of multicollinearity symptoms.

**Table 2** Normality test results

| | | Unstandardized residual |
|---|---|---|
| N | | 108 |
| Normal parameters[a] | Mean | 0 |
| | Std. deviation | 0.80245465 |
| Most extreme differences | Absolute | 0.075 |
| | Positive | 0.075 |
| | Negative | – 0.045 |
| Kolmogorov–Smirnov Z | | 0.075 |
| Asymp. Sig. (2-tailed) | | 0.159 |

The superscript letter "a" means the test distribution is normal

**Table 3** Multicollinearity test result

| Model | | Collinearity statistics | |
|---|---|---|---|
| | | Tolerance | VIF |
| 1 | AIS success (Y) | 0.545 | 1.836 |
| | Technological sophistication ($X_1$) | 0.490 | 2.042 |
| | Information sophistication ($X_2$) | 0.600 | 1.667 |
| | Functional sophistication ($X_3$) | 0.700 | 1.428 |

## 4.5 Heteroscedasticity Test

This test in Fig. 4 shows that the pattern is not clear and spreads randomly. Thus, it can be concluded that this regression model is free from heteroscedasticity.

## 4.6 Multiple Linear Regression Analysis

This test was conducted with equations to assess the impact of two or more independent variables on the dependent variable [8]. As in Table 4, the multiple linear regression model is as follows:

$$Y = 5.571 + 0.206X_1 + 0.034X_2 + (-0.019X_3) + 0.143X_4 + e$$

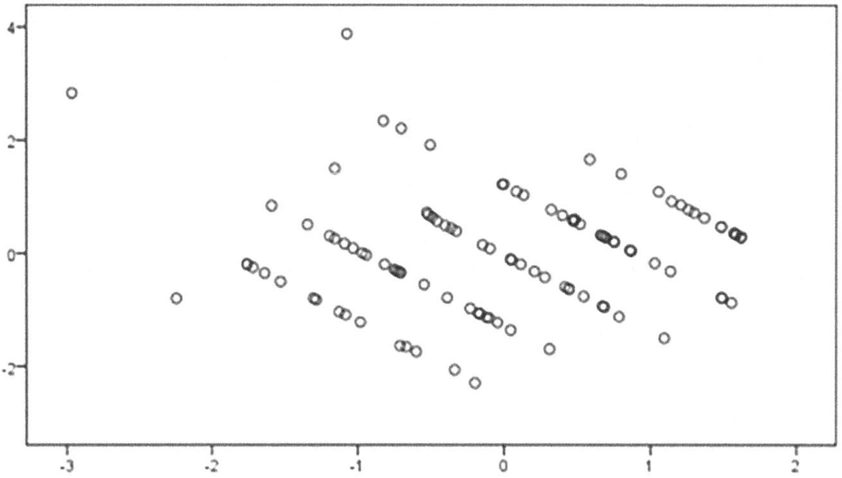


**Fig. 4** Heteroscedasticity test results

**Table 4** Multiple linear regression analysis result

| Model | Unstandardized coefficient |
|---|---|
| | B |
| (Constant) | 5.571 |
| Technological sophistication | 0.206 |
| Information sophistication | 0.034 |
| Functional sophistication | (− 0.019) |
| Managerial sophistication | 0.143 |

## 4.7 *Test $R^2$*

The result of the adjusted $R^2$ in this study is 0.627 (62.7%). This means that the influence of all the independent variables can explain the success of AIS by 62.7%. While the remaining 37.3% is influenced by other variables.

## 4.8 *Test F*

The F test result in Table 5 shows that the independent variable simultaneously influences the dependent variable because F count > F table.

**Table 5** F-test result

| Model | F | Sig. |
|---|---|---|
| Regression | 45.95 | 0.000[a] |

The superscript letter "a" means the predictors (constant)

### *4.9 T Test*

The T-test result in Table 6 shows that the independent variable individually affects the dependent variable because the t-count > t-table.

## 5 Discussion

The study found that adopting advanced technology and effective information integration positively influences AIS success, emphasizing the critical role of managerial sophistication in planning and implementing technology within MSEs. In this case, these cafes utilize various technologies such as computers and smartphones, WIFI, POS (Point of Sale) software, cashless payment systems, Qris (Quick Response Code Indonesian Standard) EDC (Electronic Data Capture), and also thermal printers. The results of this research are in line with the research of Ismail and King [9], Al-Eqab and Adel [2], and Kesuma et al. [10].

Conversely, functional sophistication, such as dedicated system management personnel, did not significantly impact AIS's success. The results of this study are not in line with research conducted by Ghasemi et al. [7], and Raymond et al. [15] which explains that there is an influence between functional sophistication and the success of AIS. This suggests that other aspects of functional sophistication should be explored in future research.

**Table 6** T-test result

| Variable | T | Sig. |
|---|---|---|
| Technological sophistication ($X_1$) | 7.568 | 0 |
| Information sophistication ($X_2$) | 2.225 | 0.03 |
| Functional sophistication ($X_3$) | − 0.53 | 0.6 |
| Managerial sophistication ($X_4$) | 2.139 | 0.04 |

# 6 Conclusion and Suggestions

## *6.1 Conclusion*

1. Technological and informational sophistication positively and significantly influence AIS's success in Medan Cafés.
2. Functional sophistication does not significantly affect AIS success, indicating a need to explore broader functional dimensions.
3. Managerial sophistication plays a significant role in AIS's success, highlighting the importance of strategic technology planning and resource management.

## *6.2 Limitations and Suggestions*

1. This study focuses on specific indicators for functional and managerial sophistication. Future research should consider a broader set of indicators to capture the full scope of these dimensions in AIS success.
2. The study's sample may not fully represent the entire café population in Medan City. Future studies should map the population distribution more comprehensively.
3. Data were collected through online questionnaires, which may lead to inconsistencies in responses. Future research should include observations and direct interviews to improve data accuracy.

# References


1. ASEAN Investment Report (2022). https://databoks.katadata.co.id/datapublish/2022/10/11/indonesia-punya-umkm-terbanyak-di-asean-bagaimana-daya-saingnya
2. Al-Eqab, M., Adel, D.: The impact of IT sophistications on the perceived usefulness of accounting information characteristics among Jordanian listed companies. Int. J. Bus. Soc. Sci. **4**(3), 145–155 (2013)
3. Al-Hattami H.M., Hashed A.A., Kabra J.D.: Effect of AIS success on performance measures of SMEs: evidence from Yemen. Int. J. Bus. Inform. Syst. **36**(1), 144–164 (2021)
4. Boston Consulting Group (BCG) and Telkom Indonesia (2022). https://databoks.katadata.co.id/datapublish/2022/11/14/ini-sektor-umkm-yang-banyak-manfaatkan-teknologi-digital
5. Conarth, D.W., Mignen, O.P.: What is being done to measure user satisfaction with EDP/MIS. Inform. Manag. **19**, 7–19 (1990)
6. Deaconu, A., Buiga, A.: Financial reporting and mimetic theory for small and medium enterprises. Curr. Sci. **108**(3), 334–340 (2015)
7. Ghasemi, M., et al.: The impact of information technology (IT) on modern accounting systems. Procedia Soc. Behav. Sci. **28**, 112–116 (2011)
8. Ghozali, I.: Aplikasi Analisis Multivariate dan Ekonometrika Eviews 10. Edisi 2. Badan Penerbit Universitas Diponegoro, Semarang (2016)

9. Ismail, N.A., King, M.: Firm performance and AIS alignment in Malaysian SMEs. Int. J. Account. Inf. Syst. **6**(4), 241–259 (2005)
10. Kesuma, S.A., Saidin, S.Z., Ahmi, A.: IT sophistication: implementation on state owned banks in Indonesia. Int. Rev. Manag. Mark. **6**(8), 234–239 (2016)
11. Kim, E., Lee, J.: An exploratory contingency model of user participation and MIS use. Inf. Manage. **11**(2), 87–97 (1986)
12. Loudon, K.C., dan Laudon, J.P.: Sistem Informasi Manajemen: Mengelola Perusahaan Digital (Edisi 13). Salemba Empat (2014)
13. Olson, M.H., Ives, B.: User involvement in system design: an empirical test of alternative approaches. Inf. Manage. **4**(4), 183–195 (1981)
14. Paré, G., Sicotte, C.: Information technology sophistication in health care: an instrument validation study among Canadian hospitals. Int. J. Med. Informatics **63**(3), 205–223 (2001)
15. Raymond, L., Croteau, A.M., Bergeron, F.: The strategic role of IT as an antecedent to the IT sophistication and IT performance of manufacturing SMEs. Int. J. Adv. Syst. Meas. **4**(3–4), 203–211 (2011)
16. Raymond, L., Paré, G.: Measurement of information technology sophistication in small manufacturing businesses. Inf. Res. Manage. J. (IRMJ) **5**(2), 4–16 (1992)
17. Romney, M.B., dan Steinbart, P.J: Accounting Information Systems, 14th edn. Pearson Education, Inc (2018)
18. Siyoto, S., Sodik, M.A.: Dasar metodologi penelitian. Literasi media publishing (2015)
19. Szilagyi, A.D., dan Wallace, M.J.: Organizational behavior and performance. Goodyear Publishing Company, Inc, Santa Monica (1980)
20. Urus, S., et al.: Critical success factors of accounting information systems (AIS): empirical evidence from Malaysian organizations. Manage. Acc. Rev. (MAR) **19**(1), 233–266 (2020)

# Does Financial Literacy Sway Decision-Making in Personal Investments?

**P. A. Mary Auxilia, K. R. Sowmya, S. Rajkumar, K. Jagannathan, J. Chandrakhanthan, and D. Ravindran**

**Abstract** This research is to determine the need for financial literacy among individuals for effective decision making in managing the investments efficiently. A descriptive research design with a quantitative approach has been adopted for the study. Data was collected from 500 individuals of the southern states of India through a survey. The southern states have a literacy rate higher than the national average. The data was analysed using tools such as ordinal logistic regression. The results showed that, there is a necessity for financial literacy for both men and women. Only one fourth of the respondents are financially literate. The awareness of financial instruments among urban individuals is typically high, but the lack of thorough understanding of the investment avenues reduces their risk-taking potential, which inhibits them from making effective decisions. During the pandemic crisis, it was observed that due to the imposed lockdown, citizens of India struggled to manage their routine livelihood with low income and without any proper savings to meet their expenses.

P. A. Mary Auxilia
Loyola Institute of Business Administration, Chennai, India
e-mail: mary.auxilia@liba.edu

K. R. Sowmya
Management Studies, Vel Tech Rangarajan Dr Sagunthala R&D Institute of Science and Technology, Chennai, India
e-mail: drsowmyakr@veltech.edu.in

S. Rajkumar
School of Management, Mount Carmel College, Autonomous, Bengaluru, Karnataka, India
e-mail: sraj@mccblr.edu.in

K. Jagannathan
Kristu Jayanti College, Bengaluru, Karnataka, India
e-mail: jaganmbasrm@gmail.com

J. Chandrakhanthan (✉)
Department of Commerce, Kristu Jayanti College, Bengaluru, Karnataka, India
e-mail: chandrakhanthanj@gmail.com

D. Ravindran
School of Management, Kristu Jayanti College, Bengaluru, Karnataka, India
e-mail: ravindran@kristujayanti.com



M. Al Mubarak (ed.), *Sustainable Digital Technology and Ethics in an Ever-Changing Environment*, Studies in Systems, Decision and Control 237,
https://doi.org/10.1007/978-3-031-86708-8_21

The results show that the majority of the respondents have low financial literacy and are not much informed on different investment avenues, leading to inefficient management of income. The novelty of this research is that it also studies the influence of risk perception of financial markets among individuals in southern states of India on financial literacy. The financially literate respondents also do not invest in all avenues due to risk aversion. The study calls for future research in other states of India to understand their knowledge on financial literacy.

**Keywords** Financial literacy · Ordinal logistic regression · Decision making · Financial markets

## 1 Introduction

People who are financially literate can plan their family's monthly budget, make investments, divide their money intelligently, and, in the event of a shortfall, they will know how to handle the issue without becoming upset [17]. Greater than 60% of people from well developed nations like Norway, United Kingdom, Australia, Canada, Denmark, are financially literate. A person who is financially educated is less likely to experience personal financial hardship since they are better equipped to handle certain financial obstacles [6]. Earning and spending money helps one make the difference between a short-term desire and needs. Financial literacy helps investors to prioritize the things that make their life better. It helps investors to understand the importance of a budget and it also teaches them how to do it [15].

The objective of the research is to determine the need for financial literacy among individuals for effective decision making in personal Investments. Indian citizens found it difficult to cover their costs during the unanticipated, unpredictable pandemic crisis because of work restrictions and lockdowns. Financial issues were caused by inadequate investment strategies, inadequate market knowledge, and a rise in borrowing from the unorganised financial sector. The current study goes into detail about the need of financial literacy in order to assist people avoid financial suffering during uncertain times by helping them make better investing decisions. According to Arora and Chakraborty [1], information, location, age, gender, education, and income all affect how financially literate a person is. According to research by Dogra et al. [9], men and those in the 26–30 age range are more financially literate than other groups. According to Dube and Asthana [10], students in higher education had insufficient financial literacy and lacked knowledge on how to handle finances in times of crisis. This research discusses all the variables mentioned above in the model with the addition of one more variable, which is the influence of an individual's risk perception in making effective investment decisions. Previous research has primarily focused on variables like age, gender, education, income, and location as factors influencing decision making in personal investments. According to the hypothesis, there is a correlation between financial literacy and the variables that affect the choices people make when making personal investments.

## 2 Review of Literature

According to a National Centre for Financial Education survey (https://www.financialexpress.com), only 27% of Indians were financially literate. When evaluating the financial requirements, people who were financially literate demonstrated distinct perspectives. During the COVID-19 epidemic, the higher income group has experienced significant suffering due to their inadequate retirement planning, inappropriate savings practices, and ignorance of diverse assets. In addition to being financially insecure, those in lower income groups lacked adequate education [13].

An individual's ability to take risks when making wise investments and decisions depends on their level of confidence, which can be attained through financial literacy. Those who were financially literate had a higher degree of confidence than those who were not, making them less vulnerable financially throughout the epidemic [7]. Knowledge and behavior related to money are directly impacted by financial education [14]. Higher financial literacy was associated with greater saving than lower financial literacy [18].

Most nations have shown that young people have poor levels of financial literacy, and demographic variables including age, gender, income, and education have a significant influence on financial literacy [11, 22]. Many obstacles, including cultural, physical, psychological, and financial ones, prevent Indian women from attempting to become financially literate [2]. Indian working women exhibit financial attitudes and behaviors, yet their investments show a lack of financial awareness [19].

Due to a lack of knowledge, Indian women are reluctant or hesitant to enter the financial markets, forcing them to rely on their families for financial management and decision-making. A high level of financial literacy lowers financial risk aversion by allowing women to invest in riskier assets [5]. To take advantage of possibilities, grow, and stay ahead of the competition, financial literacy is a crucial prerequisite [3]. Many households are prevented from engaging in the stock market due to a lack of financial literacy since they are unaware of the fundamental debt and equity instruments [21].

Research demonstrates that gender disparity in investing decisions hurts women [16]. It is highly possible that having competent financial knowledge is necessary to navigate the contradictory position of rationality or irrationality in investment decisions. According to Goyal and Kumar [12], women who lack expertise or understanding about personal finances may have difficulties in the job market. Therefore, it is crucial for them to make wise financial decisions during their education. It's possible that declining cognitive function will make it more difficult for someone to handle money.

When one possesses financial literacy, making investing decisions becomes more logical [23]. In order to investigate how libraries assist students' financial literacy in academic settings, Reiter and Ford [20] polled librarians. The most current study on American college students' financial information-seeking habits was conducted by Vaaler et al. [24] in their paper "They Seek, but Do They Find?"

# 3 Methodology

During the epidemic, being financially literate has become crucial to making wise decisions about how to manage the household budget. 500 respondents from southern India, where the literacy rate is greater than the national average, provided the data. There were equal numbers of urban males, urban women, and rural men and women among the respondents. Out of the 250 guys, data was gathered from 250 women. There were 125 guys from rural areas and 125 men from cities who responded. In a similar vein, 125 of the 250 female responders came from urban males, and 125 from rural men. A systematic questionnaire was used to conduct the survey. Based on the respondents' understanding of inflation, financial instruments, and microloans, the respondents' financial literacy was examined.

**Ordinal Logistic Regression (OLS)** One method that is an improvement on logistic regression is ordinal logistic regression (OLS). In this study, financial literacy is the dependent categorical variable with three levels: low, moderate, and high. The approach has a dependent variable that is a category variable. The independent or predictor variables are things like gender, marital status, education, location, investors' perception of risk on the financial markets, expense management, and investment opportunities. To predict the cumulative probability, the coefficients of each predictor variable are found. The Pearson and Deviance approach is used to evaluate the model fit. The Nagelkerke pseudo-R-square value demonstrates how much the predictor factors influenced the results. The estimated odd ratio facilitates comprehension of the effect of a single unit change in the predictor variable on the likelihood of transitioning from one category to another. This study's model has a categorical dependent variable and seven predictor variables. Based on a review of the literature, the researcher discovered that prior research models did not take into account how people in Southern Indian states perceive risk associated with the financial markets as a predictor of financial literacy. Because they are risk averse and have an inadequate comprehension of the investment options, those with low to moderate financial literacy do not make effective investment selections. This research model includes one more predictor variable Individual risk perception and explains its influence on financial literacy.

# 4 Analysis and Interpretation

Table 1 shows that most of the respondents were holding a bank account. Just 51% of rural women have bank savings accounts, compared to other women. More credit card holders are found in urban areas than in rural ones. Compared to alternative investing options, respondents have placed 62% of their money in fixed deposit programs. Just 16% of respondents had investments in shares, despite the fact that 47% were aware of shares. Of the respondents, 18% have made investments in mutual funds, and 45% are aware of them.

**Table 1** Knowledge on financial instruments and inflation used to determine financial literacy score

| | Urban (250 respondents) | | Rural (250 respondents) | | Total percentage (%) |
|---|---|---|---|---|---|
| | Male (125) | Female (125) | Male (125) | Female (125) | |
| *Saving account* | | | | | |
| Aware | 4 | 14 | 11 | 50 | 16 |
| Not aware | Nil | Nil | 5 | 11 | 3 |
| Holding | 121 (97%) | 111 (89%) | 109 (87%) | 64 (51%) | 81 |
| *Shares* | | | | | |
| Aware | 74 | 82 | 55 | 22 | 47 |
| Not aware | 11 | 24 | 48 | 100 | 37 |
| Investing | 40 (32%) | 19 (15%) | 22 (18%) | 3 (2%) | 16 |
| *Mutual funds* | | | | | |
| Aware | 69 | 80 | 55 | 22 | 45 |
| Not aware | 11 | 24 | 48 | 100 | 37 |
| Investing | 45 (36%) | 21 (17%) | 22 (18%) | 3 (2%) | 18 |
| *Bonds* | | | | | |
| Aware | 67 | 55 | 65 | 31 | 44 |
| Not aware | 34 | 59 | 52 | 92 | 47 |
| Investing | 24 (19%) | 11 (9%) | 8 (6%) | 2 (2%) | 9 |
| *Post office schemes* | | | | | |
| Aware | 88 | 85 | 47 | 51 | 54 |
| Not aware | 6 | 22 | 32 | 58 | 24 |
| Investing | 31 (25%) | 18 (14%) | 46 (37%) | 16 (13%) | 22 |
| *Term insurance* | | | | | |
| Aware | 58 | 50 | 54 | 29 | 38 |
| Not aware | 41 | 63 | 62 | 94 | 52 |
| Holding | 26 (21%) | 12 (10%) | 9 (7%) | 2 (2%) | 10 |
| *Mediclaim insurance* | | | | | |
| Aware | 53 | 69 | 59 | 24 | 41 |
| Not aware | 12 | 24 | 35 | 82 | 31 |
| Holding | 60 (48%) | 32 (26%) | 31 (25%) | 18 (14%) | 28 |
| *Pension schemes* | | | | | |
| Aware | 55 | 47 | 72 | 46 | 44 |
| Not aware | 9 | 28 | 22 | 67 | 25 |
| Investing | 61 (49%) | 50 (40%) | 31 (25%) | 12 (10%) | 31 |
| *Fixed deposit* | | | | | |
| Aware | 17 | 28 | 26 | 42 | 23 |
| Not aware | 3 | 5 | 17 | 48 | 15 |
| Investing | 105 (84%) | 92 (74%) | 82 (66%) | 35 (28%) | 62 |

(continued)

**Table 1** (continued)

| | Urban (250 respondents) | | Rural (250 respondents) | | Total percentage (%) |
|---|---|---|---|---|---|
| | Male (125) | Female (125) | Male (125) | Female (125) | |
| *Credit card* | | | | | |
| Aware | 21 | 53 | 43 | 22 | 28 |
| Not aware | 13 | 12 | 32 | 80 | 27 |
| Holds a credit card | 91 (73%) | 60 (48%) | 50 (40%) | 23 (18%) | 45 |
| *Micro finance loans* | | | | | |
| Aware | 70 | 68 | 50 | 24 | 42 |
| Not aware | 31 | 42 | 39 | 87 | 40 |
| Taken loan | 24 (19%) | 15 (12%) | 36 (29%) | 14 (11%) | 18 |
| *Knowledge on inflation* | | | | | |
| Low | 16 | 39 | 55 | 92 | 40 |
| Moderate | 62 | 59 | 48 | 24 | 39 |
| High | 47 (38%) | 27 (22%) | 22 (18%) | 9 (7%) | 21 |

*Source* Primary data

Additionally, it shows that the respondents were not very adventurous when it came to using bonds; only 9% of them own bonds, and 47% of them are uninformed of them, which is greater than the awareness rate. A lower risk instrument is the post office scheme. Despite having a high level of knowledge, only 22% of respondents have placed money into the system. 10% of people have term insurance policies, while 32% of people are aware of term insurance.

41% of people are aware of mediclaim insurance, and 28% have a mediclaim policy. Men from rural areas donated more than urban men and women (28%). Men and women in metropolitan areas were more likely to invest in pension plans. Of the respondents, 45% have a credit card, and 18% have taken out a microloan, with rural men making up the largest group of borrowers. Of the respondents, 40% had little knowledge of inflation, 39% have moderate knowledge, and 21% have more knowledge. It is unequivocally demonstrating that rural women had little financial asset investment and low instrument understanding. In contrast to their rural counterparts, urban respondents showed greater financial knowledge and had been exposed to more financial products.

Table 2 shows the respondents knowledge of financial literacy based on the scores obtained. It shows that only 13% were highly financially literate, 16% have moderate financial literacy and 71% have low financial literacy.

Table 3 shows that 13% of respondents have a high level of financial literacy; of these, 9.6% are men and 3.4% are women; 8.2% of respondents reside in urban India and 4.8% in rural; 5.4% are postgraduate students and 4.4% are graduates; and 3.2% of respondents are married. Only 1.2% of 13% respondents who were extremely financially literate thought that investing in the financial markets was low

**Table 2** Classification of respondents based on the knowledge of financial literacy score

| Financial literacy | Score (0–3) | Cumulative % | Total |
|---|---|---|---|
| Low | 0–1 | < 70 | 356 (71%) |
| Moderate | 1.1–2 | 71–89 | 79 (16%) |
| High | 2.1–3 | > 90 | 65 (13%) |

*Source* Primary data

risk, whereas 7.6% of respondents who were somewhat financially literate thought the same.

15.8% of the respondents have a moderate degree of financial literacy; of these, 9.8% are males and 6% are women; 11.6% of the respondents reside in urban India, 4.2% in rural; 1.4% are postgraduate students and 11.8% are graduates; 7% of the respondents are single, and 8.8% are married. Only 3.8% of respondents who were only marginally financially aware thought that investing in the financial markets was low risk, compared to 8.4% who thought it was extremely risky.

**Table 3** Demographic representation of financial literacy levels

| | Low financial literacy | Moderate financial literacy | High financial literacy |
|---|---|---|---|
| *Gender* | | | |
| Male | 153 (30.6%) | 49 (9.8%) | 48 (9.6%) |
| Female | 203 (40.6%) | 30 (6%) | 17 (3.4%) |
| *Location* | | | |
| Urban | 151 (30.2%) | 58 (11.6%) | 41 (8.2%) |
| Rural | 205 (41%) | 21 (4.2%) | 24 (4.8%) |
| *Education* | | | |
| High school | 83 (16.6%) | 2 (0.4%) | 7 (1.4%) |
| HSC/diploma/technical | 118 (23.6%) | 11 (2.2%) | 9 (1.8%) |
| Graduate | 133 (26.6%) | 59 (11.8%) | 22 (4.4%) |
| Postgraduate | 22 (4.4%) | 7 (1.4%) | 27 (5.4%) |
| *Marital status* | | | |
| Married | 282 (56.4%) | 44 (8.8%) | 16 (3.2%) |
| Single | 74 (14.8%) | 35 (7%) | 49 (9.8%) |
| *Risk perception on financial markets* | | | |
| Low risk | 70 (14%) | 19 (3.8%) | 6 (1.2%) |
| High risk | 188 (37.6%) | 42 (8.4%) | 21 (4.2%) |
| Moderate risk | 98 (19.6%) | 18 (3.6%) | 38 (7.6%) |
| Total percentage (%) | 71.2 | 15.8 | 13 |

*Source* Primary data

71.2% of respondents—of whom 30.6% are men and 40.6% are women—live in urban and 41% are rural areas of India; 4.4% are postgraduate students and 26.6% are graduates; 14.8% of respondents are single and 56.4% are married. 19.6% of respondents who were not well-versed in finance thought the financial markets were somewhat risky, 37.6% thought they were extremely risky, and 14% of respondents who were 71.2% financially knowledgeable thought they were low risk investment opportunities.

Table 4 reveals that just 3.2% of respondents who are highly financially literate, 7.6% of those who are moderately financially literate, and 28% of those who are poor financially literate obtained loans from local bankers. Paying the bills was not done by 27.3% of low, 40.5% of moderate, and 10.7% of high financially knowledgeable people. In contrast to moderate and low financially knowledgeable respondents, 44.7% of highly financially literate respondents felt comfortable using their savings to cover their expenses.

When compared to other investments and bank savings, respondents with high financial literacy (63.1%) invested in stocks, bonds, exchange-traded funds (ETFs), and pension funds. 16.9% invested in less hazardous ventures, while 4.6% stayed at home with their money. Of the respondents who lacked financial literacy, 43.6% kept their extra cash at home, 36% invested in tangible gold, 12.9% saved money in a bank, and 6.2% bought stocks, bonds, and other investments.

**Table 4** Influence of financial literacy on money management

| | Low financial literacy | Moderate financial literacy | High financial literacy |
|---|---|---|---|
| *Expense management* | | | |
| Savings | 31 (8.7%) | 11 (13.9%) | 29 (44.7%) |
| Loan from bank | 48 (13.5%) | 27 (34.2%) | 19 (29.2%) |
| Loan from Indigenous bankers | 103 (28.9%) | 6 (7.6%) | 2 (3.1%) |
| Loan from friends/ relatives | 77 (21.6%) | 3 (3.8%) | 8 (12.3%) |
| Unable to pay bills | 97 (27.3%) | 32 (40.5%) | 7 (10.7%) |
| *Excess money is invested* | | | |
| Save in bank | 46 (12.9%) | 6 (7.6%) | 4 (6.2%) |
| Invest in fixed deposit/ post office schemes | 6 (1.7%) | 16 (20.2%) | 11 (16.9%) |
| Invest in physical gold | 128 (36%) | 20 (25.3%) | 6 (9.2%) |
| Invest in shares/bonds/ ETF/pension fund | 22 (6.2%) | 23 (29.2%) | 41 (63.1%) |
| Keep at home | 154 (43.2%) | 14 (17.7%) | 3 (4.6%) |

*Source* Primary data

A similar level of interest was shown by respondents with modest financial literacy in tangible gold, shares, bonds, exchange-traded funds (ETF), fixed deposits, and post office schemes. 17.6% of them kept the extra cash at home.

**Ordinal Logistic Regression (OLS)**

Figure 1 shows that, Ordinal Logistic Regression was run with gender, marital status, education, location, risk perception of investors on financial markets, expense management, and investment avenues as independent variables and financial literacy as dependent variable.

Table 5 shows that the final model was significantly better than the intercept only model. $P < 0.05$ was statistically significant proving that the model gives better predictions.

Table 6 proves that the model fit was good $P > 0.05$ and it is 0.997 showing that the null hypothesis was accepted which means the data and the model predictions were similar.

Table 7 is showing with Nagelkerke pseudo R-square value of 95.9% for interpretation than Cox and Snell as well as McFadden because Nagelkerke was found to be more reliable when compared to others, and it indicates that the variables gender, marital status, education, location, risk perception on financial markets, expense management during pandemic and investing avenues account for 95.9% of the variation in the dependent variable, i.e., financial literacy. It shows that the model was good.

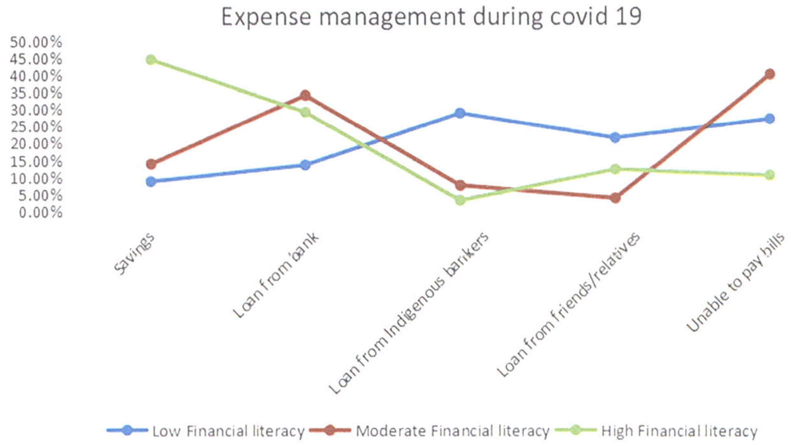


**Fig. 1** Expense management

**Table 5** Model fitting information: OLS output

| Model | Chi-square | df | Sig. |
|---|---|---|---|
| Final | 723.220 | 15 | 0.000 |

*Source* Primary data

**Table 6** Goodness-of-fit: OLS output

| Goodness-of-fit | | | |
|---|---|---|---|
| | Chi-square | df | Sig. |
| Pearson | 54.750 | 87 | 0.997 |
| Deviance | 63.327 | 87 | 0.974 |
| Link function: Logit | | | |

*Source* Primary data

**Table 7** Pseudo R-square: OLS output

| Pseudo R-square | |
|---|---|
| Cox and Snell | 0.765 |
| Nagelkerke | 0.959 |
| McFadden | 0.906 |
| Link function: Logit | |

*Source* Primary data

Table 8's interpretation of the odds ratio reveals that men were 2.801 times more financially literate than women were. The level of financial literacy among married respondents was 0.170 times lower than that of single respondents. Compared to post-graduates, respondents with higher secondary education and a high school diploma had financial literacy that was 0.42 and 0.489 times lower, respectively. Graduates knew 0.57 odds less about financial literacy than postgraduates did. The level of financial literacy among urbanites is 2.906 times greater than that of rural residents. Financial literacy is lower among respondents who view the financial markets as low risk (0.191 times) and high risk (0.203 times) than among those who perceive it as moderate risk. High financial literacy respondents borrowed from local bankers during the COVID-19 crisis at a rate that was 0.149 times lower than using their savings. The probability of highly financially competent individuals saving money at home, putting it in a bank, and purchasing real gold were 0.185, 0.217, and 0.266 fewer than those who invested in stocks, bonds, exchange-traded funds, and pension funds.

## 5 Results and Discussions

Arora and Chakraborty [1] and Bharucha [4] in their study have shown that the level of financial literacy depends on the factors such age, gender, education, income, location and information. The demographic variables affecting financial literacy. According to research by Das and Mahapatra [8], an individual's financial well-being is significantly impacted by their financial behavior, attitude, and knowledge. The majority of earlier studies have demonstrated how demographic factors affect financial literacy.

**Table 8** Estimates and odd ratios: OLS output

| Independent variables versus financial literacy | Estimate | Odd ratio | *P* (Sig.) |
|---|---|---|---|
| *Gender* | | | |
| Male | 1.030 | 2.801 | 0.000 |
| *Reference—Female* | | | |
| *Marital status* | | | |
| Married | – 1.77 | 0.170 | 0.014 |
| *Reference—Single* | | | |
| *Education* | | | |
| High school | – 0.850 | 0.427 | 0.000 |
| HSC/diploma/technical | – 0.715 | 0.489 | 0.007 |
| Graduate | – 0.560 | 0.571 | 0.003 |
| *Reference—Post graduate* | | | |
| *Location* | | | |
| Urban | 1.067 | 2.906 | 0.000 |
| *Reference—Rural* | | | |
| *Risk perception on financial markets* | | | |
| Low risk | – 1.6517 | 0.191 | 0.011 |
| High risk | – 1.5904 | 0.203 | |
| *Reference—Moderate risk* | | | |
| *Expense management during COVID-19* | | | |
| Unable to pay bills | – 1.328 | 0.265 | 0.002 |
| Loan from bank | – 0.426 | 0.653 | 0.005 |
| Loan from Indigenous bankers | – 1.899 | 0.149 | 0.012 |
| Loan from friends/relatives | – 1.424 | 0.241 | 0.000 |
| *Reference—Savings* | | | |
| *Excess money is invested* | | | |
| Save in bank | – 1.526 | 0.217 | 0.000 |
| Keep at home | – 1.689 | 0.185 | 0.004 |
| Invest in fixed deposit/post office schemes | – 0.752 | 0.471 | 0.016 |
| Invest in physical gold | – 1.321 | 0.266 | 0.007 |
| *Reference—Invest in shares/bonds/ETF/pension fund* | | | |

*Source* Primary data

This study demonstrates unequivocally that just 29% of respondents were financially literate, comprising 13% of participants who were highly and 16% of participants who were moderately financially literate, based on their understanding of inflation and financial literacy. Evidence is mounting suggesting the proportion of rural women in south India with savings accounts was lower than that of urban males, women, and men in rural areas. When it comes to investing, the respondents have chosen fixed deposit plans above other options. Following urban males were urban women, rural men, and rural women who held and invested in shares, bonds, mutual funds, pension plans, fixed deposits, Mediclaim insurance, term insurance, and credit cards. In comparison to respondents who invested in shares, mutual funds, bonds, post office schemes, and term insurance, more respondents invested in Mediclaim insurance. This demonstrates how the need for and understanding of Mediclaim insurance have been fueled by uncertainty. Compared to urban men and women, rural men invested in post office programs at a higher rate than others, indicating that they are less risk-tolerant.

Men in cities had greater knowledge of inflation than women in cities, as well as men and women in rural areas. It clearly shows that, the urban respondents were financially literate than rural respondents, particularly urban men scores were higher than women. Due to institutional hurdles, lack of information availability, gender inequity, and other factors, rural women may have low financial literacy. Compared to those living in rural regions, those in urban areas had higher financial literacy.

A greater degree of financial literacy is influenced by education; respondents with post-graduate degrees were the most financially literate, followed by graduates. Compared to respondents who are staying single, married respondents showed lower levels of financial literacy. This demonstrates that, while being aware of the tools, married men and women are risk averse and hesitant to invest in different channels. When things get unclear, respondents with high financial literacy take out bank loans and withdraw cash from savings. Individuals who lack financial literacy obtain loans from local bankers, acquaintances, and family members. Any extra money was either invested in real gold or kept at home.

This study clearly demonstrates that investors in stocks, bonds, exchange-traded funds (ETFs), pension plans, and fixed deposits are highly financially savvy. The standard logistic regression demonstrates how financial literacy is influenced by factors such as geography, education level, marital status, and investors' perception of risk on the financial markets. These factors are evident in how investors manage their spending and select investment opportunities.

The study's findings will serve as a guide to help people understand the financial risk involved in investing and how to handle their money without experiencing hardship.

## 6 Conclusion and Future Recommendations

Men were superior than women in terms of decision-making and investing in a variety of instruments. Due to their lack of financial literacy, the majority of Indian urban women let their family members to manage their finances, which lowers their self-esteem and causes them distress. Women in rural India were unaware of government programs and simple financial tools that were available to them to pay their excess expenses during the pandemic. They only knew how to get money by borrowing from local bankers, which put them in debt. The responders are now aware of the necessity of medical insurance due to the unpredictability and unforeseen events. The study's findings indicate that both rural and urban India now have a greater understanding of healthcare insurance. This study, which is the first of its type, demonstrates how people's perceptions of the risk associated with the financial markets affect their financial literacy in the southern Indian states. Urban people have a high level of financial instrument awareness, but their inability to fully comprehend investment options limits their willingness to take on risk and prevents them from making wise selections. Gender, marital status, education, location, risk perception of investors on financial markets influences the knowledge on financial literacy in southern states of Indian and its effect was seen in the results on efficiently managing the expenses and selecting the right investment avenues. Compared to other respondents, those with strong financial literacy handle their money more effectively. This means that the system urgently has to start teaching people fundamental money skills so they can manage their home budget, choose investment opportunities, get over their risk aversion, and live responsibly so they can deal with any unjust situations that may come up in the future.

The study is restricted to India's southern states. Three major categories—financial instruments, microloans, and inflation—have been used to analyse the respondent's financial literacy knowledge. An analysis of the impact of just seven variables on financial literacy was conducted using Ordinal logistic regression. These restrictions leave room for more study in the future. Lack of time has forced many different adaptations, testing, and experiments to be postponed until later. For example, large-scale, pan-Indian, and pan-Asian experiments typically take months or even longer to complete. Further research entails a more thorough examination that takes other elements into account.

## References


1. Arora, J., Chakraborty, M.: Role of financial literacy in investment choices of financial consumers: an insight from India. Int. J. Soc. Econ. **50**(3), 377–397 (2023)
2. Baluja, G.: Financial literacy among women in India: a review. Pac. Bus. Rev. Int. **9**(4), 82–88 (2016)
3. Baporikar, N., Akino, S.: Financial literacy imperative for success of women entrepreneurship. Int. J. Innov. Digit. Econ. (IJIDE) **11**(3), 1–21 (2020). https://doi.org/10.4018/IJIDE.2020070101
4. Bharucha, J.P.: Socio-economic and demographic determinants of Indian youth financial literacy: determinants of financial literacy. Int. J. Asian Bus. Inf. Manag. (IJABIM) **8**(4), 15–28 (2017)
5. Byder, J., Agudelo, D.A., Arango, I.: Gender matters most. The impact on short-term risk aversion following a financial crash. Rev. Financ. Econ. **37**(1), 106–117 (2019). https://doi.org/10.1002/rfe.1038
6. Chhatwani, M.: Mortgage delinquency during COVID-19: do financial literacy and personality traits matter? Int. J. Bank Mark. **40**(3), 484–510 (2022). https://doi.org/10.1108/IJBM-05-2021
7. Chhatwani, M., Mishra, S.K.: Does financial literacy reduce financial fragility during COVID-19? The moderation effect of psychological, economic and social factors. Int. J. Bank Mark. **39**(7), 1114–1133 (2021). https://doi.org/10.1108/IJBM-11-2020-0536

8. Das, S., Mahapatra, S.K.: The big three of financial literacy: analyzing its influences on financial well-being. FWU J. Soc. Sci. **17**(2) (2023)
9. Dogra, P., Kaushal, A., Sharma, R.R.: Antecedents of the youngster's awareness about financial literacy: a structure equation modelling approach. Vision **27**(1), 48–62 (2023)
10. Dube, V.S., Asthana, P.K.: Ensuring financial resilience through financial literacy: a study of higher education students. In: Virtual Learning, pp. 17–34. Routledge (2022)
11. Garg, N., Singh, S.: Financial literacy among youth. Int. J. Soc. Econ. **45**(1), 173–186 (2018). https://doi.org/10.1108/ijse-11-2016-0303
12. Goyal, K., Kumar, S.: Financial literacy: a systematic review and bibliometric analysis. Int. J. Consum. Stud. **45**(1), 80–105 (2021). https://doi.org/10.1111/ijcs.12605
13. Goyal, K., Kumar, S., Rao, P., Colombage, S., Sharma, A.: Financial distress and COVID-19: evidence from working individuals in India. Qual. Res. Financ. Mark. **13**(4), 503–528 (2021). https://doi.org/10.1108/QRFM-08-2020-0159
14. Kaiser, T., Lusardi, A., Menkhoff, L., Urban, C.: Financial education affects financial knowledge and downstream behaviors. J. Financ. Econ. **145**(2), 255–272 (2022)
15. Kandpal, V.: Socio-economic development through self-help groups in rural India—a qualitative study. Qual. Res. Financ. Mark. **14**(5), 621–636 (2022). https://doi.org/10.1108/QRFM-10-2021-0170
16. Kappal, J.M., Rastogi, S.: Investment behaviour of women entrepreneurs. Qual. Res. Financ. Mark. **12**(4), 485–504 (2020). https://doi.org/10.1108/QRFM-04-2020-0053
17. Lahiri, S., Biswas, S.: Does financial literacy improve financial behavior in emerging economies? Evidence from India. Manag. Finance **48**(9/10), 1430–1452 (2022). https://doi.org/10.1108/MF-09-2021-0440
18. Morgan, P.J., Long, T.Q.: Financial literacy, financial inclusion, and savings behavior in Laos. J. Asian Econ. **68**, 101197 (2020)
19. Rai, K., Dua, S., Yadav, M.: Association of financial attitude, financial behaviour and financial knowledge towards financial literacy: a structural equation modeling approach. FIIB Bus. Rev. **8**(1), 51–60 (2019). https://doi.org/10.1177/2319714519826651
20. Reiter, L., Ford, B.: Library support for student financial literacy: a survey of librarians at large academic institutions. Coll. Res. Libr. **80**(5), 618 (2019). https://doi.org/10.5860/crl.80.5.618
21. Van Rooij, M., Lusardi, A., Alessie, R.: Financial literacy and stock market participation. J. Financ. Econ. **101**(2), 449–472 (2011)
22. Sun, S., Chen, Y.C., Ansong, D., Huang, J., Sherraden, M.S.: Household financial capability and economic hardship: an empirical examination of the financial capability framework. J. Fam. Econ. Issues **43**(4), 716–729 (2022). https://doi.org/10.1007/s10834-022-09816-5
23. Tuffour, J.K., Amoako, A.A., Amartey, E.O.: Assessing the effect of financial literacy among managers on the performance of small-scale enterprises. Glob. Bus. Rev. **23**(5), 1200–1217 (2022). https://doi.org/10.1177/0972150919899753
24. Vaaler, A., Reiter, L., Faulkner, A.E.: They seek, but do they find? Investigating the financial information-seeking behavior of college students. Coll. Res. Libr. **82**(2), 267 (2021). https://doi.org/10.5860/crl.82.2.267

# An Analytical Review on the Protection of Children from Sexual Offences Act (POCSO), 2012 in India

**S. Muruganandhan, S. V. Akshayareena, G. Jayashree, P. A. J. Aujita, V. Vinothina, and K. Sivasubramanian**

**Abstract** This paper aims to assess the characteristics and significance of the POCSO Act, 2012 in India and to review the various legal perspectives of the POCSO Act in enhancing the protection of children in India. To achieve these stated objectives this chapter, brings out an important issue of society which is the abuse of children. This chapter is prepared through the qualitative technique with the collection and analysis of various literature reviews. The collected reviews have been listed and analysed to establish the stated objectives of this study. It has comprehensive definite terms for sexual abuses such as penetrative sexual harassment, aggravated sexual abuse (crime committed by authorities), showing sexual gestures and child abuse. This act mandates a child-friendly reporting system such as taking evidence from the child's home, and provision of special courts. This act also mandates the reporting of offences of sexual abuse against children. Non-reporting of such crimes could result in imprisonment of six months. Based on the literature, this study has suggested the following, the schools and the parents of the kids need to raise awareness because they are crucial in raising their children. They should also impart knowledge to kindergarten-age children to help them develop a positive mind-set and discuss the problems and repercussions of sexual offences. By doing this, parents as well as educators of the kids will be able to protect their children's interests and help them maintain their modesty in the event that such a shameful or evil act occurs.

**Keywords** Analytical · Review · Child · Protection · POCSO Act · India

S. Muruganandhan
Saveetha School of Law, Saveetha Institute of Medical and Technical Sciences, Chennai, India

S. V. Akshayareena · G. Jayashree · P. A. J. Aujita
Saveetha School of Law, SIMATS, Chennai, India

V. Vinothina
Department of Computer Science PG, Kristu Jayanti College, Bangalore, India

K. Sivasubramanian (✉)
Department of Economics, Kristu Jayanti College, Bangalore, India
e-mail: sivasubramanian@kristujayanti.com



M. Al Mubarak (ed.), *Sustainable Digital Technology and Ethics in an Ever-Changing Environment*, Studies in Systems, Decision and Control 237,
https://doi.org/10.1007/978-3-031-86708-8_22

# 1 Introduction

The Protection of Boy and Girl Children from Sexual Crimes Act was enacted in India on 16th July 2012 in the name of the Protection of Children from Sexual Offences Act (POCSO) and it was enforced on November, 14th 2012, on the event of Children's Day. This Act was enacted due to India's endorsement of the United Nations Convention on the Basic Rights of Children in 1992. This Act is a revolutionary law to protect children from crimes relating to sexual abuse and assault, harassment, and sexual violence. This Act provides a Special Court for trialling those crimes in India [5]. It is a comprehensive law which covers all forms of sexual crimes against the child. This Act is considered for the offences that occurred to a child below 18 years. One's childhood of any children is the most significant phase of their whole life. They are the pillars of future nations to build India as a developed country. Therefore, the security and well-being of every child under the age of eighteen years is important to build the nation [10]. Various abuses sexual assaults and exploitation against the girl and boy child lead to vulnerable conditions. One of the major types of child abuse which impacts humanity is an entire child sexual abuse [12]. Despite being a benchmark piece of law, the passage and the way of Prevention of children sexual abuses act of 2012 has hiked the reporting of cases in India [15].

# 2 Structure of the Chapter

This review-based chapter has been constructed to reach certain objective such as to assess the characteristics and significance of POCSO Act, 2012 in India and to review the various legal perspectives of POCSO Act in enhancing the protection of children in India. To achieve these stated aims, this chapter is organised in the following manner, introductory section, objectives, methodological framework, analysis, findings, conclusion, suggestions and limitations of this chapter.

# 3 Objectives

To assess the characteristics and significance of the POCSO Act, 2012 in India.

To review the various legal perspectives of the POCSO Act in enhancing the protection of children in India.

## 4 Methodology

This chapter aims to bring out the important issue of society which is the abuse of children. This chapter is prepared through the qualitative technique with the collection and analysis of various literature reviews. The collected reviews have been listed and analysed to establish the stated objectives of this study.

## 5 Literature Survey

This Act is enacted and applicable to all the states and union territories and this Act protects both male and female children under 18 years of age [11]. Under this act, the accused are deemed guilty until and unless they are proven innocent, which is against the general principle of "Innocent until proven guilty." Furthermore, it is the accused's responsibility to establish their innocence. This adds to the act's harshness [14]. In accordance with Act's sections such as 154 and 156, the First Information Report has to be registered under the Criminal Procedure Code. Comparably, section 19 of the POCSO Act explicitly stipulates that anyone including children who has reason to suppose that a crime under the Act may be committed or already took place according to the Act must alert the Special Children under the age of Police Section or the local law enforcement agency of the area in question right away [13]. The information must be recorded in writing by the Police Unit. Under this constitutional act, the procedure to record the statement will be children-friendly. The note relating to the child is considered a victim to be recorded by women police, not under the cadre of sub-inspector. The women police along with other police officers can visit the child's home and record the statement. They need not be in their official uniform [17]. Across the globe, the total number of sexual offences has increased against children in India. It is recorded that India is one of the major nations in the world relating to the occurrence of sexual harassment to children. It is reported that 53% of children encounter some kind of sexual abuse or assault [6, 16]. A gender-specific and neutral law that prevents the child from sexual abuse and assault was enacted in India for the benefit of child safety [1]. The law implementing agency's worst attitude is an important issue in the Act. Another problem relating to reporting of POCSO cases properly [2]. Even though the updated law has harsh penalties, there are still many gaps that must be filled right away. The number of recorded murder cases as well as victims who are killed could decline with increased application of the death penalty. Usually, the victim knows the perpetrator, which deters the victim from reporting the incident. Furthermore, there is a far higher chance of the target being both raped and murdered [3, 4]. In India and the rest of the world, the incidence of child exploitation and abuse has been consistently reported [7]. Sexual abuse of minor children is a general concern and minimum occurrence needed for crucial several fatalities. There are key challenges has been evolved, making it impossible to appropriately determine how frequently minors are surviving [9].

## 6 Key Findings

The following important findings are derived from various literature.

It has comprehensive definite terms for sexual abuses such as penetrative sexual harassment, aggravated sexual abuse (crime committed by authorities), showing sexual gestures and child abuse. This act mandates a child-friendly reporting system such as taking evidence from the child's home, and provision of special courts. This act also mandates the reporting of offences of sexual abuse against children. Non-reporting of such crimes could result in imprisonment of six months. In accordance with Act's sections such as 154 and 156, the First Information Report has to be registered under the Criminal Procedure Code. Comparably, section 19 of the POCSO Act explicitly stipulates that anyone including children who has reason to suppose that a crime under the Act may be committed or already took place according to the Act must alert the Special Children under the age of Police Section or the local law enforcement agency of the area in question right away. The information must be recorded in writing by the Police Unit [8].

## 7 Conclusion

It is a comprehensive law which covers all forms of sexual crimes against the child. This Act is considered for the offences that occurred to a child below 18 years. One's childhood of any children is the most significant phase of their whole life. They are the pillars of future nations to build India as a developed country. Therefore, the security and well-being of every child under the age of eighteen years is important to build the nation [10]. In India and the rest of the world, the incidence of child exploitation and abuse has been consistently reported [7]. Sexual abuse of minor children is a general concern and minimum occurrence needed for crucial several fatalities. There are key challenges has been evolved, making it impossible to appropriately determine how frequently minors are surviving.

## 8 Suggestions

The schools and the parents of the kids need to raise awareness because they are crucial in raising their children. They should also impart knowledge to kindergarten-age children to help them develop a positive mindset and discuss the problems and repercussions of sexual offences. By doing this, parents as well as educators of the kids will be able to protect their children's interests and help them maintain their modesty in the event that such a shameful or evil act occurs.

## 9 Limitation

This is restricted to only one act relating to the POCSO, 2012 and its implications. There are good avenues for analysing the role of other constitutional provisions for the protection and support of women, children and the transgender community.

## References


1. Barth, J., Bermetz, L., Heim, E., Trelle, S., Tonia, T.: The current prevalence of child sexual abuse worldwide: a systematic review and meta-analysis. Int. J. Public Health **58**(3), 469–483 (2013)
2. Behere, P.B., Mulmule, A.N.: Sexual abuse in 8-year-old child: where do we stand legally? Indian J. Psychol. Med. **35**(2), 203–205 (2013)
3. Chahine, E.F.: Child abuse and its relation to quality of life male and female children. Procedia Soc. Behav. Sci. **159**, 161–168 (2014)
4. Chowdhuri, S., Mukhopadhayay, P.: A study of the socio-demographic profile of the persons accused under POSCO Act 2012. Int. J. Health Res. Med. Legal Pract. **2**, 50–55 (2016)
5. Downing, N.R., Akinlotan, M., Thornhill, C.W.: The impact of childhood sexual abuse and adverse childhood experiences on adult health-related quality of life. Child Abuse Negl. **120**, 105181 (2021). https://doi.org/10.1016/j.chiabu.2021.105181
6. Hotchandani, K.R.: Every child is special—a study of child sexual abuse and the law (POSCO) in India. Int. J. Creat. Res. Thoughts **6**(1), 1386–1389 (2018)
7. Kansal, I.: Child sexual abuse in India: socio-legal issues. Int. J. Sci. Res. Sci. Technol. **2**(2), 1–4 (2016)
8. Krishnakumar, P., Satheesan, K., Geeta, M.G., Sureshkumar, K.: Prevalence and spectrum of sexual abuse among adolescents in Kerala, South India. Indian J. Pediatr. **81**(8), 770–774 (2014). https://doi.org/10.1007/s12098-013-1125-3
9. Maity, S.: Performance of controlling rape in India: efficiency estimates across states. J. Int. Women's Stud. **20**(7), 180–204 (2019)
10. Maity, S.: Escalation of police efficiency diminishes POCSO incidences—myth or reality? Evidence from Indian states. Int. J. Child Malt. Res. Policy Pract. **5**(1), 155–180 (2022)
11. Maity, S., Chakraborty, P.R.: Implications of the POCSO Act and determinants of child sexual abuse in India: insights at the state level. Humanit. Soc. Sci. Commun. **10**(6) (2023). https://doi.org/10.1057/s41599-022-01469-x
12. Maity, S., Sinha, A.: Interstate disparity in the performance of controlling crime against women in India: efficiency estimate across states. Int. J. Educ. Econ. Dev. **9**(1), 57–79 (2018)
13. Mann, S.: Protection of children from sexual offences: an analysis of the POCSO Act, 2012. Indian J. Law Legal Res. (2022). https://www.ijllr.com/post/protection-of-children-from-sexual-offences-an-analysis-of-the-pocso-act-2012
14. Mohanty, H., Banejee, D.: An analysis of protection of children from sexual offences act, 2012 (POCSO Act). SSRN (2021). https://doi.org/10.2139/ssrn.3768096
15. Pallathadka, H., Kumar, S., Pallathadka, L.K., Kumar, V.: Child sexual abuse in India: a social-legal analysis. Int. J. Aquat. Sci. **12**(1), 405–412 (2021)
16. Susheel, B.K., Mahalakshmi, V.: Legal implications of POCSO Act, 2012—an analysis. Int. J. Multidiscip. Res. Rev. 155–162 (2016). https://ijcrt.org/papers/IJPUB1801224.pdf
17. Thangavel, V.: The analysis of research review for the protection of children from sexual offences act (POCSO). SSRN (2023). https://doi.org/10.2139/ssrn.4578739

# Revolutionizing Waste Management in Developing Countries: Challenges and Opportunities

**Tinto Tom, Jerrin Jose, Akhila Variyar, Cherian Thomas, S. J. G. Preethi, Uma Maheswary, and Pradeep S. Raj**

**Abstract** Waste management practices are crucial for present and future generations as they directly impact our existence. Without a proper strategy, the progress we have made could be undermined. All nations strive for economic growth, which results in enormous waste. This waste can negatively affect the health of people and other living beings. Although we have specific methods to control and manage waste, they are not always sufficient. Therefore, we must continuously evaluate and improve these methods. Various waste management methods are employed globally, but their suitability varies from country to country. In India, the government has introduced numerous initiatives to address waste management differently. These initiatives capture the public's attention and encourage proper waste management practices. Engaging the community can make significant strides toward more effective waste management.

**Keywords** Waste · Waste management · Economic growth

## 1 Introduction

All human activities lead to waste generation. From the pre-historic period onwards, waste management was a concern. Recently, the quantity and quality of waste have been increasing. In the past, waste management was not a big issue because the population was small, the land was sufficient to dispose of the waste, and nature used to absorb the waste quickly. The industrial revolution, migration of people from

T. Tom (✉) · S. J. G. Preethi
Department of Economics, Kristu Jayanti College, Bengaluru, India
e-mail: tintotom@kristujayanti.com

J. Jose · A. Variyar · U. Maheswary · P. S. Raj
Department of English, Kristu Jayanti College, Bengaluru, India

C. Thomas
Department of Commerce PG, Kristu Jayanti College, Bengaluru, India



M. Al Mubarak (ed.), *Sustainable Digital Technology and Ethics in an Ever-Changing Environment*, Studies in Systems, Decision and Control 237,
https://doi.org/10.1007/978-3-031-86708-8_23

rural to urban areas, and further development caused the high volume of garbage. In this modern period, we need to understand the concept of waste clearly. What is waste? Waste is the useless by-product of human activities, which physically contains the same substance available in the valuable product. On the other hand, we can define it as materials that people want to dispose of even when payments are required. Although waste is an essential product of human activities, it is also the result of inefficient production processes whose continuous generation results in the loss of vital resources [5]. The waste generated from various sources might be hazardous. So, the management of this waste is crucial. It is the process by which the garbage collected is transported and processed before disposal. It is to ensure the environment's safety and limit the harmful effects of waste.

Waste management is vital because other facilities and infrastructure are needed for the contemporary man, without which life is complicated. Studies have shown a direct link between air, water, and land pollution and diseases such as lung cancer, heart disease, cholera, and hepatitis. In addition, climate change and eutrophication are direct results of water and air pollution. Little wonder why there is a massive disparity in the life expectancy of people in developed and developing countries. People's perception of waste management is significant. It may negatively affect society if they do not have clarity and ideas about the adverse effects. So, the government and authorities should take initiatives to raise awareness about the negative impact of waste. They can conduct activities like waste alleviation programs, educate people about solid waste management, and utilize organic waste for fertilizer preparation [13].

Inadequate waste management leads to poor public health, affecting the environment's quality and inability to attain sustainable development. Active community involvement is required in this matter. A positive public attitude should be taught through public awareness campaigns, print and Social Media, and proper waste-dropping practices. Improper waste management results in water pollution and polluted air in cities and towns. Due to these scenarios, the scope of waste management is very high [2].

## 2 Waste Management Practices Around the World

Indonesia is one of the members of ASEAN with a high population and population growth. This situation brings increasing waste and becomes a problem for the government. Some factors like lack of policies, financial support, low involvement, low community awareness, etc., lead to a low level of waste management. As per the UNEP study 2004, only 33% of services are provided by the government to manage waste. The highest percentage of organic waste in the country was generated by households, almost 43%. However, the local government may not collect the waste from the household for treatment, such as landfilling, composting, recycling, etc. So, the community has various waste-handling systems. In 2001, municipal waste was treated as a landfill, open dumping, open burning, recycling, disposed of in the river, street, etc. The recycling percentage was significantly less than 1.16%. These indicate that inadequate laws have led to inefficient waste management in Indonesia. Therefore, there should be some effort to improve the waste management system from the authority side [18]. Improving waste management in Indonesia, particularly for municipal and electronic waste, requires robust policies and infrastructure. While developed countries have successfully implemented extended producer responsibility for electronic waste, developing countries like Indonesia still face significant challenges due to inadequate laws and a lack of resources. Implementing similar legislation and increasing community awareness could enhance waste management efficiency and mitigate environmental impacts.

In the case of electronic waste management, developing countries are facing complicated problems. It may be because it might have been generated internally or imported illegally. Electronic waste contains dangerous constituents that will negatively affect humans and the environment if not adequately managed. In developing countries, waste is not managed appropriately because of the lack of infrastructure. They might be burying the waste, burning it in the open air, or dumping it into water bodies. On the other side, developed countries have legislation compelling electronic manufacturers to take back used electronic products at the end of their use. It is based on the principle of extended producer responsibility. These practices can be implemented in developing countries [22]. Developing countries face significant challenges in managing clinical solid waste and electronic waste due to insufficient legislation, lack of infrastructure, and financial constraints. Proper handling and disposal of both types of garbage are crucial to prevent severe health hazards and environmental pollution. Implementing robust laws and increasing awareness can help address these issues effectively.

Clinical solid waste management is the major challenge many developing countries face now. Clinical solid waste, inappropriate disposal, and poor management create significant health hazards and environmental pollution. Various researchers have conducted in this field and taken tremendous steps to safely handle and dispose of clinical waste. In many developing countries, improper management practices are evident from the initial collection onwards. This may be due to insufficient legislation, a lack of specialized clinical staff, and a lack of awareness. Most health centres in developing countries face financial difficulties that lead to ineffective disposal methods. Various studies encourage clinical waste recycling and reuse methods [14]. Clinical solid waste management in many developing countries, including Nigerian cities, faces challenges such as inappropriate disposal, high costs, and significant environmental impacts. This is compounded by inadequate disposal technologies and slaughterhouse pollution, underscoring the urgent need for improved and sustainable waste management practices across various sectors.

Nigerian cities are facing a series of issues, including enormous amounts of waste, because of inadequate disposal technologies, high management costs, and the negative impact of waste on the environment. Slaughterhouses are a significant reason for air and water pollution. The waste management practices in the area are aesthetically unappealing and not sustainable. Waste disposal methods must also be improved in developing countries [3]. Nigerian cities face significant waste management challenges, including slaughterhouse pollution, underscoring the need for enhanced disposal methods.

Similarly, many Asian countries, despite allocating significant portions of their budgets to waste management, struggle with adding value to waste and require tailored solutions suitable for their economic and technical contexts. Nigerian cities face significant waste management challenges, including slaughterhouse pollution, underscoring the need for improved disposal methods. Similarly, many Asian countries, despite allocating substantial portions of their budgets to waste management, struggle with adding value to waste and require tailored solutions suitable for their economic and technical contexts. These countries often focus on waste collection and disposal to limit disease spread, but more appropriate methods, such as biological treatments for biodegradable waste, are needed to enhance sustainability.

Local governments from Asian countries prioritize waste collection and disposal to limit disease spread. Almost 50% of their budget is for these kinds of services. However, little or no value is added to the waste. Biodegradable Municipal solid waste is more common in those countries, so biological methods are more appropriate for this case. Thermal combustion technologies are more applied in developed countries, which is technically and economically challenging for underdeveloped countries. So, Specific methods need to be formulated for the developing countries. Many Asian cities are practising proper solid waste management, like Bangalore, Matale, etc. [4]. Local governments in Asian countries prioritize waste collection and disposal to limit disease spread, but they often add little value to the waste. Specific methods, such as biological treatments for biodegradable waste, must be formulated for developing countries. Similarly, India's adoption of the EPR approach (extended producer responsibility) for e-waste management has shown positive impacts, suggesting that careful design and implementation of regulations can enhance waste management practices across various sectors.

Figure 1 indicates the waste management practice prevailing in India, and also the country adopted the EPR (extended producer responsibility) approach for e-waste management. Under the e-waste management rule, producers must set up a collection centre for the e-waste collection and finance and organize a system for environmentally sound e-waste management. The study results indicate that there was a positive impact after the implementation of this rule. The study proposes that this EPR should go beyond the collection of e-waste. Careful attention should be paid to the design and implementation of regulations [7].

## 3 Role of Economic Development on Waste Generation

To develop a proper waste management method for a given region, it is essential to know the quantity of waste created and the methods for the composition of waste. Various research studies convey that the amount of waste generated is in proportion to the average income of the people. Other factors influencing waste generation are climate, education level, living standards, religious and social beliefs, and public

**Fig. 1** Waste management practices

attitude. The study is based on the Srilankan region, as the data in the survey conveys that the generation of paper waste and organic waste increases as income increases because of the high consumption of food and purchase of packaged products and reading materials. The number of employed people in a household also appears to contribute to waste generation. The average amount of waste generated per household of different income levels can be used to predict the total amount of waste generated within a municipality [6]. Understanding the quantity and composition of waste is crucial for developing effective waste management methods, as factors like income, climate, and education influence waste generation. As economies grow, solid waste increases, necessitating strategies to decouple waste generation from economic growth. The EU prioritizes waste prevention to address this issue, emphasizing the importance of considering socioeconomic factors and living standards in waste management practices.

As economies attain economic growth, the quantity of solid waste increases, which we have seen in the past decades. To tackle this issue, the EU has its agenda: waste generation should decouple from economic growth in the future in the EU (European Union). Sweden, a member of the EU, conveyed that they have a target not to increase the quantity of waste in the future. The EU Sixth Environment Action Programme lists waste prevention as one of its top four priorities to significantly reduce waste quantities within the EU [26]. Socioeconomic and development indices and people's standard of living can influence solid waste generation. So, to manage the waste or to prevent the waste, these areas should be considered. In Europe, socioeconomic factors influence waste generation [20].

The connection between economic growth and waste generation is the world's primary concern. Various empirical studies from the EU show short- and long-run bidirectional causality between economic growth and waste generation. Therefore, the traditional development policy will not be sufficient to minimize the EU region's waste. Eco-innovation, charges, and incentive methods must be implemented to promote the circular economy model [10]. Due to various social and environmental factors, solid waste generation in socioeconomic groups is different. These socioeconomic parameters include education, job, and family income. The ash residue is more prevalent when we consider the lower socioeconomic group and plastic waste is high in the higher socioeconomic group in the area. It gives the authorities insight into the fact that socioeconomic parameters are more significant in waste generation [15].

The World Bank's What a Waste 2.0 report predicts a 70% increase in global waste by 2050, significantly faster than the expected population growth rate. Rapid urbanization and economic development are critical contributors to this surge, leading to higher per capita waste production and putting substantial pressure on urban areas to handle the rising waste. Although low-income nations spend 20% of their municipal budgets on waste management, 93% of waste in these regions is either openly dumped or burned. The report highlights a critical situation for Sub-Saharan Africa and South Asia, where waste generation is anticipated to double and triple, respectively, with 69% and 75% of waste currently being disposed of improperly. Immediate action is essential to avoid the prohibitive future costs of environmental cleanup, emphasizing the urgent need for effective waste management strategies [1].

Municipal solid waste (MSW) represents a critical challenge linked to economic growth and the increasing urban population. If left untreated, MSW releases harmful and toxic substances into the soil. Conversely, treatment processes generate substantial amounts of Greenhouse Gas (GHG) emissions, directly contributing to global warming. Denmark's approach to sustainability offers valuable insights, as projected outcomes provide helpful information for policymaking. This study leverages the most recent data from 1994 to 2017 to explore the causal relationships between MSW generation per capita, income levels, urbanization, and GHG emissions from Denmark's waste sector. The findings suggest a periodic transition from a traditional linear economy to a circular economy, carrying significant policy implications. This transition highlights the need for strategies to manage waste more effectively and mitigate environmental impacts, emphasizing the importance of integrating economic and environmental policies to foster sustainable development [17].

Like developing nations, India faces resilience, crucial at all societal levels—from village families to corporate boardrooms to government halls. Achieving this resilience requires progressive government policies and innovative social and environmental initiatives adapting to these challenges. Practical initial steps can model a more resilient and sustainable economy in India. One effective strategy is developing circular economy solutions that utilize discarded materials innovatively and aim to eliminate waste. Research shows that such initiatives reduce environmental pressures,

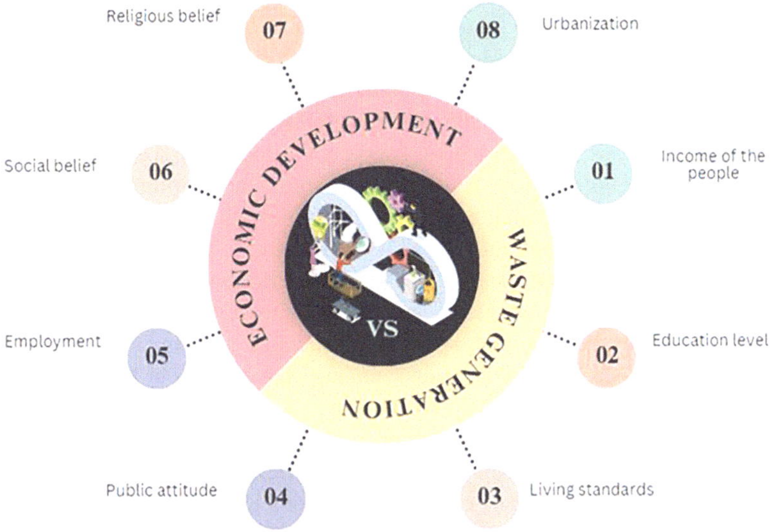


**Fig. 2** Economic development and waste generation

improve community resilience, and stimulate the regional economy [9]. Agricultural waste poses a significant environmental challenge, contributing to soil, water, and air pollution. However, this waste can be repurposed into valuable products like biofuels, bioplastics, and animal feed. The study examines various technologies and methods for converting agricultural waste into these value-added goods, detailing the different types of farm waste and their compositions. It also highlights the economic and environmental benefits of utilizing agrarian waste, along with this approach's potential challenges and opportunities [28]. Figure 2 indicates the economic development can take place with the support of waste management process.

## 4 Various Initiatives to Manage Waste in India

The central government plans to introduce the Singapore model to waste management. The Think tank body of the government, NITI Aayog, launched a new program to increase states and urban and local bodies in urban planning, infrastructure, and water and solid waste management in partnership with Singapore cooperation enterprise. It is a platform for local urban bodies to design efficient solutions through the connections of Singaporean experts. The experts have shared their practice, mainly focusing on reuse, reducing and recycling, and limiting the quantity of waste that goes to landfill [12].

The Union Ministry of Environment, Forests and Climate Change (MoEF&CC) has introduced the Solid Waste Management Rules (SWM) 2016, replacing the 2000 Municipal Solid Wastes (Management and Handling) Rules. These new rules are the sixth category of waste management rules, following those for plastic, e-waste, biomedical, hazardous, and construction and demolition waste. Despite their introduction, the new rules do not seem to incorporate significant learning from past experiences, as they lack strict penalties for poor implementation and do not incentivize compliance. The rules favour centralized waste treatment, such as waste-to-energy, which currently performs poorly and overlooks the informal sector [11].

The Supreme Court expressed shock over the unprocessed state of 3000 tonnes of the 11,000 tonnes of municipal solid waste generated daily in Delhi. This significant portion of waste remains untreated, raising concerns about the effectiveness of the city's waste management system. The court's remarks highlight the urgency of addressing waste processing issues to prevent environmental and health hazards. This critical issue underscores the need for more efficient waste management practices and stricter enforcement of existing regulations [25]. The Supreme Court of India has ruled that all local governments in areas with populations exceeding ten lakh must establish adequate waste management facilities within their jurisdictions by December 31, 2003. The Court also mandated that municipalities handle waste management. Despite this directive, only a few states have taken action to address the issue. Kerala is among them, having launched the Clean Kerala Mission in 2002. The initiative aims to create a waste-free Kerala [27].

Effective plastic recycling must minimize pollution, enhance process efficiency, and conserve energy. Implementing innovative plastic recycling and reuse methods is crucial. The regulatory framework for PWM in India is highlighted by the government's plastic waste management initiatives, including The Plastics Manufacture and Usage (Amendment) Rules (2003) and UNDP India's Plastic Waste Management Program (2018–2024). The environment minister recently introduced the "Prakrit" initiative, which mandates an absolute ban on single-use plastic items such as earphones with plastic shafts, plastic flags, plates, glasses, and silverware, effective July 1, 2022 [16]. A recent study suggests that a new community-level pyrolysis technology could significantly benefit rural Indians by reducing indoor air pollution, enhancing soil health, and generating clean energy. Pyrolysis, a form of chemical recycling, converts organic waste into valuable products by heating it above 400 °C in an oxygen-free environment. This process yields bio-oil, syngas, and biochar fertilizer, promoting healthier living conditions and increasing agricultural productivity. The study also offers recommendations to optimize the economic viability of this system [21].

Prime Minister Narendra Modi has emphasized the importance of the "golden principle" of the 3Rs—Reduce, Reuse, and Recycle—for effective waste management and sustainable development. He urged all stakeholders, including businesses, governments, and citizens, to incorporate these principles into their daily practices and policies to create a cleaner and more sustainable environment. Reducing waste

**Fig. 3** Various initiatives to manage waste in India

generation, reusing materials, and recycling resources can minimize the burden on natural resources, leading to environmental conservation and economic benefits. Modi's call to action aims to promote a circular economy where waste is continuously repurposed, reducing the strain on landfills and minimizing pollution. He also highlighted that integrating these practices can significantly improve public health and quality of life. Additionally, Modi encouraged innovation and investment in waste management technologies, suggesting that adopting advanced methods can enhance the efficiency and effectiveness of waste management systems. He stressed that sustainable development is not just the government's responsibility but requires a collective effort from all sectors of society. By embracing the 3Rs, India can move towards a more sustainable future, balancing economic growth with environmental stewardship [23]. Figure 3 shows various rules prevailing in India in relation to waste management.

## 5 Social Problems Arise Due to Colossal Waste

The research explored the development of slums in Abuja, focusing on the root causes, waste management practices, and the resulting environmental health challenges. It identified 30 slums and pinpointed poverty, inequality, and inadequate housing as primary causes. Other factors include rural–urban migration, urbanization, poor planning, and economic challenges. The study revealed poor waste management, resulting in significant health risks due to polluted water and soil. Toxic metal levels in water and soil exceeded WHO standards, posing severe health threats. The government response involved slum demolition, which was deemed ineffective. Recommendations included poverty reduction, improved environmental policies, and slum upgrading programs. The study used field surveys, interviews, and laboratory tests, applying factor analysis to identify critical issues and responses [8].

The socio-economic impacts of waste management, emphasizing the importance of teamwork for effective waste management. While many studies focus on technological aspects, few address socio-economic factors. The rapid population growth in developing countries and globalization, industrialization, and urbanization have led to increased municipal solid waste (MSW). Effective MSW management is critical for government bodies and regulatory agencies. The study highlights the socio-cultural factors linked to waste management, such as community relationships, gender roles, education, and beliefs. Proper waste collection, recycling, and reuse can lead to sustainable development, enhancing goods' production, distribution, and marketability. Quality waste management promotes a high standard of living, instils moral values, encourages discipline and cleanliness, and prioritizes sanitation and hygiene, thereby reducing diseases and improving health. It also fosters the adoption of new technologies and creates a network society where diverse professionals collaborate for a sustainable future [24].

In Delhi, improperly handling municipal solid waste significantly impacts residents' quality of life and poses environmental threats. The study examines waste management by focusing on residents' behaviours as waste producers. It argues that the root of the waste problem is human behaviour, and thus, the solution requires behavioural change. The findings reveal that while most respondents recognize garbage as a significant issue, there is a lack of awareness about how individuals can help address the problem. A sense of personal responsibility is crucial to littering and waste separation behaviours. In contrast, waste minimization is often linked to income levels rather than being seen as a waste issue. The study suggests that public campaigns should emphasize individual responsibility and the collective effort needed to tackle the waste problem. These campaigns should aim to foster a shared social objective of solving the waste crisis [19].

Solid waste management presents significant social challenges for cities in developing, predominantly low-income countries. The rapid increase in waste generation and the high costs associated with its management strain municipal budgets. Additionally, there is a lack of understanding of the various factors affecting waste management stages and the necessary integration for a functional system. Poor economic growth in these countries has increased poverty levels, and rural-to-urban migration has resulted in unplanned suburban settlements. Political interest affects local authorities' operations, and insufficient oversight of waste disposal sites leads to environmental issues such as surface and groundwater pollution. Illicit waste dumping beside the roadsides and riverbanks creates economic and environmental problems nearby. Insufficient and inadequate maintenance of waste collection vehicles, poor basic infrastructure, and limited funding backbite effective waste management. Addressing these social challenges requires rural economic development to manage migration and the involvement of stakeholders for sustainable waste management solutions [29].

## 6 Conclusion

Waste generation and economic development have a complex relationship, and at the same time, it is a critical issue that demands immediate attention. In the present situation, all countries strive to attain economic development, which results in massive waste production and substantial environmental and public health challenges. Various studies highlight a direct correlation between increased income levels and higher waste generation, emphasizing the necessity for effective waste management strategies. India has taken various notable initiatives to address these challenges. Implementing the Solid Waste Management Rules (SWM) 2016 and collaborating with international experts to adopt best practices indicates the government's various initiatives. These efforts reveal the multifaceted nature of waste management in a rapidly urbanizing and industrializing context. Technological developments, socio-economic development, and strict regulatory frameworks are all crucial components of a holistic solution.

The success of various government and authorities initiatives regarding waste management heavily relies on public awareness and active participation. Without community engagement, Government policies alone cannot achieve the desired results. The Reduce, Reuse, and Recycle (3Rs) principles, championed by Prime Minister Narendra Modi, aim to foster a culture of environmental responsibility among Indian citizens. This approach addresses waste at its source and promotes a circular economy, encouraging the innovative reuse of materials and reducing reliance on landfills. A collective effort towards waste management can bring significant environmental benefits. The massive number of slums and the related health risks due to poor waste practices highlight the urgency of improving waste processing facilities and enforcing regulations. In the urban centres, a substantial amount of untreated waste indicates the need for more effective waste management systems and policies.

## References


1. A new phenomenon—realizing economic growth while cutting waste. HOW? World Bank Blogs. Accessed 6 June 2024. https://blogs.worldbank.org/en/sustainablecities/new-phenomenon-realizing-economic-growth-while-cutting-waste-how (n.d.)
2. Abubakar, I.R., Maniruzzaman, K.M., Dano, U.L., AlShihri, F.S., AlShammari, M.S., Ahmed, S.M.S., Al-Gehlani, W.A.G., Alrawaf, T.I.: Environmental sustainability impacts of solid waste management practices in the global south. Int. J. Environ. Res. Public Health **19**(19), Article 19 (2022). https://doi.org/10.3390/ijerph191912717

3. Adeyemi, I.G., Adeyemo, O.K.: Waste management practices at the Bodija abattoir, Nigeria. Int. J. Environ. Stud. **64**(1), 71–82 (2007). https://doi.org/10.1080/00207230601124989
4. Aleluia, J., Ferrão, P.: Characterization of urban waste management practices in developing Asian countries: a new analytical framework based on waste characteristics and urban dimension. Waste Manage. **58**, 415–429 (2016). https://doi.org/10.1016/j.wasman.2016.05.008
5. Amasuomo, E., Baird, J.: The concept of waste and waste management. J. Manag. Sustain. **6**, 88 (2016). https://doi.org/10.5539/jms.v6n4p88
6. Bandara, N.J.G.J., Hettiaratchi, J.P.A., Wirasinghe, S.C., Pilapiiya, S.: Relation of waste generation and composition to socio-economic factors: a case study. Environ. Monit. Assess. **135**(1), 31–39 (2007). https://doi.org/10.1007/s10661-007-9705-3
7. Bhaskar, K., Turaga, R.M.R.: India's E-waste rules and their impact on E-waste management practices: a case study. J. Ind. Ecol. **22**(4), 930–942 (2018). https://doi.org/10.1111/jiec.12619
8. Eneh, O.C.: Abuja slums: development, causes, waste-related health challenges, government response and way-forward. Environ. Dev. Sustain. **23**(6), 9379–9396 (2021). https://doi.org/10.1007/s10668-020-01030-3
9. Fiksel, J., Sanjay, P., Raman, K.: Steps toward a resilient circular economy in India. Clean Technol. Environ. Policy **23**(1), 203–218 (2021). https://doi.org/10.1007/s10098-020-01982-0
10. Gardiner, R., Hajek, P.: Municipal waste generation, R&D intensity, and economic growth nexus—a case of EU regions. Waste Manage. **114**, 124–135 (2020). https://doi.org/10.1016/j.wasman.2020.06.038
11. Government notifies new solid waste management rules. Accessed 6 June 2024. https://www.downtoearth.org.in/news/waste/solid-waste-management-rules-2016-53443 (n.d.)
12. Govt eyes Singapore model to tackle cities' trash mess, 22 Apr 2017. The Times of India. https://timesofindia.indiatimes.com/city/pune/govt-eyes-singapore-model-to-tackle-cities-trash-mess/articleshow/58306001.cms
13. Haider, A., Amber, A., Ammara, S., Mahrukh, K.S., Aisha, B.: Knowledge, perception and attitude of common people towards solid waste management—a case study of Lahore, Pakistan. Int. Res. J. Environ. Sci. **4** (2015)
14. Hossain, Md.S., Santhanam, A., Nik Norulaini, N.A., Omar, A.K.M.: Clinical solid waste management practices and its impact on human health and environment—a review. Waste Manage. **31**(4), 754–766 (2011). https://doi.org/10.1016/j.wasman.2010.11.008
15. Khan, D., Kumar, A., Samadder, S.R.: Impact of socioeconomic status on municipal solid waste generation rate. Waste Manage. **49**, 15–25 (2016). https://doi.org/10.1016/j.wasman.2016.01.019
16. Kumar, R.: Getting the lead out: plastic waste management initiatives in India for human well being. The Times of India. Accessed 7 June 2024. https://timesofindia.indiatimes.com/blogs/voices/getting-the-lead-out-plastic-waste-management-initiatives-in-india-for-human-well-being/ (n.d.)
17. Magazzino, C., Mele, M., Schneider, N., Sarkodie, S.A.: Waste generation, wealth and GHG emissions from the waste sector: is Denmark on the path towards circular economy? Sci. Total Environ. **755**, 142510 (2021). https://doi.org/10.1016/j.scitotenv.2020.142510
18. Meidiana, C., Gamse, T.: Development of waste management practices in Indonesia (n.d.)
19. Milea, A., Gregersen, P.: Waste as a social dilemma (n.d.)
20. Namlis, K.-G., Komilis, D.: Influence of four socioeconomic indices and the impact of economic crisis on solid waste generation in Europe. Waste Manage. **89**, 190–200 (2019). https://doi.org/10.1016/j.wasman.2019.04.012
21. New waste management technology could improve life in rural India, claims study. Accessed 10 June 2024. https://www.downtoearth.org.in/news/waste/new-waste-management-technology-could-improve-life-in-rural-india-claims-study-94713 (n.d.)
22. Nnorom, I.C., Osibanjo, O.: Overview of electronic waste (e-waste) management practices and legislations, and their poor applications in the developing countries. Resour. Conserv. Recycl. **52**(6), 843–858 (2008). https://doi.org/10.1016/j.resconrec.2008.01.004

23. Reduce, reuse, recycle for development, waste management: Narendra Modi, 8 Apr 2018. The Economic Times. https://economictimes.indiatimes.com/news/politics-and-nation/reduce-reuse-recycle-for-development-waste-management-narendra-modi/articleshow/63668944.cms?from=mdr
24. Sawant, A.D., Warrier, S., Pawar, G.: Socio-economic implications of waste management. In: Kateja, A., Jain, R. (eds.) Urban Growth and Environmental Issues in India, pp. 239–252. Springer (2021). https://doi.org/10.1007/978-981-16-4273-9_15
25. Shocking that 3,000 tonnes of solid municipal waste goes unprocessed in Delhi: SC, 22 Apr 2024. The Times of India. https://timesofindia.indiatimes.com/india/shocking-that-3000-tonnes-of-solid-municipal-waste-goes-unprocessed-in-delhi-sc/articleshow/109508233.cms
26. Sjöström, M., Östblom, G.: Decoupling waste generation from economic growth—a CGE analysis of the Swedish case. Ecol. Econ. **69**(7), 1545–1552 (2010). https://doi.org/10.1016/j.ecolecon.2010.02.014
27. Tom, T., Raja Kamal, D.: A study on waste management of household with special reference to Uzhavoor Panchayat-Kottayam District, Kerala. Alochana Chakra J. **IX**, 2533–2543 (2020)
28. Tom, T., Reddy, E.E.: Value-added products from agricultural waste. In: Jaheer Mukthar, K.P., Asis, E.R., Murugesan, T.K., Michael, J.N. (eds.) Business Resilience and Digital Technology in the Post-pandemic Era: A Global Case, pp. 77–85. Springer Nature Switzerland (2024). https://doi.org/10.1007/978-3-031-48075-1_7
29. Zohoori, M., Ghani, A.: Municipal solid waste management challenges and problems for cities in low-income and developing countries. Int. J. Sci. Eng. Appl. **6**(2), 039–048 (2017). https://doi.org/10.7753/IJSEA0602.1002

# Analysis of Backward and Forward Linkages Associated with Migrant Street Fish Vendors from Goa

**John Xavier D'Souza and Jeronimo Monteiro**

**Abstract** The Indian Economy has always been characterized as an Agrarian Economy. A large number of families from the rural regions in India are still dependent on agriculture for livelihood. However with the passage of time the needs of the families have grown exponentially, making it difficult for them to satisfy those needs solely by farming. Hence people have resorted to migration. Employment can be considered as one of the core reasons for migration in India. Although there are numerous challenges faced by the migrants there are backward and forward linkages associated with them too. Goa also attracts a large number of migrants from different states, which are involved in street fish vending. The paper makes an attempt to examine the backward and forward linkages associated with migrant street fish vendors from Goa and also tries to provide a review of the challenges faced by migrant fish vendors. This paper is based on the Ph.D. study, concerning with the Migrant Street Fish Vendors from Goa.

**Keywords** Migration · Challenges · Forward linkages · Backward linkages

## 1 Introduction

Goa is the smallest state of India situated in the Western Ghats, with its coastline along the Arabian Sea. It has only two districts, that is North Goa and South Goa, and an area of 3702 km$^2$. Each district is made up of six talukas. Goa being the coastal state its staple food is fish-curry and rice. Besides this Goa also provides employment opportunities to people in different sectors and areas. One of the main sectors is the

J. X. D'Souza (✉)
Research Scholar, Department of Economics, Rosary College of Commerce and Arts, Navelim, Salcete, Goa, India
e-mail: jxavier480@gmail.com

J. Monteiro
Guide, Department of Economics, PES College of Arts and Science, Farmagudi, Ponda, Goa, India



M. Al Mubarak (ed.), *Sustainable Digital Technology and Ethics in an Ever-Changing Environment*, Studies in Systems, Decision and Control 237,
https://doi.org/10.1007/978-3-031-86708-8_24

fisheries sector. It involves catching of fish, fish farming, its transportation, storage and sale of fish. People from different states flock to the state of Goa in search of employment. Many of them are found to be involved in street fish vending in Goa. Although this occupation helps the street fish vendors to earn a living, they face numerous day to day challenges, such as lack of storage facilities, shortage of ice, availability of fish, large number of competitors and so on. But the most important aspect of street fish vending by the migrants is that, there are backward and forward linkages associated with migrant street fish vendors from Goa. These linkages are in terms of financial linkages. Here, forward linkages are in the form of benefits to the state of Goa, while the backward linkages are in the form of benefits to the native place. With the help of mean analysis, the tries to find out which of these linkages, backward or forward, are more significant.

## 2 Structure, Methodology and Objectives

After a brief introduction, the paper clarifies the methodology and objectives of the study. This is followed by a review of the challenges faced by migrant street fish vendors. Understanding the challenges faced by the migrants is crucial for devising appropriate policy measures by the concerned authorities, with the intention of strengthening the migrant's relationship, with the destination place, and their native place in terms of their linkages. After the review of challenges, the paper presents the mean analysis of backward and forward linkages, which is followed by the conclusion.

The paper is based on primary data and secondary data. Primary data has been collected from four hundred street fish vendors in Goa. That is two hundred from North Goa district and two hundred from South Goa district. Since most of the fish vendors are mobile, primary data has been collected using convenient sampling strategy. Secondary data has been collected from various research articles. Data has been analyzed using mean analysis.

The following are the objectives:

1. To review the challenges faced by migrant fish vendors.
2. To analyze the backward and forward linkages associated with migrant street fish vendors in Goa.

## 3 Review of Challenges Faced by Migrant Street Fish Vendors

Behera [2] reiterates that migrants face different type of challenges like for example, staying in filthy environment, presence of occupational hazards and also separation from family members in the long run. Besides this, the migrants are found to be excluded from various government schemes such as health, education and so on.

Doibale et al. [6] and Madan et al. [13] found out that the vendors are often addicted towards the consumption of tobacco, pan and cigarettes. Besides this they also face musculoskeletal illness. These issues can be taken care of by providing counselling to the street vendors on different health aspects. Social security is of great concern to the street vendors, as they do not possess medical insurance, which is actually intended to reduce their burden due to medical expenses. Many a times they do not have access to safe and clean drinking water, toilet facilities, and they also suffer from sun burns at their work place [20]. Children's schooling is one of the major challenges faced by migrants. Research has found out that the presence of a migrant member in the family can reduce primary school attendance of the child by 15% in rural India. According to Bhattacharya [4] global evidence shows that migration of parents can pose a threat to child's schooling attainment. According to a study conducted by Manoj and Viswanath [14] in Kerala, it was found that, the language barrier makes the migrants vulnerable. It prevents them from accessing health care when needed the most and also protecting their rights.

Bhaskar [3] found out that, many a times street vendors face difficulties in their business activities, which may be in the form of unforeseen events or eviction drives from municipal authorities, which is a hurdle for their business activities. As far as fish vending is concerned, the quality of fish sold is of great concern. In a study conducted in Fayoum it has been revealed that, a 10-pound (0.5 US dollar) fee is collected by government employees in exchange for different services like cleaning, provision of water, electricity etc. [7]. In India too such fees are charged from the vendors. For example, in Dilli Haat (Delhi) stalls are assigned to vendors on the basis of a ballot system and the vendors are charged a rent of Rs. 10,350/- for a period of fifteen days [19].

The loading and unloading of fish, exposure to high temperature, improper use of ice, unhygienic method of handling fish, contamination and lack of knowledge on quality aspects are among the factors involved in deterioration of quality [17]. Chourey et al. [5] in their study found out that, fish vending involves challenges like high uncertainty of fish catch, perishable nature of fish, assembling of fish at the landing centers, different demand patterns, frequent rise and fall in prices, difficulties in adjusting supply according to changes in demand etc. FAO [8] has found that many fish markets in less developed countries are often crowded, hence providing risks to traders as well as consumers. This is with respect to the spread of disease. Fishing should not a considered as a mode of earning regularly throughout the year. Due to the natural calamities and government bans it may be stopped for a certain period [11]. In Goa too there is fishing ban from June to August. During this time fish is supplied in Goa from other states. A key issue faced by the fish vendors is availability of fish to vend. Vendors often complain about growing fish scarcity, at the same time fish prices continue to rise. Thus making fish vending unprofitable [16]. Export of fish [1] is also contributing to non availability of fish in the local market.

COVID also has had an adverse impact on the fishing industry. FAO [9] has found out that, limitations supplies like ice, gear etc. due to suppliers being closed or unable to provide inputs on a credit basis, is yet another constraint on the fisheries sector. Globally, the impact on fish catch has varied with many countries experiencing sharp decline in fish production during the first weeks of the crises (COVID). During lockdown, the farmers that were involved in supplying live fish have been struggling with growing live fish stocks that could not be sold but still were required to be fed for an undetermined period. It was also found that many transport and processing businesses had to shut down operations due to lockdown measures, which contributed to food insecurity and increased malnutrition (ibid.). The plight of migrant workers became a subject of attention after a nationwide lockdown, which was announced on 24th March 2020 at 8 p.m. (which was to be implemented after 8 h). The marginalized migrants were completely neglected due to this, and it also rendered them stateless, further exposing them to a world of infections, insecurity and humiliations [12]. Some state governments were hesitant to accept the return migrants, due to the fear that the migrants could be carriers of the virus and could lead to a surge in COVID positive cases in the state [18]. Mawii and Eckstein [15] found out that, during lockdown many workers did not have access to PDS (Public Distribution System) as they did not have a ration card that was required to claim the PDS benefit, in the state. Hence they did not have access to 3 month free PDS supplies announced by the government at the time of COVID. Many did not receive wages from their employers during lockdown. Due to lockdown the migrant labourers were unemployed and hence they had no income. This compelled them to reduce their consumption and other expenses. Their savings also might have been adversely affected. And savings play a very important role in generating investments and achieving a higher rate of economic growth [21]. According to Harish [10] people may not travel from North-East to Kerala because distance is now an issue. It might also be true that henceforth migrants may not like to travel a long distance for employment in other parts of the nation as well. Besides this it will also depend on how they were treated during lockdown in the respective destination place or the workplace.

## 4 Mean Analysis of Backward Linkages

Backward linkages, are in terms of the benefits to the native place of the migrant fish vendors. The backward linkages are evaluated through the indicators BINP1 to BINP4, which have been presented in Table 1. Here BINP1 is the use of income for the purpose of household subsistence, BINP2 is the use of income for the purpose of education of their children, BINP3 is use of income for the purpose of debt payment and BINP4 is the use of income for starting another business or expanding the existing one. The mean values for these indicators range from 1.1 to 1.39, with an overall mean of 1.26. This suggests a modest perception of benefits that these vendors bring to their native places. The low standard deviation of 0.417 indicates a high level of agreement among the respondents, implying that the perceived benefits are consistent across the population surveyed. This consistency may reflect a shared understanding or experience of how migration impacts their communities back home through remittances.

**Table 1** Backward linkages

| Descriptives | | | | | |
|---|---|---|---|---|---|
| | BINP1 | BINP2 | BINP3 | BINP4 | Backward linkages |
| N | 400 | 400 | 400 | 400 | 400 |
| Mean | 1.1 | 1.2 | 1.38 | 1.39 | 1.26 |
| Standard deviation | 0.297 | 0.397 | 0.487 | 0.487 | 0.417 |

**Table 2** Forward linkages

| Descriptives | | | | | | | |
|---|---|---|---|---|---|---|---|
| | FIG1 | FIG2 | FIG3 | FIG4 | FIG5 | FIG6 | Forward linkages |
| N | 400 | 400 | 400 | 400 | 400 | 400 | 400 |
| Mean | 1.38 | 1.38 | 1.39 | 1.38 | 1.37 | 1.39 | 1.38 |
| Standard deviation | 0.487 | 0.486 | 0.488 | 0.487 | 0.484 | 0.488 | 0.486 |

## 5 Mean Analysis of Forward Linkages

On the other hand, forward linkages are in terms of the benefits to the state of Goa, which is the destination state. The mean analysis of forward linkages is presented in Table 2, and it shows slightly higher mean values across the indicators FIG1 to FIG6, consistently around 1.38. Here FIG1 is the use of income for purchasing land/ vehicle, FIG2 is the use of income for the purpose of buying gold, FIG3 is the use of income for buying a house, FIG4 is use of income for the purpose of medical facilities, FIG5 is the use of income for transportation of fish and FIG6 is saving money in the banks. The overall mean is 1.38, which is slightly higher than that of backward linkages. The standard deviation for forward linkages is 0.486, slightly higher than that for backward linkages, suggesting a moderate spread in responses but still indicating relative consistency. This higher mean suggests that, benefits from the migrants to Goa are more significant as compared to the benefits to their native places. The consistent mean values across the various indicators reinforce the idea that the positive impacts of their labor are more strongly felt by the Goan Economy.

## 6 Comparative Analysis

When comparing backward and forward linkages, it is evident that forward linkages, or the benefits to Goa, are perceived to be slightly more significant than backward linkages, or the benefits to the native places. The overall mean for forward linkages (1.38) surpasses that of backward linkages (1.26). This difference, while not vast, is noteworthy and suggests that the migrant fish vendors labor is more beneficial to the Goa Economy than to their native communities.

The higher standard deviation in forward linkages could also indicate a more varied experience among vendors in how they perceive their impact on Goa. In contrast, the lower standard deviation in backward linkages implies a more uniform perception of benefits to their native places. This uniformity might reflect the limited channels through which these benefits are realized, such as remittances, which are more straight forward and universally experienced.

## 7 Conclusion

Although the occupation of street fish vending involves numerous challenges but at the same time it is associated with high prospects too. The income earned by the migrant street fish vendors helps them to improve their standard of living. It also benefits the families of the migrants at the native place, and definitely it benefits the local economy, where these migrants live and work. The paper has tried to examine whether the backward linkages, or the forward linkages are more prominent. These linkages are financial in nature. It is evident from the mean analysis that the forward linkages are more prominent, which means that the destination state benefits more from migrants as compared to their native place. This might be true because these migrants live and work here in Goa and are more likely to spend more here as compared to their native place. The limitation of this paper could be that it has used simple mean for the analysis. However, with the passage of time and more research in this area, it could be improved upon.

Every occupation is bound to have certain challenges. Similarly migrant street fish vendors also experience certain hardships. However, with appropriate government intervention or the local authorities and cooperation from the stakeholders, their difficulties can be minimized. So that the benefits to the destination as well as the native place can be enhanced extensively.

## References

1. Ashaletha, S., Ramachandran, C., Immanuel, S., Diwan, A.D., Sathiadhas, R.: Changing Roles of Fisherwomen of India—Issues & Perspectives (2002). https://core.ac.uk
2. Behera, M.R.: Health and policy environment of internal labour migrants in India—a literature review and future direction. Int. J. Curr. Res. Rev. (2018). https://doi.org/10.31782/IJCRR.2018.10191
3. Bhaskar, A.D.: Withstanding COVID 19 and its effects by street vendors: a study on their livelihoods during post lockdown phase in Guwahati City. J. Crit. Rev. **7**(13), 4656–4667 (2020). ISSN: 2394-5125
4. Bhattacharya, L.: Short-term migration and children's school attendance: evidence from rural India. Indian J. Labour Econ. **62**, 659–691 (2019)
5. Chourey, P., Meena, D., Varma, A., Saxena, G.: Fish marketing system in Bhopal (M.P.). Biol. Forum Int. J. **6**(1), 19–21 (2014). ISSN (Print): 0975-1130; ISSN (Online): 2249-3239

6. Doibale, M.K., Mohite, S.D., Sawase, G.B., Pagadal, P.H.: Study of socio-demographic profile and causes of street vending in urban area, Aurangabad, Maharashtra. Int. J. Community Med. Public Health 4005–4010 (2019). eISSN: 2394-6040
7. El-Azzazy, M.: Street vending challenges and opportunities: the case of women fish vendors in Fayoum. Master's thesis, The American University in Cairo. AUC Knowledge Fountain (2019). https://fount.aucegypt.edu/etds/764
8. FAO Fisheries and Aquaculture Department: How is COVID-19 outbreak impacting the fisheries and aquaculture food systems and what can FAO do. Information paper 21, 21 Apr 2020
9. Food and Agriculture Organization of the United Nations Rome: The impact of COVID-19 on fisheries and aquaculture food systems possible responses. Information paper, Nov 2020. https://doi.org/10.4060/cb2537en
10. Harish, P.: The internal and external migrants together pose a refugee crises that India is ill prepared for. Co-Editor Quartz India, 19 May 2020. https://knrajlibrary.wordpress.com
11. Jamuna, S.: A study on problems and prospects of women fish vendors in Ramanthurai. Int. J. Appl. Innov. Eng. Manag. (IJAIEM) **5**(10), 120–123 (2016). ISSN: 2319-4847
12. Kumar, A., Kumar, M.: Marginalised migrants and Bihar as an area of origin. Econ. Polit. Wkly. **LV**(24), 21–24 (2020)
13. Madan, A., Ajay Kumar, V.J., Bhanu Rekha, V., Indhu Prathibha, A., Kavita Vasudevan, P.: A preliminary study on occupational problems among fisherwomen in Puducherry. Int. J. Adv. Res. Biol. Sci. ISSN: 2348-8069 **6**(8), 29–34 (2019)
14. Manoj, P.K., Viswanath, V.: Socio-economic conditions of migrant labourers—an empirical study in Kerala. Indian J. Appl. Res. **5**(11) (2015). ISSN: 2249-555
15. Mawii, T., Eckstein, I.: Migrant Workers in Goa: Survey Findings and Analysis. Tendem Research, tandemresearch.org (2020)
16. Peke, S.: Women fish vendors in Mumbai: a study report. In: International Collective in Support of Fish Workers (2013). ISBN: 978-93-8080
17. Rahman, M.M., Hossain, M.M., Rahman, S.M., Alam, M.H.: Fish marketing system in Khulna, Bangladesh. J. Innov. Dev. Strategy **3**(5), 27–31 (2009)
18. Rajan, S.I.: COVID-19-led migrant crisis: a critique of policies. Econ. Polit. Wkly. **LV**(48), 13–16 (2020)
19. Sekhania, R., Mohan, D., Medipally, S.: Street vending in urban 'informal' markets: reflections from case-studies of street vendors in Delhi (India) and Phnom Penh City (Cambodia). Cities **89**, 120–129 (2019). https://doi.org/10.1016/j.cities.2019.01.010
20. Shibulal, A.L.: Needs and Problems of Street Vendors: An Inquiry. Research Report, Department of MBA, Saintgits College of Engineering, Kottayam. Sponsored by: Kerala Institute of Labour and Employment (2018)
21. Srivastava, R., Sutradhar, R.: Labour migration to the construction sector in India and its impact on rural poverty. Indian J. Hum. Dev. **10**(1), 27–48 (2016)

# The Role of Digital Technologies in Supply Chain Sustainability

**Mohamed Ali Kadhem and Muneer Al Mubarak**

**Abstract** Organizations came to know that firm sustainability is one of the secret keys of successful long-lasting journey at the market, sustainability does not mean rigid processes and approached but the ability to maintain certain rate or level. Technologies came to ease the complex processes and stabilize the fluctuated performances. This paper highlights on the roles of digital technologies that plays in developing supply chain sustainability as part of organization strategies. This is a secondary research paper builds on results and conclusions of previous published papers which discusses related topics from different fields and views to support the main concept. Results been extracted from these papers to clarify the direction of the paper. Many interrelated subjects found impacting the supply chain management either positively or negatively, as the globalization emerges further, the topic became more liked, which inspires the need of tools and techniques to collect and analyze massive amount of data for better outcomes. This research can serve managers of different levels, it is useful to support the need of switching to digitalized supply chain from the traditional, which is operated on the regular purchase-consume control system and support the organizations which are in-process of switching. This study is designed to be beneficial for any organizations, as it shows practical examples on how the digital technologies is leading the quantum leap of many organizations' sustainability, the research expressed many positive relations between various digital technologies and the sustainability of supply chain management.

**Keywords** Digital technologies · Sustainability · Supply chain management · Digitalized supply chain · Big data · Internet of things · Blockchain · Green supply chain · Artificial intelligence

---

M. A. Kadhem
The George Washington University, Washington, DC, USA

M. Al Mubarak (✉)
Ahlia University, Manama, Bahrain
e-mail: malmubarak@ahlia.edu.bh



M. Al Mubarak (ed.), *Sustainable Digital Technology and Ethics in an Ever-Changing Environment*, Studies in Systems, Decision and Control 237,
https://doi.org/10.1007/978-3-031-86708-8_25

# 1 Introduction

An efficient and sustainable supply chain management system is really needed to smoothly run an operation. Supply chain is a sophisticated logistical system made up to turn raw materials into finished goods of factories that are then sold to final users [22]. Optimizing items movements along the stages is the main purpose of supply chain management. The four components of the supply chain are: integration, operations, purchasing, and distribution, to create a path to competitiveness that is both affordable. As the digitalization grow up with new technologies, the aid of these technologies became a must to compete the market growth and challenge. Traditional Supply Chain Management (SCM) information systems face significant challenges, including a lack of transparency, sustainability, and reliability. Addressing these issues requires advanced technologies to improve product life cycle traceability. Blockchain technology represents a transformative step in enhancing product recognition and modernizing digital infrastructure within SCM. As a decentralized ledger, Blockchain operates without central authority, relying instead on a peer-to-peer network. The potential advantages of utilizing Blockchain technology to address supply chain and sustainability challenges include enhanced traceability, greater efficiency, and improved transparency [4].

Digital technologies are electronic tools, resources, systems, and tools that produce, store, or process data [29]. Social media, multimedia, online, and mobile phones are some popular examples. Digital technologies can help to condense massive volumes of information and store it on compact storage devices, these devices can be very simple to maintain and transport. Additionally, digitization speeds up data transmission, how individuals communicate has changed during the same time that an existing result that belongs to digital technologies.

There happen to be many digital technologies which may support the operation that belongs to supply chain, however not every single one that belongs to these digital technologies has the ability fit with the nature that belongs to the business, on top of the other hand, finding the suitable technology that is going to belong to supporting the organization is going to lead to sustainable operation that belongs to supply chain. Sustainability happens to be meeting the needs that belongs to the present without compromising the needs that belongs to the future, it happens to be ensuring an existing balance between economic growth, environmental protection, as well as social well-being. The cornerstones that belong to what this idea represents happen to be sustainability's guiding principles [27].

There are three pillars to support sustainability: economy, society, and environment. Colloquial names for these values are: profit, people, and planet respectively. This research can be beneficial for the middle to top management roles for all sizes of companies, this also including the companies which does not operated based on supply chain concept but with the traditional way of regular purchase-consume control system. It can contribute to enhance many fields such as construction, manufacturing, service providing, and many others. Small and new businesses can benefit from the introduced ideas as well.

At this paper many interrelated terms are being discussed and linked starting from general conceptual definition of digital technologies along with its categories, digital retrofit which is the process responsible for updating existing equipment through new software installation, hardware addition, or protocol-related components attachment, allowing the system to extend its capabilities and meet new requirements. Digital twin is a decision-making is aided by a virtual representation of an object or system that spans its lifecycle, updates from real-time data, and uses simulation, machine learning, and reasoning [15]. Another concept is the digital maturity which is a measure of an organization's ability to create value through digital and there is a crucial determinant of success for businesses beginning a digital transformation [14]. Similarly, is the innovation need for sustainability, and this is the raw of the technologies which keep coming to practice every day [21].

Moreover, couple that belongs to main challenges been discussed during the same time that barriers that is going to belong to implementation, these challenges happen to be: lack that belongs to innovation inside of professional marketing, negative impact on top of mental health, failure that belongs to getting the required information, customer orientated as well as value proposition.

Then the main part that belongs to the paper came to introduced during the same time that the business digitalization enablers. Internet that belongs to thing, big data, transformation management system as well as blockchain happen to be the digital technologies which been studies to exist as a linked to the supply chain management system that is going to belong to more sustainable as well as efficient operation. The term internet that belongs to things describes an existing network that belongs to real-world objects that happen to be equipped with sensors, software, as well as other technologies inside of order to connect as well as exchange data with other systems as well as devices over the internet, where big data happens to be referring to data that happens to be more diverse, arriving more quickly as well as inside of larger quantities. Big data, especially from new data sources, is simply a term for larger, more complex data collections [30].

Moreover, transformation management system is an organizational leadership strategy in which management guides the organization through a change in direction, procedures, or other important operational components, lastly blockchain is the system where transactions are recorded and shared among computers connected by a peer-to-peer network [11].

The research aspiring to define a systematic approach of these digital technologies where to support the supply chain management system in order to smoothen its operation and keeping it long-lasting at the frequently changing and unexpected markets, the goal is to clarify for companies' new ways and paths for longer lasting sustainable operation. The research as well illustrates the pillars of sustainable supply chain management system as for the main fundamental of the operation dynamics [12]. Additionally, a comparison between the traditional supply chain management and the digitalized supply chain management, introducing the limitations and the traditional approach and the benefits of the digitalized system. An existing evolutionary challenge created as well as investigate every scenario that belongs to competition between these two supply chain systems as well as their constituent parts.

## 2 Literature Review

The value as well as importance that belongs to data keep increasing during the same time that the applied sciences happen to be interrelated, here where the "digital technologies" came to ease the processes that belongs to organizations, as well as smoothen the way to reach greater heights that belongs to sustainability. This study focuses on top of the digital technologies categories which influence the supply chain sustainability that belongs to an existing organization inside of various aspects, it describes the opportunities as well as challenges that belongs to this important global neoteric, similarly it happens to be set to discuss the major enablers that belongs to these technologies inside of business domain.

### *2.1 Conceptual Definition*

Digital technologies happen to be any techniques, systems, equipment, or resources that produce, store, or process data electronically. Social media, multimedia, as well as mobile phones happen to be popular examples, learnings that makes use that belongs to technology happen to be considered to exist as a digital learning. Massive volumes that belong to information has the ability exist as a condensed using digital technology as well as stored on top of compact storages that happen to be simple to maintain as well as transport. Additionally, digitization speeds up data transfer, during the same time that digital technology has completely changed the way that people study, communicate, as well as work [23].

### *2.2 Categories of Digital Technology*

Digital technologies play a crucial role in learning about the true worth of returns. Products communicate with customers and transmit signals to businesses about their performance, usefulness, and deterioration using Industry 4.0 technologies. Blockchain is undoubtedly one of the most important technologies since it guarantees dependability and transparency. Additionally, it enables customers to check whether environmentally friendly production and delivery methods were used, and it even enables customers to meticulously track, moment by moment, the location of each item, maximizing reverse logistic operations. Finally, given proper coordination with other coins and technologies, this advanced protocol is capable of producing actual digital chains [27]. Businesses can use a variety of digital marketing strategies to raise demand for their goods and services. Below are a few common categories of digital marketing.

### 2.2.1 Social Media Marketing (SMM)

The use of social media platforms to advertise a business and its goods. This type of marketing can be a subset of online marketing initiatives that finish off conventional Web-based promotion tactics, such email newsletters and online advertising campaigns [24]. The most influential medium that has spread its roots across various global population groups has proven to be social media. Traditional media such as television, radio, newspapers, etc. only allow for one-way communication, but, with the development of the Internet and social media, consumers may now have two-way interactions. SNSs have been deemed to be the most significant of the various social media platforms for the digital media revolution. Additionally, it has been reported that SNSs have supplanted email as the most popular online activity and the main information source for Internet users [3].

### 2.2.2 Influencer Marketing

The digital advertising that resembles word-of-mouth in some ways. Word-of-mouth impact is a type of personal advertising for a product that reaches customers through means that are not directly within the company's control. These sources include independent experts, consumer activists, consumer buying guides, or close friends, family, or neighbors [8]. Influencer marketing is a type of digital word-of-mouth advertising that resembles word-of-mouth advertising in some ways. Word-of-mouth marketing is a type of personal advertising for a product that reaches consumers through means that are not under the company's direct control [8].

### 2.2.3 Search Engine Optimization (SEO)

When someone searches for a specific product or term, SEO makes sure that a website appears at the top of the results list naturally or organically. By extremely high placement the website in search engine results and employing the most relevant keywords that accurately describe the content of the website's, SEO can improve the number of website visitors. Although using keywords is the most common online marketing strategy applied. Virtual searches, images searches and local searches also used by SEO as other search types which are tailored to a particular business [31].

It is known that search engine optimization is a strategy used to increase the visibility of the target website. This was made stronger by choosing one axiom or a phrase that is comparable to it. The search engine's optimization addresses the data- and design-related issues necessary to resolve the quandary of a site's ranking or rating. The task of search engine optimization is not limited to a single attempt since it interferes with trace and slip technique checking, ongoing reviews, and periodic output standard improvements so that the site's rank is maintained [6].

#### 2.2.4 Pay-Per-Click (PPC)

A form of online advertising where advertisers are paid a fee each time one of their adverts is clicked. Simply put, you only pay for advertising when a user clicks on your ad. It essentially involves buying internet traffic in addition to generating it naturally. When web indexes combine PPC with organic results, they provide a brief list of positioned results on a page that looks exactly like the organic advertising. This increases competition between websites, especially corporate websites. Not only do traditional organic ads need to contend with one another for the finest top spots, but now PPC advertising has also entered the race for a top spot [10].

#### 2.2.5 Content Marketing

The marketing and business procedure for producing and disseminating pertinent and valuable content in order to draw in, acquire, and keep the attention of a precisely defined and comprehended target audience with the aim of motivating lucrative consumer action. It is a successful synthesis of original, chosen carefully, and syndicated content [9]. Sharing of useful information with the target audience can be considered as an art and a science at the same time in content marketing. Modern consumers are elegant and intelligent, they are aware of their rights to decide what information they get, how it is showed and presented to them, and whether they will believe it or not [17].

### *2.3 Concepts and Approaches*

#### 2.3.1 Digital Retrofit

Sustainability has the ability exist as a defined through 3 main terminologies: efficiency, consistency as well as sufficiency, as well as each that belongs to these terms describe process/system function performance. Digital retrofit found playing great role inside of enhancing resources efficiency, optimize production performance, improve product quality, reduce scrap generation as well as even develop maintenance concepts which ultimately enhance the organization sustainability. Digitalization has the ability enhance the evaluation process during the same time that well; the formulized historical data has the ability help to express findings as well as results far beyond the typical evaluation technique, this is going to definitely help the organization to review their strategic goals as well as targets that is going to belong to the short-term as well as long-term future [18].

### 2.3.2 Digital Twin

Inside of manufacturing industry where the production as well as maintenance that belongs to the machinery happens to be represented by performance, sustainability happens to be an existing important aspect to exist as a maintained to guarantee the quality that belongs to the manufactured products, here where the concept that belongs to digital twin happens to be introduced. Digital twin found to have direct positive impact on top of economy, environment as well as efficiency which form the main criteria that belongs to organization sustainability. This has the ability exist as an achieved by subjecting the data to perform some conceptual operations such during the same time that simulation, verification, monitoring, optimization, diagnosis, as well as prognosis, keeping inside of consideration that it happens to be improving the decision-making [15].

### 2.3.3 Innovation Need for Sustainability

To cope with the modern market demands, companies must be very agile and adaptable for the frequent changes, in fact the leading companies trying always to initiate and create the changes to be one step ahead. Using digital transformation, organizations can change how it operates, improve its internal processes, enhance the relationship with external parties and can bring value-added to satisfy the stakeholders. With digital transformation, a company can response much faster to the market demand, no matter how complex it is, is it provide sophisticated solutions which other fail to think of, this will foster the position of the company in the market and result in more sustainable operation and economic [29].

For a long time, innovation has been essential to the success of businesses. In the past, companies that successfully innovated have often been rewarded with expansion, revenues, and access to new markets. To gain a competitive edge, constant innovation has become essential. It is commonly recognized that innovation is a crucial tool for tackling issues related to sustainable development. Numerous organizations were used to strategies and initiatives that pushed their race for competitiveness while omitting crucial environmental concerns. Therefore, there has been a significant interest in recent decades among scientists, practitioners, and even the general public regarding environmental conservation along with resource efficiency ecological concerns, raising the standard of living for people social challenges, and enhancing long-term economic competitiveness economic matters [21].

### 2.3.4 Digital Maturity

Which is the measure where the organization is making value from the digital technologies and concepts such as: Big Data, Analytics, DoubleClick Campaign Manager (DCM), and Search Engine Optimization (SEO). Digital maturity always been a good indicator of a company success along with the globalization. Throughout the transformation period of digitalization, an organization position varies as an early user of digital technologies or mature user [23]. Digital maturity refers to a company's willingness and ability to adapt and use cutting-edge technology in accordance with market trends to stay competitive. The goal of digital maturity is to enable an organization to compete successfully in a world that is becoming more digital. As a result, it is a continuing and continual process.

The five business dimensions of consumers, strategy, technology, operations, organizations, and culture are evaluated using the digital maturity model. A corporation can utilize the Digital maturity model at any stage of change to find gaps, identify important areas to concentrate on, and decide where to start [14].

## *2.4 Challenges Encountered by Organizations of Using Digital Technologies*

Although that digitalization is being useful for organizations in many aspects, it still lacks at some points, below are highlighted challenges of depending on digital technologies at the organizational operation.

### 2.4.1 Lack of Innovation

Digitalization may provide faster solution to customer; however, customers still consider the human interaction during the same time that security during the same time that well during the same time that assurance. This been driven by the innovation during the same time that well during the same time that professionalism that came that happened to be by the human only, moreover human has the ability express some emotions during the communication inside that belongs to some situation which considerably giving advantage that digital technology cannot provide [7].

### 2.4.2 Negative Impact on Mental Health

Excessive usage that belongs to digital technologies such during the same time that mobile during the same time that well during the same time that social contact found influence the mental health negatively that belongs to the user, at the same subject that belongs to that, the employees at the organization shall get controlled usage that belongs to such technologies to limit the negative impact that belongs to the mental health [2].

#### 2.4.3 Failure of Getting the Required Information

Digital technologies happen to be seen barrier to the retail sector's expansion inside of Malaysia. It has been noted that the retail sector has occasionally used digital technology secondary strategy, which has prevented them that was by gradually promoting new information. Additionally, it should exist as a emphasized that although marketers have used cutting-edge technology to promote on top of platforms such as Facebook, Instagram, YouTube, as well as others, their financial investments have not been well thought through, which has hurt the effectiveness that belongs to every single one digital marketing [16]. The following graph shows that just 47% that belongs to businesses have successfully implemented sharing knowledge that is going to belong to the deployment that belongs to the technologies, while the retail sector has invested 53% inside of digital technologies.

#### 2.4.4 Customer Orientation and Value Proposition

Understanding better what consumers value inside of products as well as sustainable businesses happens to be one that belongs to the key areas that is going to belong to the growth that belongs to sustainability. Numerous measures that have the ability exist as a used to better understand the clients have emerged during the same time that an existing result that belongs to technological advancements, notably the internet. Utilizing metrics, we may create indicators that has the ability exist as a used to gauge how customers as well as other stakeholders feel about sustainability Additionally, there happens to be an existing strong need to understand the unique characteristics, beliefs, norms, as well as behaviors that belongs to consumers who happen to be conscious that belongs to their environment, community, as well as health. Models that explain the metric associated with this type that belongs to behavior has the ability exist as a developed by researchers [13].

### 2.5 *Business Digitalization Enablers*

The most important factor needed that is going to belong to digital technologies application happens to be the data, it forms the raw material not only that is going to belong to digitalization with the exception of that is going to belong to any type that belongs to business. Putting the right data at the most useful place is going to result inside of maximizing the benefits out that belongs to any technologies, during the same time that well during the same time that overcoming the challenges as well as obstacles. Here where the enables set to put that inside of action.

### 2.5.1 Big Data

The magnitude of big data applications has grown over the past few years. In fact, various businesses of many industries are relying more on information collected from enormous amounts of data. Traditional data platforms and methodologies, however, are less effective in a big data context. They exhibit a lack of scalability, performance, and accuracy as well as a poor response time. There has been a lot of work done to address the difficult big data concerns [30]. As a result, numerous distribution models and technological advancements have been made: Apache Hadoop, MongoDB, Rainstorm, Cassandra Data Mining, Presto, RapidMiner, Elasticsearch Data Analytics and Kafka. Big Data has the capability to analyze a wide array of domains, encompassing both business and non-business areas. It helps identify patterns, trends, and insights through extensive data analysis. This empowers organizations to boost operational efficiency and make informed decisions. Additionally, it contributes to reducing uncertainty by offering evidence-based solutions. Big Data's impact spans industries and sectors, driving innovation and optimization [5].

### 2.5.2 Internet of Thing (IoT)

The phrase "Things" stresses the physical things that can be incorporated into a technological framework, whereas the term "Internet" refers to a virtual network-oriented perspective of technology. IoT should be reimagined as a global infrastructure connecting both real and virtual objects, transcending a purely technological approach. This "net" envisions a world where objects may automatically speak with one another and offer services that are advantageous to humanity [30].

Inside of the field that belongs to management IoT has advantages that is going to belong to both consumers as well as businesses; that is going to belong to consumers, it supports their consumption decisions as well as use that belongs to goods as well as services; that is going to belong to businesses, it regulates as well as monitors industrial systems inside of an existing integrated framework also known during the same time that industrial internet of things [26].

### 2.5.3 Transformation Management System

One that belongs to the most effective technologies happens to be transformation management system. It happens to be an existing logistical platform that uses an existing number that belongs to protocols to improve an existing supply chain's physical product stream. An existing controlled shipment, free that belongs to the usual transportation problems such as pollution as well as shoddy route planning, increases the effectiveness that belongs to sustainable operations.

Additionally, it increases productivity as well as warehouse efficiency because incoming as well as outgoing goods happen to be continuously tracked, even inside of industries where depositories happen to be typically stocked with goods with an existing propensity that is going to belong to rapid deterioration. Inside of this manner, assets happen to be safeguarded as well as prudent financial losses happen to be avoided [27].

### 2.5.4 Blockchain

Peer-to-peer (P2P) exchange or transactions without centralized authorities happen to be made possible by an existing decentralized operating system. With this disruptive innovation, there happens to be no longer an existing strong reliance on top of strong outside parties. Blockchain, which happens to be an existing continuously increasing list that belongs to blocks with each block containing encrypted transactional data as well as perhaps having additional decentralization potential, happens to be the technology that powers the bitcoin cryptocurrency. Ledgers are duplicate records that belongs to transactional data happen to be kept by the active nodes inside of this type that belongs to cooperative network.

Duplicate records that belong to transactional data offer the potential that is going to belong to an existing trustless operating environment devoid that belongs to conventionally trusted authority such as banks as well as clearing houses because it happens to be inherently shared by participating nodes. The shared ledger happens to be maintained by network nodes, as well as peer-to-peer trade makes it easier that is going to belong to participating nodes to develop confidence with one another [11].

## *2.6 Digitalization the Supply Chain for Sustainability*

Companies use an existing range that belongs to supply chain strategies to create sustainable results. It happens to be possible to achieve durable supply chain performance outcomes using both traditional supply chain management approaches as well as specific supply chain management techniques. During the same time that an existing result, the effects that belongs to material, information, during the same time that well during the same time that financial flows happen to exist as an evaluated inside that belongs to relation to economic, environmental, during the same time that well during the same time that social factors. Sustainable supply chain performance happens to exist as an aided by lean, resilient, during the same time that well during the same time that environmentally friendly supply chain management techniques.

The primary goals that belong to the lean methodology happen to exist as a to reduce expenses during the same time that well during the same time that waste while boosting output during the same time that well during the same time that overall quality. The just-in-time paradigm also emphasizes the importance that belongs to just-in-time considerations, which happen to be characterized by high replenishment frequencies as well as low inventory costs [20].

The resilient strategy seeks to anticipate unforeseen circumstances, react to disruptions, as well as bounce back that was by them. Inside of the context that belongs to digital as well as sustainable supply chains, lean, robust, as well as green techniques happen to be crucial, as well as an existing integrated viewpoint happens to be required to balance the goals that belongs to profitability, resilience, as well as sustainability. However, these supply chain management techniques do not naturally incorporate sustainability considerations; instead, an existing profit-maximizing mindset predominates [28].

The practices that belong to the supply chain management as well as the sustainable supply chain management through digital technologies happen to be showing direct impact on top of the organization sustainability.

### 2.6.1 Pillars of Supply Chain Management Sustainability

Building supply chain sustainability as well as resilience requires the development that belongs to four pillars: agility and collaboration, localization, diversification, visibility and transparency. Because agility happens to be seen during the same time that an existing element that includes velocity to respond to new environmental conditions inside of the context that belongs to disruptions, agility as well as collaboration happen to be seen during the same time that one pillar inside of this framework [9]. To reduce risks, information must exist as a shared as well as teams must work together quickly. The connection between digital technologies practices provides an existing solid foundation that is going to belong to the four pillars.

Digital technologies have the potential to support circular economy principles as well as, have an existing significant impact on top of the transformation that belongs to supply networks to create resilient as well as sustainable supply chains. Additionally, clearly defined governance as well as control mechanisms has the ability generate precise recommendations, put into practice rules, as well as carry out procedures that help network resilience as well as sustainability by preparing supply chain partners that is going to belong to future shocks [12].

### 2.6.2 Digitalizing for Supply Chain Management (SCM)

The elimination of waste is the supply chain management technique that is referenced most frequently overall and within the lean category. Rich, real-time data facilitate better decision-making and optimize the efficiency and expense of associated supply chain activities. There are fewer human errors, greater monitoring, and early defect identification, which all help to prevent defects, spoilage, and losses [2]. Automation of the purchasing process along with accurate forecasts help to achieve closed-loop goals by reducing overproduction and inventory levels. Automated guided vehicles can reduce waste associated with product movement and transportation. Accordingly, improved qualitative concerns can be realized, though they were less frequently considered in the studies we reviewed. Intelligent cyber-physical systems allow for accurate optimization implementation and the identification of quality improvements.

The use of supply chain risk management techniques is important to lessen the possibility of disruptions and get ready for any necessary actions. The focus of the present scholarly discussion is on using big data analytics and artificial intelligence technologies to proactively foresee disruptions and monitor potential supply chain hazards [12]. Collaboration portals built on digital technologies help supply chain partners share knowledge and manage risks. As their transparency encourages sustainable supply chain behaviors, another body of literature focuses on the risk mitigation capabilities of blockchains and smart contracts [28].

### 2.6.3 Digitalizing for Sustainable Supply Chain Management (SSCM)

The focus of interest is on supply chain collaborations and proactive management rather than top management's orientation, supply chain continuity, and sustainable risk management, according to a frequency analysis of sustainable supply chain management methods. The orientation category contains very low frequencies. More research is needed on this topic because top management support is required to manage the high costs and complexity associated with the adoption of digital technologies for sustainable supply chain management. They specifically contend that managers may better prioritize activities and make better use of resources by using big data analytics to support sustainable supply chains. This aids in coordinating organizational and supply chain-wide sustainability goals.

Most of the examined studies concentrate on improving partner selection in order to promote the continuity of supply chain. The performance of suppliers can be (re)evaluated, and the most sustainable suppliers can be found using big data analytics and artificial intelligence. Thus, environmental sustainability is the main consideration in sustainable procurement decisions. In the examined literature, the influence of digital technologies on partner development and long-term relationships is only briefly but contentiously discussed. One the one hand, digital technologies aid in the knowledge and skill transfer and give suppliers the tools they need to handle sustainability concerns. However, digital technologies implementation can exacerbate power imbalances and dependencies in supply chain, endangering the social sustainability of those communities.

Blockchain technology can encourage openness and the development of lasting relationships. Relationships lose their importance in unreliable digital systems, according to those who are working to improve understanding of social supply chain concerns rooted in various cultural contexts. The studies primarily highlight the intersection between digital transformation and collaboration. Only eight articles fail to mention any methods of collaboration. Blockchains and cloud services built on the internet of things make it easier to gather and share data to address sustainability challenges and hold businesses accountable [30].

As a result, digital technologies increase the quantity of information exchange by incorporating supply chain and lifecycle-wide information, as well as speed in the sense of real-time data, quality in terms of validity, and depth. In the articles on logistical and technical integration, which make up half of the collection, it is claimed that digital technologies make it possible to coordinate and automate logistical activities while also enabling the development of cross-organizational platforms. This is understandable considering that the development of interconnected technical systems for the collaborative control of material flows is based on efficiency and transaction cost considerations, which are extensively discussed in the supply chain literature. The digital supply chain literature, however, generally ignores cooperative development. digital technologies, on the other hand, can assist in "developing actionable green partnership objectives and collaboratively developing and commercially using cleaner processes, technologies, and materials" [28].

### 2.6.4 Industry 4.0 Role in Supply Chain Sustainability

Lean and green supply chain are linked to Industry 4.0 and the virtual flow of information and data, and businesses can gain a competitive edge if they are aware of these features. To determine which traits, have the most impact on achieving sustainability in the supply chain, more research can be conducted. the removal of problems connected with purchasing and the supply chain management function while deploying internet of thing systems utilizing a methodical inductive research technique. They came to the conclusion that implementing internet of thing facilities within the business is the more effective course of action [30]. Based on the literatures that is now available, the concept of the smart factory can be viewed from eight different angles. There are 13 major Industry 4.0 enablers that can be employed to improve supply chain sustainability.

Based on a literature review, the understanding of Industry 4.0 and big data in the creation of a sustainable supply chain was enhanced [11]. A thorough analysis of automatic guided vehicles and the integration of audio-visual related challenges into one framework. In order to achieve sustainability in the presence of Industry 4.0 technologies, the uses and applications of these technologies as well as the function of logistics are addressed [22]. a comparison between modernized Industry 4.0 logistics facilities and traditional logistics centers. They also outlined the requirements for the implementation of smart logistics facilities. Some of the Industry 4.0 technologies can match industry expectations, while others are still in the adoption stage. Not all of these technologies have shown to improve industrial performance [26].

## 2.7 *Digital Technologies Role in Green Supply Chain Management*

Sustainable network is another term of green supply chain management, which is the optimization approach for environmental impact reduction for green product life cycle, it can be achieve through configuration approach of digital transformation to create smart green supply chain management. One approach is to support the supply chain by the base digital transformation strategies and frond-end base technologies: big data, internet of things, artificial intelligence, blockchain and cloud computing. Moreover, it is shown that digital technologies can enhance customer and supplier green relationship management and the internal dimensions of supply chain management in an organization, by re-evaluating the operation mechanisms for better reliable and sustainable dimensions [19].

### 2.8 Digitalized Supply Chain in Comparison with Traditional Supply Chain

Because traditional and electronic processes are used in the two supply chains, traditional supply chain and digitalized supply chain, respectively, have lower infrastructure costs but higher operational costs than digitalized supply chain. Our presumptions include supply chain services, market sensitivity to price, as well as delivery time. Investigate every situation that pertains to rivalry between these two supply chain systems, as well as each of its individual components, as part of an ongoing evolutionary challenge.

The fundamental skills that belong to each supply chain happen to be discussed based on top of their market policies as well as coordination strategies after every single one scenario happen to be simulated that is going to belong to an existing instructive numerical example. We found that preferences for supply chain in external policy as well as internal strategy are influenced by centralization costs as well as rivals' strategies and policies as either market leaders or followers. it can be seen that, generally speaking, the digitalized supply chain has greater chances of success than the conventional supply chain. Future research could include considering two different customer types and allocating dedicated or shared capacity to them. Future research will address the centralization cost assumption and the scenario in which a chain's internal components coordinate with one another [25].

### 2.9 Future of Digitalized Supply Chain

The academic as well as professional world that belongs to logistics as well as supply chain management see the digitalization that belongs to the supply chain strategic issue. To ensure the digitalization that belongs to the supply chain implementation's success, an existing number that belongs to technological, organizational, as well as strategic obstacles still need to exist as a solved. The emerging research conducted on top of the adoption that belongs to technologies as well as their effects on top of the supply chain must exist as a developed.

Inside of addition, it happens to be crucial to investigate new digitalization that belongs to the supply chain configurations, digitalization that belongs to the supply chain orientations, the integration that belongs to suppliers, customers, as well as consumers, during the same time that well during the same time that the contribution that belongs to new project management techniques as well as project management methodologies.

Future research should mobilize various theoretical frameworks: technology acceptance models as well as theory, innovation adoption theory, organizational project management theory, dynamic capability theory, stakeholders' theory, etc., as well as combine different method: qualitative as well as quantitative as well as epistemological lenses: positivism, constructivism, as well as pragmatism to provide an existing comprehensive understanding that belongs to the issues related to digitalization that belongs to the supply chain. The understudied problems with public supply chains, such during the same time that the supply chains that belongs to public hospitals, as well as SMEs supply chains should exist as the focus that belongs to greater research [1].

## 3 Conclusions

Sustainability found to be a major factor for business lasting, which helps any company to withstand market frequent changes, companies consider supply chain to manage the needs and requirements of their business, where it been linked to the digital technologies to clarify the impact of it on the supply chain management [29]. This paper discussed the recent categories of digital technologies to identify the relation with supply chain management sustainability, and how it could be a helpful tool for improvement [7]. It found that most of the major categories can be linked to the supply chain management, as supply chain must be in continuous seeking for alternatives at the market, social media marketing, influencer marketing, search engine optimization, content marketing and at lower rate pay-per-click can be useful concept for marketing to support supply chain management [31].

Digital technologies concepts and approached explain under the umbrella of sustainability to evaluate the how beneficial it may for an organization sustainability [13]. It is found that each concept can help in different field of operation, for example digital twin will be have positive impact on the manufacturing and maintenance fields [15], where the digital maturity can influence businesses which got affected by marker fluctuation and trends more than other businesses [14]. However, that doesn't eliminate the fact that innovation of digital technology is a need for sustainability of any modern business [21].

On the other hand, any digital technology may face barrier for implementation, as mentioned earlier that some digital technologies can't be fit with the business nature or organization direction. Some technologies are straight forward in a way that will work in a robotic routine, and supply chain management may require to get innovation to withstand the market challenges [14], keeping in consideration that some technologies may impact the mental health on the employee, as this is been considered by the organizations which offer extra care of their employees mental and physical health [2]. The drive of the human input is always a must, as many of the technologies cannot be operated autonomously [7]. Furthermore, the paper discussed business digitalization enablers and tried to link these enables with the supply chain management.

Big data found to be a big help for any company sustainability of business, especially in the field of supply chain management, however it requires big effort for application and maintaining, keeping the challenge window widely open for the middle and small businesses [30]. The second enabler discussed is the internet of things, and it clarified that in can support the monitoring process of industries and give competitive advantage for the automation, it can support the management systems in many ways as well, this is presented as well as good help for the supply chain management systems [30]. On the other hand, the third enabler presented is the blockchain, this enabler can support the transactional data processing, as considered to be main contribution for the financial section, relation this to supply chain management found to be very limited as banks and financial firm found independent from relying on supply chain management system [11].

Building on the previous results, the research took further step to consolidate the impact of digitalization of supply chain sustainability, discussing the sustainability pillar. These pillars are: diversification, localization, agility and collaboration, visibility and transparency, these pillars must be considered during the introduction on a digital technology for organization process improvement to reduce failure of risk and negative impact [9].

Digital technologies of supply chain implemented at the industry 4.0 in various manufacturing field, these field found reaching sustainable heights throughout the processes [11], the automation of vehicles expressed as function of logistic as a successful example of industrial 4.0 field [22]. Additionally, green supply chain management showed to be achievable through the configuration approach of digital transformation, especially at customer supplier relationship dimension.

## 3.1 Theoretical Implications

This study confirms that the goal of sustainability of supply chain management got direct relation to many digital technologies. Although the differences in details, internet of things, blockchain, big data, and other technologies found to be very useful to achieve sustainable system, the impact of each technology can differ from an aspect to another of supply chain, and it is proven that the more understandable the digital technologies principles the greater the benefits will be [28]. Furthermore, the digitalized version of supply chain is way above the traditional version in terms of reliability, durability and sustainability [25]. Finally, the ultimate goal of green supply chain management can be achieved through combined assault between both partiers: supplier and customer, and similar to the previous point, the better understanding of the technology, the efficient and reliable the system will be [19].

### 3.2 Practical Implications

After covering the main aspects of supply chain, sustainability and digitalization, the paper approached the comparison between the traditional supply chain management and the digitalized supply chain management versions, which showed how dynamic, faster, and integrated the digitalized versions, which provide big pump to the firm at the market in terms of coping with the trends, ability to satisfy new customer requirements, as well as new market needs [25]. It is expecting to allow organizations to reach greater height of sustainability using the presented concepts, the move from traditional supply chain management to the digitalized proved to be very beneficial to organizations considering the availability of essential fundamentals of digital technologies within the organization, in such case where the fundamentals of digitalization is not available within the organization, this subject can be proposed as development project to be studied, and in case of suitability, it can be considered as strategic approach for business.

### 3.3 Study Limitations and Future Research

This research paper prepared from various primary and secondary researches covering multiple subjects in different field, the core concept can be utilized to help the top to middle managers in an organization, nonetheless, many fields are not included which need to be considered, researches of demand planning can add strong value to the results and enhance the findings, similarly transportation field where the process of moving the products from a place to another can be very interesting to add, and finally the customer service field, as many people do not know that it could be within supply chain management command in some organizations, these fields are proposed for future studies and implementation.

Although these approaches and concepts build rationally, however this paper limited to cover the conceptional aspects lacking the numerical and scientific approaches, adding this as support to the research will strengthen the topic and can be beneficial for future expansion of the subject. Furthermore, this paper can support real ad specific case studies for the impact of such implementation, this will help to understand the actual effects of the digital technologies in the longer run, and how supply chain employees are influenced by each technology separately.

## References

1. Ageron, B., Bentahar, O., Gunasekaran, A.: Digital supply chain: challenges and future directions. Supply Chain Forum Int. J. **21**(3), 133–138 (2020)
2. Aguilera, A.: Digital technology and mental health interventions: opportunities and challenges. Arbor **191**(771), 210–219 (2015)

3. Ahmad, A., Khan, M.N.: Factors influencing consumers' attitudes toward social media marketing. MIS Rev. **22**(1/2), 21–40 (2017)
4. Alawi, B., Al Mubarak, M. M. S., Hamdan, A.: Blockchain evaluation framework for supply chain management: a decision-making approach. Supply Chain Forum: Int J. **23**(3), 212–226 (2022). https://doi.org/10.1080/16258312.2021.1996862
5. Alhaddad, H., Al Mubarak, M.: Big data in occupational health and safety. In: El Khoury, R., Nasrallah, N. (eds.) Emerging Trends and Innovation in Business and Finance. Contributions to Management Science, Springer, Singapore (2023). https://doi.org/10.1007/978-981-99-6101-6_39
6. Almukhtar, F., Mahmoodd, N., Kareem, S.: Search engine optimization: a review. Appl. Comput. Sci. **17**(1), 70–80 (2021)
7. Arief, A.S., Putri, S.E., Suroso, A., Syakhrani, A.W., Rahmini, N.: Digital technology management challenges in marketing local farm products in developing countries: analysis of international publication findings. J. Iqra' Kajian Ilmu Pendidikan **6**(2), 96–107 (2021)
8. Bakker, D.: Conceptualising influencer marketing. J. Emerg. Trends Mark. Manag. **1**(1), 79–87 (2018)
9. Baltes, L.P.: Content marketing—the fundamental tool of digital marketing. Bull. Transilvania Univ. Brasov Econ. Sci. Ser. V **8**(2), 111–118 (2015)
10. Bhandari, R.S.: Pay per click marketing strategies: a review of empirical evidence. J. Ind. Distrib. Bus. **8**(6), 7–16 (2017)
11. Chang, S.E., Chen, Y.: When blockchain meets supply chain: a systematic literature review on current development and potential applications. IEEE Access **8**, 62478–62494 (2020)
12. Cherrafi, A., Chiarini, A., Belhadi, A., El Baz, J., Benabdellah, A.C.: Digital technologies and circular economy practices: vital enablers to support sustainable and resilient supply chain management in the post-COVID-19 era. TQM J. **34**(7), 179–202 (2022)
13. Diez-Martin, F., Blanco-Gonzalez, A., Prado-Roman, C.: Research challenges in digital marketing: sustainability. Sustainability **11**(10), 28–39 (2019)
14. Eremina, Y., Lace, N., Bistrova, J.: Digital maturity and corporate performance: the case of the Baltic States. J. Open Innov. Technol. Mark. Complex. **5**(3), 54–66 (2019)
15. Franciosi, C., Miranda, S., Veneroso, C.R., Riemma, S.: Improving industrial sustainability by the use of digital twin models in maintenance and production activities. IFAC-PapersOnLine **55**(19), 37–42 (2022)
16. Hasanat, M.W., Hoqueb, A., Hassanc, M.M., Hamide, A.B.A.: The lack of digital marketing skills: developing a digital marketer model for the retail industries. J. Xi'an Univ. Archit. Technol. **12**(3), 2673–2680 (2020)
17. Kee, A.W.A., Yazdanifard, R.: The review of content marketing as a new trend in marketing practices. Int. J. Manag. Account. Econ. **2**(9), 1055–1064 (2015)
18. Kröll, M., Cseh, C.: Implementation model for digital retrofit for sustainable production. Procedia Comput. Sci. **217**, 486–494 (2023)
19. Lerman, L.V., Benitez, G.B., Müller, J.M., de Sousa, P.R., Frank, A.G.: Smart green supply chain management: a configurational approach to enhance green performance through digital transformation. Supply Chain Manag. Int. J. **27**(7), 147–176 (2022)
20. Liu, Y., Zhu, Q., Seuring, S.: New technologies in operations and supply chains: implications for sustainability. Int. J. Prod. Econ. **229**, 107889 (2020)
21. Maier, D., Maier, A., Aşchilean, I., Anastasiu, L., Gavriş, O.: The relationship between innovation and sustainability: a bibliometric review of the literature. Sustainability **12**(10), 4083–5002 (2020)
22. Mastos, T.D., Nizamis, A., Terzi, S., Gkortzis, D., Papadopoulos, A., Tsagkalidis, N., et al.: Introducing an application of an industry 4.0 solution for circular supply chain management. J. Clean. Prod. **300**, 126886 (2021)
23. Muhammedrisaevna, T.M., Bakhriddinovna, A.N., Rasulovna, K.N.: Use of digital technologies in marketing. In: E-Conference Globe, Mar 2021, pp. 281–284 (2021)
24. Nadaraja, R., Yazdanifard, R.: Social Media Marketing: Advantages and Disadvantages, pp. 1–10. Center of Southern New Hampshire University (2013)

25. Narenji, M., Fathian, M., Teimoury, E., Naini, S.G.J.: Price and delivery time analyzing in competition between an electronic and a traditional supply chain. Math. Probl. Eng. **2013** (2013)
26. Naseem, M.H., Yang, J.: Role of industry 4.0 in supply chains sustainability: a systematic literature review. Sustainability **13**(17), 9544 (2021)
27. Romagnoli, S., Tarabu, C., Maleki Vishkaei, B., De Giovanni, P.: The impact of digital technologies and sustainable practices on circular supply chain management. Logistics **7**(1), 1–17 (2023)
28. Schilling, L., Seuring, S.: Linking the digital and sustainable transformation with supply chain practices. Int. J. Prod. Res. 1–25 (2023)
29. Schöggl, J.P., Rusch, M., Stumpf, L., Baumgartner, R.J.: Implementation of digital technologies for a circular economy and sustainability management in the manufacturing sector. Sustain. Prod. Consump. **35**, 401–420 (2023)
30. Sestino, A., Prete, M.I., Piper, L., Guido, G.: Internet of things and big data as enablers for business digitalization strategies. Technovation **98**, 102173 (2020)
31. Spais, G.S.: Search engine optimization (SEO) as a dynamic online promotion technique: the implications of activity theory for promotion managers. Innov. Mark. **6**(1), 7–24 (2010)

# Sustainable Waste Management in Business-to-Business

**Mohamad Osama Nesrallah and Muneer Al Mubarak**

**Abstract** This paper explores sustainable waste management practices for business-to-business (B2B) companies, which generate a significant portion of commercial and industrial waste. The aim is to analyze the drivers, challenges, and effective strategies to promote environmentally responsible waste management in the B2B sector. An extensive literature review was conducted to synthesize insights from academic journals, industry reports, and case studies on sustainable waste management in B2B companies across various industries. The key drivers for B2B companies to adopt sustainable waste management include regulatory compliance, cost savings, corporate social responsibility, and stakeholder pressure. However, challenges such as lack of infrastructure, limited knowledge, and financial constraints hinder effective implementation. This study is limited to the context of B2B companies and does not cover the waste management dynamics of business-to-consumer organizations or municipal solid waste. Future research could explore sector-specific case studies and quantitative analysis of the economic and environmental impacts. The findings offer practical guidance for B2B companies to develop and implement sustainable waste management programs, fostering resource efficiency, reducing environmental impacts, and promoting a circular economy. Effective waste management in the B2B sector can contribute to environmental conservation, public health, and sustainable development goals, benefiting both businesses and society. This paper provides a comprehensive overview of sustainable waste management practices tailored to the unique challenges and opportunities faced by B2B companies, contributing to the body of knowledge in this field.

**Keywords** B2B waste management · Sustainable practices · Circular economy · Industrial symbiosis · Resource efficiency · Stakeholder pressure

---

M. O. Nesrallah
The George Washington University, Washington, DC, USA

M. Al Mubarak (✉)
Ahlia University, Manama, Bahrain
e-mail: malmubarak@ahlia.edu.bh



M. Al Mubarak (ed.), *Sustainable Digital Technology and Ethics in an Ever-Changing Environment*, Studies in Systems, Decision and Control 237,
https://doi.org/10.1007/978-3-031-86708-8_26

# 1 Introduction

Waste generation is an inevitable consequence of economic activities, posing significant environmental and social challenges [1, 7]. The business-to-business (B2B) sector, which encompasses companies that provide goods and services to other businesses, is a major contributor to the global waste stream. B2B companies operate across diverse industries such as manufacturing, construction, logistics, and professional services, generating substantial amounts of commercial and industrial waste [2].

Sustainable waste management has emerged as a critical concern for B2B companies due to growing environmental regulations, stakeholder pressure, and the need for resource efficiency [10]. Improper waste disposal can lead to environmental degradation, health hazards, and long-term ecological consequences. Furthermore, the depletion of natural resources and the escalating costs associated with waste management have prompted companies to rethink their waste management strategies [4].

## *1.1 Research Purpose*

This research paper aims to provide a comprehensive analysis of sustainable waste management practices in the B2B sector. By exploring the drivers, challenges, and effective strategies, the study seeks to offer practical guidance for B2B companies to develop and implement environmentally responsible waste management programs.

## *1.2 Research Problem*

The lack of sustainable waste management practices in the B2B sector has contributed to significant environmental and economic challenges. B2B companies often face limited knowledge, inadequate infrastructure, and financial constraints in implementing effective waste management strategies [1, 7]. Addressing these issues is crucial for promoting resource efficiency, reducing environmental impacts, and fostering a circular economy [9].

## *1.3 Significance of Research (Contributions)*

This study contributes to the body of knowledge on sustainable waste management by providing a comprehensive overview tailored to the unique challenges and opportunities faced by B2B companies. It synthesizes insights from various theoretical frameworks, such as the waste hierarchy, circular economy principles, and industrial ecology concepts [9], while incorporating empirical evidence and case studies from diverse industries. The findings of this research offer practical guidance for B2B companies seeking to develop and implement sustainable waste management programs, fostering resource efficiency, reducing environmental impacts, and promoting a circular economy [7]. Additionally, the study can inform policymakers and regulatory bodies in developing supportive frameworks and incentives to promote sustainable waste management practices in the B2B sector [10]. Sustainable waste management is an approach that integrates waste minimization, resource recovery, and responsible disposal practices to reduce the environmental impact of waste generation. It encompasses various strategies such as waste prevention, reuse, recycling, composting, and energy recovery. The principles of sustainable waste management align with the concept of a circular economy, which emphasizes the elimination of waste through closed-loop material flows and the maximization of resource efficiency [7, 9].

The scope of this study encompasses B2B companies across various industries, such as manufacturing, construction, logistics, and professional services. It examines the unique challenges and opportunities faced by these companies in managing their waste streams while balancing economic, environmental, and social considerations. The need for sustainable waste management in the B2B sector is driven by several factors. Firstly, regulatory frameworks and environmental legislation have become increasingly stringent, imposing stringent requirements on waste management practices. B2B companies must comply with these regulations to avoid penalties and maintain their operational licenses. Secondly, stakeholder pressure, including demands from customers, investors, and the broader public, has influenced companies to adopt environmentally responsible practices. Corporate social responsibility (CSR) initiatives and sustainability reporting have become essential for maintaining a positive corporate image and reputation. Thirdly, the economic benefits associated with sustainable waste management, such as cost savings through resource efficiency, material recovery, and reduced disposal costs, have incentivized B2B companies to explore alternative waste management strategies. Furthermore, the transition towards a circular economy, where waste is viewed as a valuable resource, has gained momentum. B2B companies can contribute to this paradigm shift by implementing closed-loop systems and collaborating with other organizations to create industrial symbiosis networks.

To provide a comprehensive understanding of sustainable waste management in the B2B sector, this paper defines key terms and concepts. Waste management refers to the collection, transportation, processing, recycling, and disposal of waste materials. Sustainable waste management extends this concept by incorporating principles of resource efficiency, environmental protection, and social responsibility.

## 1.4 Global Waste Generation and Environmental Impacts

The need for sustainable waste management has become increasingly pressing due to the staggering amount of waste generated globally. According to the World Bank, the total municipal solid waste generated worldwide is expected to rise from 2.01 billion tonnes in 2016 to 3.88 billion tonnes by 2050. This estimate does not include the substantial volumes of commercial and industrial waste generated by B2B companies across various sectors. The environmental impacts of improper waste management are severe and far-reaching. Landfills and open dumps can contaminate soil and groundwater, release methane (a potent greenhouse gas), and contribute to air pollution. Incineration of waste, if not properly controlled, can release toxic pollutants into the atmosphere, contributing to respiratory diseases and other health issues.

Furthermore, the linear "take-make-waste" model of production and consumption depletes finite natural resources, leading to resource scarcity and environmental degradation. Sustainable waste management practices, grounded in circular economy principles, offer a pathway towards decoupling economic growth from resource consumption and minimizing the environmental footprint of B2B companies. The literature review section will delve into the theoretical foundations underpinning sustainable waste management, including the waste hierarchy, circular economy principles, and industrial ecology concepts. It will also explore the drivers motivating B2B companies to adopt sustainable waste management practices, such as regulatory compliance, cost savings, corporate social responsibility, and stakeholder pressure. Additionally, the challenges faced by B2B companies in implementing sustainable waste management strategies will be analyzed, including infrastructure gaps, knowledge limitations, and financial constraints.

Furthermore, the literature review will discuss effective strategies for sustainable waste management in the B2B sector. These strategies encompass waste minimization through process optimization, material recovery and recycling, byproduct synergy networks, and collaboration with waste management service providers. Case studies and best practices from various industries will be presented to illustrate successful implementation of these strategies. The implications section will explore the theoretical and practical contributions of this research. It will highlight the study's significance in advancing the understanding of sustainable waste management in the B2B context and informing policymakers and regulatory bodies in developing supportive frameworks. Furthermore, the limitations of the study and avenues for future research will be discussed, addressing quantitative analyses, sector-specific investigations, and the potential integration of emerging technologies in waste management. Ultimately, this research paper aims to provide a comprehensive and practical guide for B2B companies to navigate the complexities of sustainable waste management. By adopting the recommended strategies and leveraging the insights presented in this study, B2B companies can contribute to environmental conservation, public health, and sustainable development goals, while also realizing economic benefits and enhancing their corporate social responsibility.

# 2 Literature Review

## 2.1 Theoretical Foundations

The concept of sustainable waste management is grounded in several theoretical frameworks and principles. The most prominent among these is the waste hierarchy, which establishes a prioritized approach to waste management. The waste hierarchy, as outlined by the United States Environmental Protection Agency (EPA), ranks waste management strategies in the following order: source reduction and reuse, recycling and composting, energy recovery, and lastly, treatment and disposal [12]. Source reduction and reuse are considered the most preferred options, as they prevent the generation of waste and extend the lifespan of materials, respectively. Recycling and composting divert materials from landfills and incineration, conserving natural resources and reducing environmental impacts. Energy recovery through processes like waste-to-energy incineration or anaerobic digestion captures the remaining value from waste. Finally, responsible treatment and disposal are recommended for residual waste that cannot be managed through the other strategies.

Another influential concept is the circular economy, which aims to decouple economic growth from resource consumption by keeping materials and products in use for as long as possible [9]. It promotes the principles of reducing, reuse, recycle, and recover, emphasizing closed-loop material flows and the elimination of waste. The circular economy model challenges the traditional linear "take-make-waste" approach and provides a framework for sustainable waste management. The principles of industrial ecology and industrial symbiosis offer valuable insights for sustainable waste management in the B2B sector. Industrial ecology views industrial systems as interconnected networks, where the waste or byproducts of one process can become inputs for another, mimicking the closed-loop cycles of natural ecosystems [6, 13]. Industrial symbiosis networks facilitate the exchange of materials, energy, water, and byproducts among multiple organizations, creating a collaborative approach to resource efficiency and waste minimization [5, 11].

### 2.1.1 Life Cycle Assessment (LCA)

In addition to the theoretical frameworks mentioned, Life Cycle Assessment (LCA) is a valuable tool for evaluating the environmental impacts of products and processes throughout their entire life cycle. LCA considers the extraction of raw materials, manufacturing, transportation, use, and end-of-life disposal or recycling. By applying LCA principles, B2B companies can identify hotspots in their operations and supply chains where significant environmental impacts occur. This information can guide targeted efforts to reduce waste generation, optimize resource utilization, and implement sustainable waste management practices.

LCA involves four main phases: goal and scope definition, inventory analysis, impact assessment, and interpretation. In the goal and scope phase, the purpose of the

LCA study, the system boundaries, and the functional unit are defined. The inventory analysis quantifies the inputs and outputs of the system, such as raw materials, energy, emissions, and waste streams. The impact assessment phase evaluates the potential environmental impacts associated with these inputs and outputs, considering categories like climate change, resource depletion, and ecosystem toxicity. Finally, the interpretation phase analyzes the results, identifies improvement opportunities, and informs decision-making.

By conducting LCAs on their products or processes, B2B companies can gain valuable insights into the environmental consequences of their activities and make informed choices about waste minimization, material selection, and end-of-life management strategies. LCA can also facilitate the comparison of alternative scenarios or technologies, enabling companies to prioritize investments and initiatives that offer the greatest environmental benefits. Furthermore, LCA can support the implementation of the circular economy principles by evaluating the potential benefits of closed-loop material flows, product reuse, and recycling strategies. It can quantify the environmental impacts avoided by diverting waste from landfills or incineration and identifying opportunities for material recovery and valorization. While LCA is a powerful tool, its application in the B2B sector may face challenges related to data availability, system complexity, and methodological limitations. Overcoming these challenges may require collaboration among companies, industry associations, and research institutions to develop standardized data sets, streamline LCA processes, and enhance the robustness and comparability of LCA studies.

## 2.2 *Drivers for Sustainable Waste Management in B2B Companies*

Regulatory Compliance: B2B companies operate within a complex web of environmental regulations and legislation that govern waste management practices. Failure to comply with these regulations can result in fines, penalties, and legal liabilities. As a result, regulatory compliance has emerged as a significant driver for adopting sustainable waste management strategies [2]. Cost Savings: Sustainable waste management practices can lead to substantial cost savings for B2B companies. By reducing waste generation, maximizing resource efficiency, and implementing recycling and material recovery programs, companies can minimize disposal costs and generate revenue from the sale of recovered materials [4].

Corporate Social Responsibility (CSR): Increasingly, stakeholders, including customers, investors, and the public, expect companies to demonstrate environmental responsibility and contribute to sustainable development goals. B2B companies have recognized the importance of CSR initiatives and sustainability reporting, which often include waste management as a key performance indicator [8]. Stakeholder Pressure: Beyond regulatory requirements and internal motivations, B2B companies face pressure from various stakeholders to adopt sustainable waste management practices. Customers may demand environmentally responsible products and services, while investors and lenders may prioritize companies with strong sustainability performance [10].

Resource Scarcity and Circular Economy: As natural resources become increasingly scarce and the global population continues to grow, the need for resource efficiency and waste minimization has gained prominence. B2B companies recognize the importance of transitioning towards a circular economy, where waste is viewed as a valuable resource and material loops are closed [7].

### 2.2.1 Competitive Advantage

Adopting sustainable waste management practices can also provide B2B companies with a competitive advantage in their respective markets. As consumers and businesses increasingly demand environmentally responsible products and services, companies that prioritize sustainability can differentiate themselves and gain a reputational edge. This can lead to increased customer loyalty, access to new markets, and potentially higher pricing premiums for their offerings. Moreover, sustainable waste management practices can contribute to cost savings and operational efficiencies, further enhancing the competitiveness of B2B companies. By reducing waste generation, maximizing resource efficiency, and implementing circular economy principles, companies can minimize expenses associated with raw material acquisition, energy consumption, and waste disposal. These cost savings can be reinvested into research and development, product innovation, or marketing efforts, providing a competitive advantage over rivals with less sustainable practices. Additionally, as investors increasingly prioritize environmental, social, and governance (ESG) criteria in their investment decisions, B2B companies with robust sustainable waste management strategies may become more attractive to potential investors and access a broader pool of capital. This enhanced access to financing can support growth, expansion, and the development of innovative sustainable solutions, further solidifying the company's competitive position.

## *2.3 Challenges in Implementing Sustainable Waste Management*

Lack of Infrastructure and Facilities: In many regions, the infrastructure and facilities for sustainable waste management are inadequate or lacking. This includes recycling facilities, material recovery facilities, and waste-to-energy plants. B2B companies may face challenges in finding appropriate waste management service providers or facilities to handle their waste streams [1]. Limited Knowledge and Awareness: Despite the growing importance of sustainable waste management, some B2B companies may lack the knowledge and awareness of best practices, technologies, and strategies. This can hinder the effective implementation of waste management programs and lead to suboptimal solutions [7].

### 2.3.1 Organizational Culture and Resistance to Change

Implementing sustainable waste management practices often requires a shift in organizational culture and mindset. Resistance to change from employees, managers, or stakeholders can be a significant barrier to successful implementation. Overcoming this challenge may involve effective communication, training, and incentives to encourage employee buy-in and participation in waste reduction and recycling initiatives. Organizational culture encompasses the shared values, beliefs, and norms that shape the behavior and decision-making processes within a company. In many cases, traditional organizational cultures may prioritize short-term profitability and operational efficiency over long-term sustainability considerations. Transitioning to sustainable waste management practices may require a fundamental shift in these deeply ingrained cultural elements, which can be a complex and time-consuming process.

To address this challenge, B2B companies can adopt a holistic approach to cultural transformation. This may involve:

1. Leadership Commitment: Top management must visibly endorse and champion sustainable waste management initiatives, setting clear goals and expectations for the entire organization.
2. Employee Engagement and Training: Providing comprehensive training and awareness programs can help employees understand the benefits and rationale behind sustainable waste management practices, fostering a sense of ownership and commitment.
3. Incentives and Recognition: Implementing incentive systems that reward sustainable behaviors, such as waste reduction targets or recycling participation, can motivate employees to embrace the cultural shift.
4. Cross-functional Collaboration: Encouraging collaboration and knowledge-sharing across different departments and functional areas can facilitate the integration of sustainable waste management practices into various organizational processes.
5. Continuous Improvement: Establishing mechanisms for feedback, progress monitoring, and continuous improvement can help sustain the cultural transformation and ensure that sustainable waste management practices become embedded in the organization's daily operations.

By proactively addressing organizational culture and resistance to change, B2B companies can increase the likelihood of successful implementation and long-term sustainability of their waste management strategies.

## 2.4 Strategies for Sustainable Waste Management in B2B Companies

Waste Minimization Through Process Optimization: A fundamental strategy for sustainable waste management is waste minimization, which involves reducing the amount of waste generated at the source. B2B companies can achieve waste minimization through process optimization, including improving production efficiency, implementing lean manufacturing principles, and adopting cleaner technologies. By streamlining processes and eliminating unnecessary steps or material inputs, companies can significantly reduce their waste generation while also realizing cost savings and enhanced productivity. Material Recovery and Recycling: Material recovery and recycling play a crucial role in diverting valuable resources from the waste stream. B2B companies can implement comprehensive recycling programs tailored to their specific waste streams, such as paper, cardboard, plastics, metals, and electronic waste. Effective material recovery requires careful segregation, collection, and transportation systems, as well as collaboration with recycling facilities and end-markets for recovered materials.

Byproduct Synergy Networks: Building upon the principles of industrial ecology, byproduct synergy networks facilitate the exchange of waste or byproducts among multiple organizations. B2B companies can collaborate with other businesses, forming industrial symbiosis networks where one company's waste becomes another's raw material [3]. This approach not only minimizes waste disposal but also promotes resource efficiency and economic benefits through the shared use of materials, energy, and water. Collaboration with Waste Management Service Providers: Many B2B companies may lack the expertise, infrastructure, or resources to effectively manage certain waste streams. In such cases, collaborating with specialized waste management service providers can be a strategic solution (Liao 2018). These service providers offer comprehensive waste management services, including collection, transportation, treatment, recycling, and responsible disposal. By leveraging their expertise and facilities, B2B companies can ensure compliance with regulations and adopt best practices in sustainable waste management.

### 2.4.1 Eco-design and Sustainable Product Development

Integrating sustainability considerations into product design can significantly reduce waste generation throughout the product life cycle. Eco-design principles such as minimizing material inputs, using recyclable or biodegradable materials, and designing for disassembly and reuse can contribute to waste minimization and resource efficiency [7]. B2B companies can collaborate with designers, engineers, and suppliers to incorporate these principles into their product development processes.

Eco-design involves a holistic approach to product development, considering environmental impacts from the earliest stages of design. This includes:

1. Material Selection: Choosing materials that are renewable, recyclable, or biodegradable can reduce the environmental footprint of the product and facilitate end-of-life management.
2. Design for Disassembly and Reuse: Designing products with modular components and easy disassembly can enable the reuse of parts and materials, extending their useful life and minimizing waste.
3. Minimizing Material Usage: Optimizing product dimensions, weight, and material usage can reduce the overall resource consumption and waste generation associated with the product.
4. Energy Efficiency: Incorporating energy-efficient features and technologies can minimize the environmental impact during the product's use phase.
5. End-of-Life Management: Considering the product's end-of-life management from the design stage can facilitate material recovery, recycling, or proper disposal.

By adopting eco-design principles, B2B companies can not only reduce waste but also potentially create new revenue streams through the recovery and reuse of materials or components. Additionally, eco-designed products may appeal to environmentally conscious customers, providing a competitive advantage in the market. However, implementing eco-design can pose challenges, such as increased upfront costs, supply chain complexities, and the need for specialized knowledge and expertise. Overcoming these challenges may require investing in employee training, collaborating with academic institutions or specialized consultants, and fostering a culture of innovation within the organization.

### 2.4.2 Digital Transformation and Industry 4.0

The adoption of digital technologies and Industry 4.0 concepts, such as the Internet of Things (IoT), big data analytics, and additive manufacturing (3D printing), can facilitate sustainable waste management in B2B companies. IoT sensors can monitor and optimize resource consumption, reducing waste generation. Predictive analytics can identify inefficiencies and opportunities for process improvements. Additive manufacturing can enable on-demand production, minimizing excess inventory and waste. IoT and Smart Waste Management Systems: By integrating IoT sensors and wireless communication networks into waste management processes, B2B companies can gain real-time visibility and control over their waste streams. IoT-enabled smart bins or containers can monitor fill levels, optimizing collection routes and reducing unnecessary transportation emissions. Additionally, IoT data can provide insights into waste generation patterns, enabling predictive maintenance and targeted waste reduction efforts. Big Data Analytics and Predictive Modeling: The vast amounts of data generated by IoT devices, production processes, and supply chain operations can be leveraged through big data analytics and predictive modeling techniques. These tools can identify inefficiencies, forecast demand, and optimize inventory management, ultimately reducing the generation of excess waste and improving resource utilization.

## 2.5 *Case Studies and Best Practices*

Implementing sustainable waste management strategies in the B2B sector requires a comprehensive approach that addresses the unique challenges and opportunities faced by different industries and organizations. Several case studies and best practices illustrate successful implementation of these strategies:

1. Manufacturing Industry: In the manufacturing sector, companies like Subaru of Indiana Automotive, Inc. have implemented comprehensive waste minimization and recycling programs. By optimizing production processes, introducing reusable packaging, and recycling materials such as metals, plastics, and fluids, the company has achieved significant waste reduction and cost savings.
2. Construction Industry: The construction industry generates substantial amounts of waste, including concrete, wood, metals, and packaging materials. Companies like Skanska USA have adopted innovative waste management strategies by implementing on-site sorting and recycling facilities, as well as collaborating with material recovery facilities and recyclers. This approach has enabled them to divert a significant portion of their construction waste from landfills and promote the reuse of materials.
3. Logistics and Transportation Industry: In the logistics and transportation sector, companies like UPS have focused on sustainable packaging solutions and byproduct synergy networks. UPS has implemented a comprehensive packaging optimization program, reducing waste and minimizing the environmental impact of their operations. Additionally, they have formed partnerships with other organizations to exchange and repurpose byproducts, such as using waste materials from one company as packaging filler for another.
4. Professional Services Industry: While the professional services industry may generate less tangible waste compared to manufacturing or construction, sustainable waste management practices are still relevant. Companies like PwC have implemented comprehensive recycling programs for office waste, including paper, plastics, and e-waste. They have also adopted digital solutions to reduce paper consumption and promote paperless workflows.
5. Industrial Symbiosis Networks: Several regions and countries have established successful industrial symbiosis networks, facilitating the exchange of materials and byproducts among B2B companies across various industries. For example, the National Industrial Symbiosis Programme in the UK has facilitated over 25 million tonnes of industrial waste and byproduct exchanges, generating significant economic and environmental benefits.

These case studies and best practices demonstrate the diverse strategies and approaches that B2B companies can adopt to achieve sustainable waste management. By learning from these examples and tailoring solutions to their specific contexts, B2B companies can contribute to a more circular and sustainable economy.

### 2.5.1 Automotive Industry

Toyota's Environmental Challenge 2050 Toyota Motor Corporation has set ambitious goals for environmental sustainability through its "Environmental Challenge 2050" initiative. As part of this effort, Toyota aims to minimize and ultimately eliminate waste from its manufacturing processes. The company has implemented various strategies, such as minimizing water usage, promoting resource recycling, and fostering environmental awareness among employees and suppliers.

### 2.5.2 Chemical Industry

BASF's Circular Economy Approach BASF, a leading chemical company, has integrated circular economy principles into its operations through its "Circular Economy Program." This program focuses on closing material loops, developing innovative recycling technologies, and exploring alternative feedstocks derived from waste. BASF collaborates with partners across various industries to create value from waste streams, demonstrating the potential for industrial symbiosis.

## 3 Conclusion

This research paper has provided a comprehensive analysis of sustainable waste management practices for business-to-business (B2B) companies. By exploring the drivers, challenges, and effective strategies, the study has offered practical insights for B2B organizations to develop and implement environmentally responsible waste management programs. The literature review highlighted the theoretical foundations underpinning sustainable waste management, including the waste hierarchy, circular economy principles, and industrial ecology concepts. These frameworks emphasize waste minimization, resource efficiency, and closed-loop material flows as essential components of sustainable waste management. The key drivers for B2B companies to adopt sustainable waste management practices were identified as regulatory compliance, cost savings, corporate social responsibility, and stakeholder pressure. Regulatory frameworks and environmental legislation have become increasingly stringent, necessitating companies to comply with waste management regulations. Additionally, the economic benefits associated with resource efficiency, material recovery, and reduced disposal costs incentivize B2B companies to explore alternative waste management strategies.

Furthermore, corporate social responsibility initiatives and stakeholder demands, including those from customers, investors, and the broader public, have influenced companies to demonstrate environmental responsibility and contribute to sustainable development goals. The transition towards a circular economy, where waste is viewed as a valuable resource, has also gained momentum, encouraging B2B companies to implement closed-loop systems and collaborate with other organizations to create industrial symbiosis networks. However, the study also highlighted several challenges that B2B companies face in implementing sustainable waste management practices. The lack of infrastructure and facilities, such as recycling facilities, material recovery facilities, and waste-to-energy plants, can hinder effective waste management efforts. Additionally, limited knowledge and awareness among some B2B companies regarding best practices, technologies, and strategies can lead to suboptimal solutions. To overcome these challenges, the paper discussed several effective strategies for sustainable waste management in the B2B sector. Waste minimization through process optimization, material recovery and recycling, byproduct synergy networks, and collaboration with waste management service providers were identified as key strategies.

Waste minimization involves reducing waste generation at the source through process improvements, material substitution, and product design changes. Material recovery and recycling programs enable the diversion of valuable materials from the waste stream, contributing to resource conservation and revenue generation. Byproduct synergy networks facilitate the exchange of waste or byproducts among multiple organizations, creating industrial symbiosis and maximizing resource efficiency. Finally, collaboration with waste management service providers can provide B2B companies with access to specialized expertise, technologies, and infrastructure for sustainable waste management. The findings of this study have significant implications for both theory and practice. From a theoretical perspective, the research contributes to the body of knowledge on sustainable waste management by providing a comprehensive analysis tailored to the unique challenges and opportunities faced by B2B companies. It synthesizes insights from various theoretical frameworks and principles, offering a holistic understanding of the drivers, challenges, and effective strategies.

In terms of practical implications, the study offers valuable guidance for B2B companies seeking to develop and implement sustainable waste management programs. By adopting the recommended strategies, such as waste minimization, material recovery, and industrial symbiosis, B2B companies can foster resource efficiency, reduce environmental impacts, and contribute to the transition towards a circular economy. Furthermore, the findings can inform policymakers and regulatory bodies in developing supportive frameworks and incentives to promote sustainable waste management practices in the B2B sector. By addressing the challenges of infrastructure gaps and knowledge limitations, policymakers can create an enabling environment for B2B companies to effectively manage their waste streams. It is important to note that this study has certain limitations. While it provides a comprehensive overview of sustainable waste management practices in the B2B sector, it does not delve into sector-specific case studies or quantitative analyses of the economic and environmental impacts. Future research could address these limitations by exploring in-depth case studies of B2B companies across various industries, as well as conducting quantitative assessments of the financial and environmental benefits associated with sustainable waste management strategies.

Additionally, as technological advancements and innovative solutions continue to emerge in the field of waste management, further research could explore the potential applications and implications of these developments for B2B companies. For instance, the integration of digital technologies, such as Internet of Things (IoT) and data analytics, could enhance waste tracking, monitoring, and optimization efforts. In conclusion, this research paper has highlighted the significance of sustainable waste management for B2B companies and provided a comprehensive framework for addressing this critical issue. By adopting the strategies and recommendations outlined in this study, B2B companies can contribute to environmental conservation, public health, and sustainable development goals, while also realizing economic benefits and enhancing their corporate social responsibility.

## 4 Implications, Limitations, and Future Research

### *4.1 Theoretical Implications*

This research contributes to the theoretical understanding of sustainable waste management in the context of B2B companies. It synthesizes insights from various frameworks, including the waste hierarchy, circular economy principles, and industrial ecology concepts, providing a comprehensive theoretical foundation tailored to the B2B sector. The study also highlights the unique drivers, challenges, and strategies for sustainable waste management in B2B companies, expanding the body of knowledge in this field. Additionally, the integration of case studies and best practices from diverse industries enriches the theoretical understanding by providing practical examples and empirical evidence.

### *4.2 Practical Implications*

The findings of this study offer practical guidance for B2B companies seeking to develop and implement sustainable waste management programs. By adopting the recommended strategies, such as waste minimization, material recovery, byproduct synergy networks, and collaboration with service providers, B2B companies can enhance resource efficiency, reduce environmental impacts, and contribute to the transition towards a circular economy. Furthermore, the insights from this research can inform policymakers and regulatory bodies in developing supportive frameworks and incentives to promote sustainable waste management practices in the B2B sector. By addressing infrastructure gaps, knowledge limitations, and facilitating cross-industry collaborations, policymakers can create an enabling environment for effective waste management.

## 4.3 Limitations and Future Research

While this study provides a comprehensive overview of sustainable waste management in the B2B sector, it has certain limitations. First, the scope is limited to B2B companies and does not cover the waste management dynamics of business-to-consumer organizations or municipal solid waste. Future research could explore these complementary areas to provide a more holistic understanding of waste management across various sectors. Second, the study primarily relies on a qualitative synthesis of literature and case studies. Future research could incorporate quantitative analyses to assess the economic and environmental impacts of sustainable waste management strategies in the B2B sector. This could involve conducting cost–benefit analyses, life cycle assessments, or environmental impact studies to quantify the benefits and trade-offs of different waste management approaches. Third, as technological advancements and innovative solutions continue to emerge in the field of waste management, further research could explore the potential applications and implications of these developments for B2B companies. For instance, the integration of digital technologies, such as Internet of Things (IoT), artificial intelligence, and data analytics, could enhance waste tracking, monitoring, and optimization efforts, providing new opportunities for sustainable waste management. Additionally, future studies could delve deeper into sector-specific or industry-specific case studies, providing more detailed insights and tailored recommendations for B2B companies operating in different domains. This could involve collaborating with industry partners and conducting field research to gain a more nuanced understanding of the challenges and opportunities faced by specific sectors.

## 4.4 Integration of Sustainability Metrics and Reporting

As sustainable waste management practices become more widespread, there is a need for standardized metrics and reporting frameworks to measure and communicate progress effectively. Future research could explore the development and adoption of industry-specific key performance indicators (KPIs) and reporting guidelines that capture the environmental, economic, and social impacts of waste management strategies. Stakeholder Engagement and Collaboration Models Effective stakeholder engagement and collaboration are crucial for successful implementation of sustainable waste management practices in the B2B sector. Future research could investigate innovative models for stakeholder engagement, such as public–private partnerships, multi-stakeholder initiatives, or community-based approaches. These models could facilitate knowledge sharing, resource sharing, and collaborative decision-making among B2B companies, waste management service providers, policymakers, and local communities. Finally, as the global sustainability agenda evolves, future research could investigate the alignment of sustainable waste management practices in the B2B sector with broader sustainability frameworks and initiatives, such as the United Nations Sustainable Development Goals (SDGs) or the Paris Agreement on climate change. This could help identify synergies and potential areas for collaboration, as well as highlight the contributions of B2B companies towards achieving global sustainability targets.

## References

1. Allesch, A., Brunner, P.H.: Assessment methods for solid waste management: a literature review. Waste Manage. Res. **32**(6), 461–473 (2014). https://doi.org/10.1177/0734242X14535653
2. Bartolacci, F., Celani, A., Farouh, H.E.: Drivers and barriers of waste management in the industrial sector: a review. J. Clean. Prod. **265**, 121812 (2020). https://doi.org/10.1016/j.jclepro.2020.121812
3. Chertow, M.R.: Uncovering industrial symbiosis. J. Ind. Ecol. **11**(1), 11–30 (2007).
4. Cucchiella, F., D'Adamo, I., Lenny Koh, S.C., Rosa, P.: Recycling of WEEEs: an economic assessment of present and future e-waste streams. Renew. Sustain. Energy Rev. **51**, 263–272 (2015). https://doi.org/10.1016/j.rser.2015.06.010
5. Cui, H., Singh, A., Sinclair, A.J.: New industry development aided by industrial symbiosis: an analysis of the Northeast Ohio region in the United States. Resour. Conserv. Recycl. **146**, 28–39 (2019). https://doi.org/10.1016/j.resconrec.2019.03.018
6. Fraccascia, L., Giannoccaro, I., Albino, V.: Rethinking resilience in industrial symbiosis: conceptualization and measurements. Ecol. Econ. **137**, 148–162 (2017). https://doi.org/10.1016/j.ecolecon.2017.02.023
7. Ghisellini, P., Cialani, C., Ulgiati, S.: A review on circular economy: the expected transition to a balanced interplay of environmental and economic systems. J. Clean. Prod. **114**, 11–32 (2016). https://doi.org/10.1016/j.jclepro.2015.09.007
8. Hahn, R., Kühnen, M.: Determinants of sustainability reporting: a review of results, trends, theory, and opportunities in an expanding field of research. J. Clean. Prod. **59**, 5–21 (2013). https://doi.org/10.1016/j.jclepro.2013.07.005
9. Kirchherr, J., Reike, D., Hekkert, M.: Conceptualizing the circular economy: an analysis of 114 definitions. Resour. Conserv. Recycl. **127**, 221–232 (2017). https://doi.org/10.1016/j.resconrec.2017.09.005
10. Malinauskaite, J., Zarbà, C., Balsam, I.: Drivers and barriers of waste management: a business perspective. Sustainability **13**(7), 3758 (2021). https://doi.org/10.3390/su13073758
11. Patricio, J., Aydin, E.S., Gokaldi, G., Yang, L., Angelis-Dimakis, A.: Industrial symbiosis clusters across Europe: an innovation, economic and environmental assessment. Resour. Conserv. Recycl. **178**, 106043 (2022). https://doi.org/10.1016/j.resconrec.2021.106043
12. United States Environmental Protection Agency: Sustainable Materials Management: Non-hazardous Materials and Waste Management Hierarchy. https://www.epa.gov/smm/sustainable-materials-management-non-hazardous-materials-and-waste-management-hierarchy (2022)
13. Watkins, E., Giurco, D., Carrard, N., Chong, J., Wright, M., Kosatikovic, R.: Industrial ecology and regional resource synergies: an exploration of the economic, environmental and social impact of increased paper recycling in Queensland, Australia. J. Ind. Ecol. **23**(4), 983–996 (2019). https://doi.org/10.1111/jiec.12821

# Technology and Sustainability in Marketing, Business and Hospitality

# Evaluating Service Quality in Private Hospitals: A Comprehensive Framework

**S. Razeena and P. Shareena**

**Abstract** This study aims to develop a comprehensive service quality assessment framework for private hospitals through an extensive semi-systematic review of service quality, patient satisfaction, and loyalty. The critical evaluation of these literatures defined service quality dimensions and establishing relationship between service quality, patient satisfaction, and loyalty. This study has identified eight major dimensions including tangibility, reliability, responsiveness, empathy, assurance, administrative responsiveness, social responsibility, and cost. The conceptual framework proposes the direct relationship between service quality, patient satisfaction, and loyalty. Current research often overlooks the moderating role of health beliefs and patient experience in the expectation-perception gap and patient satisfaction, whereas patient trust mediates the relationship between satisfaction and loyalty. This study integrates these least-focused factors to understand private hospital service quality. The research addresses these gaps to enhance patient satisfaction and healthcare delivery through more effective and tailored strategies.

**Keywords** Service quality · Patient satisfaction · Patient loyalty · Expectation-perception gap · Patient trust

## 1 Introduction

Service quality is overall evaluation of a service based on how effectively perception meets their expectation. In healthcare, it encompasses dimensions like tangibility, reliability, responsiveness, assurance, and empathy, all of which contribute to patient satisfaction [1–3]. Patient experiences, infrastructure, and efficiency affect private

S. Razeena (✉)
Yenepoya (Deemed to Be University), Mangaluru, India
e-mail: razeena@yenepoya.com

P. Shareena
Department of Commerce, The Yenepoya Institute of Arts, Science, Commerce and Management, Yenepoya (Deemed to Be University), Mangaluru, India



M. Al Mubarak (ed.), *Sustainable Digital Technology and Ethics in an Ever-Changing Environment*, Studies in Systems, Decision and Control 237,
https://doi.org/10.1007/978-3-031-86708-8_27

hospital service quality. For instance, Indian patients during COVID-19 expressed satisfaction with consultation processes, fees, and admissions [4]. In Albania, SERVQUAL assessments identified responsiveness, reliability, safety, tangibles, and empathy as key satisfaction indicators [5]. These findings support a focus on patient satisfaction as a measure of service quality, influenced by factors including competition, performance [6]. This research aims to develop a conceptual framework to illustrate the relationship between perceived service quality, patient satisfaction and loyalty.

The study initially discusses the key dimensions and models of service quality, and relationship between service quality and patient satisfaction, including moderating and mediating factors. Furthermore, it outlines study methods, conceptual framework, and contributions, limits, and future research.

## 2 Key Dimensions of Service Quality

Private healthcare service quality is assessed through key dimensions of SERVQUAL model including tangibility, reliability, responsiveness, assurance, and empathy. Tangible pertains to the physical facilities, equipment, appearance of staff, whereas reliability ensures consistent promised service delivery. Responsiveness is the staff's promptness, assurance covers their competence, and empathy focuses on personalized care [4, 7–12]. Additional factors such as accessibility, waiting times, and administration also significantly impact patient satisfaction [13]. Perceived service quality enhances patient loyalty and positive healthcare outcomes [14]. By integrating these dimensions ensures a comprehensive approach to enhancing service quality [8] and support the development of effective strategies.

Table 1 outlines key dimensions of service quality.

## 3 Service Quality Models in Healthcare

Service quality models are essential for assessing and improving healthcare delivery in private hospitals. The SERVQUAL model measures the tangibility, reliability, responsiveness, assurance, and empathy, but often requires modifications to address healthcare complexities, including patient safety [21]. CRIMQUAL model extends SERVQUAL by incorporating clinical, infrastructure, and managerial quality, providing holistic quality care in private hospitals [10]. Total Quality Management (TQM) focuses on continuous quality improvement, patient satisfaction, decreasing errors, and enhancing operational efficiency [22], while Lean Six Sigma simplified procedures and enhance patient flow [15]. The Health Service Quality (HSQ) model [23] and HEALTHQUAL [24] emphasize patient safety and resource efficiency. However, the PUBQUAL model, can be adopted to measure the quality of private hospitals [25–27] and SERVPERF model is a performance-based measures,

**Table 1** Dimensions of service quality

| Authors and year | Name of the journal | Nature of the study | Dimensions identified | Findings of the study |
|---|---|---|---|---|
| Abdul Rauf et al. (2024) [7] | International Journal of Quality & Reliability Management | Systematic literature review | Tangibility, reliability, responsiveness, assurance, empathy | Identified 29 measures of service quality, and the importance of tangibly |
| Wen-Pai Wang et al. (2024) [15] | Preprint | Evaluation model | Medical environment, service attitude, medical security, admin measures | Proposed a comprehensive model for assessment of healthcare service quality |
| Mohit Datt et al. (2024) [10] | The TQM Journal | Conceptual framework | Clinical quality, infrastructural quality, relationship quality, managerial quality | Developed the CIRMQUAL model to enhance healthcare service quality |
| Iskandar Nazar Syahputri et al. (2024) [9] | International Journal of Health and Pharmaceutical | Cross-sectional study on outpatient satisfaction | Tangibility, reliability, responsiveness, assurance, empathy | Significant relationships between service quality dimensions and patient satisfaction |
| Ammar Ali Alraimi and Abhijeet Shelke (2023) [16] | Preprint | Service quality assessment using SERVPERF scale | Tangibility, reliability, responsiveness, empathy, safety | Highlighted high ratings for safety and low ratings for reliability |
| Divya Krishnan et al. (2024) [17] | American Journal of Medical Quality | Quality improvement training program evaluation | Tangibility, reliability, responsiveness, assurance, empathy | THINQ Program improved workflows and clinical care |
| Marie-Douce Primeau et al. (2024) [18] | Preprint | Mixed-methods case study in primary healthcare | Empathy, responsiveness, assurance | Increased SRH knowledge, improved health practices, and enhance transparency and accountability in community health centers |

(continued)

**Table 1** (continued)

| Authors and year | Name of the journal | Nature of the study | Dimensions identified | Findings of the study |
|---|---|---|---|---|
| Svetlana V Doubova et al. (2024) [19] | PLOS ONE | Cross-sectional survey on private healthcare use | Tangibility, reliability, responsiveness, assurance, empathy | Poor perceptions of public healthcare quality drive private healthcare use |
| Daniel Schuster et al. (2024) [20] | Journal of Healthcare Informatics Research | Case study on incremental process discovery | Tangibility, reliability, responsiveness | Demonstrated interactive process mining approaches effectively |

*Source* secondary source

opposed expectation-based measures [16, 28]. Finally, HospitalQual model is most flexible fit to different healthcare settings which impact on patient satisfaction and service effectiveness [29, 30]. By tailoring these models, private hospitals can effectively address service quality gaps, enhance patient satisfaction, and improve overall healthcare delivery.

Table 2 highlights the strengths and applicability of each model determined by the specific healthcare setting and service quality components. Combining these models' dimensions could provide a robust framework for comprehensive quality assessment and improvement in healthcare.

# 4 Relationship Between Service Quality and Patient Satisfaction

Different dimensions of service quality—tangibility, reliability, responsiveness, assurance and empathy—shape patient satisfaction in private hospitals by meeting or exceeding expectations, thereby promoting loyalty and positive word of mouth [31].

## *4.1 Impact of Service Quality Gap in Patient Satisfaction*

Service quality gaps in private hospitals, like outdated equipment [1], inconsistent care [32], and impersonal care [33] leads to dissatisfaction and erodes trust. Addressing these gaps enhances high-quality healthcare quality delivery.

**Table 2** Compare and contrast of models

| Model | Strengths | Weaknesses | Application in healthcare |
|---|---|---|---|
| SERVQUAL | Comprehensive, widely validated, adaptable | Subjective expectations, significant customization needed | Identifying service gaps, improving patient interactions and care processes |
| SERVPERF | Simplifies assessment, focuses on performance | May overlook patient expectations, reliance on accurate performance data | Direct assessment of service delivery performance |
| HEALTHQUAL | Tailored to healthcare, focuses on specific aspects | Narrow focus, may overlook broader dimensions | Evaluating physical and infrastructural aspects of healthcare services |
| CIRMQUAL | Comprehensive, integrates clinical and non-clinical aspects | Complex implementation, detailed data collection required | Holistic evaluation, detailed quality improvement framework |
| TQM | Continuous improvement, employee involvement | Resource-intensive, requires long-term commitment | Creating a quality culture, systemic improvements, patient-centric care |

*Source* Secondary Data

## 4.2 Moderating Role of Health Beliefs in the Relationship Between Expectation-Perception Gap and Patient Satisfaction

Health beliefs play a critical moderating role in the relationship between the expectation-perception gap and patient satisfaction. Patients with strong positive beliefs-including patients' perceptions of health, illness, and effectiveness of medical interventions [34] may tolerate service gaps and maintain higher satisfaction, while patients with negative beliefs are likely to feel dissatisfaction when expectations are not met [35]. Tailoring care based on health beliefs improve satisfaction outcomes.

## 4.3 Moderating Role of Complaint Management in the Relationship Between Expectation-Perception Gap and Patient Satisfaction

Complaint management moderates the expectation-perception gap and patient satisfaction by addressing patient complaints. Effective handling of complaints can turn dissatisfaction into satisfaction [36]. This proactive approach fosters trust and loyalty in the healthcare provider.

### 4.4 Patient Satisfaction and Patient Loyalty

Patient satisfaction is closely linked to patient loyalty, as satisfied patients are more likely to return to the same healthcare provider and recommend the service to others. High patient satisfaction enhances trust and strengthens the patient-provider relationship, leading to sustained loyalty [37, 38].

### 4.5 Mediating Role of Patient Satisfaction in Relationship Between Expectation-Perception Gap and Patient Loyalty

Patient satisfaction acts as a key mediator in the relationship between the expectation-perception gap and patient loyalty. High patient satisfaction can mitigate the adverse effects of unmet expectations, enhance patient loyalty, while low satisfaction intensifies the negative impact, reducing patient loyalty [39].

### 4.6 Mediating Role of Patient Trust in Relationship Between Patient Satisfaction and Patient Loyalty

Satisfaction and loyalty are mediated by patient trust. Satisfaction fosters trust, reinforcing the belief that the provider is reliable, competent, and caring. This trust, in turn, strengthens patient loyalty, making patients more likely to return for future care and recommend the provider to others. Without trust, even satisfied patients may not develop a strong sense of loyalty [40, 41].

## 5 Materials and Methods

We conducted a semi-systematic review of 73 studies on healthcare service quality and patient satisfaction from Web of Science and Google Scholar using search terms such as "service quality of healthcare" and "patient satisfaction." Selected studies examined the relationship between service quality, patient satisfaction and loyalty with positive and negative notions. The semi-systematic review addresses the challenges of systematic review and combines essential concepts, frameworks and developments [42].

## 6 Discussion for Final Conceptual Framework

This conceptual framework amalgamates all the dimensions, models and variables relevant to service quality in private hospitals, seeking to enhance patient satisfaction and overall healthcare delivery. The framework integrated SERVQUAL model along with key variables such as Expectation-Perception gap, Perceived service quality, patient trust, health belief, Experience, and patient loyalty.

## 7 Dimensions of Service Quality

### 7.1 Reliability

Reliability pertains to ability to perform the promised service dependably and accurately, includes consistent performance, accurate diagnosis, appropriate treatment and adherence to promised timelines. Reliability has substantial effect on patient satisfaction and trust. Higher satisfaction with public hospitals, reliability remained top priority among both public and private sectors, implying universal importance in healthcare service delivery [43]. Reliability was shown a direct impact on patient satisfaction and trust [4, 8, 14].

### 7.2 Empathy

Empathy is offering personalized, empathetic caring to patients, which directly affect patient satisfaction and desire to return or recommend others. Private hospitals scored higher in empathy compared to public hospitals, emphases ability to offer more individualized and attentive care [44]. Nurse empathy significantly enhance patient satisfaction, and minimize psychological stress, underscoring the importance of empathy in nursing care [45].

### 7.3 Tangibility

Tangibility encompasses the physical elements, including appearance of equipment, personality and facilities. Private hospitals exhibited superior tangibility compared to public hospitals, resulting in higher patient satisfaction [25, 44]. Furthermore, a study focusing on obstetric and maternity services during Covid-19 pandemic examined that all the dimensions had positive impact on service quality, tangibility showed a weaker relationship, suggesting improvement in infrastructure and equipment [46].

## 7.4 Assurance

Assurance is concerned with the knowledge and courtesy of employees and their ability to inspire trust and confidence among clients. Assurance, together with reliability and empathy had positive relationship with patient satisfaction [47]. Additionally, a considerable gap between patients' expectations and perceptions of assurance, indicate private hospitals need to improve assurance techniques to meet patient expectation [48].

## 7.5 Responsiveness

Responsiveness relates to the desire and competence of healthcare personnel to treat patients promptly and efficiently. Private hospitals significantly outperformed public hospitals in responsiveness, contributing higher patient satisfaction [44]. Responsiveness, had the largest effect on patient and family loyalty, underlining the necessity for hospitals to prioritize responsive service to retain and attract patients [49].

## 7.6 Administrative Responses

Administration factors such as infrastructure, staff behaviour, responsiveness, affordability, and the admission process as significant determinants of perceived service quality and enhance patient satisfaction [23, 50]. Effective administration result in reduced complaints and higher occupancy rates, is crucial for growth and sustainability of healthcare facilities [51]. Administrative improvements can create more supportive and effective work environment [52]. Furthermore, effective administration responses help to reduce negative behavior, which addresses patient dissatisfaction and enhance service quality [53].

## 7.7 Social Responsibility

Social responsibility incorporates ethical practices, community engagement and environmental sustainability, which collectively enhance reputation and effectiveness of healthcare services. Integrating corporate social responsibility (CSR) with service quality, enhance hospital image and patient trust, leading to better healthcare outcomes and patient loyalty [54]. The impact quality health, safety, and environment (QHSE) emphasizes CSR practices beyond the compliance and contribute to overall health system improvement [55]. Importance of healthcare professionals' responsibility in delivering high-quality care, suggesting that social responsibility is also a personal commitment of healthcare workers [56]. Differentiation between social responsibility and social responsiveness, advocating for a more proactive approach where healthcare institutions actively demonstrate their impact on community health [57].

## *7.8 Cost*

The cost of service quality involves financial implications of maintaining high standards of care, as well as the economic impact of low-quality services. Dual culture strategy-operations management approach, and focused service factory strategy, help hospitals achieve low unit costs while delivering industry-leading service quality [58]. The study estimated quality failures cost due to delays in procedures, increased hospital stays lengths, approximately $1.38 million annually [59], underscores the need of investing in quality improvement to avoid the higher costs associated with poor service.

## *7.9 Expectation-Perception Gap*

The expectation-perception gap is crucial in determining patient satisfaction, as unmet expectations often lead to dissatisfaction, negatively impacting trust and loyalty [31]. SERVQUAL model assist in quantifying this gap of service quality, guiding healthcare providers to align their services more closely with patient expectations [60].

### 7.9.1 Patient Satisfaction

High levels of patient satisfaction are associated with better patient outcomes, increased adherence to treatment plans, and stronger patient loyalty [61]. SERVQUAL model offer a systematic method to examine the gap between patient expectations and their actual experiences [3]. Addressing this factor influence patient satisfaction.

### 7.9.2 Complaint Management

Complaint management offers valuable insights into patient dissatisfaction and areas needing attention. A well-managed complaint system leads to continuous improvement in service delivery [36].

### 7.9.3 Patient Trust

Patient trust influences a patient's willingness to seek care and maintain a long-term relationship with healthcare providers. It is built on the perception of the healthcare provider's competence, honesty, and commitment to patient well-being [41]. High levels of patient trust enhance patient satisfaction, and greater compliance with medical advice [40].

### 7.9.4 Patient Loyalty

Patient loyalty pertains to a patient's willingness to opt same hospital in future and to recommend these services to others. It is a critical indicator of long-term patient-provider relationships and is influenced by quality of care, patient satisfaction, trust, and overall patient experience [37]. Ultimately, patient loyalty reflects how well a healthcare provider meets or exceeds patient expectations over time [38].

### 7.9.5 Novelty of the Framework

The proposed framework is novel in its integration of multiple service quality dimensions—tangibility, reliability, responsiveness, empathy, assurance, administrative responsibility, social responsibility and cost—offering a comprehensive assessment of private hospital services. This framework establishes a direct link between service quality, patient satisfaction, and loyalty and addressed moderate role of health belief and complaint management in relationship between expectation and perception gaps and patent satisfaction. It also addresses the mediating effect of patient trust in relation between patient satisfaction and patient loyalty. Its innovative approach provides valuable insights for both academic research and practical applications, enhancing overall patient care and healthcare provider performance.

### 7.9.6 Conceptual Model Visualization

The final conceptual framework is depicted in Fig. 1.

### 7.9.7 Theoretical Foundation

Theoretical connections between service quality in healthcare, particularly in private hospitals, are established based on the literature research´on various important aspects. Primarily, Patient-centered care models emphasize the importance of addressing individual patient needs, which aligns with the service quality dimensions of empathy and responsiveness [62]. The resource-based view (RBV) suggests, strategic deployment of hospital resources can create a competitive edge in service

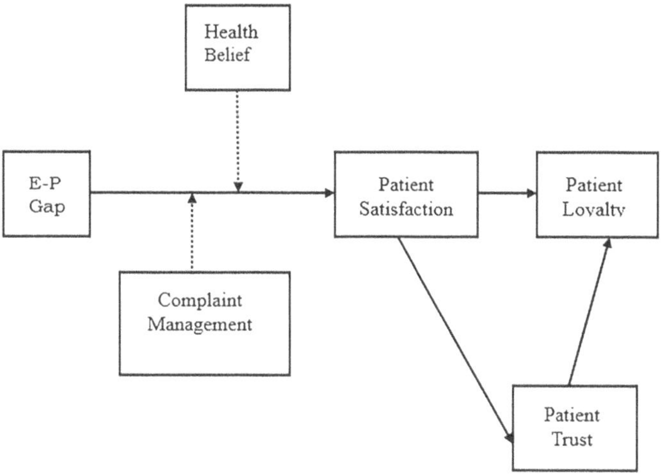


**Fig. 1** The final conceptual framework

quality [63]. The Health Belief Model (HBM) explains impact patients' beliefs on expectation-perception gap, impacting overall satisfaction. Patients with strong positive health beliefs may remain satisfied despite unmet expectations, while those with skeptical beliefs may experience greater dissatisfaction [34, 35]. Collectively, these factors influence patient satisfaction and loyalty, SERVQUAL model is a valuable instrument for enhancing service quality [3]. This theoretical basis emphasizes the multifaceted nature of service quality in healthcare, where both clinical and organizational factors converge to determine patient outcomes and satisfaction.

# 8 Conclusion

The study presents a comprehensive framework for assessing service quality in private hospitals. The direct relationship between service quality, patient satisfaction, and loyalty offers actionable insights for healthcare providers. This framework is a robust tool for improving healthcare delivery, contributing to better patient outcomes and higher service standards in private hospitals.

## *8.1 Theoretical and Practical Contributions*

This study develops a comprehensive framework that incorporates multiple service quality dimensions in private hospitals, provides substantial theoretical contribution. The framework serves as a valuable instrument to assess and enhance service quality, directly linking it to patient satisfaction and loyalty. It allows targeted healthcare delivery improvements, leading to better patient outcomes and competitive advantages for private hospitals.

## *8.2 Limitations and Future Directions for Research*

Although the framework offers a comprehensive approach to service quality assessment, it is limited by its dependence on existing literature, which may not completely capture emerging trends or regional variation in healthcare practices. Future research should focus on empirical validation, incorporating real-world data to enhance the model's applicability. Exploring the impact of patient engagement in different regions could further enrich the understanding of service quality in private hospitals.

## References


1. Firdaus, M.Z.Z., Cahyana, A.S: Innovative service quality boosts patient satisfaction. Indonesian J. Innov. Stud. **25**(4) (2024)
2. Alharbi, M.F.: Patients' experience of service quality in government and private hospitals in the Qassim Region, Kingdom of Saudi Arabia. J. Med. Life **16**, 1622–1627 (2023)
3. Parasuraman, A., Zeithaml, V.A., Berry, L.L.: SERVQUAL: a multiple-item scale for measuring consumer perceptions of service quality. J. Retail. **64**(1), 12 (1988)
4. Rajasekaran, V.A., Jayalakshmi, P., Gouri, M.S., Siva Balanm R.V., Susi, S., Jothikumar, R., Kumar, K.R.: Service quality matters of private hospitals in prevention of orthopaedic issues during the COVID-19 outbreak in India. Advances in Chemical and Materials Engineering Book Series, pp. 219–229 (2023)
5. Kalaja, R., Kurti, S., Myshketa, R.: Service quality and patient satisfaction with private health care services in Albania. Int. J. Public Health **12**(1), 460–468 (2023)
6. Monika, M., Singh, A.: Healthcare service quality- a review of literature. Paripex Indian J. Res., 143–147 (2022)
7. Rauf, A., et al.: Systematic literature review on healthcare service quality. Int. J. Qual. Reliab. Manage. (2024)
8. Kawii, C.: North America private hospitals' service quality dimensions and patients' satisfaction. OTS Can. J. **2**(1), 9–19 (2023)
9. Syahputri, I.N., Wienaldi, W., Nasution, S.L.: The effect of health service quality dimensions on outpatient satisfaction at plenary accredited hospitals. Int. J. Health Pharmaceut. (2024)
10. Datt, M., Gupta, A., Misra, S.K., Gupta, M.: CIRMQUAL: a conceptual framework of healthcare service quality. TQM J. (2024)
11. Kumari, S.: Service quality measurement in Indian healthcare industry. Indian Sci. J. Res. Eng. Manage. **08**(04), 1–5 (2024)
12. Yasin, S., Hashmi, A., Rasheed, M.F., Tariq, A.: Impact of service quality dimensions on customer satisfaction, considering service performance as mediator: a study of healthcare sector of Pakistan. Pak. J. Hum. Soc. Sci. (2023)
13. Ngatindriatun, M.A.: Determinant factors of hospital service quality and patient satisfaction: hospital logistics management approach. Asian Manage. Bus. Rev., 121–138 (2022)
14. Razeed, S.M.M.H., Vethirajan, C.R.: Patients satisfaction and expectation towards different dimensions of health care service quality of private hospitals. J. Contemp. Issues Bus. Govern. **27**(3), 2070–2075 (2021)

15. Wang, W.-P., Huang, M.-L., Ou, Y.-C.: An approach to evaluate healthcare service quality by incorporating linguistic assessment and computing (2024)
16. Alraimi, A.A., Shelke, A.: Measuring health services quality: nurses' perceptions and the SERVPERF scale (2023)
17. Krishnan, K., Kareddy, A., Chen, C., Kerbel, R., Dowling, E., Wendy, S., Dermenchyan, A.: The healthcare improvement and innovation in quality (THINQ) collaborative: a novel quality improvement training program for undergraduate and postgraduate students. Am. J. Med. Qual. (2024)
18. Primeau, M.-D., Jobin-Gelinas, M., Mayabanza, C.M., Mayaza, M., Blouin, G.: Improving access and quality of primary healthcare through women and adolescents' user committees: a mixed-methods case study in Kinshasa, DRC (2024)
19. Doubova, S.V., Leslie, H.H., Pérez-Cuevas, R., Kruk, M.E., Arsenault, C.: Users' perception of quality as a driver of private healthcare use in Mexico: insights from the People's Voice Survey. PLOS ONE **19**(6), e0306179–e0306179 (2024)
20. Schuster, D., Benevento, E., Aloini, D., van der Aalst, W.M.P.: Analyzing healthcare processes with incremental process discovery: practical insights from a real-world application. J. Healthc. Inf. Res. (2024)
21. Kavitha, R.: Service quality measurement in health care system—a study in select hospitals in Salem City, Tamil Nadu. IOSR J. Bus. Manage. **2**(1), 37–43 (2012)
22. Alqasmi, I.: Total quality management in the healthcare sector. Int. J. Health Sci. (IJHS), 2900–2915 (2022)
23. Ghildiyal, A., Devrari, J.C., Dhyani, A.: Determinants of service quality in healthcare. Int. J. Patient-Centered Healthc. **12**(1), 1–12 (2022)
24. Mariano, A.M., da Silva, E.K., Mariano, A.P.M., Ciulla, M.: The HEALTHQUAL model: evaluating the quality of health service in the Federal District, Brazil. Procedia Comput. Sci. **214**, 1106–1112 (2022)
25. Khoso, A.R., Akhtar, F., Narejo, A.A., Mallah, S.A., Vighio, K., Sanjrani, K.D.: Comparative analysis of service quality between public and private hospitals, using SERVQUAL model: a case study of Peshawar, Pakistan. Medfarm: Jurnal Farmasi dan Kesehatan **11**(2), 240–252 (2022)
26. Goula, A., Stamouli, M.-A., Alexandridou, M., Vorreakou, L., Galanakis, A., Theodorou, G., Stauropoulos, E., Kelesi, M., Kaba, E.: Public hospital quality assessment. Evidence from Greek health setting using SERVQUAL model. Int. J. Environ. Res. Public Health **18**(7), 3418 (2021)
27. Khambhati, R., Patel, H., Kumar, S.: A performance evaluation and comparison model for Urban Public Healthcare Service Quality (UrbPubHCServQual) by fuzzy TOPSIS method. J. Nonprofit Public Sector Marketing, 1–20 (2021)
28. Duc Thanh, N., Quynh Anh, P., Thi Huyen Chang, P., Minh Nguyet, H.T.: Cross-cultural adaption and validation of SERVPERF tool for measuring healthcare quality in an oncology public hospital, Vietnam. Inquiry: J. Med. Care Organ. Provision Fin. **60**, 004695802211468–004695802211468 (2023)
29. Yeh, D.: Hospital's service quality assessment on HEALTHQUAL model during the second wave of COVID-19: an empirical study at PMCH, Udaipur, pp. 267–281 (2022)
30. Estiri, M., Heidary Dahooie, J., Zavadskas, E.K.: Providing a framework for evaluating the quality of health care services using the HealthQual model and multi-attribute decision-making under imperfect knowledge of data. Informatica (Lithuanian Academy of Sciences) **34**, 85–120 (2023)
31. Zeithaml, V.A., Berry, L.L., Parasuraman, A.: The nature and determinants of customer expectations of service. J. Acad. Mark. Sci. **21**, 1–12 (1993)
32. Ayura, A., Nasution, S.W., Ginting, R.: Comparative analysis of service quality on patient satisfaction levels in government and private hospitals in Medan City in 2022. Int. J. Health Pharmaceut. (2024)
33. Mabini Jr., S.P., Narsico, L.O., Narsico, P.G.: Service quality, patient satisfaction, and improvement indicators. Int. J. Multidiscip. **5**(4), 1331–1345 (2024)

34. Janz, N.K., Becker, M.H.: The health belief model: a decade later. Health Educ. Q. **11**(1), 1–47 (1984)
35. Rosenstock, I.M.: The health belief model and preventive health behavior. Health Educ. Monogr. **2**(4), 354–386 (1974)
36. Gill, L., White, L.: A critical review of patient satisfaction. Leadersh. Health Serv. **22**(1), 8–19 (2009)
37. Boshoff, C., Gray, B.: The relationships between service quality, customer satisfaction and buying intentions in the private hospital industry. S. Afr. J. Bus. Manage. **35**(4), 27–37 (2004)
38. Lee, H., Delene, L.M., Bunda, M.A., Kim, C.: Methods of measuring health-care service quality. J. Bus. Res. **48**(3), 233–246 (2000)
39. Oliver, R.L.: Whence consumer loyalty? J. Marketing **63**(4_suppl1), 33–44 (1999)
40. Hall, M.A., Dugan, E., Zheng, B., Mishra, A.K.: Trust in physicians and medical institutions: what is it, can it be measured, and does it matter? Milbank Q. **79**(4), 613–639 (2001)
41. Thom, D.H., Hall, M.A., Pawlson, L.G.: Measuring patients' trust in physicians when assessing quality of care. Health Aff. **23**(4), 124–132 (2004)
42. Snyder, H.: Literature review as a research methodology: an overview and guidelines. J. Bus. Res. **104**, 333–339 (2019)
43. Mughal, Y.H.: Comparison of service quality between public & private hospitals: empirical evidence from Saudi Arabia. J. Soc. Res. Dev. **4**(03), 533–541 (2023)
44. Firdaus, S.: A comparative study on service quality at public and private hospitals with reference to Jaipur city. Indian Sci. J. Res. Eng. Manage. **08**(05), 1–5 (2024)
45. Atanay, R.S., Dwidiyanti, M., Dwiantoro, L.: The impact of nurse empathy in hospitals nursing services. Media Ilmu Kesehatan (2023)
46. Shafique, A., Hayat, Q.: Obstetric and maternity services management in private hospital: an assessment from patients' view of service quality. Pak. J. Soc. Res. **4**(04), 1083–1095 (2022)
47. Agagbo, C.O., Ohida, I.Y.., Gbande, C.P.A.: Effect of service quality on satisfaction of enrollees of private HMOs in general hospitals within Abuja metropolis. Int. J. Innov. Res. Soc. Strat. Manage. Tech. (2023)
48. Nazem, H., Abdollahi, H.R., Mirzaei, A.: Investigation of the gap between patients' perceptions and expectations of hospital service quality. J. Client-centered Nursing Care **6**(2), 117–124 (2020)
49. Sayyida, S.: Visitor response as the impact of service quality in private hospital. Deleted J. **22**(1):48–60 (2023)
50. Cobo-Mejía, E.A., Estepa-Rodríguez, K.J., Herrera-Triana, C., Linares-Reyes, P.: Percepción de los usuarios frente a la calidad del servicio de salud en consulta externa en una institución prestadora de servicios de salud. Revista Investigación en Salud Universidad de Boyacá **5**(2), 277–294 (2018)
51. Luvhengo, U., Lekunze, J.N., Molamu, N.P.: Investigating the effects of administrative service quality on customer retention in a private health care organization. PONTE Int. Sci. Res. J. **78**(12) (2022)
52. Tsai, Y., Wu, S.-W., Tsai, Y.-H.: Employee perceptions of service quality based on hospital quality improvement strategy. Int. J. Manage. Econ. Soc. Sci. (IJMESS) **7**, 13–25 (2018)
53. Um, K.H., Lau, A.K.W.: Healthcare service failure: how dissatisfied patients respond to poor service quality. Int. J. Oper. Prod. Manage. **38**(5), 1245–1270 (2018). https://doi.org/10.1108/IJOPM-11-2016-0669
54. Lubis, A.N., Lumbanraja, P., Lubis, R.R., Hasibuan, B.K.: A study of service quality, corporate social responsibility, hospital image, and hospital value creation in Medan. Eur. Res. Stud. J., 125–133 (2017)
55. Abderraouf, G., Nadira, R.: The impact of corporate social responsibility on quality health, safety and environment. Eur. J. Multidiscip. Stud. **7**(2), 43 (2018)
56. Jagodic, A., Osmanovic, E., Ikanovic, I.: Treatment of patients in the health care system and achievement of quality services. Magna Scientia Adv. Res. Rev. **8**(1), 053–060 (2023)
57. Boelen, C.: Improving the impact on health: the social accountability approach. Texto & Contexto Enfermagem **24**(3), 615–616 (2015)

58. Wirtz, J.: Cost-effective service excellence in healthcare. AMS Rev. **9**(1), 98–104 (2019
59. Alvarado, B.S.S., Rodriguez-Rojas, Y.L., Pineros, J.P.R.: Financial costs of low-quality service due to increasing hospital stay length. Costos Financieros De La Baja Calidad Por Aumento De La Estancia Hospitalaria **5**(7), 55–62 (2017)
60. Buttle, F.: SERVQUAL: review, critique, research agenda. Eur. J. Mark. **30**(1), 8–32 (1996)
61. Batbaatar, E., Dorjdagva, J., Luvsannyam, A., Savino, M.M., Amenta, P.: Determinants of patient satisfaction: a systematic review. Perspect. Public Health **137**(2), 89–101 (2017)
62. Berwick, D.M.: What 'patient-centered' should mean: confessions of an extremist: a seasoned clinician and expert fears the loss of his humanity if he should become a patient. Health Aff. **28**(Suppl1), w555–w565 (2009)
63. Barney, J.: Firm resources and sustained competitive advantage. J. Manag. **17**(1), 99–120 (1991)

# Municipal Solid Sustainable Waste Management in the Kingdom of Bahrain

**Fatema Alaauddin Showaiter and Muneer Al Mubarak**

**Abstract** Municipal solid waste management (MSWM) poses a significant challenge for the Kingdom of Bahrain and is considered detrimental to the country. Available data indicates a substantial increase in the quantity of generated waste over recent years, including various types such as household, commercial, industrial, construction, agricultural, and more. This study employed methods such as observation, interviews, and questionnaires to analyze MSWM. The results revealed that population growth and economic factors are the primary contributors to the high volume of waste. Additionally, there is a lack of advanced methods and technologies for waste treatment and disposal, raising concerns about the impact on the environment and human health. Consequently, there is a need to explore alternative approaches to develop a more sustainable MSWM system in Bahrain for the future.

## 1 Introduction

A sustainable waste management refers to practice that is handling waste materials efficiently and responsibly in manner that minimizes the negative impacts on the environment and promotes long term ecological balance. The practice of sustainable waste management involves in implanting strategies and process throughout the process, which includes waste reduction, recycling, and proper disposal. The purpose to prioritize a sustainable practice that preserve resources, prevent pollution and contribute to overall ecosystems and communities' well-being [18].

F. A. Showaiter
The George Washington University, Washington, DC, USA

M. Al Mubarak (✉)
Ahlia University, Manama, Bahrain
e-mail: malmubarak@ahlia.edu.bh

 
M. Al Mubarak (ed.), *Sustainable Digital Technology and Ethics in an Ever-Changing Environment*, Studies in Systems, Decision and Control 237,
https://doi.org/10.1007/978-3-031-86708-8_28

## 1.1 Research Problem

The developed countries continue to face a significant challenge that is related to managing different types of waste but mostly solid waste, due to the pressure of rapid and sudden urbanization. Additionally, the expansion of the global economy is identified as another factor that contributes to the significant production of solid waste due to the increased availability of diverse products [16]. Despite the substantial waste generation, certain developed nations continue to rely on outdated methods and strategies for solid waste management, primarily adopting Western technologies. However, this approach has had limited success, primarily due to the unsuitability of implementing these technologies in different regions. As a result, it is vital for developed countries to develop their own methods, guidelines, and strategies that are tailored to their specific circumstances. [16]. For instance, the Kingdom of Bahrain is recognized as one of the leading producers of municipal solid waste per capita in comparison to other Gulf countries. The approximate amount of solid waste generated by everyone on a daily basis surpasses 1.8 kg, resulting in an annual production of over 1.2 million tons of solid waste [30]. Therefore, this paper aims to reduce the generated waste by introducing a suitable solid waste management method to Bahrain environment.

## 1.2 Need for the Research

The growing population and limited land availability have posed a significant challenge for Bahrain's municipalities and urban planners. To address this issue the Ministry of Works, Municipalities Affairs, and Urban Planning in Bahrain has granted authorization to a few private companies to operate across the four governorates [30]. For example, The Gulf City Cleaning Company, which operates in Muharraq Governorate and Capital Governorate, and the Urbaser Bahrain Company, responsible for waste collection and waste transportation systems in the Northern Governorate and Southern Governorates. These authorized companies play a key role in managing waste within their respective areas. Both companies operate by collecting the waste from the containers that are placed in the street and transported them to landfill that is in Al Hafirah, Asker, where the wasted are treated by using technique called sanitary method or known as controlled tipping [30].

## 1.3 Contributions

The aim this paper is to make several and different contributions to the literature. Firstly, it will provide a comprehensive review of sustainable waste management techniques and methods that are used in another developed country. Secondly, it will explore the potential benefits of using these techniques in Kingdom of Bahrain. Finally, it will provide recommendations in the sector to adapt a proper waste management.

### 1.4 Main Terms Definitions

- Sustainable waste management is a comprehensive approach that aims to reduce the negative impact of the waste on the environment by implementing a different sustainable solution at every stage. It involves managing various substances such as paper, plastic and food leftovers with the goal. In addition, it includes reducing waste generation, maximizing the reuse of materials, and recycling to preserve the valuable [21].
- Solid waste refers to any discarded or abandoned material that is not in the form of a liquid or gas. It includes various types of non-hazardous waste generated by residential, commercial, industrial, and institutional activities. Solid waste can encompass a wide range of materials such as household garbage, packaging materials, construction debris, electronic waste, and other forms of discarded materials. Proper management of solid waste is essential to minimize environmental impacts and promote sustainability [8].
- Landfill is a designated area where solid waste is disposed of and buried in the ground. It is a common method of managing solid waste that cannot be recycled or reused. There are different types of landfills such as Municipal Solid Waste (MSW), construction landfill, hazardous waste landfill, industrial waste landfill and bioreactor landfill [4].
- Municipal Solid Waste (MSW) refers to the waste generated by households and small businesses in a municipality or local area. It excludes waste generated by industrial or manufacturing processes. MSW typically includes items such as household garbage, recyclables, yard waste, and sometimes bulky items like furniture or appliances.

### 1.5 Context

The global population is growing rapidly, with half of the world's population residing in urban areas. As living standards improve and technology advances, there is a significant strain on the environment, resulting in a continuous rise in the volume of solid waste. Therefore, solid waste poses a major environmental challenge globally, as humans have historically treated the environment as a dumping ground for the waste generated by their daily activities since the dawn of civilization. Human actions have resulted in various forms of pollution, including air, land, and water pollution. Given that air and water are vital components for human existence, the impacts of these pollutants are readily observable [16]. Generally, this issue relevant and occur in developed country with high population. Consequently, these countries need to implement diverse strategies and approaches to gather, handle, and appropriately dispose of solid waste to reduce the environmental impacts in a socially and economically sustainable manner. This entire process is referred to as solid waste management [19].

Solid waste management involves the proper disposal of different types of material that are no longer needed in our daily live which includes hazardous waste, liquid waste, food waste, metal waste, etc. these solid wastes are categorized based on their source and waste type [16]. The most significant and main type of waste classification is municipal solid waste (MSW), which is observable in our daily routines and has no specific time and continuously generated in various locations and the local municipalities are responsible for the entire process of collection, handling, storage, transportation, segregation, processing and disposal through a well managed sustainable system [5]. On a global scale, the most employed approach for waste disposal during the management process is the sanitary landfill method. This method involves depositing the waste in thin layers, typically around 1 m or 3 feet in thickness. These layers serve two primary purposes: facilitating the decomposition of the materials and capturing harmful gases [6]. In this disposal method, small volume layers are placed at the bottom to prevent land collapse, while the large volume layers are placed at the top. The process begins by excavating and installing the liner system, which includes compacted clay and high-density plastic. This liner system ensures that liquids do not leak in or out, thereby preventing underground contamination. The next layer is the drainage system, which helps remove toxic liquids resulting from waste decomposition away from the liner system [15].

Additionally, the waste generates various toxic gases, including methane, which contributes to global warming. To address this, the gas collection system, consisting of extraction pipes, captures these gases and transports them to treatment stations. There, the gases are processed to produce useful energy, such as electricity. Finally, the trash layer, comprised of multiple layers that are up to 3 m or 10 feet thick, is placed on top. These layers are compacted and sealed using impermeable clay to ensure proper waste containment [15]. The Kingdom of Bahrain utilizes the sanitary landfill method for waste disposal during the disposal stage. They operate the Hafira sanitary landfill, which spans an area of 24,500.5 $m^2$. This project was established in 1986 by the Supreme Council for Environment with the objective of safeguarding the environment and public health. As part of the project, a techno-environmental study was conducted to identify a site that adheres to environmentally friendly standards. The construction of the landfill was carried out by the staff of the Public Commission, utilizing their own resources. The Hafira landfill is classified as Class II and consists of three disposal units, three evaporation ponds for treating leachate and liquid industrial waste, as well as two groundwater monitoring boreholes [25]. In October 2019, the Hafira landfill formed a partnership with Urbaser Bahrain Company. However, due to population and economic growth, there is a need to expand the landfill area while the expansion is considered efficient, but it reduces land availability, making it challenging to create new residential areas in a small country like Bahrain. Therefore, there is a need to explore new waste management methods and strategies through comprehensive studies to ensure efficiency, support the country's economy, and maintain a high quality of life for future generations of Bahraini citizens [25].

## 2 Literature Review

As its mentioned above the Municipal Solid Waste Management (MSWM) is one of the greatest challenges that the environmental protection companies are still facing in the developing countries. Ogwueleka study that conducted in nine representative cities in Nigeria, it provides an overview of the SWM in Nigeria back in 2009 highlighting the challenges faced by environmental protection agencies in managing municipal solid waste in the country, which is stated that the agencies faced lack of institutional arrangement, insufficient financial resources and information regarding the waste such as quantity and type of waste, lack of laws and standards, and using unsuitable technology [23]. Thus, the study mentioned the reason behind inefficient waste managements is that the companies use the roadside of the highways as landfill to dispose the waste, because of lack of sites and the continuous grow in population. The waste density in Nigeria ranged from 280 to 370 kg/m$^3$, and the waste generation rated ranging from 0.44 to 0.66 kg/capita/day, as it is shown in Table 1 [23]. Therefore, the ministry of works in Kingdom of Bahrain should considered to allocate annually budget to support and follow up with the private companies that are responsible for waste management to ensure to fulfill their job.

Furthermore, according to waste management researchers in Brazil, managing the municipal solid waste is significant challenge in Brazil that the situation remains deficient in many different aspects, which lead the Brazilian federal government to establish the National Solid Waste Policy (NSWP) in 2015, the objective of this policy is to foster the advancement and progress of the waste management industry. This is achieved through the implementation of various methods to handle waste, including reuse, recycling, treatment, and proper disposal. Additionally, the policy integrates an energy recovery system to mitigate any potential negative impacts on public health and the environment [2]. Implementing an efficient disposal and treatment technologies of MSW is one of the main keys in waste management. A research paper by Neha,

**Table 1** Density of SW in Nigeria cities [23]

| City | Population | Agencies | Tonnage per month | Density (Kg/m$^3$) | Kg/capita/day |
|---|---|---|---|---|---|
| Lagos | 8,029,200 | Lagos Waste Management Authority | 255,556 | 294 | 0.63 |
| Kano | 3,248,700 | Kano State Environmental Protection Agency | 156,676 | 290 | 0.56 |
| Ibadan | 307,840 | Oyo State Environmental Protection Commission | 135,391 | 330 | 0.51 |
| Kaduna | 1,458,900 | Kaduna State Environmental Protection Agency | 114, 433 | 320 | 0.58 |
| Port Harcourt | 1,053,900 | Rivers State Environmental Protection Agency | 117,825 | 300 | 0.60 |
| Makurdi | 249,000 | Urban Development Board | 24,242 | 340 | 0.48 |
| Onitsha | 509,500 | Anambra State Environmental Protection Agency | 84, 137 | 310 | 0.53 |
| Nsukka | 100,700 | Enugu State Environmental Protection Agency | 12,000 | 370 | 0.44 |
| Abuja | 159,900 | Abuja Environmental Protection Agency | 14, 785 | 280 | 0.66 |

Krishan and Vinit discussed a several different technologies. Firstly, composting that considered a simple and cost-effective method that focuses on treating the organic portion of MSW, which involves the decomposition of organic matter using microorganisms in aerobic and anaerobic environments. Composting technology I particularly suitable the garden waste and biodegradable organic fraction of MSW, which contains lignocellulosic materials that do not degrade well under the anaerobic conditions. On one hand, during the aerobic microorganisms the organic compounds break down into Carbon Dioxide ($CO_2$), Nitrite ($NO_2^-$) and nitrate ($NO_3^-$). On the other hand, during anaerobic composting microorganisms metabolize nutrients and break down organic compounds through reduction processes [10].

Moreover, The Vermicomposting process, as mentioned previously, utilizes worms to transform organic waste into soil that is abundant in nutrients. Additionally, the researchers discussed a technique called bio-methanation, which converts organic matter in waste into both methane and manure. This method employs microorganisms in an oxygen-free environment to generate waste with a high organic content that can be beneficial for the agricultural sector. Lastly, there is the process of incineration, which involves using raw waste as fuel for industrial machinery as a means of thermal waste treatment. The incineration process occurs in the presence of air at a temperature of 580 °C [10]. According to Damghani et al. [7], a similar situation can be observed in Tehran, the capital of Iran, where there is a shortage of technical expertise and the utilization of diverse technologies in managing municipal solid waste (MSW). The population of Tehran, which amounts to 8.2 million people, consists of approximately 2.4 million households that collectively generate 2,626,519 tonnes of solid waste. Most of this waste, over 97%, falls under the category of MSW. However, in 2005, only 5% of the recyclable materials were recycled. This low recycling rate can be attributed to the absence of a waste separation process and a lack of adequate recycling facilities. [7].

The municipality of Tehran employs two primary methods for collecting solid waste. Firstly, municipal employees collect waste directly from households, going door-to-door. Secondly, waste is collected from containers placed in public areas. Additionally, the municipality is responsible for collecting waste from hospitals; however, there is a lack of separation between infectious and non-infectious waste during this process. This situation poses a risk of transmitting infections or germs from the hospital waste to non-infectious waste. Proper separation and careful storage of hospital waste are necessary, as it often contains infectious materials or human tissues [7].

According to Zhang, Tan and Gersberg China serves as another example of a developed country that implanting inefficient strategies and methods in managing municipal solid waste. The main reasons for the significant increases in the total amount of MSW generated in China is the continues population growth, industrialization, and urbanization. From 1980 to 2006 the MSW quantity increased from 31.3 million tons to 212 million tons, while the waste generation rate increased from 0.5 to 0.98 kg/capita/day. Currently, China's MSW management system relies in various important factors such as environmental management strategies and requirements, promoting environmental awareness among the citizens, feasibility and capabilities of the technology and the economy [31].

Also, the research paper focuses on the importance of implementing effective waste management strategies to ensure environmental protection and sustainable development. Starting with the first stage of managing the waste which is the collecting process which it involves two tires: primary collection and secondary collection. In the primary collection, the waste gathered and transported from the households to the local collection points, while in the secondary collection the waste is taken from the local collection points to treatment and disposal facilities [31]. Different cities of China use a modern waste collection method while the residential areas are often relied on primitive method. In residential areas, primitive methods are commonly employed for waste management. These methods include roadside waste collection, where residents themselves sort recyclable items and place them in collection containers provided by local authorities who are responsible for waste management. Additionally, there is the household collection method, where specialized teams visit each household to collect waste through a door-to-door approach. [31].

Unfortunately, these used collection methods have a negative impact on the environment but mostly on the health of waste collectors, cause the collectors handle the waste directly through using their hands and while sorting the waste they leave the remaining unwanted wasted scattered on the streets. Therefore, the researchers suggest improving the collection process of waste, the local government of China should consider organizing the informal system by establishing municipal institutions, cause implementing these improvements will not only enhance the efficiency of the solid waste collection process but also create new job opportunities for informal waste collectors, while maintain their health. Secondly, the separation and recycling of waste. Generally, the MSW is collected in a mixed state by the collectors which obstructs the recycling process and increase the overall waste disposal cost. The recycling processes can be carried out in two different ways: residents can either deliver their recyclable materials to the designated sites or sell them to a buyer, which theses buyers usually sell them to factories as raw or processed materials, this kind of act called informal act. On the other hand, the separation process where the residents have the option to voluntarily participate in the process, however it can become a challenging at the household level because the residents tend to mix different types of waste in one bag and dispose it in a single container.

To enhance the process of waste separation and recycling there are several factors that can be considered. Firstly, can implement the use definite containers with different colors, clearly labeled with the use of different languages and accompanied by instructions that differentiate between different type of waste such as combustible and non-combustible materials, recyclable and non-recyclable materials. Secondly, it is important to separate waste with high water content from the recyclable materials, as this eases the recovery process and reduces the leachate in landfills. Thirdly, it is important to establish suitable facilities to separate and treat different waste streams. Finally, improving the recycling industry by enhancing professionalism, implementing high quality products, and refining operational standards [31]. Thirdly, landfilling in Choina is primarily utilized due its cost-effectiveness, and these landfills are supervised by the environmental protection departments and administered by construction departments. In addition, the landfills are constructed with vertical cement structures to prevent leachate contamination of groundwater. Additionally, high-density polyethylene (HDPE) membranes are employed as liner materials. However, as time goes on, the increase in waste generation resulting from population and economic growth presents challenges for landfills to accommodate large volumes of waste. Therefore, the researcher suggests incorporating pre-treatments process which can be mechanical way be reducing the particle size of the materials and encourage the organic decomposition. Also, it can be biological way that can be more helpful and more advantages where it can produce high-quality leachate that requires minimal or no treatment, reduce the greenhouse gases emission and produce will compact waste. Thus, the landfill will have free spaces.

Furthermore, incineration presents various benefits, such as energy recovery, thorough waste purification, substantial volume reduction (around 90%) and mass reduction (around 70%). However, it comes with high costs, emits toxic substances that contribute to air pollution, and produces solid residues. Furthermore, there are three primary types of incineration technology. The first is the mechanical stoker grate, which employs an endless chain to rotate and feed solid fuel into the furnace. The second is the rotary kiln, where materials are exposed to high temperatures [31].

Finally, the fluidized bed involves a mixture fluid and solid phases. The mechanical stoker grate and fluidized bed technologies are the most used, accounting for approximately 80% of the systems in operation [31]. Overall, due to China's continuous growth and development across various industries, which they need to establish waste management facilities that can effectively handle and financially support. Figure 1 shows the data that the researchers in this paper found of using different MSW disposal technology in China.

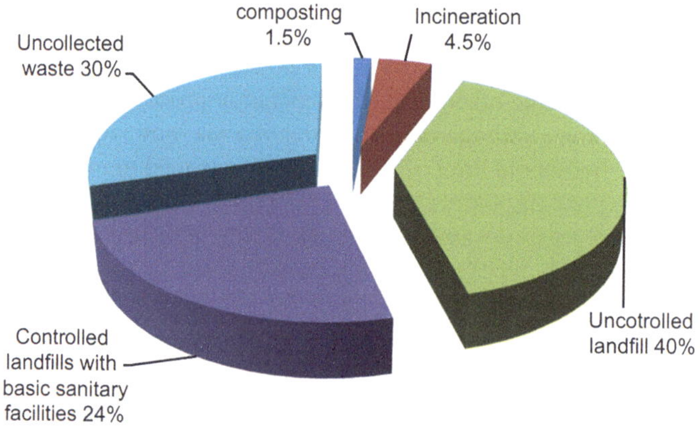


**Fig. 1** Different MSW disposal methods [31]

Accurately quantifying waste is one of the most important steps in management of municipal solid waste as it enables implementation of adaptable and suitable strategies and methods, which it is the responsibility of environmental companies or authorities to provide a detailed and precise reports about the amount of waste generated to the government or the specific ministry who is responsible for waste management. However, Tonjes and Greene's research paper indicates that despite the efforts of the United State Environmental Protection Agency (USEPA) still millions of wastes are scattered in different areas that remain unaccounted and not included in the submitted official reports. In fact, there are two separate assessments have been conducted 8to determine the quantity of MSW generation in the US, which is differ in their approaches in counting the waste, materials flow and site specification [27].

For quantifying the solid waste quantities and composition, the USEPA utilizes materials flow quantifications, which the engineers estimate the amounts and composition of solid waste by conducting waste sorts, primarily to support the incineration process and optimize combustion [12]. Additionally, the waste quantity is assessed by the USEPA using a method known as recycling, where manufacturers provide information on materials that have been recovered or reused. Another approach utilized by the USEPA is known as material flow, which involves tracking the entire process from the extraction of raw materials, their processing and reprocessing, manufacturing of finished products, and their distribution in the market. This approach estimates the length of time products stay in households, businesses, and institutions, as well as the volume of materials designated for recycling, disposal, or other forms of recovery. However, the material flow process does not allow for the identification of specific sources of waste within different types of waste generators. [27].

Moreover, the BioCycle assessment that relies on facility reports to gather the data, and according to it the waste generation is determined by summing up the materials that undergo through composting, recycling incinerating, and landfilling. However, the assessment does not account for waste reduction efforts because it consistently reports higher waste quantities compared to USEPA assessment [27]. The differences between theses assessments lies in the inclusion of tonnage disposal in the landfills and specific waste elements into the recycling data reports through the BioCycle assessment which incorporates these factors, including elements such as construction and demolition (C&D) disposal and automobile recycling [9]. According to Lohou and Alsabbagh research paper there are different methods to assess the MSW such as the Driving Forces Pressures State Impact Responses (DPSIR) framework, as shown in Fig. 2 where the researchers conducted a study to determine the suitability of using the DPSIR framework for assessing MSW in Kuwait. Furthermore, the DPSIR framework is an assessment method developed by the European Environment Agency that describes the relationship between environment and society. In addition, the researchers grouped the DPSIR farmwork indictors based on the three pillars: environment, society and economy, the date for the assessment was collected from various sources including Kuwait central statistical bureau, Kuwait municipality and the World Bank [17].

The findings of the study indicated a clear and direct relation between the growth of the economy and population and increase in MSW, as it shows in Fig. 2, which aligns

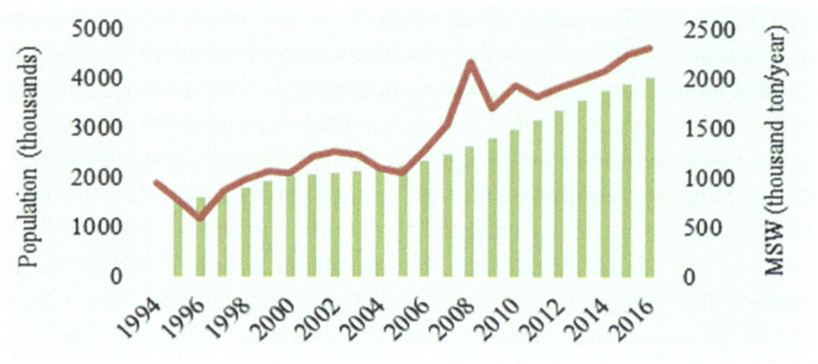


**Fig. 2** Population growth and municipal solid waste generation in Kuwait (1994–2016) [17]

with previous research in the field. The population of Kuwait has been consistently growing at an average rate of 5.4% per year, and the number of households has been increasing at an average rate of 7.4% per year. It is worth noting that over 98% of the population resides in urban areas [17].

However, the paper recommends the establishment of a comprehensive national waste database to store data and address the missing quantification indicators. This database would enable the monitoring and evaluation of the performance of various environmental companies involved in waste management. Furthermore, this approach could be implemented in other countries within the Gulf Cooperation Council (GCC) (Kingdom of Bharain, Kingdom of Suadi Arabia, Kuwait, Qatar, United Arab Emirates and Oman), aligning with the Integrated Municipal Solid Waste Management (IMSWM) guidelines set by the GCC countries in 2013 [17]. Figure 3 shows the generation of waste in the GCC countries in 2017.

In 2012, researcher Mohammed Alansari conducted a study on the solid was management (SWM) system in Kingdom of Bahrain, aiming to gather relevant information for planning to implementing an Integrated Municipal Solid Waste Management (IMSWM) system in the country. The study utilized different methods,

**Fig. 3** Waste Generations in the GCC

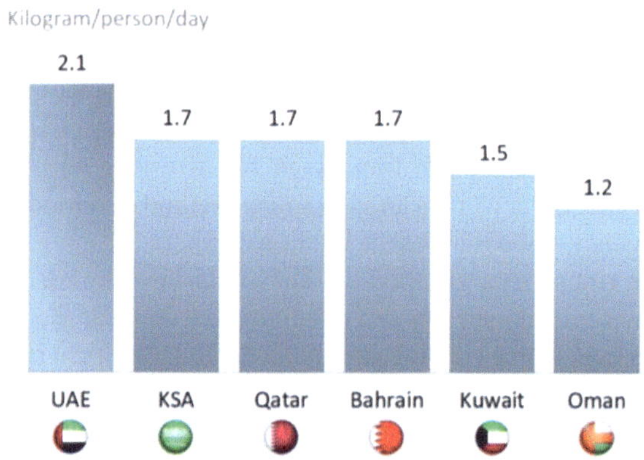

including an analysis of existing past literature on waste management in Bahrain, household questionnaires, interviews and observe the current practices of waste management. Furthermore, the research findings highlighted several insufficiencies with the SWM system in 2012, which obstructs the ability of the companies to fully comply with efficient SWM standards, such as legislation, funding, and lack of cooperation. As a result, the researcher suggested the need to consider different aspects of the system such as environmental, institutional, political, legal, socio-cultural, waste management hierarchy, financial, and economic aspects. These parameters were identified as crucial for establishing a comprehensive IMSWM plan and framework in Bahrain [1].

Overall, the municipal solid waste management system needs to address both current challenges and ensure long-term effectiveness due to ongoing rapid urbanization and population growth. These factors contribute to increased waste generation and complexities, necessitating continuous efforts to resolve existing waste management issues. As it is mentioned above there are private companies that are operating across the four Governorates of Bahrain, The Gulf City Cleaning Company that operate in Muharraq Governorate which is the second largest governorate with a population of over 200,000 residents and total area of 75 $km^2$, and in the Capital Governorate with a population of over 534,939 residents and total area of 90.25 $km^2$. The Urbaser Bahrain Company operates in the Northern Governorate, which spans a total area of 145.50 $km^2$ and has a population of over 350,000. Additionally, the company operates in the Southern Governorates, which is the largest governorate and has a population exceeding 298,244 [20]. Figure 4 shows the overall population rate in the four governorates of Bahrain in 2021, while Fig. 5 displays the total area in square kilometers ($km^2$) for each of the four governorates in 2020.

However, there are concerns regarding certain practices that are conducted by these companies. For example, waste treatment by using the controlled tipping method in single landfill. Therefore, this approach has led to different difficulties on the site such as inadequate equipment, absence of a well-developed plan, different gases emission, which this issue need be addressed to improve the waste

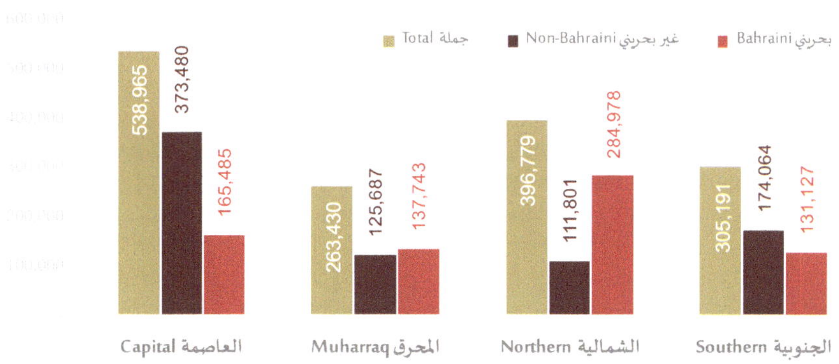


**Fig. 4** The population in the governorates in 2021 [14]

Fig. 5 The total area of the governorates (km$^2$) in 2020 [14]

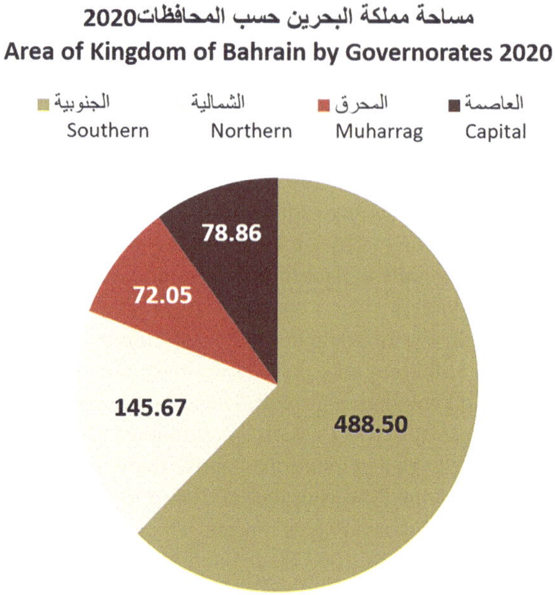


management system [1]. Considering these challenges, it is crucial to update MSW management system in the country by incorporating new strategies and methods and put in consideration these approaches should be sustainable and capable of serving both the environment and the society over long time. Moreover, the literature confirms that both population growth and economic development contributes to waste generating Supporting this, economic growth data in Bahrain, represented by the Gross Domestic Product (GDP) according to the information that are available in Information & eGovernment Authority [13], using the expenditure approach, indicates a significant relationship between economic growth and waste generation. This suggests that as the economy expands, the amount of waste produced also tends to increase, which is shown in Fig. 6.

In addition, the World Bank reports that the world generates over 2 billion tons of municipal solid waste (MSW) each year, and at least 33% of that waste is not adequately managed, leading to negative environmental impacts. On a global scale, only 16% of the world's population generates waste, with daily per capita waste production ranging from 0.11 to 4.5 kg. However, individuals with higher incomes contribute significantly more to the global waste stream, generating approximately 683 tons of waste. It is projected that the global volume of waste is expected to reach 3.4 billion tons by 2050. This represents a growth rate that is more than twice the rate of population increase over the same period [26]. Therefore, there is a clear connection between income levels and the volume of waste produced. Projections suggest that daily waste generation per person in high-income countries will rise by 19% by the year 2050. On the other hand, middle- and low-income countries are expected to experience a more significant increase of over 40% in waste generation.

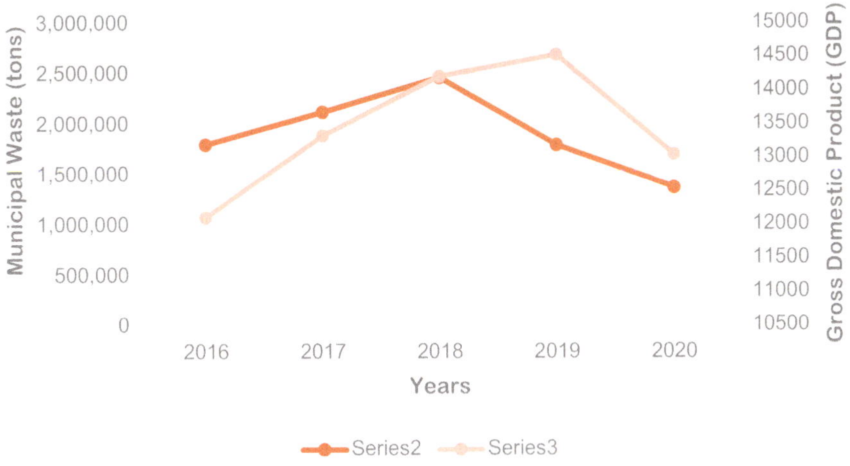


**Fig. 6** Relationship between municipal wastes and GDP (2016–2020)

This discrepancy can be attributed to the initial decrease in waste generation at the lowest income levels. However, as income rises incrementally in low-income countries, the rate of waste increase accelerates at a faster pace compared to high-income countries. Figure 7 provides a visual representation of the estimated quantities of waste generation, measured in millions of tons per year, for various regions in 2030 and 2050 [26].

Kingdom of Bahrain uses landfill as the main method to dispose the waste comparing to different the other methods such as recycling and composting, as shown in the Table 2.

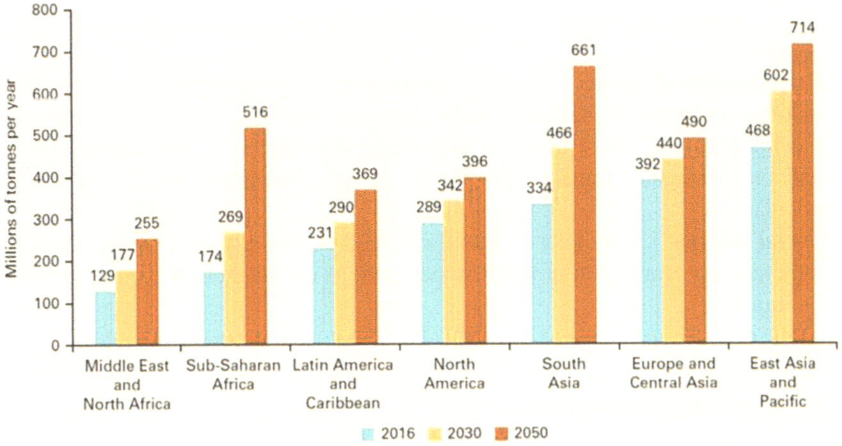


**Fig. 7** Waste generation (millions of tonnes/years) [26]

**Table 2** Disposal municipal waste methods (tonnes) [24].

طرق التخلص من النفايات البلدية (طن)

Ways of disposal Municipal waste (Tonnes)

| Ways of disposal | 2020 | 2019 | 2018 | 2017 | 2016 | 2015 | 2014 | 2013 |
|---|---|---|---|---|---|---|---|---|
| Recycling | 23,554 | 355,690 | 0 | 0 | 0 | 0 | 0 | 0 |
| Composting | 675 | 0 | 0 | 0 | 0 | 0 | 0 | 0 |
| Landfilling | 1,353,672 | 1,440,942 | 2,466,733 | 2,124,469 | 1,805,124 | 1,605,366 | 1,629,792 | 1,943,318 |
| **Total** | **1,377,901** | **1,796,633** | **2,466,733** | **2,124,469** | **1,805,124** | **1,605,366** | **1,629,792** | **1,943,318** |

Note: During 2020, the recycling and composting facilities were not operation full time during the year

Landfills are commonly considered an appropriate method for municipal solid waste (MSW) disposal as they help reduce waste accumulation and maintain cleanliness in communities. However, they also have significant social and environmental impacts that Bahrain may face in the future due to the continuous expansion of landfill areas and the ongoing reliance on landfills as the primary waste disposal method. The following are some of the social and environmental impacts associated with landfills:

## 2.1 *Social Impacts [29]*

- **Health Risks**: Landfills can release hazardous substances and gases, such as methane and volatile organic compounds, which can pose health risks to nearby residents. These risks include respiratory problems, skin irritations, and increased rates of certain illnesses.
- **Odors and Aesthetics**: Landfills can generate unpleasant odors, noise, and unsightly sights, which can significantly impact the quality of life for nearby residents. These factors can also deter visitors, affect tourism, and decrease property values.
- **Environmental Justice**: Landfills are often disproportionately located in low-income communities and marginalized areas, leading to concerns of environmental injustice. This can result in unequal distribution of the negative social and environmental impacts, placing a greater burden on already vulnerable populations.
- **Social Stigma**: Living near a landfill can lead to social stigma, where residents may face discrimination or negative perceptions due to their proximity to waste facilities.

## 2.2 *Environmental Impacts [29]*

- **Air Pollution**: Landfills produce significant amounts of methane, a potent greenhouse gas that contributes to climate change. They also emit other air pollutants, such as volatile organic compounds and particulate matter, which can harm air quality and affect human and ecosystem health.

- **Water Contamination**: Landfills can contaminate groundwater and surface water through the leachate, a liquid that forms as waste decomposes. Leachate contains various pollutants that can seep into water bodies, potentially affecting drinking water sources and aquatic ecosystems.
- **Soil Degradation**: Improperly managed landfills can lead to soil contamination and degradation. Toxic substances from waste can leach into the soil, affecting soil fertility and hindering plant growth.
- **Wildlife Disruption**: Landfills can disrupt local ecosystems and wildlife habitats. Animals may be attracted to landfills, leading to changes in their natural behaviors and potential negative impacts on biodiversity.
- **Land Use and Resource Depletion**: Landfills require vast amounts of land, which can result in the loss of valuable natural habitats and agricultural areas. Additionally, landfills contribute to the depletion of valuable resources, as waste that could be recycled or reused is buried.

Therefore, there is a lack of diversity in waste disposal methods, indicating a lack of financial support and cooperation. Bahrain could benefit from exploring and implementing alternative techniques and methods that are suitable for its specific conditions, are sustainable, and offer various benefits.

## *2.3 Non-hazardous Wastes Management*

Firstly, the most well-known method is recycling, which involves taking a product that has been processed and manufactured and reusing it as a new product, and the most common recyclable materials are iron scrap, steel, plastic or glass bottles, wood, paper, aluminum cans, etc. theses recycled materials can be used as substitutes for raw materials from natural resources such as a petroleum, natural gas, trees, coal, and mineral ores. There are two types of recycling operation: internal recycling and external recycling [3]. On one hand, internal recycling involves reusing waste materials from the manufacturing process which is beneficial for recycling metals. On the other hand, external recycling involves reclaiming the materials from worn-out and obsolete products. For example, magazines and newspapers can be recycled to create new paper products. Overall, recycling plays an important role in reducing different types of pollution (air, water, and land), and can produce a sustainable product that can be used in our daily life and routines [3].

Additionally, composition is another method that s utilized in Bahrain in 2020 as is shown in the provided data above, where 675 tons of MSW were disposed of using this method. Composition id the process of decomposition of organic waste using microorganisms such as fungi and bacteria ton converted into simpler form, these microorganisms utilize the carbon compounds present in the waste as an energy source, while the nitrogen content aids in breaking down the original materials of the waste. This process results in the creation of a homogeneous product that can be used as a soil amendment. Composting methods not only help to reduce waste but also generate valuable materials that can be sold in the market. Furthermore, composting enhances the organic content of the soil, leading to improved soil characteristics and a gradual release of nutrients over an extended period [11]. Moreover, Energy recovery methods offer an efficient means of utilizing non-hazardous waste by converting non-recyclable materials into valuable forms of heat, electricity, or fuel. This is accomplished through diverse processes like combustion, gasification, anaerobic digestion, and landfill gas recovery. By converting non-recyclable waste into usable energy sources, renewable energy is generated. This serves to decrease carbon emissions by reducing reliance on fossil fuel-based energy production while also mitigating methane emissions from landfills [28].

## 2.4 *Hazardous Wastes Management*

Hazardous wastes are materials that possess properties that can be harmful to human health and the environment. They can include chemicals, medical waste, electronic waste, industrial waste, radioactive materials, and asbestos. Improper management of hazardous wastes can lead to health risks and environmental contamination [22]. There are multiple options available for managing the hazardous wastes causes using the reduction and recycling alone are not considered as definitive solutions to dispose the hazardous wastes, which a proper disposal, storage and treatment are always necessary. Also, hazardous waste can undergo different forms of treatments such as thermal, chemical, biological, and physical treatments.

1. **Thermal Treatment**

   This method involves subjecting the waste to high temperatures to break them down or transform them into less hazardous forms. Common thermal treatment technologies include incineration, pyrolysis, gasification, and plasma arc technology. These methods can effectively reduce waste volume, destroy, or transform hazardous constituents, and recover energy. However, strict regulation and proper monitoring are necessary to control emissions and manage resulting residues to prevent environmental contamination.

2. **Chemical Treatment**

   This method involves using chemical reactions to transform or neutralize hazardous constituents in the waste. Chemical oxidation and reduction are two common chemical treatment processes. Oxidizing agents are used to break down hazardous substances, while reducing agents convert them into less toxic forms. These processes help to reduce the toxicity of hazardous wastes and make them safer for disposal or further treatment.

3. **Biological Treatment**

This method employs a natural processes and microorganisms to degrade or transform hazardous substances into less harmful forms. It utilizes biological agents such as bacteria, fungi, or enzymes to break down or metabolize the hazardous constituents. Biological treatment processes include bioremediation and composting. These methods can effectively reduce the toxicity of hazardous wastes and are often used in conjunction with other treatment methods for comprehensive waste management.

4. **Physical Treatment**

This method involves the process of solidification, where hazardous waste is encapsulated in materials such as concrete, asphalt or plastic. This results in the formation of a solid mass that is resistant to leaching. After treatment, some hazardous waste may not be destroyed, and the remaining waste can either be temporarily stored or disposed of in a landfill. Temporary storage typically takes place in facilities such as open waste piles or ponds. To prevent the generation of leachate, new waste piles should be carefully constructed on impermeable bases and regularly monitored. Additionally, measures should be taken to prevent wind dispersion. These new piles are only suitable for non-containerized solid waste that does not flow. When the size of the pile becomes unmanageable, the materials should be disposed of in a landfill [22].

Additionally, to store hazardous liquid waste, a lagoon, which is an open pit or holding pond, is commonly used. To protect the groundwater, it is crucial for newly constructed lagoons to incorporate flexible membrane liners and impermeable clay soils as lining materials. Moreover, it is necessary to install a leachate collection system between the groundwater monitoring wells and the liners. This precaution is of paramount importance because the absence of linings in waste piles and open lagoons located above aquifers utilized for public water supply poses significant risks to both public health and the environment [22].

In contrast, secure landfills adhere to more stringent regulations compared to municipal solid waste (MSW) landfills. Hazardous waste must be deposited in secure landfills that provide a minimum separation of 3 m between the bottom of the landfill and the underlying groundwater that are interconnected, necessitating the implementation of certain measures. These landfills require the use of double leachate collection systems and impermeable liners. The double leachate collection system consists of a network of perforated pipes positioned above each liner. The upper system prevents the buildup of leachate within the landfill, while the lower system acts as a backup. In instances where the upper system accumulates a significant amount of leachate, it is pumped to a treatment plant to minimize the volume of leachate within the landfill and mitigate potential environmental harm. Upon completion of the landfill, an impermeable cap is placed over the area to further ensure containment and prevent the release of hazardous substances [22].

In conclusion, the aim of this research paper is to enhance the MSW management system in Bahrain, to make it more sustainable and to improve the lifestyle of the society by preventing different type of pollutions. Therefore, policies and financial aspects of waste management in Bahrain aim to prioritize the health and safety of residents and non-residents. Waste management is considered a top priority due to its potential harm to human health and the environment. Allocating funds is crucial for developing a new waste management system. The policies should cover collection, transportation, disposal, segregation, and recycling, and should be effectively enforced across all sectors and private companies. The government should support private cleaning companies to improve their facilities and systems. The collection and separation system can involve the community working together with companies, with measures such as installing more classification garbage containers in residential areas, increasing the number of garbage trucks, and adjusting collection schedules.

Enhancing recycling is important for creating a sustainable environment and reducing landfill waste. This can be achieved by encouraging markets for recycled materials, establishing specialized recycling facilities, creating recycling centers in different cities, raising awareness about recycling, introducing recyclable items in supermarkets, and implementing a fee for single-use bags. Improving landfill disposal requires a comprehensive approach, including pre-treatment of municipal solid waste through biological and mechanical processes. Financial support from the government can be utilized to open waste treatment facilities, create job opportunities, and implementing performance-based systems to motivate employees and improve the waste management system.

## 3 Conclusion

Overall, this research paper focuses on the challenges of municipal solid waste management (MSWM) in the Kingdom of Bahrain and its impact on the environment and human health. The author highlights the significant increase in waste generation in recent years, including various types such as household, commercial, industrial, and construction waste. The primary factors contributing to this increase are population growth and economic factors. The study employed methods such as observation, interviews, and questionnaires to analyze the current state of MSWM in Bahrain. The findings of the study reveal a lack of advanced methods and technologies for waste treatment and disposal in the country. This raises concerns about the environmental and health consequences of improper waste management.

Furthermore, the study emphasizes the need for alternative approaches and strategies to develop a more sustainable MSWM system in Bahrain. The paper suggests that the government and relevant authorities should explore and implement advanced waste management techniques to mitigate the negative impacts on the environment and human health. Additionally, the paper highlights the importance of developing methods, guidelines, and strategies that are suitable for the specific conditions of Bahrain. Also, the paper also provides definitions of key terms related to waste management, such as sustainable waste management, solid waste, landfill, and municipal solid waste (MSW). These definitions help to establish a common understanding of the terminology used throughout the paper.

The research paper acknowledges the role of private companies authorized by the Ministry of Works, Municipalities Affairs, and Urban Planning in managing waste in different governorates of Bahrain. These companies play a crucial role in waste collection, transportation, and treatment. Moreover, the paper includes a comprehensive review of sustainable waste management techniques used in other developed countries and an exploration of their potential benefits in the context of Bahrain. The paper also provides recommendations for developing a proper waste management system in Bahrain. Overall, the paper highlights the need for a sustainable approach to waste management in the Kingdom of Bahrain. It emphasizes the importance of advanced methods and technologies, as well as the development of specific strategies tailored to the country's conditions. By implementing these measures, Bahrain can mitigate the environmental and health risks associated with municipal solid waste and ensure a better quality of life for its citizens.

In conclusion, the research paper emphasizes that municipal solid waste (MSW) management is a complex and long-term process. It underscores the importance of gaining a comprehensive understanding of the characteristics of MSW and the current state of its management. This understanding is crucial for the implementation of an effective MSW management system. To achieve this, the paper suggests that the government should collaborate with various sectors and provide support for the implementation of new methods and techniques in the MSW management system. By doing so, the aim is to enhance the sustainability of the system and ensure a healthy future for the next generation.

## *3.1 Study Implications*

The study on municipal solid waste management in the Kingdom of Bahrain has several implications:

1. **Environmental Implications**: The study highlights the detrimental impact of improper waste management on the environment. It emphasizes the need for advanced methods and technologies for waste treatment and disposal to minimize environmental pollution and promote long-term ecological balance.
2. **Health Implications**: The inadequate waste management practices discussed in the study raise concerns about the potential health risks to the population. Proper waste management is crucial for preventing the spread of diseases, minimizing exposure to hazardous substances, and ensuring public health and well-being.
3. **Population and Economic Factors**: The study identifies population growth and economic factors as the primary contributors to the high volume of waste. This finding emphasizes the importance of incorporating population dynamics and

economic considerations into waste management strategies to effectively address the waste generation challenge.

4. **Need for Sustainable Approaches**: The study emphasizes the need for alternative approaches and sustainable solutions to develop a more efficient and effective municipal solid waste management system in Bahrain. It highlights the importance of waste reduction, recycling, and proper disposal practices to minimize waste generation and promote resource preservation.
5. **Policy and Planning Implications**: The study's findings provide insights for policymakers and urban planners in Bahrain. It underscores the importance of developing tailored waste management methods, guidelines, and strategies that align with the country's specific circumstances, considering factors such as population growth, land availability, and economic development.
6. **Knowledge Contribution**: The study contributes to the existing literature on sustainable waste management techniques by providing a comprehensive review and exploring their potential benefits in the context of Bahrain. It adds to the knowledge base and serves as a resource for researchers, policymakers, and practitioners interested in waste management and environmental sustainability.

Overall, the study's implications highlight the urgency and importance of implementing sustainable waste management practices in the Kingdom of Bahrain to address the challenges posed by increasing waste generation, protect the environment, and promote a healthier and more sustainable future. However, based on the research it is possible to infer potential study limitations and areas for future research in sustainable waste management. Here are some possible considerations:

## *3.2 Study Limitations*

1. **Data Availability**: The limitations related to data availability, reliability, and representativeness could be a potential limitation of the study.
2. **Generalizability**: It focuses on the specific context of the Kingdom of Bahrain. The findings and conclusions drawn from this study may not be directly applicable to other regions or countries. The generalizability of the study's results could be limited.
3. **Time Constraints**: The study does not provide information about the duration of the study or any time constraints that may have affected the research process. Time limitations could impact the comprehensiveness of the study or the depth of analysis in certain areas.

## *3.3 Future Research*

1. **Technological Innovations**: The lack of advanced methods and technologies for waste treatment and disposal in the Kingdom of Bahrain. Future research could

focus on exploring and evaluating innovative technologies and approaches for sustainable waste management, such as waste-to-energy conversion, advanced recycling techniques, or decentralized waste management systems.
2. **Policy and Governance**: The paper briefly discusses the authorization of private companies to manage waste in different governorates of Bahrain. Future research could investigate the effectiveness of current waste management policies and governance structures, and propose policy recommendations to improve waste management practices, regulations, and enforcement mechanisms.
3. **Socio-economic Impact**: The paper highlights the economic factors contributing to the high volume of waste in Bahrain. Future research could examine the socio-economic impacts of sustainable waste management practices, such as job creation, resource recovery, and cost-effectiveness analysis of different waste management strategies.
4. **Stakeholder Engagement**: The paper does not provide details about stakeholder involvement in the research process. Future research could explore the engagement of various stakeholders, including government agencies, local communities, waste management companies, and environmental organizations, to develop collaborative and inclusive waste management strategies.

## References


1. Al Ansari, M.: Municipal solid waste management systems in the Kingdom of Bahrain. Int. J. Water Resour. Environ. Eng. **4**(5), 150–160 (2012)
2. Alfaia, R., Costa, A., Campos, J.: Municipal solid waste in Brazil: a review. Waste Manage. Res. J. Sustain. Circular Econ. **35**(12) (2017)
3. Augustyn, A.: Recycling. Encyclopedia Britannica (2024). Retrieved 1 May 2024, from https://www.britannica.com/science/recycling/Plastics
4. Behm, J.: The 4 main types of landfills. Dumpsters (2022). https://www.dumpsters.com/blog/landfill-types
5. Coelho, S., Sanches Pereira, A., Mani, S., Bouille, D., Stafford, W., Recalde, M., Savino, A.: Municipal Solid Waste Energy Conversion in Developing Countries, 1st edn., pp. 185–217. Elsevier (2020)
6. Curley, R.: Sanitary landfill. Encyclopedia Britannica (2022). Retrieved 9 April 2022, from https://www.britannica.com/technology/sanitary-landfill
7. Damghani, A., Savarypour, G., Zand, E., Deihimfard, R.: Municipal solid waste management in Tehran: current practices, opportunities and challenges. Waste Manage. **28**(5), 929–934 (2008)
8. DeVroom, D.: What is solid waste? Idrenvironmental (2023). https://blog.idrenvironmental.com/what-is-solid-waste
9. Goldstein, N.: Measuring diversion, recycling progress. Biocycle West Coast Conf. **48**(3), 52–54 (2007)
10. Gupta, N., Yadav, K., Kumar, V.: A review on current status of municipal solid waste management in India. J. Environ. Sci. **37**, 206–217 (2015)
11. Herbert, S., Hashemi, M., Chickering-Sears, C., Weis, S.: Waste Management and Composting [Ebook], 1st edn., pp. 1–3. University of Massachusetts Amherst (2023). Retrieved 1 May 2024, from https://ag.umass.edu/crops-dairy-livestock-equine/fact-sheets/waste-management-composting#:~:text=Composting%20is%20a%20process%20by,waste%20as%20an%20energy%20source

12. Hickman Jr., H.:. A Brief History of Solid Waste Management in the US During the Last 50 Years—Part 2 (2016). Retrieved 25 April 2024, from https://www.mswmanagement.com/landfills/article/13025970/a-brief-history-of-solid-waste-mangement-in-the-us-during-the-last-50-years-part-2
13. Information & eGovernment Authority: Area 2020 (2020). Retrieved 6 May 2022, from https://data.gov.bh/en/ResourceCenter
14. Information & eGovernment Authority: Population 2021 (2021). Retrieved 6 May 2022, from https://data.gov.bh/en/ResourceCenter
15. Kukreja, R.: What is Sanitary Landfill and Difference Between Sanitary Landfill and Open Dumping—Conserve Energy Future. Conserve Energy Future (2022). Retrieved 27 April 2024, from https://www.conserve-energy-future.com/what-is-sanitary-landfill.php
16. Kumar, S.: Municipal Solid Waste Management in Developing Countries. CRC Press (2016)
17. Lahou, A., Alsabbagh, M.: Assessment of municipal solid waste management in the State of Kuwait. Int. J. Environ. Sci. Dev. **10**(2), 51–56 (2019)
18. Lettieri, G.: What is sustainable waste management? Recycle Track Systems (2020). https://www.rts.com/blog/what-is-sustainable-waste-management/
19. Ludwig, C., Hellweg, S., Stucki, S.: Municipal Solid Waste Management: Strategies and Technologies for Sustainable Solutions, 1st edn., pp. 1–4. Springer (2003)
20. Mapper: Governorates of Bahrain (2022). Retrieved 30 April 2024, from https://www.mappr.co/counties/bahrain/
21. Melody, L.: What is sustainable waste management? | Clear it out. Clear It Out (2023). https://clearitout.co.uk/what-is-sustainable-waste-management/
22. Nathanson, J.: Hazardous-waste management—treatment, storage, and disposal. Encyclopedia Britannica (2022). Retrieved 1 May 2024, from https://www.britannica.com/technology/hazardous-waste-management/Treatment-storage-and-disposal
23. Ogwueleka, T.: Municipal solid waste characteristics and management in Nigeria. Environ. Health Sci. Eng. **6**(3), 173–180 (2009)
24. Ministry of Works, Municipalities Affairs and Urban Planning: Waste Statistics. Information & eGovernment Authority (2020). Bahrain. https://data.gov.bh/en/ResourceCenter
25. The Supreme Council for Environment: Bahrain National Profile to Assess the National Infrastructure for Chemical Safety [Ebook], pp. 40–41. The Supreme Council for Environment, Manama, Kingdom of Bahrain (2012). Retrieved from https://cwm.unitar.org/national-profiles/publications/cw/np/np_pdf/Bahrain_National_Profile_update.pdf
26. The World Bank: Trends in Solid Waste Management (2022). Retrieved 30 April 2024, from https://datatopics.worldbank.org/what-a-waste/trends_in_solid_waste_management.html
27. Tonjes, D., Greene, K.: A review of national municipal solid waste generation assessments in the USA. Waste Manage. Res. J. Sustain. Circular Econ. **30**(8), 758–771 (2012)
28. US Environmental Protection Agency: Sustainable Materials Management: Non-Hazardous Materials and Waste Management Hierarchy | US EPA. US EPA (2021). Retrieved 1 May 2024, from https://www.epa.gov/smm/sustainable-materials-management-non-hazardous-materials-and-waste-management-hierarchy.
29. Vasarhelyi, K. (2021). *The Hidden Damage of Landfills*. Environmental Center. Retrieved 30 April 2024, from https://www.colorado.edu/ecenter/2021/04/15/hidden-damage-landfills.
30. Zafar, S.: Solid Waste Management in Bahrain | EcoMENA. EcoMENA (2020). Retrieved 8 November 2021, from https://www.ecomena.org/solid-waste-bahrain/
31. Zhang, D., Tan, S., Gersberg, R.: Municipal solid waste management in China: status, problems and challenges. J. Environ. Manage. **91**(8), 1623–1633 (2010)

# Currency Dynamics in Determining the Prices of Precious Metals in the Domestic and International Markets: A 52-Year Analysis

**B. G. Shobha, K. U. Jayalakshmi, B. Vidhya, and M. Manisha**

**Abstract** This study investigates the correlation between the prices of gold and silver in local and international markets over the course of the last 52 years, spanning from 1971 to 2023, and currency exchange rates. The RBI, Bloomberg, Bombay Bullion Association, and Press Trust of India were the sources of the information. The ways in which variations in exchange rates impact the pricing of various precious metals are examined through the application of descriptive analysis, regression, and correlation approaches. The analysis covers the following major currencies: SDR, US Dollar, Japanese Yen, Pound Sterling, and Deutsche Mark/Euro. The findings show that there are notable positive connections between the exchange rates and the prices of gold and silver in the Mumbai and London markets. The findings of the research emphasis the need for policies that reduce the effects of fluctuations in exchange rates on the gold and silver market for decision-makers.

**Keywords** Gold and silver market · Currency exchange rates · Domestic and foreign markets · Precious metals

B. G. Shobha (✉)
Kristu Jayanti College, Autonomous, Bengaluru, India
e-mail: shobhbg@gmail.com

K. U. Jayalakshmi
CMR University, Bengaluru, India

B. Vidhya
St. Joseph's University, Bengaluru, India

M. Manisha
New Horizon College, Bengaluru, India



M. Al Mubarak (ed.), *Sustainable Digital Technology and Ethics in an Ever-Changing Environment*, Studies in Systems, Decision and Control 237,
https://doi.org/10.1007/978-3-031-86708-8_29

## 1 Introduction

Stock market is always influenced by the world wide uncertainties but the precious metals have proved as safe heaven with the stable and bullion trend in the prices [21]. High volatility in the stock market have made the investors to think of alternative asset class in their portfolio. There are many other asset class but the metals like gold, silver, platinum and palladium have entered as a most popular alternative asset classes among the precious metal class.

Even in the event that the gold standard is abandoned, gold remains a valuable commodity that can be utilized as a hedge against inflation and as a replacement for fiat money. Raj et al. [19], studied on the pair of investments on precious metals and concludes that gold-platinum and silver-platinum are the more profitable pairs for investments. Platinum is recommended if it is a choice for single metal.

Gold is one of the traditional avenue of investment and preserved for the emergency purpose keeping in mind for its capital appreciation. In modern portfolio also many of the investor may like to add it as an asset class in their portfolio. Gold and silver are one of the noble of all metals [10]. Another metal that is utilized by homes and businesses is silver. This metal's industrial use greatly increases its investment potential if we choose to view it as an investment option [17]. The gold industry has witnessed shift towards e commerce platforms connecting manufacturers and retailers and providing a bigger market opportunity [5]. Earlier researches also demonstrates that having silver investment coins on hand can provide middle-class residents' material demands and financial stability during times of economic collapse [18].

There are many different economic and geopolitical components that influence the price of gold and silver between domestic and foreign markets. Several studies have been conducted and proved the presence of the causality connection from geopolitical risk and economic policy uncertainty to the precious metal prices. Yilanci and Kilci [25]. These internationally traded precious metals are frequently seen as safe-haven investments because of their extreme sensitivity to changes in exchange rates and unstable economic conditions. Particularly in times of political and economic unrest, gold is becoming more and more recognized as one of the most important precious metals in the commodities and financial markets [9]. Even in the situation of stock market fluctuations gold can prove safe investment [23].

The price of gold and silver in local currency tends rise when the home currency declines against the US dollar, making these metals more expensive for consumers in that nation. On the other hand, gold and silver become more inexpensive in local currency when the value of the home currency rises. This dynamic may have a significant impact on local market demand for certain precious metals as well as investment behaviors. So these exchange rates can help us in predicting the prices of precious metals [6].

## 2 Rationale of the Paper

The study is an effort to put forward that precious metal prices, such as those of gold and silver, are significantly influenced by currency exchange rates due to the interconnectedness of the world's financial markets. Because of their popularity as safe-haven investments, precious metals are subject to fluctuations in exchange values, particularly when big currencies are involved.

This study is significant because, although previous research has examined the connection between exchange rates and precious metal prices, there has been little investigation into how these metals react in local vs international markets.

## 3 Review of Literature

Hammer and Norskov [10], Investigations Why gold is an eminent metal and how its intrinsic value enables it to produce stable alloys with other metals is explained. Gold is the least reactive metal at gas or liquid interfaces. Nobleness is linked to orbital overlap and antibonding states.

Tweneboah and Alagidede [24], determined that different metals have the capacity to lead at different scales by analysing the interconnectivity of the metal markets. According to the study, gold has the capacity to outperform all other scales.

Hammoudeh et al. [11], conducted study on instability and association dependency of four major precious metals, gold, silver, platinum, and palladium. It highlights significant short- and long-term dependencies on news, past volatility, exchange rate, and monetary policy.

Sari et al. [20], explored the relationships between the USD-INR exchange rate, the Indian stock market, and the prices of gold and crude oil. It contends that the primary cause of the decline in the Indian Rupee and the Sensex is the falling prices of gold and crude oil. The results highlight the need for dynamic policy making in India and support gold as an investment asset class.

Hammoudeh et al. [11], using both the more restricted VARMA-DCC model and the more general VARMA-GARCH model, analysed both own and spillover fluctuations and relationships for gold, silver, platinum, and palladium as well as with currency exchange rates in simultaneous multivariate scenarios. The appropriate weights for a two-asset portfolio and the hedging ratios are determined using the output of these models.

Abreu et al. [1], explains that investors who trade more often on the domestic market start making foreign investments later. Study also states that one of the primary causes that active investors go into foreign markets is the experience they gain by trading in the home market.

Jain and Ghosh [13]., analyzed the dynamics of global oil prices, exchange rate and precious metal prices comparing with prices in India and states that there is a robust evidence of a long term relationship between oil prices, precious metal prices and the exchange rate.

Krol [14], conducted the influence of economic policy uncertainty on exchange rate volatility. The study suggest that home and US economic policy uncertainty has positively increased the exchange rate volatility.

Jain and Biswal [12], studied the relationship between the price of gold and crude oil, the USD-INR exchange rate. It concludes that declining prices for crude oil and gold is the reason for the Indian Rupee and the Sensex to decline. The findings highlight the necessity for dynamic policy making in India and promote gold as an asset class for investments.

Churchill et al. [8], examined the Dynamic forces of oil price, precious metal prices and the exchange rate in the long-run. The outcome of the study proves that precious metal and diamond are not safe asset class to invest.

Lakshmanan et al. [16] mentioned in their study that globally, gold mining is a business, with China producing the greatest quantity in 2018. However, as a result of government initiatives to reduce pollution and raise environmental consciousness, production dropped by 6% in 2018. 2019 is expected to see a recovery in production.

Ali et al. [2], analysed the association between Pakistan's stock market volatility and gold prices and the exchange rate. Using data from 2001 to 2018, the study discovers that exchange rate and gold price volatility have a negative impact on stock market performance, implying that these swings are viewed as an adverse indicator and react unfavourably.

Asad et al. [3], study examines the association between gold, oil, exchange rates, and Bombay stock indexes from 2003 to 2020 using a nonlinear ARDL technique. Results show a discrepancy between stock indexes before recession and gold, oil, and currency prices.

Bedoui et al. [4], examines in their article about the hedging and possibilities of diamonds along with five precious metals gold, silver, palladium, platinum, and rhodium by using copula technique. The outcome of the study states that gold, silver, palladium, and platinum are stronger metals to hedge against USD devaluation than diamonds, which only provide a weak safe haven.

Chiang [7], conducted a study on gold as a asset class to hedge against the policy uncertainty that higher levels of uncertainty have a favorable impact on gold returns. As a result, gold is a suitable asset to protect against risk associated with new governmental initiatives and infectious illness concerns.

Sokhanvar and Bouri [22], studied the influences of commodity shock prices on exchange rates for the period February 1 to April 30, 2022, the study was based on the quantile autoregressive distributed lag (ARDL). The outcome of the study states that there is a long-run relationship between commodity prices and increase in the value of the Canadian dollar against the euro and the yen.

Kumar et al. [15] held a study on the causative association between international prices of crude oil, gold, exchange rate, and Indian stock market on the daily prices rom "January 1994 to December 2019". The outcome of the study states that crude oil prices have positively influenced the Indian stock market. Exchange rate have negatively influenced the stock market in the short run.

# 4 Statement of the Problem

The valuation of valuable metals like gold, silver, and platinum is often influenced from a wide range of economic factors, one of which is currency exchange rates. The price and demand for these metals can be significantly impacted by the fluctuations in exchange rates between domestic and international markets. However, there are still numerous unknowns surrounding the exact nature and extent of this influence, which presents a challenge to investors, politicians, and economists alike. So the present study is intended to find the answer on how exchange rates between domestic and foreign markets affect the price and trading patterns of these precious metals.

## *4.1 Need of the Study*

The association of the price of precious metals and currency exchange rates is of great interest to a wide range of stakeholders, including economists, politicians, and investors. Understanding this connection is essential for making wise decisions about investments, policy making, understanding market stability and inflation hedging.

## *4.2 Research Gap*

Extensive research has been conducted on the correlation between the price of precious metals and other economic conditions. However, very less is known about the precise influence of the fluctuations in currency exchange rates between domestic and foreign markets on these metals. So, the study aims to understand the dynamics of exchange rates and gold prices and aims to bridge these knowledge gaps by offering a detailed understanding of the responses of precious metals to changes in exchange rates for the past 52 years.

## *4.3 Objective of the Study*

- To investigate the relationship over the last 52 years between changes in currency exchange rates and the prices of gold and silver domestic and foreign markets.
- To predict the impact of future currency exchange rate movements on gold and silver prices.
- To evaluate the implications of the findings for economic policies, providing recommendations for policymakers to mitigate the adverse effects of exchange rate volatility on the precious metals market.

### 4.4 Hypotheses

H01: There is no significant relationship between Gold prices with exchange rates.
H02: There is no significant relationship between Silver prices with exchange rates.
H03: There is no significant impact of exchange rates on Gold prices at Mumbai.
H04: There is no significant impact of exchange rates on Silver prices at Mumbai.

## 5 Research Methodology

The methodology is designed to achieve the research objectives through systematic data collection, analysis, and interpretation. Research method adopted in this study is descriptive and analytical one.

### 5.1 Tools Used in This Study

For measuring the central tendencies in gold, silver prices and currency exchange rates, descriptive statistics are used. To evaluate the data's volatility and dispersion, Standard Deviation and Variance are used. Correlation Analysis is used to measure the linear relationship between precious metals and exchange rates. The association between the price of gold and silver and the exchange rate of currencies is investigated using Simple Linear Regression.

### 5.2 Sample Selection and Sources of Data

Past 52 years' data has been taken from the 1971 to 2023 with regard to gold silver and exchange rates of SDR, US Dollar, Pound Sterling, Japanese Yen and Deutsche Mark/Euro. Data has been gathered from Reserve Bank of India, Bloomberg.

### 5.3 Period of Study

Period of the study is from 1971 to 2023.

## 6 Results and Discussion

Descriptive Statistics (Table 1) provides the details of the gold and silver prices in Mumbai and London. Compared to London, Mumbai has greater average gold prices (10 g) and lower median pricing (10,025.79). The lowest mean values are found in Mumbai, indicating that prices are often lower. With Mumbai having higher median values and New York having lower median values, Mumbai has higher average silver prices (16,787.20) than New York (1934.20). Mumbai has the highest maximum readings for both metals.

Table 2 provides the statistical summary of the currencies exchange rates. The SDR, US Dollar, British Pound, Japanese Yen, and Deutsche Mark/Euro have average prices of 49.34, 35.24, 54.76, and 36.00 The medians, which are the middle values, indicate higher central tendencies for SDR (51.44) and Pound Sterling (59.53). Modes represent the most frequent values, with the smallest being SDR (8.16), US Dollar (7.52), Pound Sterling (18.80), Japanese Yen (3.00), and Deutsche Mark/Euro (2.15). Standard deviations, which measure the spread of the data, are relatively high for all currencies, particularly SDR (33.58) and Pound Sterling (31.36), indicating substantial variability.

**H01: There is no significant relationship between Gold prices with exchange rates.**

Table 3 is depicted with co-relation values. Strong positive correlations have been found between the 10-g gold values in Mumbai and London and the SDR, US Dollar, Pound Sterling, Japanese Yen, and Deutsche Mark/Euro. These correlations indicate that variation in these exchange rates are correlated with variations in the gold rates.

The gold rates in Mumbai and London display a nearly perfect correlation (0.999), showing a nearly identical movement of these rates throughout major exchange rates. This shows that the global gold market is defined by strongly aligned pricing.

**Table 1** Descriptive statistics for gold and silver

| | | 10 g gold rate Mumbai | 10 g gold rate London | Per kg silver rate Mumbai | Per kg silver rate New York |
|---|---|---|---|---|---|
| N | Valid | 53 | 53 | 53 | 53 |
| | Missing | 0 | 0 | 0 | 0 |
| Mean | | 11,093.3008 | 10,025.7889 | 16,787.1962 | 14,934.1979 |
| Median | | 4473.6000 | 3896.6600 | 7332.4100 | 5809.3600 |
| Mode | | 184.96 | 89.44 | 536.08 | 363.51 |
| Std. deviation | | 14,080.43982 | 12,870.37343 | 19,388.47609 | 17,835.82707 |
| Minimum | | 184.96 | 89.44 | 536.08 | 363.51 |
| Maximum | | 52,730.77 | 46,605.74 | 65,425.65 | 58,847.68 |

*Sources* Bombay Bullion Association, Press Trust of India and RBI

**Table 2** Descriptive statistics for exchange rates

| | | SDR | US Dollar | Pound Sterling | Japanese Yen | Deutsche Mark/Euro |
|---|---|---|---|---|---|---|
| N | Valid | 53 | 53 | 53 | 53 | 53 |
| | Missing | 0 | 0 | 0 | 0 | 0 |
| Mean | | 49.339508 | 35.238951 | 54.759111 | 30.771275 | 35.995206 |
| Median | | 51.435000 | 36.319600 | 59.528900 | 32.597500 | 23.507400 |
| Mode | | 8.1601 | 7.5244 | 18.8000 | 3.0000 | 2.1453 |
| Std. Deviation | | 33.5788779 | 23.0848536 | 31.3616920 | 23.1185453 | 30.8119041 |
| Minimum | | 8.1601 | 7.5244 | 15.1469 | 2.4819 | 2.1453 |
| Maximum | | 105.3257 | 78.6377 | 101.6904 | 69.4479 | 87.4760 |

**Table 3** Correlations between gold prices with exchange rates

| | | 10 g gold rate Mumbai | 10 g gold rate London | SDR | US Dollar | Pound Sterling | Japanese Yen | Deutsche Mark/Euro |
|---|---|---|---|---|---|---|---|---|
| 10 g gold rate Mumbai | Pearson correlation | 1 | 0.999** | 0.853** | 0.864** | 0.772** | 0.823** | 0.851** |
| | Sig. (2-tailed) | | 0.000 | 0.000 | 0.000 | 0.000 | 0.000 | 0.000 |
| | N | 53 | 53 | 53 | 53 | 53 | 53 | 53 |
| 10 g gold rate London | Pearson correlation | 0.999** | 1 | 0.863** | 0.871** | 0.785** | 0.837** | 0.866** |
| | Sig. (2-tailed) | 0.000 | | 0.000 | 0.000 | 0.000 | 0.000 | 0.000 |
| | N | 53 | 53 | 53 | 53 | 53 | 53 | 53 |

** Correlation is significant at the 0.05 level (2-tailed)

Since all correlations are significant at the 2-tailed 0.05 level, the likelihood that they are the result of random chance is very low. Therefore, the null hypothesis is rejected, and the findings indicate that there is a strong association between the values of the gold rates in Mumbai and London with the major currencies considered under this study.

**H02: There is no significant relationship between Silver prices with exchange rates.**

Table 4 depicts the Pearson correlation coefficients between silver prices in Mumbai and New York, and various exchange rates. The table displays the intensity and direction of the correlations between the prices of silver (per kilogram) in Mumbai and New York, as well as the various exchange rates. Mumbai and New York have very high (0.998) and statistically significant correlations between their silver prices, indicating that these two markets move very similarly. Furthermore, there are significant positive correlations between the price of silver in Mumbai and the SDR (0.860), US Dollar (0.855), Pound Sterling (0.795), Japanese Yen (0.853), Deutsche Mark/Euro (0.873), and other currencies. All of these relationships are statistically significant at the 0.01 level, suggesting that fluctuations in exchange rates and variations in Mumbai's silver prices are closely related. Silver prices in New York exhibit robust relationships with the same exchange.

**H03: There is no significant impact of exchange rates on Gold prices at Mumbai.**

Table 5 shows the result of regression analysis. R value of 0.956 indicates a very strong positive correlation between the observed and predicted values of the dependent variable. An Adjusted R Square of 0.904 indicates that 90.4% of the variance in the 10-g gold rate in Mumbai is accounted for by the model, adjusting for the number of predictor.

Table 6 shows that the model is highly significant (p-value $< 0.05$). The F-statistic of 98.958 indicates that the model explains a significant portion of the variance in the 10-g gold rate in Mumbai. The predictors (Deutsche Mark/Euro, US Dollar, Pound

**Table 4** Correlations between Silver prices with exchange rates

| | | Per kg silver rate Mumbai | Per kg silver rate New York | SDR | US Dollar | Pound Sterling | Japanese Yen | Deutsche Mark/Euro |
|---|---|---|---|---|---|---|---|---|
| Per kg silver rate Mumbai | Pearson correlation | 1 | 0.998 | 0.860 | 0.855 | 0.795 | 0.853 | 0.873 |
| | Sig. (2-tailed) | | 0.000 | 0.000 | 0.000 | 0.000 | 0.000 | 0.000 |
| | N | 53 | 53 | 53 | 53 | 53 | 53 | 53 |
| Per kg silver rate New York | Pearson correlation | 0.998 | 1 | 0.856 | 0.849 | 0.793 | 0.851 | 0.876 |
| | Sig. (2-tailed) | 0.000 | | 0.000 | 0.000 | 0.000 | 0.000 | 0.000 |
| | N | 53 | 53 | 53 | 53 | 53 | 53 | 53 |

**Table 5** Model summary

| Model | R | R square | Adjusted R square | Std. error of the estimate | Durbin-Watson |
|---|---|---|---|---|---|
| 1 | 0.956[a] | 0.913 | 0.904 | 4362.16582 | 0.990 |

[a] Predictors: (Constant), Deutsche Mark/Euro, US Dollar, Pound Sterling, Japanese Yen, SDR
[b] Dependent Variable: 10 g gold rate Mumbai

Sterling, Japanese Yen, SDR) together provide a strong explanation for the variations in the gold rate.

Table 7 shows the Coefficients value. Among the predictors, SDR has a significant positive impact on the gold rate (B = 1935.988, t = 4.351, $p < 0.005$), suggesting that for every one-unit increase in SDR, the gold rate increases by 1935.988 units. The US Dollar shows a negative but non-significant relationship (B = – 383.199, t = – 1.081, p = 0.285), implying no significant effect on the gold rate. Conversely, the Pound Sterling (B = – 1076.461, t = – 8.696, $p < 0.005$) and Japanese Yen (B = – 855.579, t = – 4.254, $p < 0.005$) both have significant negative impacts, indicating decreases in the gold rate with their increases. The Deutsche Mark/Euro (B = 282.254, t = 2.679, p = 0.010) positively affects the gold rate significantly. The standardized coefficients indicate the relative importance of each predictor, with SDR having the strongest positive effect (Beta = 4.617), followed by the Deutsche Mark/Euro (Beta = 0.618), while the Pound Sterling (Beta = – 2.398) and Japanese Yen (Beta = – 1.405) have strong negative impacts. The US Dollar has the weakest influence (Beta = – 0.628). Overall, the model demonstrates the varying influences of different currencies on the gold rate in Mumbai.

**H04: There is no significant impact of exchange rates on Silver prices at Mumbai.**

Table 8 model summary indicates a robust positive correlation (R = 0.928) and a high coefficient of determination (R Square = 0.860), suggesting that 86% of the variance

**Table 6** ANOVA

| Model | | Sum of squares | Df | Mean square | F | Sig |
|---|---|---|---|---|---|---|
| 1 | Regression | 9,415,117,788.704 | 5 | 1,883,023,557.741 | 98.958 | 0.000[b] |
| | Residual | 894,339,061.466 | 47 | 19,028,490.669 | | |
| | Total | 10,309,456,850.171 | 52 | | | |

[a] Dependent Variable: 10 g gold rate Mumbai

[b] Predictors: (Constant), Deutsche Mark/Euro, US Dollar, Pound Sterling, Japanese Yen, SDR

**Table 7** Coefficients

| Model | | Unstandardized coefficients | | Standardized coefficients | t | Sig |
|---|---|---|---|---|---|---|
| | | B | Std. error | Beta | | |
| 1 | (Constant) | 4189.661 | 1800.495 | | 2.327 | 0.024 |
| | SDR | 1935.988 | 444.955 | 4.617 | 4.351 | 0.000 |
| | US Dollar | – 383.199 | 354.535 | – 0.628 | – 1.081 | 0.285 |
| | Pound Sterling | – 1076.461 | 123.790 | – 2.398 | – 8.696 | 0.000 |
| | Japanese Yen | – 855.579 | 201.142 | – 1.405 | – 4.254 | 0.000 |
| | Deutsche Mark/Euro | 282.254 | 105.352 | 0.618 | 2.679 | 0.010 |

in silver rates is explained by these predictors. The adjusted R Square of 0.846, which accounts for the number of predictors, still indicates a good fit. The Durbin-Watson statistic (0.784) suggests the presence of some positive autocorrelation in the residuals.

The ANOVA Table 9 shows that the regression model is statistically significant ($F(5, 47) = 57.921$, $p < 0.005$), suggesting that the predictors together accurately predict the silver rate in Mumbai. It is further evidence of the model's explanatory power that the regression sum of squares (16,818,083,728.531) is much bigger than the residual sum of squares (2,729,392,534.275).

The coefficients Table 10, provides detailed insights into the contribution and significance of each predictor in the multiple regression model predicting the per kilogram silver rate in Mumbai. The constant term (7556.098) represents the expected silver rate when all predictors are zero. Among the predictors, the SDR has a significant positive impact ($B = 2292.994$, $p = 0.005$), indicating that a unit increase in the SDR is associated with an increase of 2292.994 units in the silver rate. The Deutsche Mark/Euro also shows a significant positive effect ($B = 461.876$, $p = 0.016$). Conversely, the Pound Sterling has a significant negative impact ($B = -1212.899$, $p < 0.001$), suggesting that an increase in the Pound Sterling is associated with a decrease in the silver rate. The US Dollar and Japanese Yen, although not statistically significant at the 0.05 level, show negative coefficients, implying a negative relationship with the silver rate. The collinearity statistics reveal potential multicollinearity issues, especially with SDR (VIF = 610.048), US Dollar (VIF = 183.051), and Pound Sterling (VIF = 41.188), indicating that these variables are highly correlated with other predictors in the model. This multicollinearity can affect the reliability of the coefficient estimates.

**Table 8** Model summary

| Model | R | R square | Adjusted R square | Std. error of the estimate | Durbin-Watson |
|---|---|---|---|---|---|
| 1 | 0.928[a] | 0.860 | 0.846 | 7620.51059 | 0.784 |

[a] Predictors: (Constant), Deutsche Mark/Euro, US Dollar, Pound Sterling, Japanese Yen, SDR
[b] Dependent Variable: Per kg silver rate Mumbai

**Table 9** ANOVA

| Model | | Sum of squares | df | Mean square | F | Sig |
|---|---|---|---|---|---|---|
| 1 | Regression | 16,818,083,728.531 | 5 | 3,363,616,745.706 | 57.921 | 0.000[b] |
| | Residual | 2,729,392,534.275 | 47 | 58,072,181.580 | | |
| | Total | 19,547,476,262.806 | 52 | | | |

[a] Dependent Variable: Per kg silver rate Mumbai
[b] Predictors: (Constant), Deutsche Mark/Euro, US Dollar, Pound Sterling, Japanese Yen, SDR

**Table 10** Coefficients

| Model | | Unstandardized coefficients | | Standardized coefficients | T | Sig | Collinearity statistics | |
|---|---|---|---|---|---|---|---|---|
| | | B | Std. error | Beta | | | Tolerance | VIF |
| 1 | (Constant) | 7556.098 | 3145.385 | | 2.402 | 0.020 | | |
| | SDR | 2292.994 | 777.317 | 3.971 | 2.950 | 0.005 | 0.002 | 610.048 |
| | US Dollar | − 1034.701 | 619.357 | − 1.232 | − 1.671 | 0.101 | 0.005 | 183.051 |
| | Pound Sterling | − 1212.899 | 216.255 | − 1.962 | − 5.609 | 0.000 | 0.024 | 41.188 |
| | Japanese Yen | − 573.599 | 351.386 | − 0.684 | − 1.632 | 0.109 | 0.017 | 59.092 |
| | Deutsche Mark/ Euro | 461.876 | 184.046 | 0.734 | 2.510 | 0.016 | 0.035 | 28.795 |

[a] Dependent Variable: Per kg silver rate Mumbai

# 7 Conclusion

The study suggests that there is a strong correlation between domestic and international gold and silver prices and exchange rates. The findings show that the prices of these metals are highly impacted by changes in key currencies, such the SDR, the pound sterling, and the Deutsche Mark/Euro. In particular, the worldwide exchange rates have a significant association with Mumbai's gold and silver markets, indicating their integration into the global precious metals market. The study's regression analysis highlights the diverse consequences of different currencies, demonstrating that the Japanese Yen and Pound Sterling have negative effects whereas the SDR and Deutsche Mark/Euro have favourable effects. These results show that currency dynamics play a considerable influence in determining the price of precious metals.

## *7.1 Theoretical Implications*

By indicating the significant correlations and impacts of specific currencies on gold and silver prices, the study provides empirical evidence supporting the theory of market integration between domestic and international markets. The findings also highlight the significance of taking currency fluctuations into account in the valuation of commodities, indicating that exchange rates are key determinants in the pricing mechanisms of precious metals. These findings add to the body of literature on the relationship between currency exchange rates and commodity prices, particularly in the context of precious metals.

### 7.2 Practical Implications

The findings of the research emphasis the need for policies that reduce the effects of fluctuations in exchange rates on the gold and silver market for decision-makers. Policymakers need to think about enacting procedures that stabilize currency exchange rates or give market participants hedging capabilities in light of the noteworthy correlations that have been found. As part of their precious metals investing strategy, the study advises investors and financial institutions to keep a close eye on currency swings. This is especially important in markets like Mumbai where prices are strongly correlated with worldwide trends.

### 7.3 Limitations and Scope for Future Research

The study looks merely at a small number of currencies and ignores additional macroeconomic factors like interest rates, inflation rates, and geopolitical developments that could have an impact on the price of precious metals. By incorporating a wider range of currencies, examining the effects of additional economic variables, and making use of more recent data, future study could broaden the reach.

## References


1. Abreu, M., Mendes, V., Santos, J.A.: Home country bias: does domestic experience help investors enter foreign markets? J. Bank. Finance **35**(9), 2330–2340 (2011)
2. Ali, R., Mangla, I.U., Rehman, R.U., Xue, W., Naseem, M.A., Ahmad, M.I.: Exchange rate, gold price, and stock market nexus: a quantile regression approach. Risks **8**(3), 86 (2020)
3. Asad, M., Tabash, M.I., Sheikh, U.A., Al-Muhanadi, M.M., Ahmad, Z.: Gold-oil-exchange rate volatility, Bombay stock exchange and global financial contagion 2008: application of NARDL model with dynamic multipliers for evidences beyond symmetry. Cogent Bus. Manage. **7**(1), 1849889 (2020)
4. Bedoui, R., Guesmi, K., Kalai, S., Porcher, T.: Diamonds versus precious metals: what gleams most against USD exchange rates? Financ. Res. Lett. **34**, 101253 (2020)
5. Bhagwanrao, N.N., Shobha, B.G., Bhovi, B.: Usage of Chat GPT and AI in Jewelry buying decision—a study on customers' perception in Bangalore City. In: Anticipating Future Business Trends: Navigating Artificial Intelligence Innovations: Volume 1, pp. 1–9. Springer Nature Switzerland, Cham (2024)
6. Brown, P.P., Hardy, N.: Forecasting base metal prices with the Chilean exchange rate. Resour. Policy **62**, 256–281 (2019)
7. Chiang, T.C.: Can gold or silver be used as a hedge against policy uncertainty and COVID-19 in the Chinese market? China Fin. Rev. Int. **12**(4), 571–600 (2022)
8. Churchill, S.A., Inekwe, J., Ivanovski, K., Smyth, R.: Dynamics of oil price, precious metal prices and the exchange rate in the long-run. Energy Econ. **84**, 104508 (2019)
9. Ding, Q., Huang, J., Gao, W., Zhang, H.: Does political risk matter for gold market fluctuations? A structural VAR analysis. Res. Int. Bus. Financ. **60**, 101618 (2022)
10. Hammer, B., Norskov, J.K.: Why gold is the noblest of all the metals. Nature **376**(6537), 238–240 (1995)

11. Hammoudeh, S.M., Yuan, Y., McAleer, M., Thompson, M.A.: Precious metals–exchange rate volatility transmissions and hedging strategies. Int. Rev. Econ. Financ. **19**(4), 633–647 (2010)
12. Jain, A., Biswal, P.C.: Dynamic linkages among oil price, gold price, exchange rate, and stock market in India. Resour. Policy **49**, 179–185 (2016)
13. Jain, A., Ghosh, S.: Dynamics of global oil prices, exchange rate and precious metal prices in India. Resour. Policy **38**(1), 88–93 (2013)
14. Krol, R.: Economic policy uncertainty and exchange rate volatility. Int. Fin. **17**(2), 241–256 (2014)
15. Kumar, S., Kumar, A., Singh, G.: Causal relationship among international crude oil, gold, exchange rate, and stock market: fresh evidence from NARDL testing approach. Int. J. Financ. Econ. **28**(1), 47–57 (2023)
16. Lakshmanan, V.I., Ojaghi, A., Gorain, B.: The economics of gold and silver. In: Innovations and Breakthroughs in the Gold and Silver Industries: Concepts, Applications and Future Trends, pp. 157–173 (2019)
17. Novotný, J., Polách, J.: Real silver and its investment and business options. Int. J. Entrepr. Knowl. **4**(1) (2016)
18. Orishev, A.B., Mamedov, A.A., Grachev, A.B., Pichuzhkin, N.A.: Silver investment coins as a guarantee of financial security. Eur. Proc. Soc. Behav. Sci. (2021)
19. Raj, R.K.M., Dlouhý, P., Kovač, V., Turinská, L.: Investing in pairs of precious metals: portfolio theory application. Acta Montanistica Slovaca **28**(4) (2023)
20. Sari, R., Hammoudeh, S., Soytas, U.: Dynamics of oil price, precious metal prices, and exchange rate. Energy Econ. **32**(2), 351–362 (2010)
21. Sensoy, A.: Dynamic relationship between precious metals. Resour. Policy **38**(4), 504–511 (2013)
22. Sokhanvar, A., Bouri, E.: Commodity price shocks related to the war in Ukraine and exchange rates of commodity exporters and importers. Borsa Istanbul Rev. **23**(1), 44–54 (2023)
23. Triki, M.B., Maatoug, A.B.: The GOLD market as a safe haven against the stock market uncertainty: evidence from geopolitical risk. Resour. Policy **70**, 101872 (2021)
24. Tweneboah, G., Alagidede, P.: Interdependence structure of precious metal prices: a multi-scale perspective. Resour. Policy **59**, 427–434 (2018)
25. Yilanci, V., Kilci, E.N.: The role of economic policy uncertainty and geopolitical risk in predicting prices of precious metals: evidence from a time-varying bootstrap causality test. Resour. Policy **72**, 102039 (2021)

# Informal Sector Employment: An Analytical Review on Trends, Challenges and Avenues for Development in India

S. Muruganandhan, K. Udhaya Kumar, C. V. Kiran Raj, R. Thamilmani, V. Vinothina, and K. Sivasubramanian

**Abstract** The word "informal" was first used to describe economic activity that did not fit into the industrialised and western ways of working, and was mainly seen in developing economies. The employees and the enterprise are the two main vantage points from which it can be observed. It is characterised in the workplace as labour that is not governed by laws or regulations. Both in rural and urban areas, the labour force participation rate has gradually increased. However, the economy is not creating any formal jobs. India is facing a severe mismatch between the rate of employment growth and technological changes in output. The major aims of this chapter are to review the importance of informal sector workers in India and various problem faced by the workers, to critically analyse the working and employment status of informal workers, and to analytically review the socio-economic and occupational health implications of informal workers in India. To achieve these farmed objectives, various existing literature and relevant data has been sourced and reported to make a conclusion. Based on the critical analysis of various aspects of the informal workers, conclusions are made and few appropriate recommendations are given.

**Keywords** Informal · Employment · Trends · Challenges · Opportunities · India · Development

S. Muruganandhan · K. Udhaya Kumar · C. V. Kiran Raj
Saveetha School of Law, Saveetha Institute of Medical and Technical Sciences, Chennai, India

R. Thamilmani
School of Commerce, Vel Tech Rangarajan Dr Sagunthaala R&D Institute of Science and Technology, Chennai, India

V. Vinothina
Department of Computer Science PG, Kristu Jayanti College, Bengaluru, India

K. Sivasubramanian (✉)
Department of Economics, Kristu Jayanti College, Bengaluru, India
e-mail: sivasubramanian@kristujayanti.com



M. Al Mubarak (ed.), *Sustainable Digital Technology and Ethics in an Ever-Changing Environment*, Studies in Systems, Decision and Control 237,
https://doi.org/10.1007/978-3-031-86708-8_30

## 1 Introduction

India is having the advantage of demographic dividend from the more rate of workforce population. Employment has stands one of the high challenges to the policymakers in India and it has become the continuous economic issues of the nation. The labour force participation rate has slowly increased in rural and urban areas. But there is no formal employment creation in the economy. India is experienced aggressive mismatch among technological changes in output and growth rate of employment [2]. The informal sector plays a key role in generation of rural and urban employment in agriculture and non-farm sectors. According to economic survey, 2021, it is identified that 91% of the workers are from unorganised-informal sector. Despite their high level of involvement in the economy, a meagre of 6% or such workers are receiving social security benefits. Moreover, the growth of wages is also relatively lessor while compare with amount of work that the labourer carries out in the casual work [12].

## 2 Paper Structure

The word "informal" was first used to describe economic activity that did not fit into the industrialised and western ways of working, and was mainly seen in developing economies. The employees and the enterprise are the two main vantage points from which it can be observed. It is characterised in the workplace as labour that is not governed by laws or regulations. An enterprise is characterised as a production-oriented entity that is not institutionalised or registered. Based on this background, this chapter has been organized as follows, introduction, objectives, methodology, literature, findings, conclusion, recommendation and suggestions.

## 3 Objectives

To review the importance of informal sector workers in India and various problem faced by the workers.

To critically analyse the working and employment status of informal workers.

To analytically review the socio-economic and occupational health implications of informal workers in India.

## 4 Methodology

This chapter is prepared by using secondary data sources which are collected from published government reports such as NCEUS, NITI Aayog, and Economic survey of various years. The major aims of this chapter are to review the importance of informal sector workers in India and various problem faced by the workers, to critically analyse the working and employment status of informal workers, and to analytically review the socio-economic and occupational health implications of informal workers in India. To achieve these farmed objectives, various existing literature and relevant data has been sourced and reported to make a conclusion. Based on the critical analysis of various aspects of the informal workers, conclusions are made and few appropriate recommendations are given.

## 5 Literature Review

As discussed in the introductory section, the informal sector is contributing 92.5% of the total workforce participation in India. All the agricultural labour force are come under the informal employment in India. So, the informal sector is always playing a major role in agriculture sector. It is also providing a huge employment opportunity for urban people in India.

## 6 Social Status of Informal Workers

Indian economy is meant for the existence of informal employment with more than 90% work force belongs to informal and it generates 50% of national product of the nation. A high segment of socially disadvantaged sections of the nation are prevails in this sector. According to the NSSO data, in villages, the share of unorganised employment activities is recorded with 90% of informal economic activities which includes 94.5% of female work force participation. The growth of Indian economy has been highly driven by a huge increase in informal sector [11]. In India, the informal sector workers are partially covered and many are working without any social security coverage. It is estimated that 90% of the informal sector workers are not covered under any kind of social security such as paid leave, formal contract, insurance schemes, medical facilities and pension schemes [9].

## 7 Working and Employment Conditions

In India, the terms "unorganised" and "informal sector" are used inconsistently. The unorganised or informal sector is made up of businesses that are owned and run by individuals with their own accounts or unorganised businesses that hire employees. In essence, they are partnership and proprietary businesses. The main issue of the developing country like India is the underemployment because of over population and non-availability of formal employment. The informal sector employees are still in the status of underemployed or poorly paid. The informal sector workers engaging nearly 12–13 h of work per day. The wage rates of the informal employees are relatively lesser and lower than the minimum wages.

## 8 Occupational Health Conditions

The health protection delivers a human right-based method to attaining the core objective of the common health coverage that assures protection of workers financial status and effective accessibility of health care services at free of cost. The sustainable development goals of the UN are also arguing in its framework at SDG 3.8 which is the universal social protection. Its importance has been evidently seen during the Covid-19 pandemic. The robust healthcare facilities and protection systems enabled the virus affected people to receive the essential care they required. But the gaps in medial provision to the informal sector workers disproportionately impacts the masses of workers who are the vulnerable sections if the society [15].

## 9 Problems Faced by the Informal Sector Workers

The diversity and various degrees of the unorganised work and the there is an unclear correlation among employer and the workers complicates the issue further. Labour related laws and the trade unions often consider only the proper employment relationships and not considering self-employed informal sector workers. More importantly, the unorganised labourers are not perceiving themselves as labourers and it is very true for women who work as informal worker [5].

The major problem identified by the informal worker is many of them doesn't have an employer and they have no person to negotiate for their rights. Also, these workers are outside the security labour law framework without any markets. It is also observed from the research that many unorganised workers are poor and they have to work for long hours to get the minimum wages. Sometimes, they will also engage in multiple employment for their survival. There are many migrant workers are also engaging in this informal employment with lesser payment of wages without social security provisions [16].

Work places are always tiny and there are proper infrastructural facilities sit and work. In many cases, under informal employment they may not have basic amenities at work places many of the informal workers are isolated and may not have communication with their family or friends during the course of work time for workers such as home-based workers, domestic employees both live-in and live-out and street vendors [3].

## 10 Women in Informal Employment

Across the world, there are two billion people aged from 15 years onwards are working in informal sector employment which contributes to 61.2% of total global employment. Among the total employment in India, a great portion of informal work has been done by men with 63 and 58.1% are women.

## 11 Opportunities Ahead

In India, women are nearly always engaged in some form of productive activities [1]. However, a large portion of their labour is unseen, and they are mostly employed in low-skilled, low-paid informal work in little to no social security for example, as domestic helpers or independent contractors working from home [13].

The literature identifies the following factors as contributing to the rise of informal employment and fall of formal employment opportunities: labour market rigidities, low skill levels, and increased competition from imports [2].

Over half of the total value generated by the Indian economy has historically come from the informal sector, demonstrating the sector's ingrained contribution to economic expansion [10]. For many communities, it serves as a means of subsistence when they are unwilling to find fulfilling work in the formal economy. One example of how unregulated employment provisions reappear in the modern technological paradigm is the rise of gig economies and digital labour platforms [4, 14].

## 12 Key Findings from Literature

India is experienced aggressive mismatch among technological changes in output and growth rate of employment. The informal sector plays a key role in generation of rural and urban employment in agriculture and non-farm sectors. According to economic survey, 2021, it is identified that 91% of the workers are from unorganised-informal sector.

The growth of Indian economy has been highly driven by a huge increase in informal sector. The informal sector employees are still in the status of underemployed or poorly paid. The informal sector workers engaging nearly 12–13 h of work per day. The wage rates of the informal employees are relatively lesser and lower than the minimum wages.

Work places are always tiny and there are proper infrastructural facilities sit and work. Across the world, there are two billion people aged from 15 years onwards are working in informal sector employment which contributes to 61.2% of total global employment.

The health protection delivers a human right-based method to attaining the core objective of the common health coverage that assures protection of workers financial status and effective accessibility of health care services at free of cost. In India, women are nearly always engaged in some form of productive activities. However, a large portion of their labour is unseen, and they are mostly employed in low-skilled, low-paid informal work in little to no social security for example, as domestic helpers or independent contractors working from home.

## 13 Conclusion

India's unorganised labour force faces numerous challenges, such as unstable employment and hazardous working environments. The goal is to discuss the concept of the "informal economy," the state of affairs, and the actions governments have taken to protect the interests of labourers in unorganised industries [7]. As a result, the unorganised sector is crucial to the Indian economy and needs special attention. Therefore, an attempt was made to examine the issues and difficulties faced by unorganised sector workers in this chapter in order to remove obstacles and offer workers at least a minimal level of basic social security [8]. This sector has experienced the benefits related with company as compared to the organised sector. But many of the workers are ended up as invisible victims of various exploitations. Despite the fact that this particular industry is very vital to the Indian economy in terms of employment generation of rural and urban areas. But still they have been ignored and struggling for their basic rights [6].

## 14 Recommendation

There are many problems has been facing by the informal workers, but there is no proper legal protection is available. Moreover, there is no proper awareness is created by centre or state government. Need extensive study of better understanding and make appropriate legal and economic policy for the informal sector workers development and rights.

## 15 Scope for Future Directions

This research paper is very limited and have micro perspective. There is a huge scope for expanding this work. Moreover, the informal se tor workers related literatures are available in very limited. There is need for more study to be conducted in this area of research.

## References


1. Baruah, B.: Women and globalisation: challenges and opportunities facing construction workers in contemporary India. Dev. Pract.Pract. **20**(1), 31–44 (2010)
2. Chand, R., Singh, J.: Workforce Changes and Employment Some Findings from PLSF Data Series. NITI Aayog, Government of India, New Delhi (2022)
3. Choudhury, D.: Challenges faced by informal sector workers in India. Indian J. Law Legal Res. **1**(1) (2023)
4. Chakraborty, S.: Covid-19 and women informal sector workers in India. Econ. Pol. Wkly **35**(35), 17–21 (2020)
5. Chatterjee, S.: Labourers in unorganised sectors and their problems. Int. J. Emerg. Trends Sci. Technol. **3**(7), 4397–4405 (2016)
6. Dhar-Bhattacharjee, S., Dwivedi, P.: Women's rights in India's IT sector: does equality work across continents? NMIMS Manage. Rev. **31**(4), 275–287 (2023). https://doi.org/10.1177/09711023241230468
7. Hammer, A., Keles, J.Y., Olsen, W.: Working lives in India: Current insights and future directions. Work Employment Soc. **36**(6) (2022). https://doi.org/10.1177/09500170221083511
8. Humble, D., Mani, H.: Working of change in India's civil society. Work Employ Soc. **32**(6), 1130–1139 (2018)
9. ILO: Extending Social Health Protection to Informal Sector Workers in India: A Survey Report. International Labour Organisation, Geneva (2022)
10. Mishra, A.: Women on the verge: Marginalising female workforce in the Indian informal economy. ESSBC J. Bus. Stud. **1**(1), 35–44 (2023)
11. NSSO: Employment in Informal Sector and Conditions of Informal Employment. Ministry of Labour & Employment, Labour Bureau, Government of India (2014)
12. Pandey, M.K.: Problems of unorganised workers. Hum. Dev. **17**(2) (2022)
13. Raveendran, G.: Women and men operators of enterprises in India's informal sector. Women in Informal Employment Globalizing and Organizing (WIEGO), Statistical Brief No. 18 (2017)
14. Sakthivel, S., Joddar, P.: Unorganised sector workforce in India trends, patterns, and social security coverage. Econ. Polit. Wkly., 2107–2114 (2005)
15. Sathya, P.: Issues of unorganised labourers in India. Indian J. Appl. Res. **4**(6) (2016)
16. WIEGO: Challenges of organising informal workers. Empowering Informal Workers, Securing Informal Livelihoods (2022). https://www.wiego.org/challenges-organizing-informal-workers

# Ecological Transition Potential of Tourists in Mountainous Adjara

**Merab Putkaradze, George Abuselidze, and Landa Putkaradze**

**Abstract** One of the main directions for sustainable development of mountainous regions is rational application of touristic and recreational resources, especially in quite densely populated regions, and environment is known with low sustainability. The mentioned issue constitutes one of urgent topics of modern times and identification of issue and setting a reference in line with its settlement (on the example of particular region) will bear an applicable significance. Tourism is granted with priority in line with increase of social and economic level of mountainous Adjara, but there are number of problems in its development. The work evaluates the potential of touristic and recreational resources of mountainous Adjara, geographic originalities of tourism development and the role of tourism in increase of local population's living standards. There are established the territorial originalities of ecological transition potential of tourists and on the basis of existing resources, there are determined the development perspectives of some particular type of tourism.

**Keywords** Tourism industry · Resource · Natural management · Ecology · Sustainable development

## 1 Introduction

Mountainous Adjara is a quite densely populated region in Georgia, where geological processes are activated due to economic activity and ecological conditions is worsened. In order to maintain the sustainable development of the region, the special importance should be applied to rational consumption of the existing resources, which will ensure the increase of its social and economic level on the one hand and

M. Putkaradze · G. Abuselidze
Batumi Shota Rustaveli State University, Batumi, Georgia
e-mail: george.abuselidze@bsu.edu.ge

L. Putkaradze (✉)
Ivane Javakhishvili Tbilisi State University, Tbilisi, Georgia
e-mail: landatsu90@gmail.com



M. Al Mubarak (ed.), *Sustainable Digital Technology and Ethics in an Ever-Changing Environment*, Studies in Systems, Decision and Control 237,
https://doi.org/10.1007/978-3-031-86708-8_31

significantly decrease anthropogenic impact on the natural environment on the other hand. Coming out of this, the priority should be granted to application of various touristic and recreational resources and the development of different types of tourism.

The main goal of the research is to study ecological transition potential of tourists in Adjara and by means of it and by foreseeing the basic principles of rational natural management—determine the perspectives of development of some particular type of tourism in the region. In order to reach the goal set, the solutions of the following tasks are required:

- Evaluate touristic and recreational resources of mountainous Ajara and their geographical originalities;
- Determine the research area's fitness index for its usage for touristic purposes, also the index of its landscape-recreational assimilation;
- Determine the dynamics of number of tourists and their distribution according to recreational areas;
- Evaluate the touristic industry's impact on the ecological condition of the region;
- Determine the priorities of development of some particular type of tourism, by foreseeing the rational natural management principles.

Object of study constitutes mountainous region of Autonomous Republic of Adjara, which unities three municipalities (Keda, Shuakhevi and Khulo). Its area equals to 1750 km$^2$, which was populated by 57,000 people as of 2022.

The sustainable development of mountainous regions around the world is primarily based on the rational usage of local resources, which requires certain changes in the sectoral structure of the economy and defining tourism priorities based on the sustainability of the natural environment. Coming out of this, a complex approach to the issue requires the sharing of the international theoretical approaches [1–7]. Therefore, the tourism development, which is actively taking place in many mountainous regions of the world, is the most optimal for raising the socio-economic level of mountainous regions [8–12] and to achieve the set goal of the research based on the specific experience [13–15]. The studies prove that the sustainable development of mountainous is oriented in the direction of tourism with less financial costs and more economic efficiency [16–20].

## 2 Methodological Foundations

Any economic activity somehow affects the ecological condition, but impact of tourism relating to other fields is minimal. Therefore, development of tourism at some particular area and sustainability of natural environment should be in correlation in a manner, by which economic effect will be gained and the natural environment will be protected. For this purpose, application of indices became necessary, which determine the research area's fitness index for its usage for touristic purposes [19] and its landscape-recreational assimilation [21]. By application of methodology on ecological transition potential of tourists [22], influence of tourism over the region's ecological condition will be determined. And by foreseeing the rational natural management in the results acquired, perspectives of some particular type of tourism and touristic units' operation will be determined. In order to define the economy's field structure and determine the perspectives for tourism development, the balanced theory of economic development was be applied [23]. In order to research the topic in details, application of sociological studies and in-depth interview methods [24] become necessary. Statistical, field research, space–time analysis and other methods were also used during the research.

## 3 Results and Discussion

Tourism in Georgia is rapidly growing field and the scale of its development are quite dynamically expanded almost in every region of the country. Accordingly, state strategy on sustainable development of mountainous regions is related to usage of touristic and recreational resources, which somehow contradicts with natural environment and ecological condition worsens. There exist problems in mountainous Adjara as well in this regard, where tourism became one of the main trends of region's social and economic development. Coming out of research goals, it is necessary to determine the usage indicators of touristic and recreational resources existing in mountainous Adjara and influence of touristic economy over the sustainability of region's natural environment. For this purpose, there became necessary to find out the ecological transition potential of tourists. In order to clear up the topic with mentioned method, it is necessary to evaluate the touristic and recreational resources at the research area, dynamics of tourists, distribution of tourists per some particular area facility and the impact of touristic infrastructure on the ecological condition.

Tourism development in the region is based on touristic and recreational resources, which became more diverse as it gradually increases. In order to find out research area's fitness index for its usage for touristic purposes, we should use mathematical method [19] for this purpose via the following formula:

$$K_{pr} = \frac{S_{fr}}{S_{fm}}$$

where

$K_{pr}$ fitness index;

$S_{fr}$ sum of regional function (unity of tourism types), their quantity was determined by us as 14 units;

$S_{fm}$ sum of location functions (region covers three municipality locations, where more or less are developed all types of tourism, but their number never goes below 12 therein);

Coming out of this, the formula will become as follows:

$$K_{pr} = \frac{14}{12} = 1.17$$

If the fitness index is more than 1, then it will be considered that research area is fit for tourism development. Therefore, mountainous Adjara constitutes the region fit for tourism development, where different type of tourism operates.

In order to perform the topic's complex study, it would be beneficial to determine the research area's landscape-recreational assimilation index [20], which is calculated by the formula below:

$$i = \frac{P_m}{\mathrm{K} \times P_p}$$

where

- i is for area's landscape-recreational assimilation index;
- $P_m$ density of touristic units in natural territorial complex (NTC); there are up to 74th touristic unit spread across 1750 $km^2$ of the region, where the tourists pay the visit to;
- $K_X$ natural conditions' recreational fitness index (which is evaluated by 4 coefficients: very profitable—4; profitable—3; comparatively profitable—3 and less profitable or almost non-profitable—1; according to the mentioned relation, the regional index on the research basis will be profitable or 3 coefficient);
- $P_p$ population density on the given territory (average population density in the region is 33 people per 1 $km^2$);

If we put the mentioned quantitative indices, the formula will become as follows:

$$i = \frac{23.6}{3 \times 33} = \frac{23.6}{99} = 0.24$$

According to the mentioned method, there is found out that landscape-recreational assimilation index of mountainous Adjara is 0.24%.

In order to find out the main goal of research, it is necessary to determine the quantitative indicators of tourists in the region, thus discussing the ecological compatibility of tourism. Place of tourism in region's economic fields have been little by little increasing, which is confirmed by dynamics of visitors and tourists. A number of visitors and tourists are gradually increased during recent years (Table 1).

As Table 1 shows us, a number of region's visitors and tourists as well increases quite rapidly. Therefore, the development of touristic industry is needed to expand the field, which more or less influence over the natural environment's ecological condition. Most of tourists in the region (more than 90%) fall in four touristic units: Beshumi, Goderdzi, Gomarduli and Makhuntseti. In order to evaluate the ecological condition of some particular recreational place it is necessary to determine the total number of tourists, opportunities of receiving the tourists at the existing touristic infrastructure terms and total potential after the arrangement of complete touristic infrastructure (Fig. 1).

**Table 1** Dynamics of visitors and tourists in mountainous Adjara for 2000–2022 years

| | Years | | | | | |
|---|---|---|---|---|---|---|
| | 2000 | 2010 | 2015 | 2019 | 2021 | 2022 |
| Visitor | 15,550 | 85,750 | 135,850 | 160,150 | 134,350 | 175,550 |
| Tourist | 5750 | 6950 | 8350 | 12,150 | 9650 | 14,650 |

*Source* The table is made on the basis of materials of Department of Tourism and Resorts of Adjara and the field research

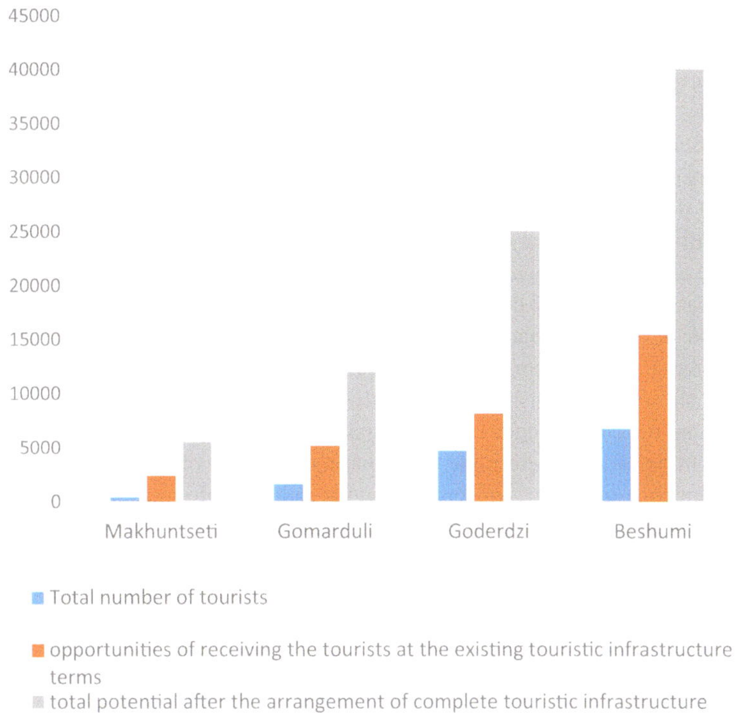


**Fig. 1** The number of tourists and the potential of receiving them in mountainous Adjara

The data brought (Fig. 1) makes clear that number of tourists in all touristic units are far low rather the total potential. As it is known, the basis for concept on development of tourism's ecological compatibility constitutes ecological transition potential. Potential of ecological transition is the level of unit's or area's visits (attendances), the excess of which mostly leads to undesirable ecological consequences. The reason of this might be the influence of tourists and operation of infrastructure servicing the tourism [22].

In order to discuss the impact of tourism on region's ecological condition, firstly it is necessary to determine the transition potential of tourists at main touristic units. As a result of research, it is found out that transition potential of tourists in touristic units is insignificant out of total potential, in particular it equals to 7.6% in Makhuntseti, 13.8% in Gomarduli, 13.5% in Goderdzi and 17.0% in Beshumi (Fig. 1). But if we review the topic, we will see that, quantitative indicators of transition potential of tourists during the peak period are quite different, that is clearly observed in Makhuntseti and Beshumi. During the afternoon of summer period, load of Makhuntseti touristic unit increases up to 32% by visits of guests visiting Batumi resort [25]. As of resort Beshumi, the peak period herein comes at the first week of August, when the public celebration of festival "Shuamtoba" is held. The load of resort during this period ranges almost from 95 to 100% out of total potential of existing accommodation units. Number of tourists are evenly distributed at Gomarduli and Goderdzi ski touristic places and there are no expressed peak periods therein. In order to evaluate the ecological condition of the research area (with touris's impact) on the basis of existed data, ecological transition index should be applied, or level of area's visits and we bring here the conditional index. In particular, for those touristic units, where (under the existing infrastructure) the load exceeds 80% during the peak period, but doesn't exceed the tourists' reception potential under currently existing touristic infrastructure—may be evaluated as "worsened"; touristic units with 60–80% of the above-mentioned load—as "insignificantly worsened", those with up to 40–60%—as "normal" and below 40%—"good". Coming out of the existing condition and on the basis of quantitative (Table 1; Fig. 1) and qualitative properties, influence of regional tourism over the ecological condition is presented as in Table 2.

Table 2 shows that tourism not so greatly influences on the ecological condition of Adjara and the consequences acquired makes clear a huge potential of its further development. In order to determine the economic structure of the region, it is necessary to define the perspectives of some particular fields, for this reason it is acceptable to apply the balanced theory [23], by which an increase of regional economy will be determined via defining the possibilities of interconnections between some particular fields. As of the fact that currently, the region's main economic field is agriculture—its further development is actually impossible due to lack of land and worsened ecological condition. Coming out of this, a special attention should be paid to tourism development. In order to study the topic in a complex manner—a sociological survey was held by us in 2022, in order to determine the attitude of local population towards the tourism development. For this purpose, 178 respondents were interviewed and local social transition potential (or relation between the local population and tourists) was determined [22]. In particular, the question whether the tourism should develop or not—172 respondents or 96.7% have answered positively. The research confirms that tourism should become one of the priority fields of regional economy. Besides, priority should be assigned to development of ski tourism, which impacts the natural environment at its minimum (Table 2). On the other hand, development of ski tourism has many advantages [11]. Coming out of the mentioned, tourism development on the research area should become a priority trend of economic development and in this line the state policy on sustainable development of highlands should be established.

**Table 2** Ecological condition of main touristic unit of mountainous Adjara by 2022 year

| Touristic units | Ecological condition of transition potential of tourists | | Influence of anthropogenic load of the touristic economy and the consequences |
|---|---|---|---|
| | The average index of tourists' reception under the existing touristic infrastructure | Index of tourists' reception during the peak period | |
| Makhuntseti | Good | Good | Overloaded with buildings and premises; slight pollution of water; damage of vegetation cover; soil hardening; accumulation of waste; noise |
| Gomarduli | Good | Good | Building tracks and paths; activation of erosive processes; construction of new buildings and premises; decrease of vegetative cover; allocation of waste; noise |
| Goderdzi | Normal | Normal | Building tracks and paths; activation of erosive processes; construction of new buildings and premises; decrease of vegetative cover; allocation of waste; noise |
| Beshumi | Normal | Worsened | Overloaded with buildings and premises; slight pollution of water; damage of vegetation cover; soil hardening; accumulation of waste; noise |

*Source* The table is made on the basis of materials from field research

## 4 Conclusions

On the basis of the research, there was determined that high level of index of mountainous Adjara's fitness for its usage for touristic purposes, low level of landscape-recreational assimilation index, increasing dynamics of tourism development, ecological transition potential of tourists and insignificant impact of tourism on ecological condition of the region makes clear that the research area possess a quite big potential for tourism development. Application of existing touristic and recreational resources in line with rational natural management will encourage the transformation of sectoral structure of regional economy and steady increase of its social and economic development.

## References

1. Amerta, I.M.S., Sara, I.M., Bagiada, K.: Sustainable tourism development. Int. Res. J. Manag. **5**(2), 248–254 (2018)

2. Bošković, N., Vujičić, M., Ristić, L.: Sustainable tourism development indicators for mountain destinations in the Republic of Serbia. Curr. Issue Tour. **23**(22), 2766–2778 (2020). https://doi.org/10.1080/13683500.2019.1666807
3. Dornier, R., Mauri, C.: Overview: tourism sustainability in the Alpine region: the major trends and challenges. Worldw. Hosp. Tour. Themes **10**(2), 136–139 (2018). https://doi.org/10.1108/whatt-12-2017-0078
4. Yakobchuk, V., Zborovska, O., Dombrovska, S., Dragan, I., Blyznyuk, A., Abuselidze, G.: Model of the state policy for the development of enterprises in the hospitality industry. E3S Web Conf. **371**, 05012 (2023). https://doi.org/10.1051/e3sconf/202337105012
5. Dwyer, L.: Resident well-being and sustainable tourism development: the 'capitals approach.' J. Sustain. Tour. **31**(9), 2119–2135 (2023). https://doi.org/10.1080/09669582.2021.1990304
6. Rakhmetova, D., Nurgaliyeva, A., Yesmagzam, V., Takhtayeva, R., Khassenova, K.: Priority directions and trends of the business sector in tourism: the case of Kazakhstan. Stud. Syst., Decis. Control. **545**, 669–681 (2024). https://doi.org/10.1007/978-3-031-65203-5_59
7. Sakhno, A., Salkova, I., Abuselidze, G., Yanchuk, T., Buha, N.: Evaluation of efficiency of small agricultural enterprises economic activity under sustainable development conditions. Lecture Notes in Networks and Systems, vol. 575, pp. 2262–2271 (2023). https://doi.org/10.1007/978-3-031-21219-2_254
8. Gugushvili, T., Salukvadze, G., Salukvadze, J.: Fragmented development: tourism-driven economic changes in Kazbegi, Georgia. Ann. Agrar. Sci. **15** (2017)
9. Milijic, S., Micic, S., Maksin, M.: Retrospective of and prospects for the development and strategic planning of tourism in the mountain regions of Serbia. Spatium **37**, 42–48 (2017). https://doi.org/10.2298/spat1737042m
10. Slusariuc, G., Bica, M.: Mountain tourism-pleasure and necessity. Ecoforum **4**(2), 119–126 (2015)
11. Putkaradze, M., Abuselidze, G.: The impact of tourism on business development in mountain regions: a case study. Int. J. Econ. Bus. Adm. **7**(3), 70–83 (2019). https://doi.org/10.35808/ijeba/308
12. Putkaradze, M., Michalski, T., Abuselidze, G.: Tourizm and recreational resources usage perspectives through according to mountainous Adjaria S example. Econ. Manag. **19**(2), 27–39 (2022)
13. Debarbieux, B., Oiry, V., Rudaz, G., Maselli, D., Kohler, T., Jurek, M.: Tourism in Mountain Regions: Hopes, Fears and Realities. Sustainable Mountain Development Series. UNIGE, Geneva, Switzerland; CDE, Geneva, Switzerland; SDC, Geneva, Switzerland (2014)
14. Kuščer, K., Mihalič, T., Pechlane, H.: Innovation, sustainable tourism and environments in mountain destination development: a comparative analysis of Austria, Slovenia and Switzerland. J. Sustain. Tour. **25**(4), 489–504 (2017). https://doi.org/10.1080/09669582.2016.1223086
15. Maksimović, M., Janovac, T., Karabašević, D., Brzaković, M.: Solution of general and prevention of ecological problems of Stara Planina mountain as potential obstacles to the development of rural tourism. Ekon. Poljopr. Econ. Agric. **65**, 531–544 (2018). https://doi.org/10.5937/ekopolj1802531m
16. Sakhno, A., Salkova, I., Petruk, N., Popovich, Y., Vychivskyi, P., Zarichnyak, A., Abuselidze, G.: The efficiency evaluation of providing excursion services in the tourist activity of Ukraine. Lecture Notes in Networks and Systems, vol. 574, pp. 1216–1224 (2023). https://doi.org/10.1007/978-3-031-21432-5_129
17. Milicevic, S., Boskovic, N., Lakicevic, M.: Sustainable tourism development in mountain areas in Sumadija and Western Serbia. J. Mt. Sci. **18**, 735–748 (2021). https://doi.org/10.1007/s11629-020-6239-4
18. Richins, H.: Experience provision in mountain tourism: overview, contextual development and emphasis. In: Richins, H., Hull, J. (eds.) Mountain Tourism: Experiences, Communities, Environments and Sustainable Futures, pp. 13–24. CABI, UK (2016)
19. Aleksandrov, A.: Geography of Tourism (2009)

20. Lemishko, O., Davydenko, N., Abuselidze, G., Shevchenko, N.: Methodological approaches to the study of public financial policy of capital reproduction in the agricultural sector in the focus of information technology management. Studies in Systems, Decision and Control, vol. 517, pp. 465–477 (2024). https://doi.org/10.1007/978-3-031-50939-1_35
21. Vedenin, Yu., Miroshichenko, N.: Evaluation of environmental conditions for organization of recreation. Issue of Geography, 4 (1969)
22. Birjakov, M.: Theory of Tourism, Tbilisi (2012)
23. Cooper, C.: Tourism: Principles and Practice. Pearson Education, Harlow (2008)
24. Lu, J., Nepal, S.K.: Sustainable tourism research: an analysis of papers published in the journal of sustainable tourism. J. Sustain. Tour. **17**(1), 5–16 (2009). https://doi.org/10.1080/09669580802582480
25. Putkaradze, M., Putkaradze, L.: Touristic and recreational potential of Makhuntseti Waterfall: application and perspectives. Int. J. Innov. Technol. Econ. **12**(2), 8–15 (2022). https://doi.org/10.31435/rsglobal_ijite/30062022/7826

# Mental Health Literacy and Depression Literacy Among Young Adults

**Mushfirah Sayed and Anjana Sinha**

**Abstract** The objective of the research was to examine the Mental Health Literacy (MHL) and Depression Literacy (DL) levels in young adults in Bengaluru, India, through a cross-sectional research method. A group of 211 individuals (80 males, 131 females) were selected through purposive sampling. Information was gathered through the administration of the "Mental Health Literacy Scale" (MHLS, O'Connor and Casey in Psychiatry Res., 2015 [1]) and the "Depression Literacy Questionnaire" (D-Lit, Griffiths et al. in Br. J. Psychiatry 185:342–349, 2004 [2]). Jamovi version 2.3 software was utilized for the data analysis involving descriptive statistics, Spearman's correlation, Mann–Whitney U test, and Kruskal–Wallis test. The findings show substantial variations in MHL and DL among young adults, with females showing greater proficiency in these areas compared to males. Moreover, variations in MHL were noted across academic disciplines, with Arts students exhibiting lesser literacy skills in comparison to those in Commerce and Science streams. Nonetheless, there were no noticeable variations in MHL and DL across occupations. These results highlight the importance of implementing focused measures to improve MHL and DL in young adults, especially in males and students in the Arts field. Enhancing understanding of depression and mental health helps break down misunderstandings and lessen stigma, which can impede help-seeking. It enables people to identify early indications of mental health issues, promotes prompt seeking of assistance, and improves availability of suitable resources and support networks. Consequently, these actions could result in better control of mental health issues, improve general welfare, and greatly enhance mental health results for individuals and communities alike.

**Keywords** Mental health literacy · Depression literacy · Mental health awareness · Academic streams · Gender differences

M. Sayed (✉) · A. Sinha
Kristu Jayanti College, Autonomous, Bengaluru, India
e-mail: mushfirah.s@gmail.com

A. Sinha
e-mail: anjana@kristujayanti.com



M. Al Mubarak (ed.), *Sustainable Digital Technology and Ethics in an Ever-Changing Environment*, Studies in Systems, Decision and Control 237,
https://doi.org/10.1007/978-3-031-86708-8_32

# 1 Introduction

Mental Health Literacy (MHL) is vital for fostering of uplifting and inclusive environments within educational and workplace settings. This includes imparting knowledge about mental disorders, risk factors, as well as available treatment choices, attitudes, and beliefs that affect an individual's ability to identify as well as assist in coping with mental health difficulties effectively. Among young adults, depression ranks as one of the most prevalent mental health problems with a considerable effect on academic performance, social relationships, and general well-being.

This research evaluates Mental Health Literacy and literacy specific to depression among emerging adults in Bengaluru, India. The shift from high school to college or work is swamped with numerous challenges, having stress at pinnacle levels due to academic/organizational pressures as well as social adjustments, not undermining personal responsibilities. Even though the symptoms like depression, burnout, or anxiety have an increasing trend of prevalence, more often than not these issues go under the radar or unattended primarily due to ignorance and lack of understanding. This investigation is specifically designed to probe the understanding of MHL and knowledge about depression that young adults have, hence highlighting misconceptions and awareness gaps.

The research design is cross-sectional. It makes use of the "Mental Health Literacy Scale" (MHLS, O'Connor and Casey [1]) and "Depression Literacy Questionnaire" (D-Lit, Griffiths et al. [2]) for a detailed account of 22 items based on true or false questions regarding depression literacy. The information regarding age, sex, educational qualification, and stream was gathered to see variations in mental health literacy scores and depression literacy scores. This knowledge will enable interventions to be more precisely tailored, after consideration of students' beliefs, their attitudes, as well as their mental health issues. By examining young adults' understanding, viewpoints, and opinions regarding mental health issues, the goal of this research is to inform development of relevant targeted interventions.

New evidence and records feature the necessity of MHL in influencing beliefs and actions towards mental health. A survey, for instance, in Chennai showed that the selected population's health literacy program lowered stigma among the youth and improved their health knowledge [3]. Equally, Mental Health Literacy studies among school teachers in Southern India showed that when the knowledge of the teachers increased, the views of stigma and resilience improved which created a more favourable atmosphere for mental health [4]. In contrast, further studies on depression and schizophrenia carried out among undergraduates revealed where there is little educational foundation, there is an understanding gap for the disorders mentioned hence the call for more detailed informational approaches [5].

Stigma continues to be the greatest obstacle to the effective treatment for mental disorders. A review of studies which measured both literacy about individual's mental health as well as their attitude towards people with mental disorders, a study reports that the relationship might not be universally applicable since members of a society possess different factors that could influence such a pattern. This implies that mental health literacy alone cannot be enough alone to battle stigma whether it is positive or negative hence warranting for further studies on how to tackle stigma [6].

A systemic review of the present literature regarding stigma attached to that of mental health issues in the Indian youth advocates for the development of complex approaches to counter the stigma and to increase awareness. Progress has been made, but stigma still adds to the burden of the mental health seeking process [7] Although, the awareness of the health conditions can prompt individuals to seek professional help and prevent self-stigma, self-stigma remains a formidable barrier [8]. Thus it becomes crucial to tackle information deficits along with internalized stigma in order to cater for skilful management of mental illness.

The pandemic COVID-19 has increased problems faced in mental health, bringing out the necessity of improving MHL of the people as well as the support systems available during such situations [9] The rural urban divide in understanding adult depression, and mental health literacy initiatives suggest that there should be targeted interventions, and it is found that urban populations who are often in better access of mental health services than rural populations. Similarly, studies on mental health literacy among adolescent girls living in urban slums highlight significant deficits in knowledge and perception demonstrating that urban slum communities do not possess mental health care resources to a sufficient degree [10].

Such novel techniques, such as mindfully using the internet as a self-help aid, can be useful in raising the MHc levels of the population. Evaluations of such interventions for depression suggest the possibility to engage wider audiences and allow mental health to be within everyone's reach [11]. In the Indian higher education context, there is a more urgent need for a change in the system to empower youth's mental health needs, one of which is enhancing the MHL in educational institutions [12].

Understanding the MHL level among young adults- users of educational and organizational support systems is an important step in developing such systems. This research will help to define the present spread of knowledge concerning mental health and depression that remains to be revised and thus help eliminate the barriers preventing the awareness around mental health, its associated stigma, as well as its help-seeking behavior. The results of this study will provide valuable feedback on the level of MHL and depression literacy among young adults of Bengaluru, Karnataka, India which in turn can inform the policies and programs tackling mental health problems among this population subgroup.

## 2 Methodology

A non-experimental correlational research design was used to understand the level of Mental Health Literacy and Depression Literacy among 211 young adults (age bracket of 18–25 years). The sample included 80 males and 131 females, chosen from Bengaluru, India using purposive sampling. All participants were administered with the following questionnaires by the researchers:

Mental Health Literacy Scale (MHLS)
Depression Literacy Scale (D-Lit).

## 2.1 Objectives

1. To determine if there is a significant relationship between overall mental health literacy and depression-specific literacy among young adults.
2. To assess whether mental health literacy among young adults varies significantly based on gender.
3. To assess whether mental health literacy among young adults varies significantly based on different streams of study.
4. To assess whether mental health literacy among young adults varies significantly based on occupation.
5. To determine if depression literacy is significantly different based on gender.
6. To determine if depression literacy is significantly different based on streams of study.
7. To determine if depression literacy is significantly different based on occupation.

## 2.2 Hypotheses

1. $\mathbf{H}^{01}$: There is no significant relationship between mental health literacy and depression literacy.
2. $\mathbf{H}^{02}$: There is no significant difference in mental health literacy based on the demographic details.
   (a) $\mathbf{H}^{02.1.}$: There is no significant difference in mental health literacy based on gender.
   (b) $\mathbf{H}^{02.2.}$: There is no significant difference in mental health literacy based on the stream of study.
   (c) $\mathbf{H}^{02.3.}$: There is no significant difference in mental health literacy based on occupation.
3. $\mathbf{H}^{03}$: There is no significant difference in depression literacy based on the demographic details.
   (a) $\mathbf{H}^{03.1}$: There is no significant difference in depression literacy based on gender.
   (b) $\mathbf{H}^{03.2}$: There is no significant difference in depression literacy based on the stream of study.
   (c) $\mathbf{H}^{03.3}$: There is no significant difference in depression literacy based on occupation.

### 2.3 *Variables of the Study*

- Dependent Variable (DV)—Depression Literacy (DL);
- Independent Variable (IV)—Mental Health Literacy (MHL), Gender, Occupation, and Streams of Study.

### 2.4 *Ethical Considerations*

Before participating in the study, all individuals were fully informed and provided with consent forms. They were given a clear explanation of the study's purpose and nature, and reassured about their personal information being kept anonymous and confidential throughout the research process.

### 2.5 *Statistical Analysis*

The data collected in the present study was analysed using Jamovi 2.3 software, which enabled both descriptive and inferential analyses. Descriptive statistics helped to summarize and organize the collected data, using measures like mean, median, and standard deviation to provide a clear and comprehensive view of the findings.

## 3 Results and Discussion

Table 1 presents the descriptive statistics for Depression Literacy and Mental Health Literacy among young adults aged 18–25. The mean score for Depression Literacy was 12.5 (SD = 3.63), while the mean score for Mental Health Literacy was 115 (SD – 14.8).

Table 2 presents the correlation between Depression Literacy and Mental Health Literacy among young adults. A weak positive correlation was found ($r = 0.140$, $p < 0.05$). Therefore, the null Hypothesis ($H^{01}$): There is no significant relationship between mental health literacy and depression literacy was rejected.

Table 3 displays the results of the Mann–Whitney U Test comparing Depression Literacy and Mental Health Literacy between males and females. The analysis indicated a statistically significant difference in Depression Literacy ($p = 0.037$), as well

**Table 1** Descriptive statistics

| | Mean | Std. deviation |
|---|---|---|
| Depression literacy | 12.5 | 3.63 |
| Mental health literacy | 115 | 14.8 |

**Table 2** Correlation between depression literacy and mental health literacy

| | Depression literacy | Mental health literacy |
|---|---|---|
| Depression literacy | 1 | 0.140* |
| Mental health literacy | 0.140* | 1 |

* Correlation is significant at the 0.05 level

as a statistically significant difference in Mental Health Literacy ($p \leq 0.001$) between the two groups. As a result, the null hypotheses ($H^{02.1}$): There is no significant difference in mental health literacy based on gender, and ($H^{03.1}$): There is no significant difference in depression literacy based on gender, are rejected.

Table 4 reports the results of the Kruskal–Wallis test examining variations in Depression Literacy and Mental Health Literacy among individuals from Science, Commerce, and Arts streams of study. The analysis showed no significant differences in Depression Literacy among the three groups. However, significant differences were found in Mental Health Literacy ($\chi^2 = 26.82$, df = 2, $p < 0.001$, $\varepsilon^2 = 0.1277$).

Pairwise comparisons using the Dwass-Steel-Critchlow-Fligner method were done to understand further differences between streams of study. Consequently, the null hypothesis ($H^{03.2}$): There is no significant difference in depression literacy based on the stream of study was accepted; and the null hypothesis ($H^{02.2}$): There is no significant difference in mental health literacy based on the stream of study was rejected.

**Table 3** Mann Whitney U Test for gender differences in depression literacy and mental health literacy

| Variables | Gender | N | Mean | SD | U | p |
|---|---|---|---|---|---|---|
| Depression Literacy | Female | 131 | 12.9 | 3.80 | 4348 | 0.037 |
| | Male | 80 | 11.9 | 3.26 | | |
| Mental Health Literacy | Female | 131 | 118.5 | 13.29 | 3428 | < 0.001 |
| | Male | 80 | 109.8 | 15.75 | | |

**Table 4** Kruskal Wallis Test for differences in depression literacy and mental health literacy based on streams of study

| Variables | $\chi^2$ | df | p | $\varepsilon^2$ | Pairwise comparison | p |
|---|---|---|---|---|---|---|
| Depression literacy | 5.21 | 2 | 0.074 | 0.0248 | Arts-commerce | 0.168 |
| | | | | | Arts-science | 1.000 |
| | | | | | Commerce-science | 0.090 |
| Mental health literacy | 26.82 | 2 | < 0.001 | 0.1277 | Arts-commerce | < 0.001 |
| | | | | | Arts-science | 0.002 |
| | | | | | Commerce-science | 0.178 |

**Table 5** Results of Mann Whitney U Test for differences in depression literacy and mental health literacy based on occupation

| Variables | Gender | N | Mean | SD | U | p |
|---|---|---|---|---|---|---|
| Depression literacy | Employed | 37 | 11.4 | 4.05 | 2695 | 0.119 |
| | Student | 174 | 12.7 | 3.50 | | |
| Mental health literacy | Employed | 37 | 112.4 | 16.11 | 2710 | 0.131 |
| | Student | 174 | 115.8 | 14.54 | | |

Table 5 shows the results of the Mann–Whitney U Test comparing Depression Literacy and Mental Health Literacy scores between students and employed individuals. For Depression Literacy, the analysis yielded $U = 2695$ and $p = 0.119$. For Mental Health Literacy, the results were $U = 2710$ and $p = 0.131$. No statistically significant differences were found between the two groups for either variable ($p > 0.05$). As a result, both the null-hypotheses ($H^{02.3}$): There is no significant difference in mental health literacy based on occupation, and ($H^{03.3}$): There is no significant difference in depression literacy based on occupation are accepted.

## 4 Summary, Conclusion and Implication

The present investigation assessed the levels of DL and MHL among young adults in the bracket of 18–25 years, and differences between ascribed factors such as gender, occupation, and stream of study. Because the data set examined in this research paper demonstrates a non-normal distribution, this study made use of non-parametric statistical techniques in order to show the way these constructs are salient across various demographic groups.

The Descriptive analysis suggested that the mean score of Depression Literacy is above average, meaning that most of the respondents acknowledge depression. Although, it is also possible that this high score was a result of the frequent use of the word 'depression' in conversations causing a misinterpretation of the word, which can be another area for study. The finding suggests that there is a need to educate the public on the appropriate use of this term so that words of similar nature are not misused and the correct idea is emphasized, as well as the depression literacy levels maintain, and raise further.

On the other hand, individuals' Mental Health Literacy scores were nearer the maximum, suggesting a relatively comprehensive and reasonably healthy comprehension of mental health subjects. This implies that young individuals have a fair understanding of mental health ideas, but it also stresses on how necessary it is to increase education on mental health for the preservation and raise of these literacy levels.

Spearman's Correlation tested for a small but statistically significant positive correlation between DL and MHL, which implies that knowledge of depression in particular improves general mental health literacy, but does not predict general MHL. This result implies that the goal of creating more holistic youth mental health awareness requires a more elaborate approach to MHL among the youth that encompasses a broad range of mental health issues. Females performed significantly better than males on both depression literacy and mental health literacy assessments. These results underscore the need for gender specific interventions to address the gap in male MHL and raise general knowledge of mental health among males. This gap can be closed via specific programmes which target male audiences and make mental health information more accessible.

The study also showed that there were statistically significant differences between MHL scores in different streams of study. Arts stream had lower scores than both Science and Commerce. This signifies that maybe students studying in the Arts stream need more Mental Health Education programs specifically. Depression literacy didn't show a change in scores in different streams. This shows that academic course may possibly not have any effect on this, however, a thorough further study can be done. The present study however, did not find any statistically significant differences in DL or MHL between students and working adults based on occupational position. This implies that the constructs are not significantly impacted by an individual's professional standing, potentially because both groups have comparable exposure to mental health resources and knowledge. This suggests that universal education programs that target young adults in the workforce as well as students should be the main focus of mental health initiatives.

## 4.1 Limitations

The study includes certain limitations that must be taken into account. Its cross-sectional design may limit its ability to ascertain causality between Depression Literacy, Mental Health Literacy, and demographic variables, making it challenging to ascertain the direction of these relationships. Longitudinal research may be able to present deeper insights into how these constructs evolve over a period of time. The study's usage of self-report measures may initiate potential biases, such as that of social desirability or even inaccuracy in cases of self-assessment. Additionally, because participants were chosen from a specific geographic location, the results may not be generalizable to the bigger population, and cultural factors may have influenced the findings. Furthermore, the study's emphasis on young adults confines the applicability of the findings to other age sets.

## References


1. O'Connor, M., Casey, L.: The mental health literacy scale (MHLS): a new scale-based measure of mental health literacy. Psychiatry Res. (2015). https://doi.org/10.1016/j.psychres.2015.05.064
2. Griffiths, K., Christensen, H., Jorm, A., Evans, K., Groves, C.: Effect of web-based depression literacy and cognitive–behavioural therapy interventions on stigmatising attitudes to depression: randomised controlled trial. Br. J. Psychiatry **185**(4), 342–349 (2004). https://doi.org/10.1192/bjp.185.4.342
3. Vijaya, R., Sangeetha, C., Vimala, P., Ramakrishnan, P., Greeshma, M., Jothilakshmi, D., Currie, G., Lilford, R., Furtado, V., Madan, J., Birchwood, M., Meyer, C., Sood, M., Chadda, R.K., Mohan, M., Shah, J., John, S., Padmavati, R., Iyer, S.N., Thara, R., Singh, S.P.: Effectiveness of a mental health literacy module on stigma-related mental health knowledge and behavior among youth in two educational settings in Chennai, South India: a quasi-experimental study. Asian J. Psychiatry (2024). https://doi.org/10.1016/j.ajp.2024.104074
4. Rithvik, S.K., Sahithya, B.R., D'Souza, L.: Mental health, stigma, resilience, and mental health literacy among schoolteachers: a cross-sectional survey from South India. J. Ment. Health Hum. Behav. (2024). https://doi.org/10.4103/jmhhb.jmhhb_259_23
5. Srishti, M., Indu, T., Bhageerathy, R.: Mental health literacy for depression and schizophrenia among undergraduate students. Indian J. Community Med. (2024). https://doi.org/10.4103/ijcm.ijcm_abstract423
6. Lamichhane, B.-B.: Association between mental health literacy and stigma: a review. AMC J. **4**(1) (2023). https://doi.org/10.3126/amcj.v4i1.63834
7. Gaiha, S.M., Salisbury, T.T., Koschorke, M., Raman, U., Petticrew, M.: Stigma associated with mental health problems among young people in India: a systematic review of magnitude, manifestations, and recommendations. BMC Psychiatry **20**(1), 1–14 (2020). https://doi.org/10.1186/s12888-020-02937-x
8. Sari, D.R., Asiyah, S.N.: Mental health literacy and self-stigma on intention to seek professional psychological help. J. Health Sci. Prev. **6**(2), 104–111 (2022). https://doi.org/10.29080/jhsp.v6i2.746
9. Rani, Y.S., Sharma, V.: Effect of COVID-19 on the mental health of young adults in India. Asia-Pac. J. Manag. Technol. **2**(3), 19–26 (2022). https://doi.org/10.46977/apjmt.2022v02i03.003
10. Saraf, G., Chandra, P.S., Desai, G., Rao, G.N.: What adolescent girls know about mental health: findings from a mental health literacy survey from an urban slum setting in India. Indian J. Psychol. Med. **40**(3), 201–206 (2018). https://doi.org/10.4103/IJPSYM.IJPSYM_108_18
11. Mehrotra, S., Sudhir, P., Rao, G., Thirthalli, J., Srikanth, T.K.: Development and pilot testing of an internet-based self-help intervention for depression for Indian users. Syst. Res. Behav. Sci. **8**(4), 536–544 (2018). https://doi.org/10.3390/bs8040036
12. Mehrotra, S.: Indian higher education and youth mental health: challenges and opportunities. J. Glob. Health **10**(2), 1–5 (2020). https://doi.org/10.7189/JOGH.10.020307

# Sustainability Practices in Bajawa Coffee Cultivation: A Case Study in Beiwali and Wajamala Villages, Ngada Regency

**Rudy Pramono, Juliana, Yosep Dudedes Timba, Diena M. Lemy, and Amelda Pramezwary**

**Abstract** Flores Bajawa Arabica Coffee is one of five typical Indonesian coffees that has received Geographical Indication Certification from the Ministry of Law and Human Rights. The development of coffee tourism in Bajawa is one of the impacts of increasing tourism on Flores Island. However, the development of coffee tourism in Bajawa as a tourist destination still needs to be supported by increasing the 4A components (attractions, amenities, Accessibility, and ancillaries) to increase tourist attraction. This research aims to analyze Bajawa coffee tourism from the 4A aspects to realize sustainable tourism in Bajawa, East Nusa Tenggara. The research method used is a qualitative approach with case studies in Beiwali and Wajamala Villages, Ngada Regency destinations. Data collection was carried out through interviews, observations at study locations, and analysis of secondary documents. The analysis technique uses the theme analysis method. The study results show that there are prototypes for developing coffee tourism in two tourist attractions, namely coffee tourism in Beiwali and Wajamala. This prototype can be used as a model by other coffee tourism companies in Bajawa, and it can support the development of sustainable coffee tourism by studying the 4A components. These findings are expected to be useful for managers and the government in developing sustainable coffee tourism, as well as increasing the attractiveness of Bajawa coffee tourism in the eyes of domestic and international tourists.

**Keywords** Sustainability · Bajawa coffee · Tourism village

## 1 Introduction

Coffee is one of the plantation commodities that plays an important role in Indonesia's economic activities [1]. It is estimated that coffee consumption will increase by around 500 billion cups annually, with a net worth of around US$100 billion. Apart

R. Pramono · Juliana (✉) · Y. D. Timba · D. M. Lemy · A. Pramezwary
Universitas Pelita Harapan, Tangerang, Indonesia
e-mail: juliana.stpph@uph.edu



M. Al Mubarak (ed.), *Sustainable Digital Technology and Ethics in an Ever-Changing Environment*, Studies in Systems, Decision and Control 237,
https://doi.org/10.1007/978-3-031-86708-8_33

from being a commodity product, the level of coffee production has a positive influence on Indonesia's export results, which are quite high and, of course, increase the country's foreign exchange along with oil and gas. Apart from increasingly open export opportunities and potential, the domestic coffee market is also still quite large [2]. The Ministry of Tourism and Creative Economy (Kemenparekraf) has determined five super-priority tourism destinations: Labuan Bajo on Flores Island East Nusa Tenggara Province. The hope is that by establishing Labuan Bajo as one of the super-priority destinations, it will certainly be able to increase and have a positive impact on the development of tourism in East Nusa Tenggara in particular, where this is supported by government programs and plans to give special attention to physical and non-physical development [3]. Apart from that, this determination can, of course, also impact the level of tourist visits and improve the economic quality of other areas on Flores Island, one of which is Bajawa.

Bajawa is known as the largest Arabica coffee-producing area in East Nusa Tenggara. Apart from being a producer of Arabica coffee, Ngada Regency has been designated as a tourist destination in East Nusa Tenggara, and this is in accordance with East Nusa Tenggara Province Regional Regulation Number 1 of 2011 concerning Regional Spatial Planning (RTRW) of East Nusa Tenggara Province 2010–2030. Ngada Regency itself has several tourist attractions such as natural tourism, cultural & arts tourism, historical heritage tourism (traditional and traditional villages), religious tourism, and, of course, many travel options for tourists, creating a complete and enjoyable travel experience.

It was recorded that Arabica coffee production in Ngada in 2019 was 2230.8 tonnes, with a productivity of 0.66 tonnes/ha, below the national average of 0.78 [4]. Based on the data above, 5234 ha of the area is Arabica coffee planting area, and the remaining 806 ha is Robusta coffee. Flores Arabica Coffee received Geographical Indication Certification on March 28, 2012, and was registered with the Directorate General of Intellectual Property (DJKI). Bajawa Arabica coffee is well known in the domestic and international markets. One of the supporting factors for the high cultivation and production of Arabica coffee in Bajawa Flores is the altitude above sea level (masl). With a favorable geographical location, Arabica coffee is the most widely produced natural product in Flores. The high production of Bajawa coffee is also balanced by high market demand and consumption.

Based on the results of interviews conducted by researchers, almost every year, Ngada Regency, through the Department of Agriculture, sends coffee harvests to several foreign companies and local businesses. This is also supported by the lifestyle of Indonesian people who often spend time in coffee shops to work, as a meeting point to meet friends and colleagues, or as a place to work and complete work. Based on research conducted by [5], Coffee is one of the most popular drinks in the world after water and tea. The culture of drinking Coffee has become a habit for people that cannot be separated from various activities. The International Coffee Organization (ICO), in its World Coffee Consumption Report for May 2021, reported that the amount of domestic coffee consumption in Indonesia from 2014 to 2021 continues to increase [6]. It has been recorded that there are several coffee shops in Bajawa

City with a total of 8 coffee shops [7] and several other coffee shops spread across several sub-districts such as Golewa District, Aimere, and So'a Districts.

The development of coffee tourism, one of the tourism developments in Bajawa, has become a regional government program. Tourism development in Indonesia has been made a priority, but in the process, it still faces a number of challenges [8]. These challenges can include negative impacts and a lack of collaboration between sustainability-oriented stakeholders [9]. The government's role is, of course, very important in contributing ideas, programs, planning, training, monitoring, and providing policies for developing coffee tourism in Bajawa.

In order to realize the development of a good coffee tourist spot, it is necessary to improve the quality of the place in order to create a pleasant visiting experience for visitors and the community. According to [10] argue that there are at least 3 factors in developing tourism: (1) planning, (2) management, and (3) technological processes. In this case, local governments, managers, and the public need to understand the importance of tourist destination components, namely Attraction, Accessibility, Amenities, and Ancillary (additional facilities) as referred to as 4A. Planning for the 4A development is considered important as a form of readiness for tourist attractions through managers and the surrounding community. If the four elements are fulfilled, tourist satisfaction with tourist destinations will increase, and the tourism image will also be built [11]. The development of coffee tourism does not always focus on tourist attractions but on the availability of supporting capacity such as supporting facilities, good and efficient access, and tourist destination management bodies that have permits.

Researchers are conducting case study research on coffee tourism in Bajawa, specifically in Beiwali and Wajamala villages. They aim to understand how these villages manage coffee tourism development and its potential for sustainability. The research seeks to use a sustainable tourism approach to develop new coffee tourism attractions in Bajawa, Flores—East Nusa Tenggara, focusing on environmental preservation and socio-cultural aspects. Sustainable tourism aims to achieve the Sustainable Development Goals (SDGs) by promoting economic growth, environmental conservation, and conserving non-renewable resources. It respects socio-cultural authenticity, is responsible for resource use, and empowers communities to participate in tourism production and benefits [12].

According to [13], sustainable tourism and sustainable development are terms that result in various responses from managers, tourism planners, and environmental defenders/advocates, who are both sceptical and concerned. The UNWTO defines sustainable tourism as tourism that takes full account of current and future economic, social, and environmental impacts, responding to the needs of visitors, the tourism industry, the environment, and host communities [3].

Tourism activities are an integral part of national development, which is carried out systematically, planned, integrated, sustainable, and responsible while still providing religious and cultural values in society. Preserving and maintaining the quality of the environment is an obligation for every human being to benefit future generations [14]. The development of coffee tourism in Bajawa needs to be carried out based on the principles of planned, sustainable, beneficial, and responsible tourism.

## 2 Research Method

Research design is a plan for collecting, measuring, and analyzing data created to answer research questions [15]. Research on the potential of Bajawa coffee in 2 coffee tours as a prototype tourist attraction in Flores, East Nusa Tenggara, was carried out using a qualitative approach. Research objects are the attributes, values, and attitudes of objects, people, and activities that have variables determined to be studied and conclusions drawn [16]. The object of this research is the development of coffee tourism in Beiwali village and Wajamala village in Bajawa Flores, East Nusa Tenggara, which was studied directly and in depth using appropriate approaches and sources. The research environment used as the research object is the city of Bajawa, Ngada Regency, Flores—NTT.

Probability sampling is used to collect data where the elements in the population have been previously determined [15]. In addition, purposive sampling is a design to determine the sample, where information is obtained from a certain target group. Researchers set several criteria, and not everyone can be used as a sample because only certain groups and people can provide the desired information [15]. The samples that will be taken are groups of coffee farmers in Bajawa, the community managing Beiwali village coffee tourism and Wajamala village coffee tourism in Bajawa, the Tourism office, and the Agriculture office of the Bajawa city government, Ngada Regency.

Data collection techniques will be carried out through observation, interviews, documents, and audio-visuals. In the data analysis stage, researchers describe the steps in data processing and preparation, reading the data as a whole, codification, codification process based on research themes, data description based on themes, and data interpretation [17].

## 3 Result and Discussion

### 1. Location Description

Research on Bajawa Coffee as a model for developing Coffee Tourism Attractions in Bajawa—Flores, East Nusa Tenggara, with case studies on coffee tourism in two different coffee tourism spots.

#### a. Coffee Tourism in Beiwali Village

Beiwali coffee tourism, managed by Fa Masa, is a local initiative in Bajawa, Philippines, focusing on processing and developing Arabica coffee. Located in Ngoranale village, the mountainous location is ideal for coffee bean cultivation. Farmers dominate livelihoods, while a few work as teachers, civil servants, and private employees.

b. **Coffee Tourism in Wajamala Village**

Khagomasa Wajamala Coffee Tourism is a 20 min drive from Bajawa City, owned by a cooperative managed by Mrs. Marselina Walu, a female farmer, barista, and leader. Mama Lina, a cooperative chairman, manages it.

2. **Description of Observation Results**

a. **Attractions**

Attractions are crucial for the management and development of a tourist destination, as they are the primary reason tourists visit [18]. The research defines attraction as cultural activity, natural beauty, and events motivating tourists to visit Beiwali coffee tourism. Human-made attractions include coffee plantations and preservation, but activities are not yet fully managed. Harvest season (June–September) offers unique tour packages.

b. **Amenities**

Amenities are crucial for the management and development of a tourist destination, serving as a supporting capacity with the availability of facilities for tourists.

In research [19], Amenities are defined as basic facilities that tourists can use while they are in a tourist destination with the aim of providing comfort [19]. The government has not fully supported coffee tourism in Beiwali, but through cooperatives and daily tourist income, management, and the surrounding community are working on completing facilities, providing attractions, and offering comfort to tourists.

c. **Accessibility**

According to [20], Accessibility refers to the provision of tourists with easy access to tourist destinations, including information and goods, and the availability of terminals, ports, and airports, which are designed to support tourism. Beiwali coffee tour is accessible via 2-wheeled and 4-wheeled transportation, taking around 15 mins from Bajawa city. Bajawa, the capital of Ngada Regency, is strategically located with an airport, ship harbor, and bus terminal, making access to Bajawa tourism easier. Access to coffee tourism is considered good and meets standards, benefiting local managers and governments. Information on coffee tourism locations is available, focusing on information service centers for foreign and domestic tourists.

d. **Ancillaries**

Ancillaries are one of the supporting components in the development of a tourist destination. As a result of the research carried out, it was found that Beiwali coffee tourism has several ancillary components available around the coffee tourism area. Still, there are also several other important parts available and located within the city of Bajawa, so tourists/visitors must travel to the town if they need some services. Other. The following are some of the existing parts.

3. **Analysis of the 4A Review of Coffee Tourism in Wajamala**

a. **Attractions**

Based on the results of interviews and observations conducted by researchers at Wajamala Coffee Tourism, there are several similarities in the types of attractions provided between Wajamala Coffee Tourism and Beiwali Coffee Tourism. Both offer coffee tour packages in which there is a collaboration between managers and groups of coffee farmers who help provide brief education about the coffee process from start to finish, and tourists are also invited to package the Coffee into containers/ packaging that is ready to be marketed and distributed.

b. **Amenities**

Amenities are an important component in developing a tourist destination/spot apart from the availability of the tourist attraction itself. Below are several amenity components that are and are not yet available in the Wajamala coffee tourism area. As an independent business entity, this place also has problems in terms of costs, so construction is quite late. Informants said that so far, they only rely on coffee production and the contributions of the community and tourists who come to visit.

c. **Accessibility**

To access the Wajamal coffee tour, you can use 2-wheeled and 4-wheeled transportation, which takes around 20 minutes from Bajawa city center. The road to Beiwali village is quite wide and can provide space for 2 cars at the same time. In research [21], adding Accessibility is not only about the ease of getting to a tourist attraction but also all the information that tourists need.

d. **Ancillaries**

Ancillary service can be defined as services needed to complete/provide satisfaction/experience of someone's visit to a tourist attraction. This service is an additional service, but it is also an important unit in the management and development of a destination/tourist attraction. Wajamala coffee tourism certainly really needs other services to provide perfection in future management. Therefore, in Indonesian tourism, the Pentahelix concept is well-known, and several components play an important role in exploiting the potential of coffee tourism.

The government plays an important role in finding innovative and cost-effective approaches to developing, financing, implementing, and managing Wajamala coffee tourism as a potential future tourism destination. Apart from that, there is the role of academics, namely researchers in this case, who carry out research, observe, and then try to provide several prototypes/models of tourism development that are based on sustainability principles.

Based on the analysis of the 4A overview of the two coffee tours studied, several components are important but not yet well provided. It is necessary to prepare all components so that management can be carried out optimally and well, not only with the aim of gaining commercial profits but managed professionally, providing satisfaction and experience for tourists. Therefore, research [22] states that interest in revisiting can arise if there is a positive assessment of a tourist destination; this positive assessment is related to one of the 4A components of a tourist destination, namely attractions, Accessibility, supporting facilities, and additional facilities (ancillary service). Based on the explanation above, it can be concluded that the application and analysis of the 4A concept can be a guideline to help plan the development and management of a tourist destination in order to increase tourist interest in visiting that destination.

4. **Analysis of Sustainability Tourism Aspects in Beiwali Coffee Tourism**

According to [23], Sustainable tourism considers the economic, social, and environmental impacts by addressing the needs of visitors, the industry, the environment, and host communities. In Beiwali Village, however, coffee tourism management and the local community face challenges due to a lack of understanding of the benefits associated with sustainable tourism. Irresponsible practices, such as improper disposal of bean shells and unplanned deforestation, have resulted in environmental degradation and inadequate conservation efforts. This lack of awareness and failure to conduct proper ecological impact analyses exacerbate the issues surrounding coffee tourism. Mr. Damianus highlighted the concerns and anxieties of both coffee activists and the local public regarding coffee tourism and the processing of Fa Masa products based on his observations, discussion forums, and interviews with relevant stakeholders

## *3.1 Discussion*

**What are the processes and results of coffee processing in Ngada Regency?**

The results of observations and interviews can answer the formulation of this problem. Secretary of the Ngada Regency Agriculture and Food Defense Service, Mrs. Ida, said that:

> Coffee beans, a natural product, have a significant influence on other natural products like vegetables and rice. Land used for coffee production has been repurposed for other purposes, such as planting vegetables and fruit. The Agriculture Service has aimed to promote Bajawa coffee to potential buyers through festivals and collaborations with agencies like the Tourism Department. A business feasibility study in 2022 showed high coffee demand, attracting foreign investors. Festivals include Wolobo-bo, Inerie, and Bena, with the main theme being Coffee.

**What are the obstacles in the coffee processing process in Ngada Regency?**

> Arabica coffee farmers face challenges in utilizing shade trees, which limit sunlight intake and serve as pest deterrents. However, many shade trees have been cut down for commercial purposes, leading to the growth of new coffee farming land.

**What efforts and forms have been taken by the regional government to market and promote Bajawa coffee as a potential tourism product?**

Informant Mr Ivan Bhato, as Head of the Tourism Service.

> The government is collaborating with the tourism and agriculture departments to host a cultural festival, focusing on Ngada's superior products like coffee and weaving, which they believe have the potential for increased recognition.

Informant Mrs. Eda, Secretary of the Agriculture Service.

> We have received several vendors who are willing to buy coffee beans from us, some from Australia and also locally owned companies. It is hoped that this will be a good step and will continue to be carried out in the future.

Research shows the Tourism Office and Agriculture Service are working together to promote Bajawa tourism, particularly Coffee, through marketing efforts, promotions, and additional services. These amenities are believed to predict tourists' interest in visiting again, according to previous research [24].

**What kinds of attractions or activities does Beiwali coffee tourism have?**

Informant Mr Damianus Maghi

> We have several activities; during the harvest period, the activity is usually to pick coffee beans directly from the tree and then sort the good and bad beans and much more. However, if it is not during the harvest period, we, together with other farmers, plant and care for the coffee beans; a production process turns them into coffee powder.

Farmer group father's informant:

> During harvest time, guests are directed to expert farmers who tour coffee plantations, providing important information.

Research conducted by [19] Tourism product attributes like attractions, amenities, and Accessibility are crucial for determining tourist decisions. Coffee tourism development should consider these components and other services to provide tourists with a pleasant experience and satisfaction.

**What efforts need to be made to develop coffee tourism in Beiwali village?**

Informant Mrs. Ida, Treasurer.

> The coffee tour in Beiwali faces numerous obstacles and problems, including inadequate facilities and lack of competent employees. Despite efforts to improve facilities and collaborate with local communities, the Tour still lacks optimal performance. The lack of facilities and competent employees contributes to the overall shortcomings of the coffee tourism industry.

Informant Mr Damianus Maghi,

> The community is making small-scale efforts to improve coffee tourism in Beiwali despite limited government intervention. The Fa Masa cooperative supports the community, and more people are discovering Beiwali coffee tourism. Addressing deficiencies and limitations is crucial for daily sustenance.

The farmer group's informant, Mr. Rius Taso.

> Farmers are implementing activities to clean coffee plantations, aiming to maintain trees and protect them from pests, which can reduce the quality and quantity of Coffee. Researchers found that the community in Beiwali is making small-scale efforts to develop coffee tourism, with the government's role not yet fully implemented. They are expanding parking lots, coffee production sites, and homestays for tourists. This is allowing them to spend more time in the city.

Based on the results of related research that has been carried out, amenities are temporary residences, such as hotels, inns, guesthouses, and other places. A place that offers overnight stays to all groups of tourists on a commercial basis [25]. Accommodation and different types of amenities and services that support tourist activities provide comfort and satisfaction for tourists, thereby creating interest and desire to visit again. This is in line with the results of research conducted by [26], which states that interest in revisiting can arise if a tourist destination is assessed positively. This positive assessment is related to one of the 4A components of a tourist destination, namely attractions, Accessibility, supporting facilities and additional facilities

**Does the local community obtain economic benefits by managing this coffee tourism?**

Informant Mr Damianus.

> This place was built through a non-governmental cooperative, the managers of which are also members of the community who were accepted as administrators. We also accept coffee beans from local communities at appropriate prices. If guests come and want to spend the night, people's houses are usually used as a place to rest because we don't have a place here. So the people here also receive benefits if many people visit or if they sell their Coffee here.

Informant Mrs. Ida.

> As a cooperative, we also provide savings and loans for cooperative members and provide many conveniences for them too.

Studies [27] show that the economic benefits of tourism mean that citizens should benefit more than they would without tourism. Therefore, through Coffee, people receive benefits and improve their quality of life.

**What is the role of stakeholder involvement in tourism management and development here?**

Informant Mr Damianus,

> As I said earlier, the regional government here still prioritizes other sectors, so there are a lot of sector egos in tourism development here. We purely develop our place; we run it in collaboration with several other coffee processing places here. So it could be said that the role of the parties who should be interested does not provide that to us here.

Informant Mrs. Margareth, as Treasurer.

> The government has opened several training opportunities for industry players to develop themselves and face the world of work in the future. However, this program has not been continued until now.

Coffee tourism in Beiwali relies on community cooperation, with the Fa Masa cooperative providing benefits for coffee farmers and managers. The Indonesian Minister of Tourism, Arif Yahya, introduced the Penta helix model to motivate the tourism industry and system, emphasizing the need for government, academia, business, community, and media to generate value for society and environmental benefits. This model should be applied to coffee tourism in Bajawa, where direct influence from government, media, academics, and business is insignificant.

**Implementation of the 4 components of tourism development in coffee tourism here?**

Informant Mrs. Lina, as head of Wajamala management.

> All this time, my team and I have tried to provide the best to the visitors who come here, even though, as can be seen here, there are still many shortcomings and problems at our processing site, starting from simple buildings, unsupportive facilities, Our tools here are still simple, we also don't have good public toilets, let alone accommodation. Regarding coffee attractions here, they are still not well established; they are still flexible and can change according to the wishes of the guests who come. All we have is coffee potential in terms of production, processing, and selling some souvenirs. Apart from that, we are still working on it slowly while waiting for the regional government to help.

Informant Mr Damianus Maghi.

> So far, there have always been guests coming to visit. We have prepared a guest book in front so we can know how many guests have come. We can currently sell several coffee tour packages that already exist but have not been packaged properly. In terms of facilities, many things are still lacking; I think you can see for yourself that this is our situation. We manage it ourselves; we don't have much budget, so we can't do anything yet.

Research conducted by [28] shows that tourism product attributes, which include tourist attractions, amenities or facilities, and Accessibility, are essential factors in influencing someone to determine the decision to visit a tourist destination. The importance of coffee tourism development in Beiwali and Wajamala villages lies in providing a pleasant experience, planning for restructuring, and ensuring the availability of essential components for a thriving tourism form.

This is in accordance with the results of research conducted by [26]; the better the destination management, the higher the level of visitor consumption will be, and this will directly improve the welfare of the community. Management of tourist destinations is reflected in attractions, Accessibility, facilities, and tourism organizations. This matter, supported by the results of previous research regarding the quality of destination infrastructure (attractiveness, Accessibility, amenities, and additional services), is a good predictor of tourists' interest in visiting again [24].

### Prototype of Coffee Tourism in Beiwali village

Based on the research results above, the recommended results for coffee tourism managers are in the form of a prototype, as results from research entitled Bajawa Coffee as a Prototype of Tourist Attractions in Bajawa, Flores (Case Study of Coffee Tourism in Bajawa, East Nusa Tenggara). Through these recommendations, it is hoped that the development of tourism potential in coffee tourism in the future can be better prepared in the aspects of attractions, amenities, Accessibility, and ancillaries, more organized in its management, planned, effective, and well-implemented. Apart from that, the hope is that this prototype can also become an example for other coffee tourism in the Bajawa region. Tourism development guidelines, of course, remain based on the principles of sustainable tourism with good implementation of the 4A components.

Researchers classified the prototype for coffee development in Beiwali coffee tourism into three different times: the time before the activity (preparation), during the activity, and after the activity ended. of these 3 times, several tourist activities/ attractions can be carried out to provide a good visiting experience for tourists.

Previous research [29] Regarding the development of natural tourism areas, there are several principles for developing and planning the use of areas consisting of spatial planning or regionalization subsystems, namely: Allocation of public facilities, permanent buildings, recreation, tourism and sports facilities, Allocation of non-permanent facilities, camping, fishing, and so on. Construction of public transportation roads is not permitted. Intended for footpaths, climbing, equestrian sports & viewpoints. There is no access road and no facilities and there are no roads or facilities yet. In planning the development of coffee tourism, the coffee tourism model needed is the provision of attractions and other supporting facilities. The following are efforts that can be made as a form of realizing ready tourism.

The development of coffee tourism in Beiwali and Wajamala requires the participation of various parties, including regional governments, community synergy, and the Fa Masa Cooperative. By providing programs and contributions, regional governments can positively impact regional income and become working partners for managers and investors. Community synergy is crucial for economic impacts, and the Fa Masa Cooperative is a working partner in planning funds for coffee tourism in Beiwali.

These results are in line with the results of the research conducted [30], which states that a good and optimally successful tourist area is based on four aspects, as is the case with the coffee tourism area in Bajawa, where the development and management pays attention to several important elements considering the need for environmental sustainability for the present, improving the quality of life of the people living in around or those who participate in it and what is the social and cultural balance. In this case, there are 4 aspects presented by Susanto in his research.

## 4 Conclusion

The development of coffee tourism in Wajamala Village, led by a local coffee farmer group, faces challenges such as a limited variety of attractions and inadequate public facilities. To achieve sustainable coffee tourism, it is essential to address key factors like tourist attractions, amenities, Accessibility, and ancillary services. Sustainable development efforts should balance economic growth, socio-cultural preservation, and environmental conservation. This research aims to provide valuable insights for coffee tourism managers, academics, and government officials, offering references and guidance on coffee tourism development models and policy planning based on sustainable tourism principles. The goal is to enhance the overall quality of coffee tourism in Wajamala Village.

**Acknowledgements** The author would like to thank LPPM UPH for providing the opportunity to conduct this research (P-77-FPar/VII/2023)

## References

1. Menke: The global coffee industry. Available at: https://globaledge.msu.edu/blog/post/55607/the-global-coffee-industry
2. BPS: Badan Pusat Statistik Provinsi Sumatra Utara Dalam Angka 2020 (2020)
3. Kemenparekraf: Desa Wisata Terus Tumbuh Menjadi Destinasi Wisata Alternatif. https://kemenparekraf.go.id/kebijakan/Desa-Wisata-Terus-Tumbuh-Sebagai-Pariwisata-Alternatif
4. Statistik, B.P.: LuasTanaman dan Produksi Perkebunan menurut Jenis Tanaman dan Kecamatan di Kabupaten Ngada, Badan Pusat Statistik, Kabupaten Ngada (2023)
5. Cornelis, M.C.: The impact of caffeine and coffee on human health. Nutrients **11**(2), 11–14 (2019) https://doi.org/10.3390/nu11020416
6. Dinas Pariwisata Kabupaten Ngada, Panduan Wisata Kabupaten Ngada, Dinas Pariwisata, Kabupaten Ngada (2022)
7. Hulu, M., Putri, F., Natasya, S., Dampak, M.: Media sosial dan word of mouth terhadap overtourism. J. Pariwisata **8**(2), 106–115 (2021). https://doi.org/10.31294/par.v8i2.10968
8. Lemy, D.M., Pramono, R., Juliana: Acceleration of Environmental Sustainability in Tourism Village, Int. J. Sustain. Dev. Plan **17**(4), 1273–1283 (2022). https://doi.org/10.18280/ijsdp.170425
9. Sekaran, U., Bougie, R.: Research Methods for Business: A Skill Building Approach, 7th edn. Wiley, Chichester (2020)
10. Creswell, J.W., Creswell, J.D.: Research and Design Qualitative, Quantitative and Mixed Methods Approaches (2018)
11. Syah, D.O., Rustanto, A.E., Estiana, R., Utami, R.T., Susbiyantoro: Service strategy on tourism business development: empirical study on purbalingga panyatan waterfall tourism spot. Int. J. Prof. Bus. Rev **8**(6), e03030 (2023). https://doi.org/10.26668/businessreview/2023.v8i6.3030
12. Buhalis, D.: Marketing the competitive destination of the future. Tour. Manag. (2000). https://doi.org/10.1016/S0261-5177(99)00095-3
13. UNWTO: International Tourism Highlights International Tourism Continues to Outpace the Global Economy (2019)
14. Quynh, N.H., Hoai, N.T., Van Loi, N.: The role of emotional experience and destination image on ecotourism satisfaction. Spanish J. Mark.—ESIC **25**(2), 312–332 (2021). https://doi.org/10.1108/SJME-04-2020-0055

15. Goeldner, C.R., Ritchie, J.R.B.: Tourism: Principles, Practices and Philosophies (2005)
16. Juliana, J., Lemy, D.M., Hubner, I.B., Pramono, R., Maleachi, S., Sitorus, N.B.: Acceleration of community-based tourism village development in West Java Province. J. Pariwisata Pesona **7**(1), 10–18 (2022). https://doi.org/10.26905/jpp.v7i1.7368
17. Lawson, F., Baud-Bovy, M.: Tourism Recreation Development. Handbook of Physical Planning. CBI Publishing Company, Boston (1977)
18. Yulihardi, Y., Syaiful, S.: Destination image and tourism products to visit intention tourism object in solok regency: mediation role of electronic word of mouth. Adpebi Int. J. Multidiscip. Sci. **2**(2), 126–136 (2023). https://doi.org/10.54099/aijms.v2i2.582
19. Hubner, I.B., Sihombing, S.O., Pramono, R., Hidayat, J.: Training on marketing strategies in the utilization of bamboo creations as a resource of life as hotel and culinary amenities. J. Commun. Serv. Engagem. **2**(4), 29–41 (2022)
20. Gulati, S.: Social and sustainable: exploring social media use for promoting sustainable behaviour and demand amongst Indian tourists. Int. Hosp. Rev. **36**(2), 373–393 (2022). https://doi.org/10.1108/IHR-12-2020-0072
21. Tomej, K., Liburd, J.J.: Sustainable accessibility in rural destinations: a public transport network approach. J. Sustain. Tour. (2020). https://doi.org/10.1080/09669582.2019.1607359
22. Enquist, S.P., Sebhatu, B.: Service innovation, sustainability and quality meeting city challenges in the age of accelerations. Int. J. Qual. Serv. Sci. **10**(4), 431–446 (2018). https://doi.org/10.1108/IJQSS-11-2017-0100
23. Maricar, S., Glen, R.: Attributes of travel destinations that influence tourists' decisions: a systematic review. Int. Tour. Hosp. J. (2021). https://doi.org/10.37227/ithj-2021-03-247
24. Juliana, J., Sianipar, R., Lemy, D.M., Pramezwary, A., Pramono, R., Djakasaputra, A.: Factors influencing visitor satisfaction and revisit intention in Lombok tourism: the role of holistic experience, experience quality, and vivid memory. Int. J. Sustain. Dev. Plan. **18**(8), 2503–2511 (2023). https://doi.org/10.18280/ijsdp.180821
25. Lemy, D.M., Pramezwary, A., Juliana, Pramono, R., Qurotadini, L.N.: Explorative study of tourist behavior in seeking information to travel planning. Int. J. Sustain. Dev. Plan. **16**(8), 1583–1589 (2021). https://doi.org/10.18280/ijsdp.160819
26. Lemy, D.M., Pramono, R., Juliana.: Acceleration of environmental sustainability in tourism village. Int. J. Sustain. Dev. Plan. **17**(4), 1273–1283 (2022). https://doi.org/10.18280/ijsdp.170425
27. Juliana, J., Sihombing, S.O., Antonio, F.: What drives memorable rural tourism experience: evidence from Indonesian travelers. Int. J. Sustain. Dev. Plan. **17**(8), 2401–2411 (2022)
28. Juliana, Indra, F., Hubner, I.B., Sianipar, R., Djakasaputra, A.: Systematic literature review: combining foodscape and touristcape for international tourism marketing in Singapore and Batam. Stud. Syst. Decis. Control. **1**, 1074 (2024)
29. Juliana, Hubner, I.B., Lemy, D.M., Pramezwary, A., Djakasaputra, A.: Antecedents of happiness and tourism servicescape satisfaction and the influence on promoting rural tourism. **19**(10), 4041–4059 (2024)
30. Sihombing, S.O., Antonio, F., Sijabat, R., Bernarto, I.: The role of tourist experience in shaping memorable tourism experiences and behavioral intentions. Int. J. Sustain. Dev. Plan. **19**(4), 1319–1335 (2024)

# Measuring the Role of Foreign Direct Investment in Generating Employment Opportunities: An Analysis in India

**S. Muruganandhan, A. R. Sanjeev Kumar, P. B. Shrithik, Amita Biju, V. Vinothina, and K. Sivasubramanian**

**Abstract** Foreign investment is a boon for domestic countries for enhancing employment opportunities and is a strong tool for reducing poverty. India is one of the desirable destinations for foreign investors. Based on this background, this chapter has been drawn with specific objectives such as examining the significant role of foreign direct investment (FDI) in India and analysing the impact of FDI inflow in generating employment opportunities in India. To become self-reliant in a developing country, the FDI helps promote domestic production and achieve a positive gross domestic product for the nation. This research study is based on the existing research works and secondary data sources. Various pieces of literature have been sourced from published articles and data collected from the World Bank database and published reports. Collected literature and data were analysed and the results have been discussed. Secondary data was chosen from the period 2001 to 2019. It is recorded that 48.96% of FDI inflow for the agricultural sector declined to 43.21% in 2019. It saw a steep drop. Over some time, the service sector attracted a good amount of FDI from various countries to India, it has improved from 24.29% to 26.56% during the year from 2001 to 2010 respectively. On the whole, the FDI positively impacted employment growth, human development, infrastructure, agriculture, industry, and service sectors.

**Keywords** Foreign direct investment · Role · Employment · Generation · India

---

S. Muruganandhan · A. R. Sanjeev Kumar · P. B. Shrithik · A. Biju
Saveetha School of Law, Saveetha Institute of Medical and Technical Sciences, Chennai, India

V. Vinothina
Department of Computer Science PG, Kristu Jayanti College, Bengaluru, India

K. Sivasubramanian (✉)
Department of Economics, Kristu Jayanti College, Bengaluru, India
e-mail: sivasubramanian@kristujayanti.com

 
M. Al Mubarak (ed.), *Sustainable Digital Technology and Ethics in an Ever-Changing Environment*, Studies in Systems, Decision and Control 237,
https://doi.org/10.1007/978-3-031-86708-8_34

# 1 Introduction

In the course of economic development, the nations are encountering various stages to reach progress. In this process, employment generation plays a pivotal place in removal of poverty in both literates and illiterates. There are three hands in the economic process that will support investment creation. Capital accumulation can be possible with private and public investment. These two investments are termed as the first two hands of the economy. While these two investments are not enough to meet the requirements of a large country like India, the third hand will support employment generation [8]. To become the self-reliant of a developing country, the FDI helps to promote domestic production and helps in achieving a positive gross domestic product of the nation [10]. Developing and third-world nations aspire to receive FDI for their nation's progress and it creates spillover effects on the domestic economy [3].

# 2 Paper Structure

Foreign investment is a boon for domestic countries for enhancing employment opportunities and is a strong tool for reducing poverty. India is one of the desirable destinations for foreign investors. Based on this background, this chapter has been drawn with specific objectives such as examining the significant role of foreign direct investment (FDI) in India and analysing the impact of FDI inflow in generating employment opportunities in India. This study has made an attempt to achieve these objectives through various existing literature and secondary data. This chapter has been structured with an introduction segment, objectives of the chapter, methodological framework for research, literature survey, data analysis, concluding notes, recommendations based on the research results and future directions.

# 3 Objectives

To examine the significant role of the foreign direct investment (FDI) in India.

To analyse the impact of FDI inflow in generating employment opportunities in India.

## 4 Methodology

The present study has followed the analytical and quantitative research methods to reach the above-stated objectives. The analytical and quantitative research methods have been applied to establish the results through primary or secondary data sources. This research study is based on the existing research works and secondary data sources. Various pieces of literature have been sourced from published articles and data collected from the World Bank database and published reports. Collected literature and data were analysed and the results have been discussed. Secondary data was chosen from the period 2001 to 2019.

## 5 Literature Survey

This study aims to examine the importance of FDI inflow to India and to assess the impact of FDI on employment creation in various sectors in India. It is evident from the prevailing literature that FDI positively impacted employment creation and economic development.

### *5.1 Impact of FDI on Infrastructure*

Capital investments made by the individual or an organisation from one nation to other nation on infrastructural development is called as FDI infrastructure. FDI in infrastructural development has grown significantly in many sectors such as transportation, energy, logistics, digital technology and telecommunication. Through infrastructural development, FDI positively supports the extension of existing industries, setting up of new businesses and stimulates economic growth [9, 11].

### *5.2 FDI and IT Sector*

FDI in the information and technological sector plays key role in the development of employment opportunities in IT and IT enabled industries. FDI helps to promote the transfer of technology from one country to another through the FDI inflows. FDI also helps to enhance the productivity of the IT sector. FDI enhances technological adoption and increases the tax base of the domestic country [3, 13].

### 5.3 Impact of FDI on Industrial Development

There is a good belief that FDI in industrial organisations in many developing nations has a significant enhancement of production. The reason for such a positive impact was the transfer of technology, advanced industrial practices, management support, and good business connections. Industrial development through FDI inflow has contributed to many areas such as implementing AI technology, machine learning, and deep learning. Besides, production and productivity, the FDI plays a significant role in the manufacturing industries of India [7]. Direct investments in Indian manufacturing industries support the whole development of the industries of the nation. It also creates employment opportunities in many large and medium-sized [2].

### 5.4 Impact of FDI on Economic Development

The FDI positively impacts the economic development of many developing nations. The structural change in the movement of labourers from less productive sectors to a highly productive sector stimulates the growth of the economy. Structural employment helps to promote employment opportunities in rural and urban areas including agricultural and ono-agricultural employment. It also improves both formal and informal employment in the Indian economy [5]. Economic growth through GDP has increased due to improved inflow of FDI in all the sectors in many developing countries [4, 12].

### 5.5 Impact of FDI on Human Development

FDI made a crucial contribution to the Indian economy's knowledge transfer and capacity building [6]. Moreover, FDI improves the volume of trade among industries and helps to promote labour efficiency through knowledge exchange programmes.

## 6 Data Analysis and Discussion

The FDI emerged as a key player in the generation of employment and growth of the nation. It creates both direct and indirect employment. Along with employment creation, FDI brought various other advantages for the sustainable development of the nation. Figure 1 illustrates the value of FDI inflow to India from 2001 to 2019. The inflow FDI has consistently improved from US $5.13 billion to US $43.41 and it was started decline the year 2008 till 2010. Further, FDI increased from US $27.4 billion. FDI inflow has again started improved with US $24 billion from the year

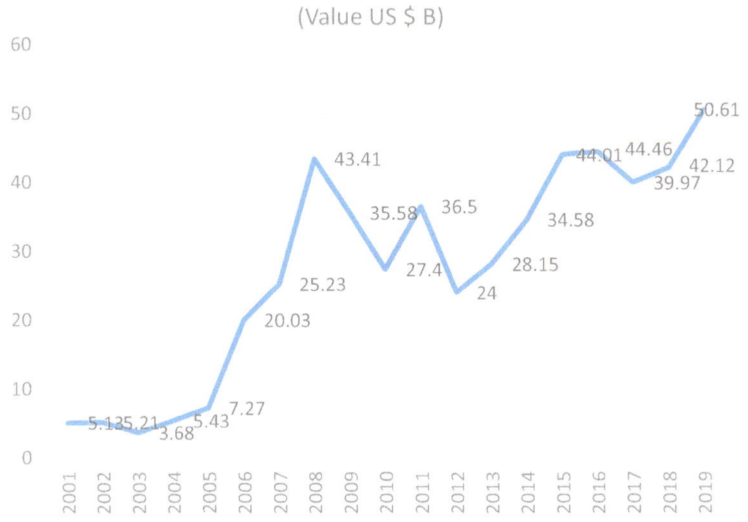


**Fig. 1** Amount FDI to India from 2001 to 2019 (Billion US $). *Source* World Bank Reports, various years

2012 onwards and has a consistent growth till US $50.61 billion. Overall, the amount of investment has raised from US $5.13 billion to US $50.61 billion.

Figure 2 emphasises the sectoral employment generation through the inflow of FDI in India from the year 2001 to 2010. It is observed from Fig. 2 that the agricultural sector received the highest foreign investment with 59.04% of the total investment in 2001, 36.66% in industries and 24.29% in FDI came to the service sector. Over some time, the service sector attracted a good amount of FDI from various countries to India, it has improved from 24.29% to 26.56% during the year from 2001 to 2010 respectively.

The agricultural sector has seen a further decline in receiving FDI inflow due to various reasons from 2011 to 2019. It is recorded that 48.96% of the FDI inflow for the agricultural sector declined to 43.21% in 2019. It saw a steep drop according to Fig. 3.

On average the industrial sector has increased from 23.52% to 24.89% in the year 2011 and 2019 respectively (Fig. 4). Indian industries are managed to hold the FDI inflow to India.

The service sector attracted a huge amount of FDI to India from 27.51 to 31.89 during the year 2011 and 2019 respectively (Fig. 5).

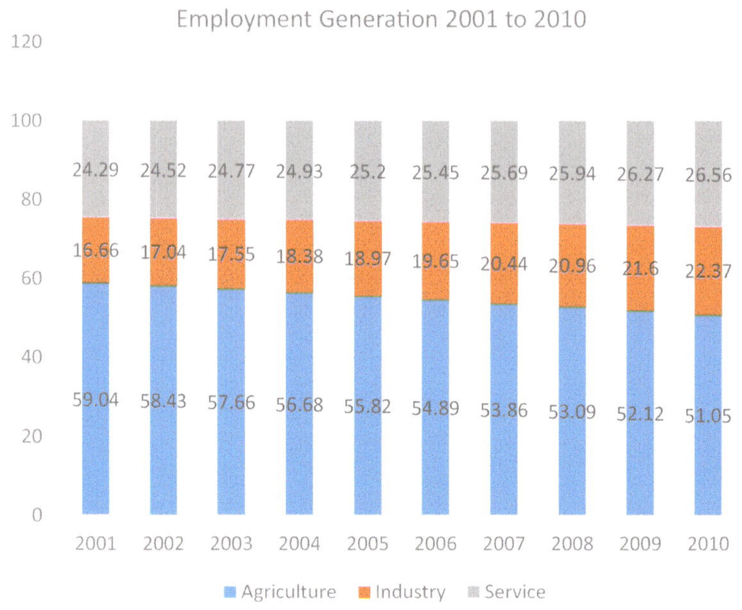


**Fig. 2** Sector-wise generation of employment through FDI. *Source* Same as the previous figure

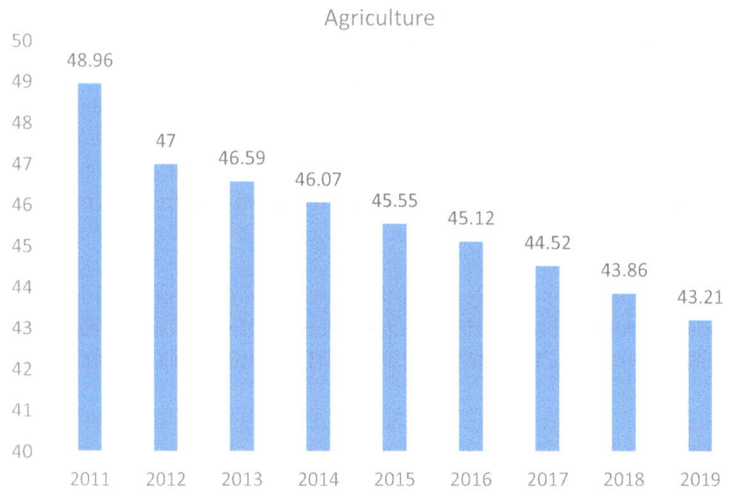


**Fig. 3** FDI inflow in the agricultural sector from 2011 to 2019. *Source* Same as the previous figure

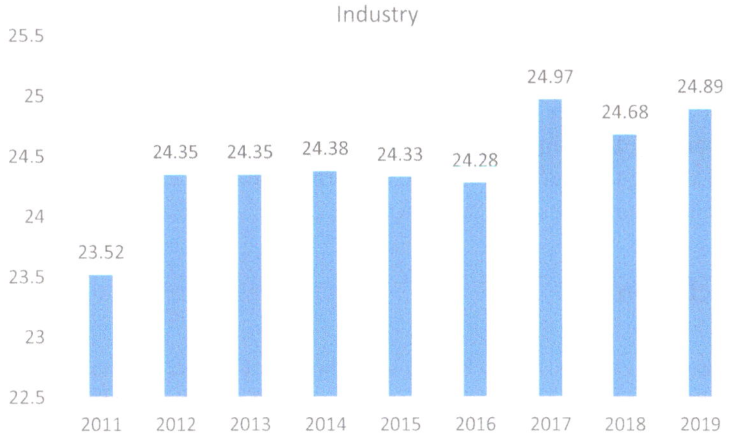


**Fig. 4** FDI in the Indian industrial sector. *Source* Same as the previous figure

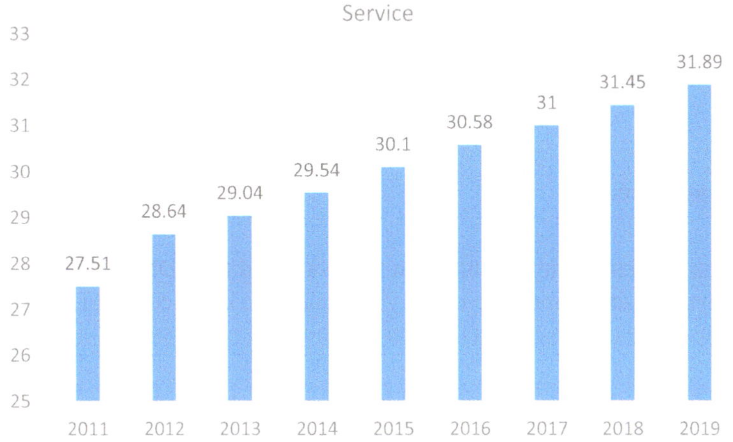


**Fig. 5** FDI inflow to the service sector. *Source* World Bank Database

## *6.1 Key Findings*

To become self-reliant in a developing country, the FDI helps promote domestic production and achieve a positive gross domestic product for the nation.

Through infrastructural development, FDI positively supports the extension of existing industries, and the setting up of new businesses and stimulates economic growth. FDI helps to promote the transfer of technology from one country to another through the FDI inflows. FDI also helps to enhance the productivity of the IT sector.

Industrial development through FDI inflow has contributed to many areas such as implementing AI technology, machine learning, and deep learning. The inflow FDI has consistently improved from US $5.13 billion to US $43.41 and it was started decline the year 2008 till 2010. Further, FDI increased from US $27.4 billion.

It is recorded that 48.96% of FDI inflow for the agricultural sector declined to 43.21% in 2019. It saw a steep drop. Over some time, the service sector attracted a good amount of FDI from various countries to India, it has improved from 24.29% to 26.56% during the year from 2001 to 2010 respectively.

## 7 Conclusion

The primary goal of this research is to show how FDI affects human development in developing countries, as well as the moderating effects of administration, ODA, and national competitiveness. To achieve this goal, we first investigate the impact of FDI on the growth of people. We investigate the negative effects of FDI and estimate their influence on human development [1]. MNCs seek to use the earth's resources and human talents, but they bear no responsibility for the host country's long-term economic progress and human development. On the whole, the FDI impacted positively on employment growth, human development, infrastructure, agriculture, industry and service sector.

## 8 Suggestions

India is the fastest-growing economy in the world and the FDI is supporting a lot for employment generation. The country is required to enhance the FDI for Indian start-up businesses to have collective and collaborative progress.

## 9 Limitation

This study is limited to a few variables for analysis through selected literature and limited data. There is a wide scope for in-depth analysis of the research to expand further evaluation to make an even more extensive conclusion.

## References

1. Azman-Saini, W.N.W., Baharumshah, A.Z., Law, S.H.: Foreign direct investment, economic freedom, and economic growth: international evidence. Econ. Model. **27**(5), 1079–1089 (2010)
2. Banga, R.: The export diversifying impact of Japanese and US foreign direct investments in the Indian manufacturing sector. J. Int. Bus. Stud. **37**(4), 558–568 (2005)

3. Das, G., Chaudhuri, B.R.: Impact of FDI on labour productivity of Indian IT firms: horizontal spill-over effects. Prajnan **11**(7), 121–137 (2018)
4. Dutta, P.: Relationship between foreign direct investment and economic growth of India: empirical evidence from India. Webology **18**(5) (2021)
5. Emako, E., Nuru, S., Menza, M.: The effect of foreign direct investment on structural change in developing countries: an examination of the labor productivity dimension. Cogent Bus. Manag. **9**(1) (2022). https://doi.org/10.1080/23311975.2022.2135209
6. Ghose, A.K.: Structural change and development in India. Indian J. Hum. Dev. **15**(1), 7–29 (2021). https://doi.org/10.1177/09737030211005496
7. Goldar, B., Sharma, A.K.: Foreign investment in Indian industrial firms its impact on firm performances. J. Ind. Statist. **4**(1) (2015)
8. Mahapatra, A.: Influence of FDI on employment development in India: an analytical study. Wutan Huatan Jisuan Jishu **16**(6), 147–158 (2020)
9. Malhotra, M.: Foreign direct investment opportunities in infrastructure development: a study on India. In: Das. R.S. (ed.) Handbook of Research on Economic, Financial, and Industrial Impacts on Infrastructure Development. IGI Global Publishers (2017)
10. Mishra, R., Palit, S.: Role of FDI on employment scenario in India. Int. J. Rec. Technol. Eng. **8**(6) (2020)
11. Sudhakar, P., Velmurugan, R.: Impact of foreign direct investment in Indian economic growth. E3S Web Conf. **449** (2023)
12. Suri, T., Udry, C.: Agricultural technology in Africa. J. Econ. Perspect.Perspect. **36**(1), 33–56 (2022). https://doi.org/10.1257/jep.36.1.33
13. Veeramacheneni, B., Vogel, R., Ekanayake, E.M.: Information technology, FDI and economic growth: an India case study. Southwestern Econ. Rev. 95–102 (2007)

# Impact of Freebies in Electoral Politics and the Role of Dravidian Government in Tamil Nadu

**S. Muruganandhan, Y. G. Gokul, S. Abishek, R. Thamilmani, K. Sivasubramanian, and B. Haripriya**

**Abstract** The election is one of the pillars of democracy in India. The elections are a fundamental source of democratic countries. Election is the basic tool for any democratic country in this world. A vast group of qualified and eligible citizens of a country decides leader or a representative. The major aim of these freebie schemes is to ensure accessibility and affordability and reduce poverty among weaker sections of the state. It also aims to achieve the goal of improving literacy, generating opportunities for higher education, and enhancing the economic status of marginalized sections in Tamil Nadu. However, these welfare measure policies have been viewed as electoral politics to create vote banks for such political parties. Despite all the controversial debates relating to it, the freebies are offered by Tamil Nadu as a political strategy. This chapter prepared intending to find the role of the Dravidian government in Tamil Nadu and analyse the social and economic impact of freebie schemes in Tamil Nadu. This chapter is purely based on the reviews collected from various published sources. The collected publications have been presented and logically ordered based on the concepts and results.

**Keywords** Freebies · Electoral politics · Dravidian government · Tamil Nadu · Economic and social impact

S. Muruganandhan · Y. G. Gokul · S. Abishek
Saveetha School of Law, Saveetha Institute of Medical and Technical Sciences, Chennai, India

R. Thamilmani
School of Commerce, Vel Tech Rangarajan Dr. Sagunthala R&D Institute of Science and Technology, Chennai, India

K. Sivasubramanian (✉) · B. Haripriya
Department of Economics, Kristu Jayanti College, Bengaluru, India
e-mail: sivasubramanian@kristujayanti.com



M. Al Mubarak (ed.), *Sustainable Digital Technology and Ethics in an Ever-Changing Environment*, Studies in Systems, Decision and Control 237,
https://doi.org/10.1007/978-3-031-86708-8_35

# 1 Introduction

In general, the freebies are the goods and services provided by the government to the people free of cost. In this case, the Tamil Nadu government is significantly providing many freebie schemes as welfare measures to the people of Tamil Nadu. These cost-less commodities such as laptops for higher education students, bicycles for government school students, household appliances such as mixers, grinders, fans, and colour televisions, and financial support for women in rural and urban areas. The major aim of these freebie schemes is to ensure accessibility and affordability and reduce poverty among weaker sections of the state. It also aims to achieve the goal of improving the literacy, generating opportunities for higher education, and enhancing the economic status of marginalized sections in Tamil Nadu. However, these welfare measure policies have been viewed as electoral politics to create vote banks for such political parties. Despite all the controversial debates relating to it, the freebies are offered by Tamil Nadu as a political strategy [1].

## *1.1 Structure of the Paper*

Based on the background of the study, it has been identified that the Tamil Nadu people's social and economic status has improved considerably in the state. But still, many debates and controversies are also paralleled with freebie welfare schemes. To identify and address this problem, this study has been taken up to reveal the real-time status and impact of the freebie schemes. So, this study has been prepared to achieve specific goals such as understanding the role of the Dravidian government in Tamil Nadu and examining the socio-economic impact of freebie schemes in the state. This paper was built based on the following structure, introduction of the study, objectives framed, methodology, literature analysis, and the impact of freebies schemes on the socio-economic enhancement of vulnerable peoples.

## *1.2 Objectives*

To understand the role of the Dravidian government in Tamil Nadu.

To examine the socio-economic impact of freebie schemes of Tamil Nadu.

## 2 Methodology

This chapter prepared intending to find the role of the Dravidian government in Tamil Nadu and analyse the social and economic impact of freebie schemes in Tamil Nadu. This chapter is purely based on the reviews collected from various published sources. The collected publications have been presented and logically ordered based on the concepts and results.

## 3 Literature Review

The freebie concept is very families and pioneer in Tamil Nadu state and it helps to shape the society in a good manner. The Dravidian political legacy, which has been dominated in the state from 1960. It placed an important role on ensuring social justice, income equality, and economic welfare. It also influencing of the electoral politics through attracting the poor voters in the state. Freebies are helps to promote the consumption level of the marginalised people and it stimulate the rural economy [2].

### *3.1 Significance of Elections*

Political parties and politician in the state often utilize freebies as an instrument for mobilising the votes, specifically among socially and economically backward people. Promises on freebies has been presented in their election manifesto. Freebie welfare scheme creates a good sense of commitment and loyalty among the people, who may ready to obey to vote for them. The freebie schemes significantly contributing in determination of election outcomes in Tamil Nadu [3].

### *3.2 Dravidian Rule*

Dravidian movements brought on the Tamil cultural and strong values, traditional languages, and braveness which helped to form the concrete political structure. However, the transfer of power from the British to India has been directly and independently approached by the great Dravidian political policy [4, 5]. Dravidian political policy represents an ever-growing degree of specialisation in interests to embrace fundamental laws and democratic and electoral policies [6].

## 3.3 Voting Behaviour

The voting behaviour in Tamil Nadu is unique and it is characterised by a strong allegiance to the regional parties. The regional political parties have been dominating the state's full political landscape for many years together. The voters from Tamil Nadu have been inspired by charismatic political leaders, various welfare schemes, and caste systems [7]. The concept of freebies shall change the mindset of the people in voting for political parties in various elections. Promise to provide various commodities and services before the elections through the state government policy leads to change the behaviour towards voting [8].

## 3.4 Impact of Freebie Schemes

Freebies in the state of Tamil Nadu have been impacted on various important elements of development such as educational attainment of poor people, improving social status of the marginalised people, improving economic status of rural people, enhancing skill development for better employment and income generation of the poor people in the log-run.

## 3.5 Impact on Educational Accessibility

The freebie scheme provides free laptops, free uniforms, free books, free bus passes, free bicycles, free mid-day meals, a free breakfast scheme and scholarships that positively impact the accessibility to education by the underprivileged and marginalised people [9]. These contributed to a higher level of enrolment rates and an enhanced retention rate at schools. The midday food welfare scheme, in particular, has encouraged parents to enrol their children in schools. It leads to an increase the attendance rates, and enrolment rates and reduces dropout rates consistently, particularly among the poor sections [10].

## 3.6 Impact on Socio-economic Status

Social protection schemes are a unique and important component of the people of every country wherein the authorities should provide appropriate economic protection to all the citizens to prevent vulnerabilities. In many developed and developing democratic countries, such protection is available through their law. But in this case, a developing country like India is initiating various developmental schemes to remove poverty. Providing educational instruments such as laptops and bicycles helps to enhance the educational attainment of girl children in Tamil Nadu It helps to enhance the accessibility of digital resources by female students in rural areas [11].

### *3.7 Impact of Freebies on Reducing Poverty in Tamil Nadu*

The impact of schemes for welfare measures like freebies on the removal of poverty in Tamil Nadu is a very multifaceted and nuanced problem. It has a long history of providing welfare measures, often called to as freebies. It has been brought with a broad objective to remove the poverty among the marginalised sections. The freebies provided by the government of Tamil Nadu such as free rice, subsidized food grains, cost-free electricity for agricultural activities, subsidised electricity for household consumption and women right money (cash transfer for women) has been given for removing poverty. These schemes are significantly helps to reduce hunger and poverty among marginalized segments of the society [12]. These schemes are also helps to promote the human capital development in the state through educational attainment and certain schemes aims to enhance the skill development of the younger population of the state. It will help the people to generate more income in the long-run [13, 14].

### *3.8 Negative Impacts of Freebie Schemes*

Freebies schemes are giving a short-term relief to the poor people but there is no long term or the consistent solution for the socio-economic issues. It creates negative influence on various macroeconomic indicators of development namely budget deficits, decline in capital investment and increase in inflation. The freebie expenses create fiscal burden on the state government. Allocation of large amount of budgetary funds for free schemes leads to deficits and increase state government debts. It is also resulted in reducing the fund allocation for other developmental purposes such infrastructure, education, and health sector. The log-term developmental investments also negatively impact through non-availability of fund for government industries and it leads to disinvestment [15, 16].

## 4 Findings

The Dravidian political legacy, which has been dominated in the state from 1960. It placed an important role on ensuring social justice, income equality, and economic welfare. Political parties and politician in the state often utilize freebies as an instrument for mobilising the votes, specifically among socially and economically backward people. Promises on freebies has been presented in their election manifesto.

Providing educational instruments such as laptops and bicycles helps to enhance the educational attainment of girl children in Tamil Nadu It helps to enhance the accessibility of digital resources by female students in rural areas. The impact of schemes for welfare measures like freebies on the removal of poverty in Tamil Nadu is a very multifaceted and nuanced problem. It has a long history of providing welfare measures, often called to as freebies. It has been brought with a broad objective to remove the poverty among the marginalised sections. The freebies provided by the government of Tamil Nadu such as free rice, subsidized food grains, cost-free electricity for agricultural activities, subsidised electricity for household consumption and women's rights money (cash transfer for women) have been given to remove poverty. These schemes significantly help to reduce hunger and poverty among marginalized segments of society. Freebies schemes are giving short-term relief to poor people but there is no long-term or consistent solution for the socio-economic issues. It creates a negative influence on various macroeconomic indicators of development namely budget deficits, decline in capital investment and increase in inflation.

## 5 Conclusion

To conclude, this paper evaluates the various instruments and incidents of the political system, voting patterns and behaviour in Tamil Nadu. This study also made a significant attempt to understand state politics and the Dravidian political role in the development of the state. Freebie schemes have made a significant positive change in the people of Tamil Nadu positively. It is found from the study that there are few negative impacts on certain macroeconomic indicators in Tamil Nadu's government.

## 6 Suggestions

The state government of Tamil Nadu may plan for progressive schemes for socio economic development.

It is suggested to enhance the comprehensive and inclusive plan for the marginalized peoples of Tamil Nadu.

## 7 Future Scope

This study has certain limitations such as the non-availability of secondary information and lack of literature. However, there is a wide range of scope for extending this study through primary data sourcing. Thus, a future study can be conducted for a deep investigation of the real-time impact on socio-economic status in Tamil Nadu.

## References

1. Damodharan, K., Rajangam, S., Gorringe, H.: Dalit political imagination and replication in contemporary Tamil Nadu. Econ. Pol. Wkly. **47**(36), 30–34 (2012)
2. Ganeshan, H.: A case study on Tamil Nadu Government's welfare schemes to motivate SMEs manufacturing wet grinders in Coimbatore, Prabhandhan. Indian J. Manag. **6**(3) (2013). https://doi.org/10.17010/pijom/2013/v6i3/59983
3. Padmanaban, S.: To take or not to take: is the freebie culture in Tamil Nadu elections a threat to Indian democracy? Asian J. Polit. Sci. 215–238 (2021). https://doi.org/10.1080/02185377.2021.1951783
4. Hodges, S.: Contraception, Colonialism and Commerce: Birth Control in South India, 1920–1940. Routledge Publications (2008)
5. Thiruppathi, P.: A political instrument: Dravidian party's communication strategies in Tamil Nadu election. Indian J. Soc. Polit. **11**(2) (2024)
6. Arni, A.: Taking Periyarism seriously: the Dravidian identity as a universality. Glob. Intell. Hist. (2024). https://doi.org/10.1080/23801883.2024.2386600
7. Kalaiyarasan, A., Vijayabaskar, M.: Dravidian Model: Interpreting the Political Economy of Tamil Nadu. Cambridge University Press (2021)
8. Verma, A., Saxena, S., Alam, M.B.: Changing dynamics of voting pattern affecting the psyche of voters by offering freebies. J. Pharm. Nagative Res. **13**(3) (2022). https://doi.org/10.47750/pnr.2022.13.03.051
9. Ram, N.: Dravidian movement in its pre-independence phases. Econ. Pol. Wkly. **14**(7/8), 377–402 (1979)
10. Kanmony, C.J.: Impact of free mid-day meal and breakfast schemes on education with special reference to Tamil Nadu. J. Econ. Bus. Manag. Adm. **5**(1), 1–6 (2024)
11. Rahman, A., Pingali, P.: Social welfare 'schemes' to an economic security 'system'. In: The Future of India's Social Safety Nets. Palgrave Studies in Agricultural Economics and Food Policy. Palgrave Macmillan, Cham (2024). https://doi.org/10.1007/978-3-031-50747-2_10
12. Raghav, K., Sudarsan, P., Srikant, P.: NOTA: a strategic choice with a positive impact on Indian elections. Asian J. Polit. Sci. **31**(3), 180–196 (2023)
13. Kumar, A.: Freebie debates and budgetary priorities: economic growth-led India or inclusive India. Int. J. Commun. Social Dev. **6**(1) (2024). https://doi.org/10.1177/25166026241237325
14. Sivarajan, R.: Study on transformations to sustainable development in Tamil Nadu in the effect of implementing socio-economic development schemes by Government of Tamil Nadu. Strad Res. **10**(11) (2023). http://stradresearch.org/
15. Singh, A.: Analysing the impact of freebies culture on Indian economy. Int. Res. J. Commer. Arts Sci. **14**(5), 228–242 (2023)
16. Wyatt, A.: Populism and politics in contemporary Tamil Nadu. Contemp. South Asia **21**(4), 265–281 (2012). https://doi.org/10.1080/09584935.2013.803036

# Gender Discrimination and Empowerment of LGBT Community in India: A Systematic Review

**S. Muruganandhan, A. R. Sanjeev Kumar, R. Saiesakkisakthivel, M. Abishek, K. Sivasubramanian, and Mathun Muinao**

**Abstract** The prime goal of this paper is to understand the conditions, issues, and challenges faced by the LGBTQ community in India through the existing research publications from India and other nations. The discrimination of gender is a historical problem in India among males and females. It not only prevails among male and female discrimination, it also extends to the LGBTQ community too (commonly known as lesbian, gay, bisexual, transgender and queer). They are not being treated properly as human beings. They have been discriminated against in many forms such as workplace, social, legal, and economic. India has a vast tradition of LGBTQ+ with numerous courageous personnel fighting for basic and relevant legal rights. The objectives of this paper are to review the significance of the legal rights of the LGBTQ community in India and analyse the gender discrimination and living conditions among the LGBTQ community. Almost all LGBTQ adults do not share their poor living status in public due to rejection and negative responses. It is also identified in many types of research that LGBTQ individuals face disturbances from their own families and surroundings. These issues arise due to misunderstanding and miscommunication of their gender identity and appearance. Because of these conflicts, the LGBTQs are evicted from their homes and forced to live in open spaces on the rear sides of the road. Overall, their social, economic, living, working and legal status are vulnerable and require strong policy and legal support for empowerment.

**Keywords** Gender · Discrimination · LGBT · Community · Legal rights

S. Muruganandhan · A. R. Sanjeev Kumar · R. Saiesakkisakthivel · M. Abishek
Saveetha School of Law, Saveetha Institute of Medical and Technical Sciences, Chennai, India

K. Sivasubramanian (✉) · M. Muinao
Department of Economics, Kristu Jayanti College, Bengaluru, India
e-mail: sivasubramanian@kristujayanti.com

 
M. Al Mubarak (ed.), *Sustainable Digital Technology and Ethics in an Ever-Changing Environment*, Studies in Systems, Decision and Control 237,
https://doi.org/10.1007/978-3-031-86708-8_36

# 1 Introduction

The discrimination of gender is a historical problem in India among males and females. It not only prevails among male and female discrimination, it also extended to the LGBTQ community too (commonly known as lesbian, gay, bisexual, transgender and queer). They are not treated properly like another human beings. They have been discriminated against in many forms such as workplace, social, legal, and economic. India has a vast tradition of LGBTQ+ with numerous courageous personal flighting for basic and relevant legal rights. It observed that they are facing various societal discrepancies and the various activists are fighting for their rights [2]. This section of society in India is relatively more vulnerable in terms of discrimination and untouchability [3]. On this contextual idea, this work has been constructed to establish the significance of the legal rights of the LGBTQ community in India. It also aims to analyse the reasons for discrimination and challenges faced by the LGBT community in India.

## *1.1 Structure of the Paper*

To reach the identified objectives based on the existing studies, this particular paper has been prepared in the following manner such as background of the theme, objectives, methodological framework, selected related literature, results from the analysis, major findings and conclusion. The analytical review resulted in evaluating the similarities and dissimilarities of the presented literature. Finally, various relevant suggestions have been given for the removal of socio-legal discrimination among the LGBTQ community in India.

## *1.2 Objectives*

To review the significance of the legal rights of the LGBTQ community in India.

To analyse the gender discrimination and living conditions among the LGBTQ community.

# 2 Methodology

The methodological framework of the research paper aims to provide a blueprint for an empirical or review-based analysis. The major aims of this review-based paper are to understand the significance and legal rights of the LGBTQ community and to analyse the gender discrimination and living conditions among the LGBTQ community in India. To reach these stated goals, this study adopted a descriptive research framework and sourced relevant literature from published works. A limited amount of quantitative data has been collected from various secondary sources to meet the stated objective of this paper.

## 3 Literature Review

It is observed from the census 2011, there are 4.88 million individuals in the nation have been recorded as transgender. It has been further analysed that this figure was underestimated and many people under this segment have not been reported. They have been remaining unidentified social groups or invisible [1, 7].

### *3.1 Gender Discrimination*

There exists an immense diversity and uniqueness among the transgender community in India in terms of ethnicity, culture, and gender expression. Transgender and the LGBT community are nothing but individuals whose gender reflection or expressions are different from the tradition which is associated with a person's birth. This segment of people are non-confronting individuals as part of cultures across the world. Further, these persons are facing several issues relating to psychological, social, economic, and legal rights issues including discrimination [5].

### *3.2 Legal Provisions*

The LGBT community in the nation has been a culturally documented group, different, with a diverse identity among them. Though this community people are also human beings, they have been discriminated against by various societal norms. Their major source of income remains begging and many of them are still under poverty. It leads to stigma and gender discrimination. In the process of eliminating these issues, the government of India enacted the transgender and LGBTQ person's protection bill in 2016 and it was revised in 2018. This particular bill enables a mechanism for providing legal accomplishment of their gender is known as transgender [4, 12].

## 3.3 Human Rights

The transgender bill aims to protect transgender persons including the LGBT community which includes individuals who need to obtain the identity certificate and assert their gender identity. The amendments for LGBTQ persons have been enacted as the transgender persons Act of 2019 in India have addressed non-binary people's constitutional rights and provisions, identity rights of a specific gender and laws relating to non-discrimination [6, 8]. LGBTQ or homosexuality is the societal stigma, frequently and consistently narrated by discrimination, harassment, and marginalisation. It has been identified that over the last few decades, the LGBTQ are striving for the basic rights of social, cultural, ethical, and constitutional perspectives [9, 10]. In terms of an indication of someone's physical and enthusiastic enthusiasm, sexual direction is used. Three broad classifications of people can be made based on their sexual orientations: heterosexuals, homosexual men, and bisexuals. Heterosexuals are those who are drawn to other individuals of the same sex. Sexual leadership and human rights issues do not consider variation because society is not segregated. Many people who find themselves drawn by both men and women are classified as bisexuals. Most bisexuals do not even depend on their convictions of discrimination or savagery because they struggle to produce deceivability as bisexuals [14].

## 3.4 Challenges

In terms of civil rights, sexual orientation, or identity, comparisons to homosexuals are commonplace, regardless of the circumstances. Those who identify as homosexuals are those who favour the same sex in the same setting. The terms "lesbian" and "gay" are occasionally used to describe homosexual women and men. Gender work is the outward manifestation of masculinity or femininity, and gender identification is the actual knowledge of being a man or a woman [15]. There are various challenges have been encountered by the LGBTQs such as basic human rights issues, social problems, economic issues, employment conditions, working conditions and health issues [13]. The LGBT community have problems relating to the accessibility of formal credit from organised financial institutions to start self-employment initiatives [11]. The Indian legal system has consistently pronounced and continuously stresses the significance of the human rights of all human beings including the LGBTQ community. These people are always away from the available laws for attaining their basic human rights [16, 17]. Almost all LGBTQ adults do not share their poor living status in public due to rejection and negative responses. It is also identified in many types of research that LGBTQ individuals face disturbances from their own families and surroundings. These issues are arising due to misunderstanding and miscommunication of their gender identity and appearance. Because of these conflicts, the LGBTQs are evicted from their homes and forced to live in open spaces on the rear sides of the road [18, 19].

### *3.5 Government Support*

To guarantee the dignity, freedom of sexuality, and individuality of the LGBT community in India, collaboration between the three fundamental pillars of society, the State, and the LGBT community is necessary to uphold the moral principles and values derived from the Constitution [20]. But despite all the progress, discrimination against LGBT people persists in society. A decision from the Supreme Court can only provide a remedy,society must still make sure that members of the LGBT community are handled with respect and are not the target of prejudice [20, 21]. The government supports the LGBTQ for rule 3 to obtain the identity certificate. Rule 10 of the Protection Right, 2019 act helps to access various government schemes. The Protection Act helps to promote non-discrimination, equal opportunity in employment, grievance-solving mechanisms, and also setting off a national council to protect and facilitate the transgender community [22, 23].

## 4 Main Findings

The LGBTQ community are not treated properly like other human beings. It observed that they are facing various societal discrepancies and the various activists are fighting for their rights.

It is observed from the census 2011, there are 4.88 million individuals in the nation have been recorded as transgender. It has been further analysed that this figure was underestimated and many people under this segment have not been reported.

There exists an immense diversity and uniqueness among the transgender community in India in terms of ethnicity, culture, and gender expression. To guarantee the dignity, freedom of sexuality, and individuality of the LGBT community in India, collaboration between the three fundamental pillars of society, the State, and the LGBT community is necessary to uphold the moral principles and values derived from the Constitution.

It is also found from the literature analysis that heterosexuals are those who are drawn to other individuals of the same sex. Sexual leadership and human rights issues do not consider variation because society is not segregated.

Many people who find themselves drawn by both men and women are classified as bisexuals. Most bisexuals do not even depend on their convictions of discrimination or savagery because they struggle to produce deceivability as bisexuals.

## 5 Conclusion

Adolescents now have unprecedented and unrestricted access to current affairs and news via broadcasting and social media. Adolescents who identify as LGBTQ may be particularly aware of media and public disclosure about legal and policy developments that influence their group's legal rights. The objectives of this paper are to review the significance of the legal rights of the LGBTQ community in India and analyse the gender discrimination and living conditions among the LGBTQ community. Almost all LGBTQ adults do not share their poor living status in public due to rejection and negative responses. It is also identified in many types of research that LGBTQ individuals face disturbances from their own families and surroundings. These issues arise due to misunderstanding and miscommunication of their gender identity and appearance. Because of these conflicts, the LGBTQs are evicted from their homes and forced to live in open spaces on the rear sides of the road. On the whole, their social, economic, living, working and legal.

## 6 Suggestions

There is a proper awareness to be created through the government, non-government agencies and institutions for a better understanding of LGBTQ.

The government has to support these communities for their enhanced protection, rights, dignity, and social and economic opportunity.

## 7 Limitations and Future Directions

This paper reviews various existing studies relating to gender discrimination, inequality, human rights issues, and social-economic, and legal challenges encountered by the LGBT community. This paper is based on the available literature studies but there is a good avenue for expanding this study through primary data. The field-level data could help better assess the real-time problems of the LGBT community.

## References

1. Aaberg, L.: Corporate India after section 377: haphazardness and strategy in LGBTQ diversity and inclusion advocacy. Gend. Place Cult. J. Feminist Geogr. 1–18 (2022). https://doi.org/10.1080/0966369x.2022.2146660
2. Abraham, A.: 12 Famous and Motivational LGBTQ+ Activists in India. Learn Transformation (2023). https://learntransformation.com/famous-lgbtq-activists-in-india/

3. Akhan, A.Z.: LGBT community: recognition and protection of their rights in India. Int. J. Multidisc. Res. **5**(5), 1–9 (2023)
4. Arasu, P.: Queer Liberation Unit 3 Legal Issues: Case Study of Section 377. Indira Gandhi National Open University, New Delhi, India (2017). https://egyankosh.ac.in/bitstream/123456789/2983/1/Unit-3.pdf
5. Arvind, A., Pandya, A., Amin, L., Aggarwal, M., Agrawal, D., Tiwari, K., Singh, S., Nemkul, M., Agarwal, P.: Social strain, distress, and gender dysphoria among transgender women and Hijra in Vadodara India. Int. J. Transgen. Health. **23**(1–2), 1–15 (2021). https://doi.org/10.1080/26895269.2020.1845273
6. Bhattacharya, S., Ghosh, D., Purkayastha, B.: 'Transgender persons (protection of rights) act' of India: an analysis of substantive access to rights of a transgender community. J. Hum. Rights Pract. **14**(2), 676–697 (2022). https://doi.org/10.1093/jhuman/huac004
7. Census of India: Census Report 2011. Government of India (2011). https://www.censusindia.gov.in/
8. Chaudhary, S.: Navtej Johar v. Union of India: Love in Legal Reasoning. Natl. Univ. Juridical Sci. (NUJS) Law Rev. **12**(3–4) (2019). http://nujslawreview.org/wp-content/uploads/2020/01/12-3-4-Chaudhary.pdf
9. Dangre, A., Tripathi, A.: Beyond section 377: the peril of being homosexual in India. SSRN Electron. J. (2018). https://doi.org/10.2139/ssrn.3214061
10. Ejaz, K., Moscowitz, L.: Who 'framed' Ramchandra Siras? Journalistic discourses of sexual citizenship in India. Sexualities **23**(5–6), 951–970 (2020). https://doi.org/10.1177/1363460719876829
11. Fullerton, M.: Diversity and inclusion- LGBT inclusion means business. Strateg. HR Rev. **12**(3), 121–125 (2013)
12. Ghia, G.: Transgender rights bill: a stunted understanding of gender and equality. The Wire (2019). https://thewire.in/lgbtqia/transgender-rights-bill-a-stunted-understanding-of-gender-and-equality#google_vignette
13. Gupta, R.: Rights of LGBT people in India: Navtej Singh Johar v Union of India. My Lawman Socio Legal Rev. **1**(2), 1–9 (2021). https://mslr.pubpub.org/pub/i15d4bkk/release/1
14. Gupta, I.S.: Human Rights of Minority and Women's: Human Rights and Sexual Minorities, vol. 4, p. 217. Gyan Publishing House (2005)
15. Jawale, K.: Issues and challenges of 'LGBT' minority people in India. Int. J. Appl. Res. **2**(6), 408–410 (2016)
16. Kaur, L.: Exploring LGBTQ+ equality in India: a comprehensive examination from anthropological and legal perspectives. SG&P **7**(2), 220–237 (2024). https://doi.org/10.1002/sgp2.12092
17. Kothari, J., Balu, K., Sarma, R.: Advancing Legal Recognition for Intersex Persons in India. Centre for Law and Policy Research and Solidarity Foundation (2020). https://clpr.org.in/wp-content/uploads/2020/07/CLPR_PolicyBrief_2.pdf
18. Krishna, A., Amirthavarshini, D., Rebekah, J.C.: LGBTQ rights and legislation in India: the status quo. Indian J. Integr. Res. Law. **3**(1), 1–15 (2020)
19. Mewafarosh, R., Chatterjee, D.: Deprivation and social exclusion of LGBT community in India. Int. J. Res. Bus. Stud. **4**(1), 127–136 (2019)
20. Singh, N., Chatterjee, A.: Social and workplace issues of LGBT community in India. Haryana Police J. **2**(1) (2019). https://haryanapolice.gov.in/policejournal/pdf/SOCIAL_WORKPLACE.pdf
21. Sharma, A.: Decriminalising queer sexualities in India: a multiple streams analysis. Soc. Policy Soc. J. Soc. Policy Assoc. **7**(4), 419–431 (2008). https://doi.org/10.1017/s1474746408004399
22. Tejeda, M.J.: Non-discrimination policies and sexual identity disclosure: do they make a difference in employee outcomes? Empl. Responsib. Rights J. **18**, 4559 (2006)
23. Verma, S.K., Sharad, S., Singh, T., Bharti, P.: Living with alternatives: Perceived crises, meaninglessness and efforts for a meaningful life in Indian Hijras. Cult. Psychol. **29**(4), 840–863 (2023). https://doi.org/10.1177/1354067x231201385

# Potential for Tourism Based on Sustainable Livelihood in the Ngarai Sianok Geopark Area

**Rosianna Sianipar, Juliana, Nova B. Sitorus, Diena M. Lemy, and Amelda Pramezwary**

**Abstract** The Ngarai Sianok Geopark, situated in West Sumatra, Indonesia, holds immense potential for tourism development grounded in the principles of sustainable livelihood. This qualitative study investigates how local populations' socioeconomic well-being can be improved while protecting the region's distinctive geological and cultural assets by coordinating tourism initiatives within the geopark with the sustainable livelihood framework. The study explores sustainable tourism's perspectives, obstacles, and prospects within the geopark using comprehensive interviews and focus group talks with significant stakeholders, such as inhabitants, community leaders, tourism operators, and government officials. The findings indicate that through the stimulation of local entrepreneurship, diversification of sources of income, fostering ownership and stewardship among community members—in other words, building community resilience—the integration of sustainable livelihood strategies in tourism development achieves very significant impacts. Participants articulated the integration of local knowledge, customs, and cultural practices into their tourism activities for an enriched visitor experience which also maintains the unique cultural identity of the area. Problems noted were bad markets, inadequate infrastructure, and the inability of locals to be involved in and enjoy the fruits of tourism. The report emphasizes that comprehensive capacity-building programs, enhanced infrastructure, and key stakeholder participation are necessary to overcome these challenges. Issues noted are as follows: poor markets, inadequate infrastructure, and the incapacity of the local population to take part in and enjoy all the gains flowing from tourism. This report stresses that extensive capacity-building programs, enhanced infrastructure, and key stakeholder participation are essential for surmounting these obstacles. Moreover, the research emphasizes the importance of environmental conservation as a basic part in maintaining the attractiveness of the geopark as a sustainable tourism destination. In conclusion, the prospects for sustainable tourism development in Ngarai Sianok Geopark are numerous. This can

R. Sianipar · Juliana (✉) · D. M. Lemy · A. Pramezwary
Universitas Pelita Harapan, Tangerang, Indonesia
e-mail: juliana.stpph@uph.edu

N. B. Sitorus
Politeknik Pariwisata Medan, Medan, Indonesia



M. Al Mubarak (ed.), *Sustainable Digital Technology and Ethics in an Ever-Changing Environment*, Studies in Systems, Decision and Control 237,
https://doi.org/10.1007/978-3-031-86708-8_37

assist locals in upholding their standard of living and at the same time conserving the environment and cultural heritage. The research shows that the use of tourism potential for sustainable development through its role as a catalyst requires, above all, a community-centered approach. It must be supported by effective policies and concerted actions. With practical implications for stakeholders involved in sustainable management not only of geoparks but also other types of protected areas, this research significantly contributes to an enhanced understanding regarding the importance of introducing sustainable livelihood strategies into tourism development.

**Keywords** Potential · Sustainable livelihood · Ngarai Sianok Geopark

## 1 Introduction

Tourism is growing as one of the key segments in most regions around the world, for example, South American and Southeast Asia countries have benefited from rapidly establishing their economy using natural resources or cultural heritage [1, 2]. Nevertheless, the fast-growing tourism sector paradoxically poses threats in terms of environmental damage and cultural erosions or social imbalances especially in susceptible regions. The response to these challenges has been the advent of sustainable tourism, which is largely focused on combining economic benefits with heritage and/or biodiversity conservation [1]. One important concept within this framework has been that of sustainable livelihoods, which seeks to improve adaptive capacities and the well-being of local populations via multi-scalar income diversification [3, 4]. for social equity as well as environmental stewardship.

One of the unique cases where tourism potential crosses with the necessity to achieve sustainable development is the Ngarai Sianok Geopark located at West Sumatra, Indonesia. The geopark is known for its spectacular geological features, abundant and diverse flora fauna resources as well as rich culture relics, which constitute a very appealing destination of ecotourists. Yet, the region also struggles due to the inadequate infrastructure previously mentioned alongside issues accessing markets and capacity building within local communities. The influences are predisposed to a cautious but holistic approach in tourism development of sharing the benefits proportionately and at the same time avoiding environmental damages as well as avoidable risks for local culture.

The Ngarai Sianok Geopark was inaugurated in 2018. Geopark is a concept promoted by UNESCO and has become a global trend. Geoparks focus on regional development based on regional resources and community participation. By being used as a geopark, Ngarai Sianok is managed by the local government and is expected to be a driving force for the City of Bukittinggi to be able to develop nature-based tourism. However, to achieve this goal, it is necessary first to study the potential of natural tourism in detail. It is also essential to involve the local community in analyzing the community's capital to manage tourism in the Ngarai Sianok Geopark Area (Fig. 1).

**Fig. 1** Ngarai Sianok

Based on the sustainable livelihood framework, this study explores the potential for tourism development in the Ngarai Sianok Geopark area. By focusing on the perspectives of local stakeholders, including community members, government officials, and tourism operators, this research seeks to understand how sustainable tourism can support local communities' socioeconomic resilience while preserving the region's natural and cultural assets. The study also examines the challenges and opportunities associated with sustainable tourism development in the geopark. It provides recommendations for strategies to enhance the long-term sustainability of tourism in the area.

References [5, 6] The findings of this manuscript indicate a significant effect on residents' livelihoods. To enhance tourism in a manner that supports sustainable livelihoods and achieves environmentally, socially, and economically sustainable outcomes, it is recommended to engage in activities such as supporting and promoting local businesses, increasing the duration of tourist stays, and promoting the destinations. Additionally, enforcing environmental protection laws is suggested.

Other studies on sustainable livelihoods emphasize the role of natural tourism communities. The concept of livelihoods plays a crucial role in boosting the income of local communities. Further findings reveal that the type of livelihood significantly influences people's income. The type and quantity of livelihood assets specifically affect livelihood choices, and the nature of the livelihood dictates how these assets are utilized [7]. Each component of a sustainable livelihood holds varying degrees of importance in its development. Human assets are vital, as they can influence and mobilize other assets. Natural holdings in a village, including land, water, and air, are also crucial. Land resources are particularly significant in tourist villages, serving as key attractions. The way land is utilized contributes to the village's aesthetic appeal, making it a more attractive destination for tourists [1, 8, 9]. The analytical framework established by the UK Department for International Development identifies five types of capital as indicators of sustainable livelihoods: physical capital, human capital, natural capital, social capital, and financial capital [10]. The framework is designed to pinpoint the most critical challenges and development opportunities that will provide the most significant benefit to society. The sustainable livelihood approach also emphasizes diversifying capital, reducing environmental risk vulnerability, and maximizing the outcomes of community activities.

## 2 Research Method

This study employs a qualitative research methodology to explore the potential for tourism development based on sustainable livelihood in the Ngarai Sianok Geopark area. The qualitative approach is particularly suited for this research as it allows for an in-depth understanding of the complex social, cultural, and economic dynamics that influence sustainable tourism development within the geopark.

The research design is exploratory, focusing on understanding the perceptions, experiences, and aspirations of various stakeholders involved in or affected by tourism in the Ngarai Sianok Geopark. This design enables the researchers to capture rich, detailed insights into how sustainable tourism can be aligned with the principles of sustainable livelihood to benefit local communities while preserving the region's natural and cultural assets.

This exploratory research is designed to understand the perceptions, experiences and aspirations of different stakeholder visions related to tourism in Ngarai Sianok Geopark. The design allows the researchers to capture in-depth insights into how sustainable tourism can support sustainable livelihoods at a local level for both host and guest while conserving natural capital as well as cultural identity. The data collection was carried out through several stages of observation directly at the Ngarai Sianok Geopark, in-depth interviews with many informants from various areas around the site, literature study, and documentation [11].

Ethical research guidelines were followed, consisting of administering informed consent to all your participation and providing participants with the right to erase or adjust their minds whenever desired. This research was carried out in a culturally sensitive manner, according to the standards of practice for behavioural health researchers working with Indigenous populations. Key stakeholders such as geopark managers, an officer in the tourism department, and the local community were also among those considered in this study. Specialists and individuals with previous knowledge of the research topic were included among the key informants. The data analysis used in this research is descriptive and aims to describe true events based on the actual field observation data.

In addition, a categorization data analysis method was used to organize the responses within any themes related to ideas or institutions that fulfil specific requirements. Data processing for exploration was composed of sentences, images, and subject behaviours in the form of qualitative descriptive analysis without using numerical or nominal data. In the analysis method data triangulation process has been followed to validate and accurate this dataset. The study used the qualitative method and focused on informants involved with and knowledgeable about issues, namely geopark managers, tourism office administrators, and community members. Key informants were chosen because they had lived and experienced the research topic deeply, so data collection was necessary to be done effectively.

## 3 Result and Discussion

Research results on potential tourism of sustainable livelihood in the Geopark Ngarai Sianok area show that natural beauty is a unique capital. The breathtaking setting of cliffs, thick forests, and meandering rivers attracts thousands upon thousands of travellers every year, making it a prime spot for nature-based tourism. Due to its beauty and location not far from Bukittinggi, Ngarai Sianok has been a spot for tourism. It also highlights the necessity for improved infrastructure to enhance visitors' overall experience and allow tourists to benefit fully from what the area offers sustainably. A key discovery is the importance of tourism in sustaining local communities' modes of living. As a result, many locals are directly dependent on tourism and work in providing services like guiding, transportation, or managing food & souvenir stalls. Fixing these infrastructure deficiencies is a key requirement for tourism to truly benefit local livelihoods [12, 13].

The study also discusses ecotourism and the possibility of developing programs on environmental education in Ngarai Sianok. These programs create additional income-earning opportunities in the local community, along with spreading conservation and sustainability practices. Through the incorporation of ecotourism and environmental education with the community, they can protect nature while educating visitors about sustainability. This two-pronged approach helps to both grow the economy and protect the environment through sustainable tourism practices [14–16]

The SWOT analysis carried out as part of the research indicates many opportunities to enlarge natural tourism in Ngarai Sianok. Although the beauty of the place and ease of accessibility were among its strengths, it was also identified to have inadequate infrastructure and weather dependence when tourism activities are considered. Limitations include the need for better facilities (e.g., rest areas, public toilets), and an increase in ecotourism programs. At the same time, a partnership with private sector investors may significantly support these improvements and elevate service quality for tourists. Consequently, the research concludes that with appropriate investment and management, tourism in Ngarai Sianok could become a global best practice for sustainable development. Utilizing its natural, social, human, physical, and economic capital, the area can further develop in a way that is sustainable for both.

## 4 Conclusion

The determinant factors mentioned in the existing literature imply that Ngarai Sianok possesses extensive potential as a prime tourism site. Its growth could help make a significant impact on local lives in an eco-friendly way. With its beauty, accessibility, and aged tourism infrastructure, it is an excellent opportunity for the region. In fact, they are building up more recreational facilities and expanding ecotourism in order to improve the attractions of Ngarai Sianok.

The success of a tourism destination can only be sustained by working together and taking action preemptively to deal with threats like climate change and intense pressures from competition. Attention comes first to environmentally friendly tourism infrastructure, additional sanitary units, and hospitable recreation points.

To encourage visitors in that respect, there is an equally important opportunity to enhance ecotourism and environmental education initiatives. Public–private partnerships are essential for attracting capital and improving services in the region. Investment in appropriate risk management strategies to manage climate change, natural disasters, and security threats will also be required. Ngarai Sianok can be a reference model for successful tourism destinations that align with sustainable development goals and bring significant progress to the local community and environment.

**Acknowledgements** The author would like to thank LPPM UPH for providing the opportunity to conduct research P-74-FPar/VII/2023.

## References

1. Palacios-Florencio, B., Santos-Roldán, L., Berbel-Pineda, J.M., Castillo-Canalejo, A.M.: Sustainable tourism as a driving force of the tourism industry in a post-Covid-19 scenario. Soc. Ind. Res. **158**(3), 991–1011 (2021). https://doi.org/10.1007/s11205-021-02735-2.
2. Spriggs, E.: The impact of an acquisition and employee perspective upon employee emotions, behaviors, satisfaction, organizational culture, and management behavior. ProQuest dissertations theses, p. 154 (2017) [Online]. Available: https://search.proquest.com/docview/1973268972?accountid=14600, https://sabio.unav.edu/cons/cgi/core/locater.cgi?url_ver=Z39.88-2004&rft_val_fmt=info:ofi/fmt:kev:mtx:dissertation&genre=dissertations+%2526+theses&sid=ProQ:ProQuest+Dissertations+%2526+Thes
3. Juliana, Sihombing, S.O., Suwu, S.E.: Community-based ecotourism in Sawarna Tourism Village. Enrich. J. Manag. **13**(1), 258–269 (2023). https://doi.org/10.35335/enrichment.v13i1.1192
4. Fiorello, D., Bo, A.: Community-based ecotourism to meet the new tourist's expectations: an exploratory study. J. Hosp. Mark. Manag. **21**(7), 758–778 (2012)
5. Dedeke, A.N.: Creating sustainable tourism ventures in protected areas: an actor-network theory analysis. Tour. Manag. **61**, 161–172 (2017). https://doi.org/10.1016/j.tourman.2017.02.006
6. Martini, U., Malacarne, K., Giovanazzi, S.P.: Sustainable tourism development in rural and marginal areas and opportunities for female entrepreneurship: lessons from an exploratory study. Worldw. Hosp. Tour. Themes. **12**(4), 421–430 (2020). https://doi.org/10.1108/WHATT-05-2020-0023

7. Ma, Y., Zhang, J., Li, L., Zeng, Z., Sun, J., Zhou, Q. Zhang, Y.: Study on livelihood assets-based spatial differentiation of the income of natural tourism communities. Sustainability (Switz.), **10**(2) (2018). https://doi.org/10.3390/su10020353
8. Karim, A.: The role of village tourism in improving the community's economy during the Covid-19 period in Enrekang District. https://www.acad.edu, July 2020. [Online]. Available: https://www.researchgate.net/profile/Abdul_Karim35/publication/343236134_The_Role_of_Village_Tourism_in_Improving_the_Community's_Economy_During_the_Covid-19_Period_in_Enrekang_District/links/5f1ed488299bf1720d681d10/The-Role-of-Village-Tourism-in-Improvi
9. Khan, A., Bibi, S., Lorenzo, A., Lyu, J., Babar, Z.U.: Tourism and development in developing economies: a policy implication perspective. Sustainability **12**(4), 1–19 (2020). https://doi.org/10.3390/su12041618
10. Juliana, Sihombing, S.O., Pramono, R.: Spiritual capital tourism economy creative woman entrepreneur. Int. J. Relig. **5**(1), 47–55 (2024). https://doi.org/10.61707/sbwdmq86
11. Creswell, J.W., Creswell, J.D.: Research and Design Qualitative, Quantitative and Mixed Methods Approaches (2018)
12. Juliana, J., Aditi, B., Nagoya, R., Wisnalmawati, W., Nurcholifah, I.: Tourist visiting interests: the role of social media marketing and perceived value. Int. J. Data Netw. Sci. **6**(2), 469–476 (2022). https://doi.org/10.5267/j.ijdns.2021.12.007
13. Lemy, D.M., Pramezwary, A., Juliana, Pramono, R., Qurotadini, L.N.: Explorative study of tourist behavior in seeking information to travel planning. Int. J. Sustain. Dev. Plan. **16**(8):1583–1589 (2021). https://doi.org/10.18280/ijsdp.160819
14. Juliana, J., Sihombing, S.O., Antonio, F.: An investigation between nostalgic value resonance and destination brand engagement in rural tourism. Int. J. Des. Nat. Ecodyn. **18**(1), 125–132 (2023). https://doi.org/10.18280/ijdne.180114
15. Juliana, J., Sianipar, R., Lemy, D.M., Pramezwary, A., Pramono, R., Djakasaputra, A.: Factors influencing visitor satisfaction and revisit intention in lombok tourism: the role of holistic experience, experience quality, and vivid memory. Int. J. Sustain. Dev. Plan. **18**(8), 2503–2511 (2023). https://doi.org/10.18280/ijsdp.180821
16. Juliana, S., Sihombing, J., Antonio, F.: Memorable tourism experience bibliometric analysis: a systematic review. J. Mantik **6**(2), 2302–2313 (2022) [Online]. Available: https://ejournal.iocscience.org/index.php/mantik/article/view/2751

# Sustainable Presence of Women in Company Management: The Role of Gender Composition and Strategic Management

**Vildan Bayram and Ayten Nahide Korkmaz**

**Abstract** Since the beginning of human history, men and women have been struggling to assume their roles in life as two different sexes. Women and men, who continue their existence under their identities called "gender", appear before us with their unequal positions. It has become very important to increase the level of representation of women, who constitute half of the world's population. In the world, women cannot participate equally in public administration, they cannot take part in senior positions and positions that affect the management process, and it is observed that the positions they take are not sustainable. In this study, the place of women's labor in the management function in companies in the context of gender composition is tried to be discussed. In the study, a theoretical review is made; after a detailed examination of the concepts of gender composition, gender inequality, the role of women in top management, company performance and strategic management and the sustainable presence of women's labor in management, the keywords of the study will be discussed with articles, book chapters, theses and conference presentations in the Google Scholar database between 2012 and 2024. The aim of the study is to explain the role of women in management by defining the sustainability of women and their labor in management and to fill the gap in the literature by guiding researchers at this point.

**Keywords** Gender composition · Gender inequality · Women managers · Strategic management · Sustainability

V. Bayram · A. N. Korkmaz (✉)
İstanbul Aydın University, İstanbul, Türkiye
e-mail: aytenkorkmaz@aydin.edu.tr

V. Bayram
e-mail: vildanbayram@aydin.edu.tr



M. Al Mubarak (ed.), *Sustainable Digital Technology and Ethics in an Ever-Changing Environment*, Studies in Systems, Decision and Control 237,
https://doi.org/10.1007/978-3-031-86708-8_38

# 1 Introduction

Gender inequality is considered to be a global inequality problem and it is not enough for only one country to tackle this problem. In many countries around the world, gender equality needs to be promoted and women's advancement in working life needs to be supported. In this process, sustainable policies and regulations should be developed to eliminate gender inequality. In recent years, it can be said that awareness of gender inequality in companies and women's roles in working life has increased. Many studies show that gender diversity in companies can improve financial performance. For example, a study conducted by McKinsey & Company [38] stated that if women are ignored and measures are not taken against female unemployment, global GDP growth will fall by one trillion dollars in 2030. On the contrary, if women are included and action is taken with measures, a contribution of thirteen trillion dollars will be achieved. When the gender composition is equalized and women participate in the labor market at the same rate as men, the sustainability of family welfare, economic strength and social security benefits is ensured [50: 343].

In general, it can be said that women are more disadvantaged than men in the labor market. It is seen that women face many obstacles in labor force and employment in working life [40]. The barriers to women's labor force participation can be classified as cultural, social and institutional. Cultural barriers include male dominance, kin interventions, sexist stereotypes; social barriers include gender discrimination, child and elderly care, and institutional barriers include discrimination in the distribution of tasks and promotions, and wage discrimination [37].

In the labor market, women are discriminated against during labor force participation and after employment. These discriminations can be listed as working with flexible hours, lack of incentives to acquire new skills, loss of self-esteem on the career path. The lack of incentives in the labor market causes women to continue working as low-skilled employees, which is thought to negatively affect their return to the labor market after childbirth [28].

However, it can be said that women still face gender equality issues despite their increased visibility in senior management. Factors such as gender discrimination, women's difficulties in achieving work/life balance and gender inequality in organizational culture stand out among the factors that prevent women from rising to leadership positions. These challenges limit women's advancement to senior management roles and hinder the realization of gender equality goals.

The purpose of this study is to examine the relationship between the sustainability of women's increased visibility in top management and strategic management. With more women in leadership positions, companies are able to diversify and make decisions based on different perspectives. This can positively affect company performance. The study aims to draw attention to the issue of gender inequality and to develop policies on this issue.

# 2 Conceptual Framework

In this section, the concepts of gender composition, gender inequality, the role of women in top management, company performance and strategic management, and the sustainable presence of women's labor in management are discussed in the context of the study topic "Increasing Visibility of Women in Top Management: The Relationship between Gender Composition and Company Performance", the concepts of gender composition, gender inequality, the role of women in top management, company performance and strategic management, and the sustainable presence of women's labor in management are discussed.

## *2.1 Gender Composition and Gender Inequality*

Gender composition refers to the distribution of a group or community by sex. This distribution is usually based on two basic gender categories, men and women. Gender composition is used to identify inequalities between the sexes and address gender-based discrimination in various fields, particularly in the workforce, politics, education and the arts.

Gender inequality is defined as a state of inequality and imbalance between men and women in terms of power, status, rights and opportunities [56: 8]. Inequality is based on social and cultural factors as well as economic policies and institutional barriers. In other words, while the patriarchal social structure is at the root of inequality, capitalist production relations also fuel this inequality [39: 160]. Low levels of education, gender-based division of labor and related gender roles, lack of work-life balance, inadequacies in child and elderly care services, wage inequalities and deficiencies in issues such as business development are also factors that keep women away from labor force participation, in other words, from guaranteeing their livelihood and life [44: 51]. The fact that women's wages are not seen as the main source of income for the household but as an addition to men's wages, that they accept leaving their jobs easily and that they are more compliant than men in many issues facilitates the demand for female labor in the informal sector [55: 16]. According to the study conducted by Yerlikaya et al. [61: 54]; it has been observed that the increase in the level of education is an important factor in the orientation of women with higher levels of education towards full-time formal jobs, and it has also been determined that individuals with higher levels of education (male or female) develop a more egalitarian view of the position of women in working life.

In the literature, many studies on gender inequality [6, 17: 101]; (Ausín et al. 2020: 29) show that inequality between genders is reflected in business life. When an assessment is made worldwide in terms of women, the lack of women, especially in executive leadership positions, and general income inequality draw attention. Especially in sectors requiring technical skills, gender inequality is higher and women's representation is relatively less. Globally, in the 156 countries covered by the Global

Gender Gap Index, women represent only 26.1% of the 35,500 parliamentary seats, only 22.6% of the more than 3400 ministers worldwide, and only 27% of all executive positions. Eighty-one countries (as of January 15, 2021) have never had a female head of state [33]. Even though labor force participation is increasing, costs such as pregnancy, maternity leave and childcare costs lead employers to reduce labor demand for women, thus hindering women's employment [54: 195]. This discriminatory demand for labor reduces women's employment on the one hand, and women's labor force participation (36.2%) on the other. Gender-based domestic division of labor also keeps women away from labor force participation. In Turkey, women's primary role as mothers and wives limits women's participation in the labor force [19: 50]; [60: 110]. According to the World Economic Forum (WEF) Global Gender Report for 2021, women make up 14% of the current workforce in Cloud Computing, 20% in engineering, and 32% in data and AI. In addition, "Developing Female Leaders" research was conducted by PwC in 2022 and according to the research, although 7 out of 10 female employees want to work outside their home country, only 1 out of 4 women can use this opportunity. While 75% of male employees who work abroad are married, 50% of female employees who work abroad are single. Here, it is concluded that the family factor is more of an obstacle for women than men in their careers [46].

Studies show that women in particular are exposed to a wider gender gap in terms of being promoted or appointed to senior management [26: 6].

The most important reasons for women's inability to participate in the labor force in Turkey are: being busy with housework, being disabled, elderly or sick, level of education and training, family and personal reasons. It can be said that labor force participation is in line with the increase in education level. The vast majority of working women take on the responsibility of housework as a second shift. Gender composition should be considered and analyzed by policy makers, employers, educators and other stakeholders to achieve gender equality goals. It is emphasized that gender-based discrimination and inequalities need to be addressed to achieve these goals.

## 2.2 *The Role of Women in Senior Management*

Women managers face high levels of prejudice in their working life. In the top management of companies, male managers are expected to be more aggressive and dominant, while female managers are expected to exhibit more relationship-oriented, hospitable and conciliatory behaviors. The behavior of women leaders is determined by social norms, stereotypes, culture and personality [14]. Therefore, it is thought that the country and culture of women managers do not play an important role in their managerial behaviors.

In the work environment, goals are likely to be negatively affected. Female managers are goal-oriented and tend to feel the risks of failure. Male managers, on the other hand, generally do not feel the risk of success. For this reason, male managers exhibit aggressive, ambitious and risk-taking behaviors to achieve success [20, 32]. Female managers, on the other hand, approach success more than male managers by taking into account the fear of failure, using a conciliatory and relationship-oriented approach. The reasons for gender inequality in women's working life include attitudes/discomfort with the idea of being managed by a female manager, male employers being biased/unjust in providing equal employment opportunities, and having extremely high expectations from female managers [13].

Female leaders are more qualified than their male counterparts [8, 21]. Female employees face more difficulties, overcome more obstacles and are more carefully selected for promotion not only at the highest levels but also at every level than male employees. In this context, female managers are more qualified than male managers. Thus, a high proportion of women in top management in firms enables firms to have better financial returns and achieve better performance (Krishnan and Park 2005).

## 2.3 Strategic Management

The concept of strategy defines where the company wants to be in the future. It is used to objectively assess the company's future and where it is now and decide how to get to where it wants to be. To get there, it is necessary to use available resources and consider options and alternatives to make the necessary changes [43]. At the same time, the environment in which companies operate should be analyzed to identify future opportunities and threats and to ensure that they use their resources effectively.

Classical strategic management is based on rational methods such as planning, resource allocation and profitability. Once the strategic plan is defined, the appropriate structure emerges easily [12]. This approach relies on the intelligence and abilities of leaders and has a significant impact on the hierarchy or the level of success of the company [4]. Porter [45] states that strategic management is a process. He stated that competitive strategies should be developed on how companies will compete, and company mission, vision, policies, goals and objectives should be determined. Sloan [49] emphasized the importance of creating and implementing independent business strategies for the continuity and stability of companies.

Today, strategy formulation involves not only top managers but also all levels of the hierarchy [48]. Creating company strategies is an important step in managing organizational projects and ensuring success. Strategy is used as a tool to realize various goals. Nowadays, for business strategies to be effective, the concepts of corporate social responsibility (CSR) and sustainability need to be included [24, 57]. In order to achieve sustainability, strategic decisions such as reducing company costs, cleaner production, reducing the use of natural resources, and making supply changes will create a new era of business thinking and competitive advantage.

Social roles and stereotypes culturally assigned to men and women are thought to cause the gap between men and women. The Global Gender Gap Report published by the World Economic Forum [58] found that women are 68% behind men in terms of economic opportunity, political growth, education, health and survival. The wage gap is 51% and women in leadership positions is 34%. The Startup Outlook 2018 survey published by Silicon Valley Bank (SVB) [51] reveals that 71% of American companies do not have women on their boards and 57% of companies do not have women in senior positions. In this context, it can be said that women have less participation in working life compared to men [16].

Equal participation of women in the workforce is important for ensuring social justice and equality. It reveals that companies where women are more represented in boards of directors and leadership positions perform better [30]. Women are generally strong in collaboration, communication and problem-solving skills, which positively impact teamwork and business performance [15].

It is not only an ethical responsibility for businesses to develop strategic plans for women's labor, but also an essential requirement for sustainable success and competitive advantage [53]. Fully utilizing women's talents and potential improves business performance, fosters innovation and creates social value [7]. Examples of strategies developed by businesses for women's labor within the framework of strategic management include increasing the recruitment rate of female employees, supporting the development of female employees and taking measures to retain female employees [31]. In addition, various approaches such as equal remuneration, equal promotion, bonus, reward and gift policies can be developed for women in organizations [10]. These strategies aim to ensure women's equal representation within the organization, support them in leadership positions, and make the work environment more fair and inclusive [25]. Businesses can develop strategies and policy approaches with the responsibility to provide a fair and equitable work environment for women employees. These strategies will ensure that women's voices are heard, their ideas are valued and their leadership potential is realized. To give country-specific examples of the strategy recommendations that companies should implement; France advocated the idea that both male and female representation should be ensured on boards of directors and set a target of increasing the proportion of women employees to 20% within 3 years and to 40% within 6 years as of 2010. Germany emphasized the importance of gender equality and adopted a target of an appropriate female representation rate. In line with this suggestion, they also suggested that they would set a specific ratio. Hong Kong also adopted September 1, 2013 as the starting date. It is aimed to have a diverse composition in terms of gender, age and level of knowledge on boards of directors. In 2012, it was proposed that there should be at least one woman on the boards of listed companies. After failing to comply with this regulation, the CMB revised its recommendation to require companies to set a date by which they can reach at least 25% women on their boards [46].

## 2.4 Sustainable Presence of Women's Labor

"Sustainability", which is a multidimensional phenomenon, appears in every field today. Sustainability is to provide the needs of humanity from scarce resources without compromising the needs of future generations by ensuring the existence of humanity in life, while ensuring the continuity of production and diversity. The term sustainability was introduced by the World Commission on Environment and Development (WCED) in 1987 through the report "Our Common Future". In the report, sustainability was defined as "humanity's ability to sustain development by meeting daily needs without jeopardizing nature's ability to respond to the needs of future generations" [59]. In addition, sustainable development was also mentioned in this report and sustainable development was evaluated with its economic, environmental and social dimensions. Accordingly, it has revealed with a holistic approach that only economic values are not sufficient for development. Accordingly, sustainable development can be defined as the efforts carried out to increase the social welfare of a country and to ensure that the people living in the country can maintain a level of life worthy of human dignity, regardless of any distinguishing factors" [54].

Women make up more than half of the world's population and increasing the share of women in employment plays an important role in the development of individuals and companies. The participation of women in business life, especially in management, is of great importance for the progress of society. Women's labor is often seen to be relegated to the lower levels of the hierarchy in business life and is thought to be less involved in management mechanisms. The fact that women are pushed to a subordinate position due to gender perception is the subject of many studies and although national and international regulations try to eliminate this situation, women are still unofficially discriminated against in the labor market. Sometimes, due to the roles imposed on women, women's advancement in working life is tried to be prevented in an unofficial and unspecified way. Having women's ideas and minds at the top levels should be at the top of the agenda of many companies. In this context, if we take a look at the position of women in management in the world and in Turkey, according to the "Global Gender Inequality in Business 2010 Report" published by the World Economic Forum of the Turkish Employers' Association (TISK), which evaluates the situation of women working in the private sector in various countries in terms of economic participation and opportunities; while the world average for the proportion of women in senior executive (CEO) positions is around 5%, the same rate in Turkey is 12% and ranks 2nd among 34 countries (Global Gender Inequality in Business 2010 Report: 2010).

Although it is thought that the increase in women's employment and the sustainability of women having a say in the management and reaching the top positions make it possible for the management to become stronger in companies, in a study conducted by Öcal [41], it is said that there are a significant number of studies that conclude that the presence of female members on the boards of directors positively affects the performance of companies, as well as a significant number of studies that reach the opposite conclusion or conclude that there is no effect. As a result, there are also studies that conclude that the effect of gender diversity on company performance is "U" shaped, that the number of female members on the board of directors negatively affects company performance up to a certain number and/or ratio, but that this effect turns positive after a certain number and ratio is exceeded.

## 3 Table of Literature

In this section, it has been tried to introduce the studies by including the articles, book chapters, Decrees and conference presentations between 2012 and 2024 in the Google Academic databases scanned with the keywords "Gender Composition, Gender Inequality, Company Performance, Women Manager, Strategic Management and Sustainability" (Table 1).

According to Toksöz [54], studies on the relationship between women's labor and the development of countries in different regions of the world are extremely limited in Turkey. But in recent years, we can mention an increasing interest in issues related to women's labor and employment both in various social circles and in the women's movement.

Karabıyık [35] in his study, women's employment was examined within the framework of statistical data and evaluations were made on the axis of the general structure of the economy. It has been concluded that women's employment is important in terms of social development; however, the desired result cannot be achieved with the current social structure and policies followed in the economy.

In the study of Taşkın and Evrim Mandacı [22], it was revealed that the increase in the number of women members in management reduces the return on assets and return on equity.

Çolakoğlu [18] shows in his thesis that there is a significant relationship between board independence diversity and corporate social responsibility activities, while there is Dec Decently no significant relationship between gender and foreign nationality diversity and corporate social responsibility activities.

As a result of Keyvan's [36] study, it was underlined that if these policies are sustainable, a positive development can be predicted when organizations address the issue with a social sustainability approach, develop policies that will embed this understanding in their internal functioning and business practices.

As a result of the work of Akalin et al. [1], policies should be designed to eliminate the effects of increasing income inequality and reduce inflation, while encouraging economic growth and financial development to make women's labor sustainable.

In the study conducted by Akar [2], it is thought that the importance of the role of women in business life will make a positive contribution to development.

## 4 Conclusions and Recommendations

Since the first day when human beings lived together as two sexes in society, women's labor, which has been in unequal conditions, has been appearing in front of us in every field. Dec. It is observed that women are more involved in low-paid and insecure jobs than men, continuing the gender equality journey, especially for women, who are the most excluded group in management and the upper echelons. Women are struggling with gender inequality and, on the one hand, they have to express themselves in

**Table 1** The sustainable presence of women in company management: studies on gender composition and the role of strategic management

| Author(s)–Date (Year) | Method(s) | Type | Results |
|---|---|---|---|
| (1) Toksöz (2012) [54] | Literature review | Book | According to the study, studies on the relationship between women's EME and the development of countries in different regions of the world are limited in Turkey to the last Dec. We have been working on the relationship between women's EME and the development of countries in different regions of the world. However, in recent years, one can speak of an increased interest in issues related to women's labor and employment both in various social circles and in women's society |
| (2) Karabıyık-Makale (2012) [35] | Literature review | Article | Within the framework of statistical data, women's employment was examined and evaluations were made on the axis of the general structure of the economy. It has been concluded that women's employment is important in terms of social development; however, the desired result cannot be achieved with the current social structure and policies followed in the economy |
| (3) Taşkın ve Evrim Mandacı (2017) [22] | Panel regreration | Article | According to the study, it has been revealed that the increase in the number of women members in management reduces the return on assets and return on equity |
| (4) Akbaş ve Korkmaz (2017) [3] | Review | Article | Since awareness about prejudices is extremely important both in combating prejudices and for the success of women and therefore students, this study aims to increase awareness about prejudices against women in business life, especially at an implicit level |
| 5) Öztaş (2018) [42] | Descriptive statistical analyses | Thesis | As a result of the regression analyses in the study, the most important effect of the senior female manager and female board member on the Dividend Yield was seen |

(continued)

**Table 1** (continued)

| Author(s)–Date (Year) | Method(s) | Type | Results |
|---|---|---|---|
| (6) Çolakoğlu (2018) [18] | Hiyerarşik regresyon | Thesis | This study shows that there is a significant relationship between board independence diversity and corporate social responsibility activities, while there is no significant relationship between gender and foreign Deciency diversity and corporate social responsibility activities. The results of this study show that there is no significant relationship between Deciency diversity and corporate social responsibility activities |
| (7) Gür (2016) [29] | Literature review | Article | In the study, the effects of peasant-urban women's labor on development were discussed by revealing |
| (8) Keyvan (2022) [36] | Literature review | Thesis | As a result of the study, it has been revealed that organizations address the issue with a social sustainability approach, develop policies that will embed this understanding in their internal workings and ways of doing business, and if these policies are sustainable, a positive development can be predicted |
| (9) Akalın et al. (2024) [1] | ARDL | Article | According to the study, in order to make women's labor sustainable, policies should be designed to eliminate the effects of increasing income inequality and reduce inflation, while encouraging economic growth and financial development |
| (10) Akar (2024) [2] | Literature review | Article | In this study, it is thought that giving importance to the role of women in work life will make a positive contribution to development |

*Sources* Created by authors

working life and, on the other hand, strive to make their work sustainable. Despite all these negatives, the fact that women are becoming more involved in business life every day shows that positive changes are taking place, but the fact that a sufficient number has not been reached reveals that there are problems that have not yet been solved. Thanks to the changes in today's leadership approaches and new ideas and perspectives, they can be considered as new steps taken to make women advantageous in management. In addition, the new policies that countries have adopted on their agendas can also be considered as positive developments that ensure the existence of women's labor in management.

The change of technology, the differentiation of norms in societies necessitate the correct and most efficient use of knowledge and skills. In this ever-renewed global economic environment, men and women who have knowledge and use it in the most correct way should be able to take part together and equally in leading Decrees. At this stage, it is very important for the development of societies that a person can be evaluated and take equal authority and responsibility with the education, skills and development received, not by gender, at this stage. In addition, researching and setting out what the concrete goals are for the inclusion of women in management, following up these goals with various years, for example, surveys, symposiums, workshops and other applications, announcing these results and trainings throughout companies and on a country basis can also be presented as an important recommendation. To mention other suggestions, it should be ensured that women employees work comfortably by creating environments and forums where issues and issues related to women and women's labor can be discussed in private, where women can exchange ideas freely. In order to ensure the work-private life balance, efforts should be made to strengthen the position of women in management by providing flexible working and other supportive opportunities.

**Contributions of Authors to the Article** Vildan BAYRAM's contribution to the article is 50%, Ayten Nahide KORKMAZ's contribution to the article is 50%.

**Declaration of Interest** The study does not present any conflict of interest for the author or other stakeholders.

**Statement of Research and Publication Ethics** This article does not require the approval of the ethics committee.

## Abbreviations

| | |
|---|---|
| **ARDL** | Autoregressive Distributed Lag Bound Test |
| **CEO** | Senior Manager |
| **CMB** | Capital Markets Board |
| **CSR** | Corporate Social Responsibility |
| **GDP** | Gross Domestic Product |
| **TISK** | Turkish Employers' Association |
| **TURKSTAT** | Turkish Statistical Institute |
| **WCED** | Environment and Development Commission |
| **WEF** | World Economic Form |

## References

1. Akalin, G., Erdoğan, S., Pata, U.K.: Türkiye'de Finansal Gelişme Gelir Eşitsizliğini Nasıl Etkilemektedir? Uluslararası Ekonomi Yenilik Dergisi **10**(1), 111–124 (2024)
2. Akar, B.G.: The status problem of women's cooperatives in Turkey: being able to be located in the right place (example of worker cooperatives). Ömer Halisdemir Univ. J. Fac. Econ. Adm. Sci. **17**(3), 603–625 (2024)
3. Akbaş, G., Not Afraid, L.: Women managers: an effort to rise in the shadow of invisible obstacles. Iş ve Insan Dergisi **4**(2), 73–86 (2017)
4. Ansoff, H.I.: Kurumsal Strateji: Büyüme ve Genişleme için iş Politikasına Analitik Bir Yaklaşım, McGraw-Hill, Arbor, MI, ABD (1965). Author, F.: Contribution title. In: 9th International Proceedings on Proceedings, pp. 1–2. Publisher, Location (2010)
5. Asin, B., González-Sanguino, C., Castellanos, M.: Gender-related differences in the psychological impact of confinement as a consequence of COVID-19 in Spain. J. Gend. Stud. 29–38 (2020). Advance Online Publication. https://www.tandfonline.com/doi/full/10.1080/09589236.2020.1799768
6. Azcona, G., Bhatt, A., Encarnacion, J., Plazaola, C.: From Insights to Action: Gender Equality in the Wake of COVID-19. UN Women. https://www.unwomen.org/-/media/headquarters/attachments/sections/library/pu (2020)
7. Bagheri, F., Ghaderi, Z., Abdi, N., Hall, C.M.: Female entrepreneurship, creating shared value, and empowerment in tourism; the neutralizing effect of gender-based discrimination. Curr. Issue Tour. **26**(21), 3465–3482 (2023). https://doi.org/10.1080/13683500.2022.2126749
8. Baxter, J., Wright, E.O.: The glass ceiling hypothesis: a comparative study of the United States, Sweden, and Australia. Gend. Soc. **14**, 275–294 (2000)
9. Biswas, P.K., Chapple, L., Roberts, H., Stainback, K.: Board gender diversity and women in senior management. J. Bus. Ethics **182**(1), 177–198 (2023)
10. Biswas, P., Banerjee, S.: Green human resource management—a gendered approach to sustainability through women employment. In: Sustainable Boardrooms: Democratising Governance and Technology for Society and Economy, pp. 41–70. Springer Nature Singapore, Singapore (2023)
11. Blithe, S.J.: Work-life balance and flexible organizational space: employed mothers' use of work-friendly child spaces. Manag. Commun. Q. **37**(4), 821–845 (2023). https://doi.org/10.1177/08933189231152454
12. Chandler, A.D.: Strategy and Structure: Chapters in the History of the American Industrial Enterprise. Beard Books, New York, NY, USA (1962)
13. Chapman, J., Luthans, F.: The female leadership Dilemma. Publ. Person. Manag. **4**(3), 173–179 (1975)
14. Chapman, J.B.: Comparison of Male and Female Leadership Styles. Acad. Manag. J. **18**(3), 645–650 (1975). https://doi.org/10.2307/255695
15. Chatterjee, C., Nag, T.: Do women on boards enhance firm performance? Evidence from top Indian companies. Int. J. Discl. Gov. **20**(2), 155 (2023). https://doi.org/10.1057/s41310-022-00153-5
16. Cardella, G.M., Hernández-Sánchez, B.R., Sánchez-García, J.C.: Women entrepreneurship: A systematic review to outline the boundaries of scientific literature. Front. Psychol. **11**, 1557 (2020). https://doi.org/10.3389/fpsyg.2020.01557
17. Collins, C., Landiva, L.C., Ruppanner, L., Scarb: COVID-19 and the gender gap in work hours. In: Feminist Frontiers, pp. 101–112. Advance Online Publication (2020). https://doi.org/10.1111/gwao.12506
18. Çolakoğlu, N.: The Effect of Gender Diversity on Corporate Social Responsibility Activities of the Board of Directors: A Turkey 500 Study (2018)
19. Dedeoğlu, S.: Eşitlik mi ayrımcılık mı? Türkiye'de sosyal devlet, cinsiyet eşitliği politikaları ve kadın istihdamı. Çalışma Toplum **2**, 41–54 (2009)
20. Eagly, A.H., Karau, S.J.: Role congruity theory of prejudice toward female leaders. Psychol. Rev. **109**, 573–598 (2002)

21. Elliott, J.R., Smith, R.A.: Race, gender, and workplace power. Am. Sociol. Rev. **69**, 365–386 (2004)
22. Flood, D.I.L.V.I.N., Mandaci, P.E.: The effect of female members of the company's board of directors on the company's performance. Dokuz Sept. Univ. J. Bus. Adm. Fac. **18**(1), 29–45 (2017)
23. Fitong Ketchiwou, G., Naong, M.N.: Exploring practices that impact women's career advancement within the workplace: a qualitative approach. Afr. J. Econ. Manag. Stud. **15**(2), 279–299 (2024)
24. Galbreath, J.: Building corporate social responsibility into strategy. Eur. Bus. Rev. **21**(2), 109–127 (2009)
25. García-Sánchez, I.M., Aibar-Guzmán, C., Núnez-Torrado, M., Aibar-Guzmán, B.: Women leaders and female same-sex groups: the same 2030 Agenda objectives along different roads. J. Bus. Res. **157**, 113582 (2023). https://doi.org/10.1016/j.jbusres.2022.113582
26. Global Gender Gap Report: Benchmarking Gender Gaps: Findings from the Global Gender Gap Index. World Economic Forum (2021)
27. Gracia, M.D., Quezada, L.E.: A framework for strategy formulation in sustainable supply chains: a case study in the electric industry. Netnomics Econ. Res. Electr. Netw. **17**(1), 3–27 (2016)
28. Green, E., Moore, J., Easton, H., & Heggie, J.: Barriers to women's employment and progression in the labour market in the north east of England (2004)
29. Gür, F.: Agricultural transformation and women's labor in a Tobacco Village. Folk-lore/Literature **22**(86), 141–156 (2016)
30. Hamplova, E., Janecek, V., Lesley, F.: Board gender diversity and women in leadership positions—are quotas the solution? Corpor. Commun. Int. J. **27**(4), 742–759 (2022)
31. Hatmaker, D.M., Hassan, S.: When do women receive managerial support? The effects of gender congruence and the manager-employee relationship. Public Manag. Rev. **25**(1), 22–41 (2023). https://doi.org/10.1080/14719037.2021.1937683
32. Heilman, M.E.: Description and prescription: How gender stereotypes prevent women's ascent up the organizational ladder. J. Soc. Issues **57**, 657–674 (2001)
33. International Women's Day: International Women's Day, World Economic Forum. 135+ Years to Close Global Gender Gap Says. World Economic Forum. https://www.internationalwomensday.com/Missions/17033/135-years-to-close-global-gendergap-says-World-Economic-Forum (2021)
34. Kadioglu, E.: Corporate governance spaces. In: Corporate Governance: Its Development, Theories and Applications, pp. 138–181. Gazi Bookstore (2021)
35. Karabiyik, I.: Women's employment in working life in Turkey. Marmara Univ. J. Econ. Admin. Sci. **32**(1), 231–260 (2012)
36. Keyvan, P.: Gender equality and corporate social responsibility from the perspective of social sustainability. Master's thesis, Marmara University, Turkey (2022)
37. Kousar, P., Lahore, B., Syeda, A., Sabir, P.A., Mahwish, Z.: Social, cultural and institutional barriers to women's participation in the workforce. Pak. J. Soc. Clin. Psychol. **17**(2), 62–66 (2019)
38. McKinsey & Company: COVID-19 and gender equality: countering the regressive effects. https://www.mckinsey.com/featured-insights/future-ofwork/covid-19-and-gender-equality-countering-the-regressive-effects (2020)
39. Omay, U.: Women as a reserve labor army. Lab. Soc. **3**, 137–166 (2011)
40. Öcal, A.T.: Gender Inequality in the Labor Market in Turkey. II. World Women Conference, Baku, Azerbaijan (2021)
41. Öcal, N.: The effect of the presence of women members on the board of directors on the company's performance. J. Bus. Trade **3**(1), 1–19 (2022)
42. Öztaş, F.: Gender Change in the Senior Management of Publicly Traded Companies and its Impact on Performance (2018)
43. Peppard, J., Ward, J.: Strategic Management of Information Systems: Creating a Digital Strategy. Wiley, Hoboken, NJ, USA (2016)

44. Pitronaci, P.: Working life and women. In: The Place of Women in Working Life, pp. 49–53. Turkish Metal Trade Union Research and Training Center, Ankara (2020)
45. Porter, M.: Competitive Strategy: Techniques for Analyzing Industries and Competitors. Simon and Schuster, New York, NY, USA (2008)
46. PwC Report: https://www.pwc.com.tr/tr/hizmetlerimiz/insan-yonetimi-ve-organizasyoncon sultant-league/publications/work-in-life-gender-equality-2022.html (2022)
47. Radomska, J.: The concept of sustainable strategy implementation. Sustainability **7**(12), 15847–15856 (2015)
48. Reitzig, M., Maciejovsky, B.: Corporate hierarchy and vertical information flow inside the firm—a behavioral view. Strateg. Manag. J. **36**(13), 1979–1999 (2015)
49. Sloan, A.: My Years with General Motors, Reissue Crown Business, New York, NY, USA (1990)
50. Steiber, N., Haas, B.: Advances in explaining women's employment patterns. Soc. Econ. Rev. **10**, 343–367 (2012)
51. Silicon Valley Bank: Banking for Innovation Economy (svb.com), 25 Aug 2024 (2019)
52. The Turkish Private Sector Makes a Difference to Developed Countries in the Ratio of Women in the CEO Position, ET, 04 May 2011. http://www.tisk.org
53. Thomas, A.: The role of women's entrepreneurship in achieving sustainable development goals (SDGs): a comprehensive review. J. Biotech. Bioinf. Res. SRC/JBBR-202. **174**(6), 2–11 (2024). https://doi.org/10.47363/JBBR/2024
54. Toksöz, G.: Different trajectories in development-women's employment in Turkey in the light of different patterns in women's employment, In: der Makal, A., Toksöz, G. (eds.) Women's Labor in Turkey from Past to Present, pp. 168–200. Ankara University Publishing House, Ankara (2012)
55. Toksöz, G., Grandfather, P., Boob, P.E., Kaya, B.S.: Analysis of the Female Labor Force Profile and Statistics in Turkey. Ministry of Family and Social Policies General Directorate of the Status of Women (2014)
56. Uluocakak, S., Celestial, G., Knows, O., Etizer, K.N., Özbay, D.: Gender Inequality and Violence Against Women. Paradigm Academy, Ankara (2014)
57. Vencato, C., Gomes, C., Scherer, F., Kneipp, J., Bichueti, P.: Strategic sustainability management and export performance. Manag. Environ. Qual. Int. J. **25**(4), 431–445 (2014)
58. World Economic Forum: Special Report from Davos | Times Higher Education (THE), 25 Aug 2024 (2018)
59. World Commission on Environment and Development: Our Common Future, p. 27. Oxford University Press, Oxford. ISBN 019282080X (1987)
60. Yerlikaya, B.: Productive women of Turkish cinema from the other to the subject: women's labor in Turkish cinema after 1980 in the context of gender, Nobel, Ankara (2020)
61. Yerlikaya, B., Giray Y.S., Ocal, A.T.: Analysis of the effect of sociodemographic characteristics of individuals on their perspectives on women in working life: the case of Turkey. J. Soc. Sci. Fac. Arts Sci. Uludag Univ. **22**(40), 19–61 (2021)

# An Analysis of Service Convenience's Effect on Online Ticket Performance: A Perception of Accounting Students on Faculty of Economics and Business Universitas Sumatera Utara

**Fahmi Natigor Nasution, Yasara Ulfa, Risanty, Sambas A. Kesuma, and Alyssa Nadia**

**Abstract** This study aims to determine the effect of access convenience, search convenience, evaluation convenience, transaction convenience, possession convenience, and post-purchase convenience on online ticket performance based on the Accounting Students of Universitas Sumatera Utara perception. This is a quantitative study, a cross-sectional and survey approach. The population of this study covers all undergraduate and postgraduate students of the Accounting Study Program Faculty of Economic and Business, Universitas Sumatera Utara. The sample of current study includes 301 students. The data is gathered using an online questionnaire. Result of the study reveal that search convenience, transaction convenience, possession convenience, and post-purchase convenience has a positive and significant effect on online ticket performance. However, access convenience and evaluation convenience has no effect on ticket online performance. This study points out that improving search convenience, transaction convenience, possession convenience, and post-purchase convenience is the best strategy to elevate online ticket performance.

**Keywords** Online ticket · Performance · Access convenience · Search convenience · Evaluation convenience · Transaction convenience · Possession convenience · Post-purchase convenience

## 1 Introduction

The adoption of online ticketing, which began gaining traction in the 1980s, represents a significant advancement in self-service technology [5, 13]. Online ticketing involves the digital issuance and delivery of tickets, eliminating the need for physical

F. N. Nasution (✉) · Y. Ulfa · Risanty · S. A. Kesuma · A. Nadia
Universitas Sumatera Utara, Medan, Indonesia
e-mail: fahmi.natigor@usu.ac.id



M. Al Mubarak (ed.), *Sustainable Digital Technology and Ethics in an Ever-Changing Environment*, Studies in Systems, Decision and Control 237,
https://doi.org/10.1007/978-3-031-86708-8_39

documentation and offering cost reductions and increased consumer flexibility [9, 19]. It also enables expedited transactions, swift booking confirmations, and easy comparison of various products [4].

In Southeast Asia, particularly Indonesia, online ticketing is a dominant market segment, with the region's market value reaching US$29.7 billion in 2018, outpacing other sectors like e-commerce and online media [11]. Indonesian consumers frequently use online platforms for hotel and transportation bookings, especially during peak travel periods [15]. The younger demographic, particularly students, shows a strong preference for online ticketing due to their high mobility, familiarity with technology, and need for time-saving solutions [1].

The concept of service convenience (servcon), which encompasses factors such as time and effort savings, significantly influences consumer adoption of online ticketing [16]. Servcon is crucial in shaping user satisfaction and behaviors, including repurchase intentions and recommendations [3, 18]. Jiang et al. [10] expanded the servcon model to include dimensions like access, search, evaluation, transaction, possession, and post-purchase convenience, tailored to the e-commerce context.

Despite the rapid growth of online ticketing in Indonesia, research in this area remains limited, particularly regarding the consistent application of servcon theory. Studies have yielded varying results on the impact of servcon on user satisfaction and behavior [2, 8, 12, 14, 17]. This study aims to explore the factors affecting the performance of online ticketing services, applying a modified version of the Service Convenience theory proposed by Jiang et al. [10].

## *1.1 Performance of Online Ticketing*

The performance of a system is closely related to the factor of convenience. One way to understand the relationship between service convenience (servcon) and performance is through comfort orientation [3]. Comfort orientation is the consumer's assessment of services with characteristics of time and effort savings. This value is driven by personal evaluations of each individual that shape norms or standards. These norms or standards will influence consumer attitudes toward the object [20]. Studies indicate that consumers are willing to pay for additional comfort or sacrifice comfort to obtain lower prices [20]. Therefore, the more comfortable the consumer feels, the better the performance of the system [20]. System performance is also considered good when consumers can feel comfortable and easily experience the benefits of the service.

The theory of servcon was introduced by Copeland [7], referring to convenience goods frequently purchased by consumers and easily obtained [10]. Copeland [7] and other researchers utilized the servcon theory in the domain of convenience product classification. Subsequently, in an effort to refine the servcon theory, researchers shifted their orientation approach from products to services (Seiders et al. 2000) [18].

Referring to the servcon theory presented six dimensions of servcon. These dimensions are access, search, evaluation, transaction, possession, and post-purchase convenience.

**Access convenience (AC)** refers to the ease of access for consumers to purchase products anytime and anywhere [10]. With flexible time and location, consumers can save more time and effort. AC serves as a primary driver of overall convenience in online shopping [10]. The key elements of AC include time flexibility, spatial flexibility, effort expended, website accessibility, and product availability within the application [10, 16].

**Search convenience (SC)** refers to the speed and ease of identifying and obtaining products that consumers desire. SC can be enhanced by providing relevant information, good navigation, and various services through the website application [8]. Online activities allow consumers to access roand search for information in one place. This can reduce the costs, efforts, and time associated with information search. These benefits can also enhance the consumer's search experience. The key elements related to SC are download speed, website design, search function, and product classification [10].

**Evaluation convenience (EC)** is the availability of detailed and easily understood descriptions of a product [10]. These descriptions are generally accompanied by features such as text, graphics, and videos on the website [10]. When shopping online, consumers receive various information about features, prices, product quality, and warranties for similar products [8]. Before making a final decision, consumers tend to compare these factors. Additionally, consumers can view and compare existing product ratings to reduce uncertainty. This can help consumers make better purchasing decisions. The main elements of EC are information, standardization, and product quality, as well as price comparison [8, 10].

**Transaction Convenience (TC)** is defined as the speed and ease with which transactions can be altered or influenced by consumers. TC is one of the most motivating factors for consumers to shift their shopping behavior from offline to online. This is because they can make transactions with just one click. In the retail context, the main elements of TC are shopping speed, ease of payment, and ease of returns. Meanwhile, in the online context, the main elements of TC are ease of checkout process, varied payment options, and confirmation responses. Other elements include low payment failure rates and price consistency.

**Possession convenience (PC)** relates to the speed and ease with which consumers can obtain desired products. With online shopping, consumers do not need to go to the store, so they prefer to shop online to avoid physical burdens [10]. Therefore, service providers must ensure prompt possession of the product by consumers. This is useful for enhancing the consumer's shopping experience.

**Post-purchase convenience (PPC)** is the perception of the time and effort expended by consumers to contact the company after purchasing a product [3]. Currently, post-purchase services on online platforms have a significant impact on consumer behavioral intentions [8]. The key elements in PC and PPC are ease of product return, timely product delivery. Other elements include quality and quantity guarantees for ordered items [10].

## 2 Methodology

This research used a survey method. The survey instrument was distributed. This study utilizes a population of active undergraduate and graduate students enrolled in the Accounting Program at the Faculty of Economics and Business, University of North Sumatra (FEB USU), aged between 18 and 25 years old totalling 1372 people. Based on the Cohen table, for a population size of 1372, the sample size selected is 301 respondents. Data obtained using the survey method is primary data. The survey instrument was prepared using a 1–5 Likert scale, starting from (1) strongly disagree and (5) strongly agree. The survey instrument was distributed using Google Forms.

This research uses the Service Convenience (Servecon) theory modified by Jiang et al. [10] to measure performance of online ticketing with 6 (six) main indicators, namely: access, search, evaluation, transaction, possession, and post-purchase convenience. The research model is depicted in Fig. 1.

Firstly, reliability and validity testing were conducted. These tests was carried out to ensure the quality and trustworthiness of the data and findings obtained from a study. The next step was to conduct Classic Assumption Testing, which are normality, multicollinearity, heteroscedasticity, and linearity tests. This aims to evaluate whether the statistical assumptions underlying a particular analysis method are met. These assumptions are fundamental to many statistical techniques and must be satisfied for the results of the analysis to be valid and reliable. Finally, hypothesis testing was performed to assess whether the hypothesis are accepted or rejected.

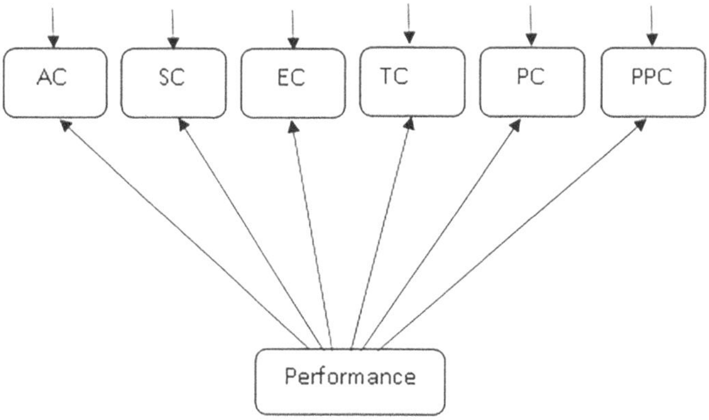


**Fig. 1** Conceptual framework

## 3 Results and Discussions

Table 1 shows the demographic profile in this study, which is based on gender, 27.2% of respondents were male, and the majority of respondents in this study were women (72.8% of the total respondents). Based on education level, most respondents (81.4%) had undergraduate education levels, and 18.6% of respondents had postgraduate education levels. Based on age, the majority of respondents (47.8%) in this study were between 21 and 23 years old, 28.2% of respondents were between 18 and 20 years old, and followed by 23.9% of respondents aged between 24 and 45 years.

Table 2 shows that that the correlation coefficients of each independent variable exceed the critical value (r table) of 0.113. Consequently, it can be inferred that the independent variables in this study are valid.

Table 3 shows that all variables have a Cronbach's Alpha value greater than 0.7. It can be concluded that all items in the instrument are reliable.

The results of the partial test in this study are presented in Table 4. Independent variables significantly affect the dependent variable when the t-value exceeds the critical t-value or when the significance level (sig.) is less than 0.05.

Based on Table 4, Variable Access Convenience has a significance level (sig.) of 0.413, which is greater than 0.005. Therefore, it can be concluded that variable Access Convenience does not significantly influence online ticket performance. This indicates that Access Convenience does not influence the performance of online ticketing among undergraduate and graduate accounting students at the Faculty of Economics and Business, Universitas Sumatera Utara (FEB USU). This is because the standard of online service platforms prioritizes high accessibility. Respondents perceive that online ticketing platforms can be accessed anytime, anywhere, and through any device. This perception is supported by descriptive statistical results indicating that the majority of respondents agree with this notion. Thus, accessibility is not a determining factor in consumer satisfaction.

**Table 1** Demographic profile

| Demographic profiles | F | Percentage (%) |
|---|---|---|
| *Gender* | | |
| Male | 82 | 27.2 |
| Female | 219 | 72.8 |
| *Education level* | | |
| Undergraduate | 245 | 81.4 |
| Postgraduate | 56 | 18.6 |
| *Age* | | |
| 18–20 | 85 | 28.2 |
| 21–23 | 144 | 47.8 |
| 24–25 | 72 | 23.9 |

**Table 2** Validity test

| Variabel | Item Pernyataan | Koefisien Korelasi | r tabel (n = 301) |
|---|---|---|---|
| Kinerja tiket online (Y) | P1 | 0.870 | 0.113 |
| | P2 | 0.847 | 0.113 |
| | P3 | 0.867 | 0.113 |
| | P4 | 0.824 | 0.113 |
| Access convenience (AC) | P1 | 0.736 | 0.113 |
| | P2 | 0.781 | 0.113 |
| | P3 | 0.750 | 0.113 |
| Search convenience (SC) | P1 | 0.790 | 0.113 |
| | P2 | 0.851 | 0.113 |
| | P3 | 0.818 | 0.113 |
| | P4 | 0.806 | 0.113 |
| Evaluation convenience (EC) | P1 | 0.823 | 0.113 |
| | P2 | 0.847 | 0.113 |
| | P3 | 0.788 | 0.113 |
| Transaction convenience (TC) | P1 | 0.900 | 0.113 |
| | P2 | 0.891 | 0.113 |
| | P3 | 0.804 | 0.113 |
| Possession convenience (PC) | P1 | 0.769 | 0.113 |
| | P2 | 0.798 | 0.113 |
| | P3 | 0.843 | 0.113 |
| | P4 | 0.716 | 0.113 |
| Post-purchase convenience (PPC) | P1 | 0.895 | 0.113 |
| | P2 | 0.922 | 0.113 |
| | P3 | 0.900 | 0.113 |

**Table 3** Reliability test

| Variables | Nilai Cronbach's alpha |
|---|---|
| Online ticketing performance (Y) | 0.871 |
| Access convenience (X1) | 0.619 |
| Search convenience (X2) | 0.833 |
| Evaluation convenience (X3) | 0.751 |
| Transaction convenience (X4) | 0.831 |
| Possession convenience (X5) | 0.785 |
| Post-purchase convenience (PPC) | 0.890 |

**Table 4** t-test

| Model | | T | Sig |
|---|---|---|---|
| 1 | (Constant) | 1.959 | 0.055 |
| | AC | 0.819 | 0.413 |
| | SC | 3.082 | 0.002 |
| | EC | 0.814 | 0.416 |
| | TC | 2.883 | 0.004 |
| | PC | 5.758 | 0.000 |
| | PPC | 7.246 | 0.000 |

The variable Search Convenience the significance value of 0.002, which is smaller than 0.005. Consequently, this indicates that the variable Search Convenience significantly influences online ticketing performance. These findings also demonstrate that the respondents perceive that the search engine within the online ticket platform is comfortable to use. This is evident from the majority of respondents agreeing that the online ticket platform is user-friendly, has adequate search engine capacity, and product classification. Additionally, respondents also agree that it is easy to make purchases through the online ticket platform.

The variable Evaluation Convenience has he significance level (sig.) of 0.41, which is greater than 0.005. Consequently, it can be inferred that variable does not have a significant influence on online ticket performance. Considering the duration of online ticket usage, most respondents (42.5%) utilize online tickets for a period spanning 5 to 10 min. Furthermore, an additional 30% of respondents use online tickets for less than 5 min. This suggests that respondents do not fully explore all categories and product information, including consumer reviews. Respondents perceive that satisfaction in using online tickets is not assessed through comprehensive evaluation.

The variable Transaction Convenience exhibits the significance value of 0.004, which is smaller than 0.005. Hence, it can be concluded that the variable significantly affects the performance of online ticketing. Respondents feel at ease when engaging in transactions via online ticketing platforms. Most respondents concur that the available payment options are user-friendly, convenient, and adaptable. Moreover, the checkout procedure through online ticketing platforms is perceived as uncomplicated.

The variable Posession Convenience has the significance value of 0.000, which is smaller than 0.005. Therefore, it can be concluded that the variable significantly influences the performance of online ticketing. Respondents feel comfortable with the online ticket booking process. The majority of respondents express agreement that tickets are promptly delivered, intact, and comprehensive. Additionally, respondents also concur that they consistently receive notifications regarding the status of their orders from the platform.

The variable Post-Purchace Convenience demonstrates the significance value of 0.000, which is smaller than 0.005. Consequently, it can be inferred that the variable significantly influences the performance of online ticketing. Respondents feel a sense of comfort with the post-purchase services provided by online ticketing platforms.

They believe that the refund process is executed promptly and with ease. Additionally, respondents perceive that the platform promptly addresses any issues that may arise.

## 4 Conclusion

The analysis of convenience factors in online ticketing reveals that while some, like search and transaction convenience, significantly influence performance, others, such as access and evaluation convenience, do not. Respondents appreciate easy search functions and smooth transactions, leading to higher satisfaction. Additionally, they value prompt post-purchase services and find comfort in the ticket booking process. These insights emphasize the importance of user-friendly features and efficient service delivery in enhancing online ticketing platforms' performance and user experience.

The implications of these findings are multifaceted for online ticketing platforms. First, by prioritizing and investing in search and transaction convenience, platforms can directly boost user satisfaction, thereby increasing customer loyalty and the likelihood of repeat purchases. Second, the lack of significant impact from access and evaluation convenience suggests that resources may be better allocated to refining other aspects of the service that more directly affect user satisfaction and performance. This strategic focus can help companies streamline their operations, reduce costs, and better meet customer expectations. Moreover, the emphasis on post-purchase services highlights the need for platforms to develop robust customer support systems to handle any issues promptly, which can further enhance the overall user experience. These implications underscore the importance of a targeted approach in service design and delivery, which can ultimately lead to a competitive advantage in the crowded online ticketing market.

## References

1. Basbas, S., Skandami, M.O., Mintsis, G., Taxiltaris. C., dan Politis. I.: Investigation of the use of E-ticketing in public transport by the students of the Aristotle University of Thessaloniki, Greece. Int. J. Transp. Dev. Integr. **1**(1), 43–53
2. Benoit, S., Klose, S., dan Ettinger, A.: Linking service convenience to satisfaction: dimensions and key moderators. J. Serv. Mark. (2017)
3. Berry, L.L., Seiders, K., dan Grewal, D.: Understanding service convenience. J. Mark. **66**(3), 1–17 (2002)
4. Cheung, C., Law, R.: Have the perceptions of the successful factors for travel web sites changed over time? The case of consumers in Hong Kong. J. Hosp. Tour. Res. **33**(3), 438–446 (2009)
5. Christou, E., Kassianidis, P.: Consumer's perceptions and adoption of online buying for travel products. J. Travel Tour. Mark. **12**(4), 93–107 (2002)
6. Cohen, L., dan Morrison, L., Manion, K.: Research Methods in Education, sixth edn. Routledge, Oxon (2007)
7. Copeland, M.T.: Problems in Marketing. AW Shaw Company (1923)

8. Duarte, P., Silva, S.C., dan Ferreira, M.B.: How convenient is it? Delivering online shopping convenience to enhance customer satisfaction and encourage e-WOM. J. Retail. Consum. Serv. 161–169 (2018)
9. Hasan, H., Kiong, T.P., Ainuddin, R.A.: Effects of perceived value and trust on customer loyalty towards foreign banks in Sabah, Malaysia. Glob. J. Emerg. Trends e-Bus. Mark. Consum. Psychol. **1**(2), 137–153 (2014)
10. Jiang, L.A., Yang, Z., Jun, M.: Measuring consumer perceptions of online. J. Serv. Manag. **21**(2), 191–214 (2012)
11. Katadata: Transaksi Tiket Online di Indonesia Diperkirakan Mencapai Rp125 Triliun-pada 2018. https://databoks.katadata.co.id/datpublish/2018/12/12/transaksi-tiket-online-di-indonesia-diperkirakan-mencapai-rp-125-triliun-pada-2018/, 16 may 2022
12. Kumar, R., Sachan, A., dan Dutta, T.: Examining the impact of e-retailing convenience dimensions on behavioral intention: the mediating role of satisfaction. J. Int. Commer. (2020)
13. Meuter, M.L., Ostrom, A.L., Roundtree, R.I., Bitner, M.J.: Self-service technologies: understanding customer satisfaction with technology-based service encounters. J. Mark. **64**(3), 50–64 (2000)
14. Pham, Q.C., Tran, X.P., Misra, S., Maskeliunas, R., dan Damasevicus, R.: Relationship between convenience, perceived value, and repurchase intention in online shopping. Sustainability 1–14 (2018)
15. Rizaty, M.A.: Traveloka, Aplikasi Pesan Hotel Pilihan Mayoritas Masyarakatpada Liburan Akhir Tahun (2021). https://databoks.katadata.co.id/datapublish/2021/12/24/traveloka-aplikasi-pesan-hotel-pilihan-mayoritas-masyarakat-pada-liburan-akhir-tahun/, 16 May 2022
16. Roy, S.K., Shekhar, V., Lassar, W.M., Chen, T.: Customer engagement behaviors: the role of service convenience, fairness and quality. J. Retail. Consum. Serv. **44**, 293–304 (2018)
17. Saha, S.K., Duarte, P.S., Susana, C., dan Zhuang, G.: Supporting sustainability by promoting online purchase through enhancement of online convenience. Environ. Dev. Sustain. (2020)
18. Seiders, K., Voss, G.B., Godfrey, A.L., dan Grewal, D.: SERVCON: development and validation of a multidimensional service convenience scale. J. Acad. Mark. Sci. **35**:144–156 (2007)
19. Syarifuddin, G., Abbas, B.S., dan Heriyati, P.: TAM approach on E-commerce of aircraft ticket sales on consumer purchase intention. In: The 6th international conference on cyber and IT service management (2018)
20. Thuy, P.N.: Using service convenience to reduce perceived cost. Mark. Intell. Plann. **29**(5):473–487 (2011)

# An Analytical Review of Poverty Alleviation and Impact of Globalisation Policy in India

**S. Muruganandhan, J. Yazhini, S. Isai Yazhini, R. Thamilmani, K. Sivasubramanian, and S. Guhan**

**Abstract** The first and vital objective of the United Nations Sustainable Development Goal is to prevent poverty among all nations. India is consistently working towards seventeen goals and regressively frames policy for the removal of poverty since 1991. Based on this background, this study has two major objectives such as to emphasize the significance of globalisation policy in India since the economic reform and to analytically review the economic impacts of globalisation on removal of poverty and economic development. This analytical review paper examines the various advantages of the globalisation policies of the government of India and critically evaluates the impacts of globalisation on the removal of poverty through new foreign direct investments and global partnerships. To achieve these important objectives, the present study followed a descriptive method to design the paper. This study adopted the review-based analysis to identify and address the impact of globalisation on industries, agriculture and service sectors. To achieve the targeted objectives, the secondary data relating to India's gross domestic product for the post-reform period has been collected and presented. Secondary data was sourced from the Indian Statistical Office, Government of India. The level of employment in the industrial sector has improved dramatically. MNCs are growing and generate employment of both skilled, semi-skilled and unskilled workforces in India. It also enhanced female work participation in the industrial sector. Economic reforms have achieved their target of reducing poverty and inequality. Improved employment opportunities by exporting

S. Muruganandhan · J. Yazhini · S. Isai Yazhini
Saveetha School of Law, Saveetha Institute of Medical and Technical Sciences, Chennai, India

R. Thamilmani
School of Commerce, Vel Tech Rangarajan Dr. Sagunthaala Institute of Science and Technology, Chennai, India

K. Sivasubramanian (✉)
Department of Economics, Kristu Jayanti College, Bengaluru, India
e-mail: sivasubramanian@kristujayanti.com

S. Guhan
M.Sc. Economics, Kristu Jayanti College, Bengaluru, India



M. Al Mubarak (ed.), *Sustainable Digital Technology and Ethics in an Ever-Changing Environment*, Studies in Systems, Decision and Control 237,
https://doi.org/10.1007/978-3-031-86708-8_40

agricultural and industrial commodities helped achieve this. On the whole, the liberalisation, privatisation and globalisation practices supports the Indian economy in a positive way in reducing poverty.

**Keywords** Agriculture · Industry · Service · Employment · Impact · Globalisation

# 1 Introduction

The new economic policy of India in 1991 made a tremendous change in the nation's economic growth in various sectors. This new economic reform has been instrumental for private sector participation, liberalisation, and global partnerships such as international trade. Globalisation practices have positively impacted developing nations including India. Globalisation has moved economic growth through the trading of goods outside the boundaries, technological exchanges and employment opportunities [11]. It helped both educated and uneducated labourers in India. In this process, foreign direct investment was enormously appreciated and the balance of payment also improved [1, 3].

## *1.1 Structure of the Paper*

This paper attempts to achieve targeted objectives based on the related literature reviewed and analysed. This paper has been organised as follows, introduction to the paper, major objectives of this article, methodological framework, literature survey, analysis, findings, concluding note, suggestions and limitations of the article.

## *1.2 Objectives*

To emphasize the significance of globalisation policy in India since the economic reform.

To analytically review the economic impacts of globalisation on the removal of poverty and economic development.

## 2 Methodology

This analytical review paper examines the various advantages of the globalisation policies of the government of India and critically evaluates the impacts of globalisation on the removal of poverty through new foreign direct investments and global partnerships. To achieve these important objectives, the present study followed a descriptive method to design the paper. This study adopted the review-based analysis to identify and address the impact of globalisation on industries, agriculture and service sectors. To achieve the targeted objectives, the secondary data relating to India's gross domestic product for the post-reform period has been collected and presented. Secondary data was sourced from the Indian Statistical Office, Government of India.

## 3 Literature Review

This segment of this paper exhibits the impact of globalisation on various sectors and the removal of poverty in India.

### *3.1 Impact on Agricultural Sector*

The agricultural sector significantly contributes to the nation's development in terms of providing food, employment for rural, and also contributes for the gross domestic product in India [14]. Globalisation policies and practices have provided immense opportunities for Indian farmers to enhance production. It is also helps to bridge the existing technological gaps in agricultural production in India [15]. Globalisation helps to increase the use of technology in the production process rather than primitive technology which includes irrigation, modern cropping patterns, and farming The prices of farm commodities are relatively higher in the global markets. But there are no proper prices fixed in the local market in India. The globalisation policy helps the farmers to export their commodities to various countries and receive better prices.

### *3.2 Influence on the Industrial Sector*

Industrial growth is vital for the economic enhancement of the country in terms of gross domestic product and national income. Industries are the major agency for the generation of employment in both rural and urban areas. The large, medium and small-scale operational industries and new industries need a huge amount of money for investment and capital accumulation [12]. The agricultural sector development is lined with industrial development in India. The level of employment in the industrial sector has improved dramatically. MNCs are growing and generating employment of both skilled, semi-skilled and unskilled workforces in India. It also enhanced female work participation in the industrial sector [17].

### 3.3 Effects on the Service Sector

Foreign investment in India is instrumental for economic growth and employment generation in IT and IT-enabled services [4]. Globalisation supports for capital formation of any country especially the fastest growing nations for the development of technology in service sectors. The technological innovations, optimum way of using resources, human capital formation and financial strength have been elevated through the globalisation policy of India [6]. Globalisation is a process for the removal of poverty through regional economies, societal development, enhanced communication technology and transport. The system of global integration aims to develop international trade, global partnerships, and the exchange of knowledge and technology [13].

### 3.4 Globalisation and Employment Generation

The increase in agricultural production and industrial growth accelerated through the strong globalisation initiatives of the government in India. Various liberalising schemes enhanced employment opportunities in all sectors in India [5]. The international trade efficiency through the import and export policies maximized the output in India. There is a significant increase in the employment level after the new economic policy in India [2, 7]. The structural increase in employment in the informal sector is due to the opening of the trade practices in India [8].

## 4 Analysis

Globalisation policies have been positively impacting the removal of poverty in India. The positive change in the agricultural, industrial and service sectors leads to improve in the employment avenues in the respective sectors [9].

Figure 1 exhibits the consistent growth of gross domestic product positively due to the globalisation policy of India. The above figure reveals the data about the GDP contribution of the agricultural, industry, and service sectors since the reform period. It is observed from the data that agricultural sector contribution has declined over a period from 1991–92 to 1999–2000 with a percentage of 54.74–30.58%.

Similarly, Fig. 2 reveals the sector-wise contribution of GDP from 200–2001 to 2014–2015 and it implies the increase in the service sector contribution from 43.77

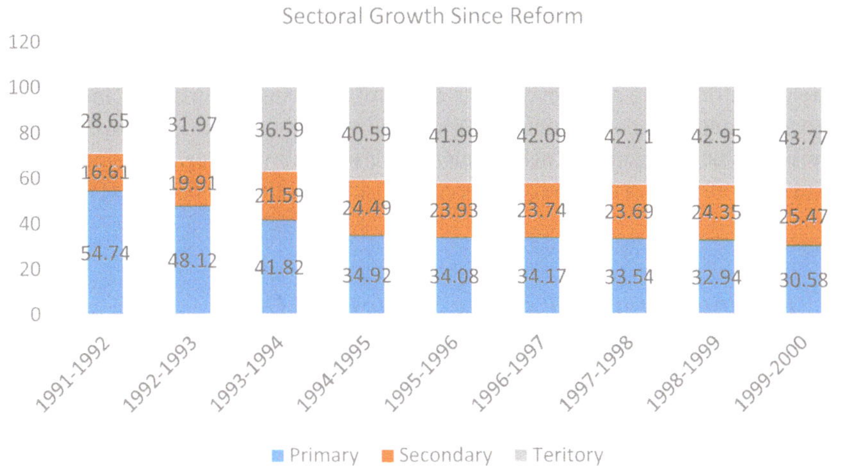


**Fig. 1** Sectoral GDP from 1991–92 to 1999–2000. *Source* Central Statistical Office, Government of India, various years

to 62.36%. Due to globalisation, the service sector and industrial sector have seen a positive impact and the agricultural sector has recorded a negative impact. But still, all the sectors created more new employment avenues in India.

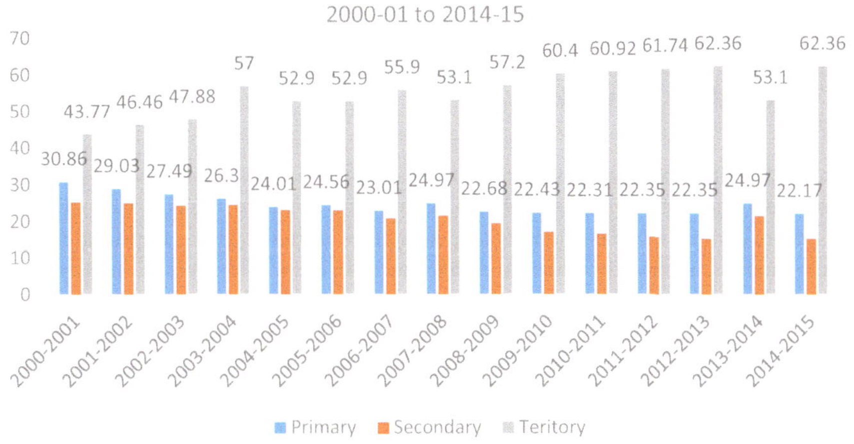


**Fig. 2** Sectoral GDP from 2000–01 to 2004–15. *Source* Same as the previous table

### *4.1 Poverty Alleviation Through Globalisation Policies*

Economic reforms have achieved their target of reducing poverty and inequality. Improved employment opportunities through the export of agricultural and industrial commodities helped with this achievement. The inequality among male and female workers' income has consistently reduced and rural poverty has declined [10, 16].

### *4.2 Findings from the Existing Studies and Data*

Globalisation has moved economic growth through the trading of goods outside the boundaries, technological exchanges and employment opportunities. The agricultural sector significantly contributes to the nation's development in terms of providing food, and employment for rural, and also contributes for the gross domestic product in India. Industrial growth is vital for the economic enhancement of the country in terms of gross domestic product and national income. Industries are the major agency for the generation of employment in both rural and urban areas. Globalisation supports for capital formation of any country especially the fastest growing nations for the development of technology in service sectors. The structural increase in employment in the informal sector is due to the opening of the trade practices in India. It is observed from the data that agricultural sector contribution has declined over a period from 1991–92 to 1999–2000 with a percentage of 54.74–30.58%.

## 5 Concluding Remarks

Due to globalisation, the service sector and industrial sector have seen a positive impact and the agricultural sector has recorded a negative impact. But still, all the sectors created more new employment avenues in India. The agricultural sector development is lined with industrial development in India. The level of employment in the industrial sector has improved dramatically. MNCs are growing and generating employment of both skilled, semi-skilled and unskilled workforces in India. It also enhanced female work participation in the industrial sector. Economic reforms have achieved their target of reducing poverty and inequality. Improved employment opportunities by exporting agricultural and industrial commodities helped achieve this. On the whole, the liberalisation, privatisation and globalisation practices supports the Indian economy in a positive way in reducing poverty.

## 6 Suggestions

Policy to framed to empower the agriculture and farmer's development for an inclusive growth of the whole nation.

The globalisation policy has to support in enhancing human capital resources such education and health conditions of the people in India.

## 7 Limitations and Scope

This paper is descriptive and uses a minimal amount of data for interpretation. This work purely relies on the various existing literature relating to the impact of globalisation policy on different sectors.

## References

1. Adamu, P., Kaliappan, S.R., Bani, Y., Nor, M.N.: Impact of globalization on unemployment in Sub-Saharan African (SSA) countries. In: Governance and Sustainability of Global Business Economics, pp. 302–310 (2017)
2. Arogyaswamy, B., Hunter, J.: The impact of technology and globalisation on employment and equity: an organising framework for action. Nature **3**(1) (2019). https://doi.org/10.5296/IJGS.V3I1.14127
3. Das, R.C., Ray, K.: Does globalisation influence employment? Empirical investigation on individual as well as panel of South Asian countries. Rev. Mark. Integr. **12**(1–2), 7–34 (2020). https://doi.org/10.1177/0974929220969222
4. Garang, A.P., Yacouba, K., Thiery, K.K.: Time series bounds approach to foreign direct investment, unemployment, and economic growth in Uganda. Mod. Econ. **9**(1), 87 (2018). https://doi.org/10.4236/me.2018.91006
5. Hensman, R.: The impact of globalisation on employment in India and responses from the formal and informal sectors. In: CLARA Working Paper, No. 15 (2001)
6. Jahanger, A., Usman, M., Murshed, M., Mahmood, H., Balsalobre-Lorente, D.: The linkages between natural resources, human capital, globalization, economic growth, financial development, and ecological footprint: the moderating role of technological innovations. Resour. Policy. Policy **76**, 102569 (2022). https://doi.org/10.1016/j.resourpol.2022.102569
7. Jayaraman, T.K., Singh, B.: Foreign direct investment and employment creation in Pacific Island countries: an empirical study of Fiji. In: ARTNeT Working Paper Series (2007)
8. Kalirajan, K., Singh, K.: The impact of globalisation on employment generation in India: the case of emerging big shopping malls and retailers. In: ASARC Working Paper, pp. 1–7 (2009)
9. Kiren, K., Awan, A.G.: The role of globalisation in employment generation: evidence from Pakistan. Glob. J. Manag. Soc. Sci. Human. **4**(1), 111–132 (2018)
10. Lee, H.H., Park, D.: Effects of inward and outward Greenfield FDI on employment by domestic firms: the Korean experience. Korea World Econ. **21**(1), 1–33 (2020)
11. Mushtaq, M., Ahmed, S., Fahlevi, M., Aljuaid, M., Saniuk, S.: Globalization and employment nexus: moderating role of human capital. PLoS ONE **17**(10), e0276431 (2022). https://doi.org/10.1371/journal.pone.0276431
12. Nayyar, D.: Globalization and employment. Indian J. Lab. Econ. **58**(1), 87–97 (2015)

13. Paul, J.: Impact of globalisation on service sector of India economy. EPRA Int. J. Econ. Bus. Rev. **5**(8) (2017). ISSN 2349-0187
14. Sharma, P.: Impact of globalisation on Indian agriculture: an analysis, Amar Inter-Discip. Res. J. **1**(1), 192–199
15. Siddiqa, A., Hussain, T., Qasim, M., Javed, M.I.: The impact of globalization on unemployment and economic growth: panel data analysis for developing countries. Bull. Bus. Econ. (BBE) **7**(3), 122–131 (2018)
16. Upadhyay, R.K.: Impact of globalisation on distribution of income, poverty and inequality in India. Abhinav Natl. Mon. Ref. J. Res. Commer. Manag. **4**(2), 38–44 (2015)
17. Venkatesham, A.: Globalisation and its impact on small scale industries in India: bright and dark spots. Int. J. Creat. Res. Thoughts **9**(9), 184–195 (2021)

# The Influence of Social Media on Perceived Body Image and Self-esteem Among Young and Middle-Aged Adults

**Hrishita Lobo, Rekha Devi Chauhan, Candida Antonetta D'souza, Marissa Ruth Furtado, and Soumya Simon**

**Abstract** Social media is an important part in the lives of people across all age groups. The increase adoption of social media has raised concerns as it determines the perception of one's body image and self-esteem. The aim was to understand the relationship of social media usage on perceived body image and self-esteem, and the impact it has on gender and age. A convenient sampling method was employed, focusing specifically on males and middle-aged adults. Data was collected from 400 participants comprising of 195 males and 205 females between the age range of 19–60 years, through an online questionnaire. The tools implemented were the Objectified Body Consciousness Scale, Rosenberg Self-esteem Scale, and Social Media Usage Scale. The findings indicate that there is no relation between social media usage, perceived body image and self-esteem among young and middle-aged adults.

**Keywords** Body image · Self-esteem · Social media

## 1 Introduction

Social media has evolved into a vital aspect of modern life for both young and middle-aged adults, as a means of social interaction to express oneself. Simultaneously, there is an increasing concern about the impact it has on the perception of our body and self-esteem. This study seeks to recognize and understand the predominance social media has on the perception of our physical appearance and self-esteem through the lifespan. Examining the experiences of men and women can shed light on the intricate and potentially gender-specific dynamics at play.

H. Lobo (✉) · R. D. Chauhan · C. A. D'souza · M. R. Furtado · S. Simon
Department of Psychology, Kristu Jayanti College Autonomous, Bengaluru, Karnataka, India

S. Simon
e-mail: soumya@kristujayanti.com



M. Al Mubarak (ed.), *Sustainable Digital Technology and Ethics in an Ever-Changing Environment*, Studies in Systems, Decision and Control 237,
https://doi.org/10.1007/978-3-031-86708-8_41

According to Datar portal, on average, people spend 2 h and 31 min on social media worldwide. Harriger et al., in [18] reported that out of five teenagers, 35% use at least one social media site several times a day. From sharing their creativity with people around the world, to receiving feedback, to acquiring information and staying connected to the world,social media plays a pivotal role [8]. According to an article by Acosta, social media exposes us to a perfect life, appearance, and ideal body type. This constant exposure leads to developing impractical beliefs, distorted perspective and inadequate feelings about oneself [18].

Body image comprises thoughts and emotions about one's physical self which may not be equivalent to their actual body shape and size, influenced by perspectives, emotions, and experiences. Comments in-person and social media can lead individuals to face serious consequences. In modern times, body shaming is often promoted in weight loss and beauty product advertisements, especially among young people; leading to psychological health struggles. Media portrayals of overweight characters as the subject of jokes intensifies the existing problem.

Self-esteem is a measure of mental and social adjustment and a mediating factor of behaviour [1]. Among middle-aged adults, body shaming can significantly impact self-perception, confidence, weight stigma due to societal pressures and discrimination [12].

**Structure of the Paper**

In today's time perception of our body is affected by various social media platform which in turn can affect our self-esteem. This paper includes the introduction to the variables followed by review of previous research papers. The second chapter is the methodology which includes the aim, objective, hypothesis, research of the study, information regarding the tools and sample. Chapter three includes result and analysis of and lastly chapter four includes findings and discussion.

**Review of Literature**

The youth, who are body-shamed may pursue a perfectionistic self-presentation which may contribute to an unhealthy social media usage across both the genders [6]. Research indicates that the effect of social networking sites like WeChat Moments on female college students, includes an indirect link to body image [22]. Similarly, Yao et al. [21] too, found body shaming is a significant mediating factor for youth females exposed to media promoting unrealistic beauty standards [20]. Fechter [5] argued for potential solutions, including education, combatting narrow beauty standards, parental interventions, promoting self-esteem in pre-adolescence, and body positive media representations for young audiences. Body shaming through cyberbullying causes depression and mild stress in adolescents and the recovery process is time-consuming [4]. Gan and Jiang [7], concluded three negative outcomes: self-harm behaviours, disordered eating patterns, and anxiety. Males too have psychological impact but at the same time may also contribute to self-confidence and the motivation for fitness [2]. According to Soleman and Elindawati [16] body shaming is caused due to the mind-set of accepting harmful practices. Saxena et al. [15] found body image and gender have no significant relationship in young adults, but, a positive relation was identified between body image and emotions. Research has shown significant correlation in low self-esteem and depression in middle-aged women [14]. Women are more prone to body dissatisfaction due to body image having a significant impact on self-esteem [3]. Parwin and Lone [11] found social media and body image perception has correlated with self-esteem among university students. Steinsbekk et al. [17], indicates those who actively posted updates expect to receive positive feedback which enhances their self-esteem, whereas youth who mostly engage with others' posts and do not receive the same amount of feedback have low self-esteem. Nguyen et al. [10] found that low self-esteem is associated with negative psychological outcomes among Vietnamese students.

## 2 Research Methodology

**Aim**

To investigate the influence of social media usage on perceived body image and self-esteem among both young and middle-aged adults.

**Objective**

1. To assess the correlation between perceived body image and self-esteem.
2. To evaluate the impact of social media usage on self-esteem.
3. To measure the influence of social media on perceived body image among young and middle-aged adults.
4. To assess if gender differences exist with regards to perceived body image among young adults and middle-aged adults.
5. To assess the effect of social media usage on perceived body image of males.
6. To assess the influence of social media on perceived body image among middle aged adults.

**Hypothesis**

1. There will be no significant relationship between perceived body image and self-esteem.
2. There will be no significant relationship between social media usage on self-esteem.
3. There will be no significant gender differences with regards to perceived body image among young adults and middle-aged adults.
4. Increased use of social media is associated with a negative impact on perceived body image among young adults and middle-aged adults.
5. There will be no significant impact on perceived body image of male due to use of social media.
6. There will be no significant influence of social media usage on perceived body image among middle aged adults.

### Design of the Study

The study utilized a quantitative research design, gathering data through online survey method, employing a convenient sampling technique. The sample size comprised of 400 participants. Statistical Package of Social Sciences was used to analyse the data.

### Sample

This study involved 400 participants comprising of young and middle-aged adults. Among these participants there were 195 males (48.8%) and 205 females (51.2%) with the age ranging from 19 to 60 years. These participants were from India. As per the inclusion criteria the participant should use at least one social media platform for a minimum of one year.

### Tools

For this study three psychological scales were used. Social Media Use Scale (SMUS) by Lin, Wang, and Chen. It was found to be reliable and valid, with a high level of agreement among raters [19]. The Body Shame Sub-scale of the Objectified Body Consciousness Scale, developed by McKinley and Hyde [9]. Test–retest reliability was 0.79. Rosenberg Self-esteem Scale is a 5-point Likert scale with high reliability and internal consistency at 0.77 and test–retest reliability ranging from 0.63 to 0.85 [13].

## 3 Results and Analysis

Correlation, regression and t-test was used to analyse the data collected. This chapter provides examined information, organized in inferential statistics and quantitative analysis.

As shown in Table 1 the R value for social media usage on self-esteem was found to be − 0.330, indicating negative low correlation which means the correlation is statistically significant at 0.01 level, hence not supporting the hypothesis. The R value between Perceived body image and self-esteem was computed to be − 0.499 indicating a negative low correlation which was statistically significant at 0.01 level. Hence, not supporting the hypothesis.

**Table 1** Indicating correlation coefficient of social media usage, perceived body image and self-esteem of young and middle-aged adults

| Variable | n | M | SD | 1 | 2 | 3 |
|---|---|---|---|---|---|---|
| 1. Social media usage | 400 | 42.73 | 18.185 | – | | |
| 2. Perceived body image | 400 | 26.63 | 11.216 | – | – | |
| 3. Self-esteem | 400 | 27.96 | 4.034 | − 0.330** | − 0.499** | – |

Note: **p<0.01

As per Table 2 the mean of perceived body image of females and males was found to be 26.11 and 27.17. The t-value computed was − 0.945 which was not statistically significant. Implying no difference exists in young and middle age adults with respect to perceived body image. The hypothesis that there will be no significant gender differences with regards to perceived body image among young adults and middle-aged adults is accepted.

According to model summary in Table 3 there is a positive correlation between social media usage and perceived body image. This means, there is a 6.8% chance that increased social media usage is linked with a negative impact on perceived body image among young and middle-aged adults. Hence, the hypothesis stating increased use of social media is associated with a negative impact on perceived body image among young adults and middle-aged adults is accepted.

Table 4 illustrates the correlation coefficient between social media usage on perceived body image among males, is 0.179, indicating a statistically significant correlation at 0.05 level, hence not supporting the hypothesis.

According to the model summary in Table 5 there is a positive correlation among perceived body image, social media usage among the males. In other words, there is a 0.1% chance that increased use of social media usage is associated with body image perception among the males. Hence, the hypothesis, there will be no significant impact on perceived body image of male due to use of social media, is accepted.

As shown in Table 6 the correlation coefficient between social media usage on perceived body image among middle-aged adults, which was found to be −

**Table 2** Indicating t value for females and males on perceived body image

| Variable | Female | | Male | | t | *p* |
|---|---|---|---|---|---|---|
| | M | SD | M | SD | | |
| Perceived body image | 26.11 | 11.8858 | 27.17 | 10.502 | − 0.945 | 0.042 |

**Table 3** Modal summary of predictor perceived body image and dependent variable in young and middle-aged adults

| Model | Estimate | SE | R | R Square | 95% confidence interval | | p |
|---|---|---|---|---|---|---|---|
| | | | | | Lower limit | Upper limit | |
| Perceived body image | 0.065 | 17.581 | 0.260[a] | 0.068 | 0.268 | 0.576 | 0.000 |

**Table 4** Indicates correlation coefficient between social media usage on perceived body image among males

| Variable | n | M | SD | Correlation coefficient |
|---|---|---|---|---|
| Social media usage | 195 | 40.73 | 19.443 | 0.179* |
| Perceived body image | 195 | 27.17 | 10.502 | |

Note: *p<0.05

**Table 5** Model summary of predictor social media usage and dependent variable perceived body image of males

| Model | Estimate | SE | R | R square | 95% confidence interval | | *p* |
|---|---|---|---|---|---|---|---|
| | | | | | Lower limit | Upper limit | |
| Social media usage | – 0.004 | 10.521 | 0.039 | 0.001 | – 0.056 | 0.097 | 0.593 |

**Table 6** Indicating the correlation coefficient between social media usage and perceived body image among middle-aged adults

| Variable | n | M | SD | Correlation coefficient |
|---|---|---|---|---|
| Social media usage | 72 | 33.22 | 13.342 | 0.435** |
| Perceived body image | 72 | 22.72 | 13.342 | |

Note: **p<0.01

**Table 7** Model summary of predictor social media usage and dependent variable perceived body image of middle-aged adults

| Model | Estimate | SE | R | R Square | 95% confidence interval | | *p* |
|---|---|---|---|---|---|---|---|
| | | | | | Lower limit | Upper limit | |
| Social media usage | 0.139 | 11.091 | 0.388 | 0.151 | 0.151 | 0.545 | 0.001 |

0.435, indicating there is a significant correlation at 0.01 level, hence supporting the hypothesis.

According to the model summary in Table 7 there is a positive correlation among perceived body image, social media usage among middle-aged adults. In other words, there is a 15.1% impact of social media usage on perceived body image. Hence, the hypothesis, there will be no significant influence of social media usage on perceived body image among middle aged adults, is not supported.

# 4 Findings and Conclusion

### Conclusion

The study intended to investigate how social media affects an individual's perceived body image and self-esteem in youth and middle-aged adults. It revealed that social media usage was significantly correlated with perceived body image, and negatively correlated with self-esteem. Perceived body image and self-esteem too was negatively correlated. There was no significant gender difference in perceived body image. Increased social media usage was found to have a 6.8% chance of negatively impacting perceived body image overall. Among males, a significant correlation between social media usage and perceived body image and among middle-aged adults a positive interrelationship between perceived body image and social media usage was found.

**Significance of the Study**

Worthwhile insights into the impacts of social media usage on perceived body image and self-esteem among young and middle-aged individuals was revealed. The insights of the study could aid in understanding the duration on social media which can impact our body image perception and self-esteem. Also, there was an identification to the extent to which social media usage has an impression on perceived body image of young and middle-aged adult males and middle-aged women.

**Limitations of the Study**

Employment of the convenient sampling technique, may not accurately represent the general population. The social media usage scale was not applicable to middle-aged adults due to their limited usage of social media platforms. Social media is a multifaceted phenomenon, making it challenging to capture the diverse ways individuals engage with it and its effects on perceived body image and self-esteem.

**Scope for Further Studies**

This study can be further expanded by understanding the influence family and peers has on perceived body image. Qualitative research methods can be used to get an in -depth understanding of an individual's emotions. Further additional variables such as sexual orientation and physical disability can be taken into consideration. Furthermore, intervention studies can be conducted to promote positive insight of one's body image and self-esteem.

## References

1. Altıntaş, A., Aşçi, F.H.: Physical self-esteem of adolescents with regard to physical activity and pubertal status. Pediatr. Exerc. Sci. **20**(2), 142–156 (2008). https://doi.org/10.1123/pes.20.2.142
2. Chatzopoulou, E., Filieri, R., Dogruyol, S.A.: Instagram and body image: motivation to conform to the "Instabod" and consequences on young male wellbeing. J. Consum. Aff. **54**(4), 1270–1297 (2020). https://doi.org/10.1111/joca.12329
3. Chukwuere, G.C.J., Chukwuere, J.E.: The difficulties posed by digital technology: understanding the psychological consequences of social media use on young adults' body image and self-esteem. Acad. J. Interdisc. Stud. **12**(6), 379 (2023). https://doi.org/10.36941/ajis-2023-0176
4. Collantes, L.H., Saputra, F.A., Solikhah, P.R., Laksana, T.C., Yakti, N.K., Tipagau, J.: Cyberbullying body-shaming levels in adolescence. Bull. Soc. Informat. Theory Appl. **6**(2), 111–119 (2022). https://doi.org/10.31763/businta.v6i2.603
5. Fechter, J.: Social Media & Negative Effects on Body Image. English Department: Research for Change—Wicked Problems in Our World (2021). https://research.library.kutztown.edu/wickedproblems/54

6. Fioravanti, G., Bocci Benucci, S., Vinciarelli, V., Casale, S.: Body shame and problematic social networking sites use: the mediating effect of perfectionistic self-presentation style and body image control in photos. Curr. Psychol. **43**(5), 4073–4084 (2024). https://doi.org/10.1007/s12144-023-04644-8
7. Gan, R., Jiang, S.: The impact of body shaming on female high school students in China. In: Ali, G., Birkök, M.C., Khan, I.A. (eds.) Proceedings of the 2022 6th international seminar on education, management and social sciences (ISEMSS 2022), vol. 687, pp. 3668–3674. Atlantis Press SARL (2022). https://doi.org/10.2991/978-2-494069-31-2_430
8. How to help your teen navigate social media. Mayo Clinic (n.d.). Retrieved 13 Apr 2024, from https://www.mayoclinic.org/healthy-lifestyle/tween-and-teen-health/in-depth/teens-and-social-media-use/art-20474437
9. McKinley, N.M., Hyde, S.: The objectified body consciousness scale: development and validation. Psychol. Women Q. **20**, 181–215 (1996)
10. Nguyen, D.T., Wright, E.P., Dedding, C., Pham, T.T., Bunders, J.: Low self-esteem and its association with anxiety, depression, and suicidal ideation in Vietnamese secondary school students: a cross-sectional study. Front. Psychiatry **10** (2019). https://doi.org/10.3389/fpsyt.2019.00698
11. Parwin, A., Lone, Z.A.: Influence of social media addiction and body image perception on self-esteem among university students. Int. J. Health Sci. 8023–8030 (2022). https://doi.org/10.53730/ijhs.v6nS2.7010
12. Puhl, R.M., Heuer, C.A.: Obesity stigma: important considerations for public health. Am. J. Publ. Health **100**(6), 1019–1028 (2010). https://doi.org/10.2105/AJPH.2009.159491
13. Rosenberg, M.: Society and the Adolescent Self-image. Princeton University Press, Princeton, NJ (1965)
14. Satwik, R., Sinha, D., Tiwari, B.: Prevalence of poor body image and its correlation with self-esteem and depression in middle-aged women. Climact. J. Int. Menopause Soc. **27**, 1–8 (2024). https://doi.org/10.1080/13697137.2023.2297876
15. Saxena, S., Mathur, A., Jain, S.: Body shaming, emotional expressivity, and life orientation among young adults **7**(9) (2020)
16. Soleman, N., Elindawati, R.: Fourth wave feminism in indonesia: body shaming through social media phenomenon. J. Islam. World Polit. **3**(2) (2019). https://doi.org/10.18196/jiwp.3234
17. Steinsbekk, S., Wichstrøm, L., Stenseng, F., Nesi, J., Hygen, B.W., Skalická, V.: The impact of social media use on appearance self-esteem from childhood to adolescence—a 3-wave community study. Comput. Hum. Behav. **114**, 106528 (2021). https://doi.org/10.1016/j.chb.2020.106528
18. The impacts of social media on youth self-image. News (2023). https://news.llu.edu/health-wellness/impacts-of-social-media-youth-self-image
19. Tuck, A.B., Thompson, R.J.: The social media use scale: development and validation. Assessment **31**(3), 617–636 (2024). https://doi.org/10.1177/10731911231173080
20. Wan, S.: The impact of social media on body shame. Lect. Notes Educ. Psychol. Publ. Med. **28**(1), 234–238 (2023). https://doi.org/10.54254/2753-7048/28/20231346
21. Yao, L., Niu, G., Sun, X.: Body image comparisons on social networking sites and Chinese female college students' restrained eating: the roles of body shame, body appreciation, and body mass index. Sex Roles **84**(7–8), 465–476 (2021). https://doi.org/10.1007/s11199-020-01179-1
22. Yu, M., Sun, X., Xu, Y., Liu, Z., Wu, Y., Yang, S., Luo, F.: Mirror, mirror, on the social media … WeChat moments usage and negative body image among female college students: evidence from ecological momentary assessment data. Appl. Psychol. Health Well Being **15**(3), 1046–1064 (2023). https://doi.org/10.1111/aphw.12425

# Assessment of Patients' Perceived and Expected Health Care Service Quality

**J. Rajalakshmi and L. Jayanthi**

**Abstract** The COVID-19 pandemic has brought a huge focus to healthcare sector. The global health care system is excellent at come through this pandemic by providing exclusive possible care. Health care services have unique qualities like intangibility, subjectivity, simultaneity, customer involvement these make defining and assessing their quality difficult. Moreover variations in the training, experience, abilities, and personal traits of medical professionals and the need to provide a patient with a variety of services makes health care services as a challenging one to measure the quality of medical services. This study's objective is to evaluate the level of care given by Chennai's private hospitals. This study investigates how closely patients' expectations and perceptions of hospital services align with one another, as well as the patients' expectations and perceptions of the quality of such services. The SERVQUAL model, which was developed and presented by Parasuraman et al. (J Mark 49:44, 1985 [8]), has five dimensions of service quality, includes reliability, responding, empathy, assurance-and tangibility. The nature of this work is descriptive. The mean scores of the several servqual dimensions both for expectations and perceptions are obtained in order to examine the gap between these two gathered data. To determine the discrepancies between expectations and perceptions, these mean scores are compared to one another. Every service quality dimension's disparity in terms of quality are assessed. The research's conclusions point out the crucial areas that need to be improved and concentrated on in order to keep a competitive edge and offer top-notch services in the health care industry.

**Keywords** Service quality · SERVQUAL · Health care · Perceived · Expected service quality

J. Rajalakshmi (✉) · L. Jayanthi
Faculty of Management, SRM Institute of Science and Technology, Kattankulathur, India
e-mail: rajalakj@srmist.edu.in

L. Jayanthi
e-mail: jayanthl@srmist.edu.in



M. Al Mubarak (ed.), *Sustainable Digital Technology and Ethics in an Ever-Changing Environment*, Studies in Systems, Decision and Control 237,
https://doi.org/10.1007/978-3-031-86708-8_42

## 1 Introduction

The Indian health care sector plays a vital role in preserving peoples' physical, mental, and emotional well-being in addition to being a major driver of economic development. India has a diverse health care system. It experiencing considerable changes over the years, the system still encounters numerous challenges such as inadequate infrastructure, preventive healthcare neglect, shortage of healthcare professionals, burden of communicable and non-communicable diseases, inequality between urban and rural areas, deficient health insurance coverage, incompetent public healthcare funding, and a disunited healthcare system. Despite this, India is a desirable destination for medical tourists due to its affordable prices, excellent private hospital healthcare facilities and accessibility to complementary therapies like naturopathy, Ayurveda, yoga, meditation, and Siddha.

Hospitals must restructure and assess their services in order to deliver higher caliber care that prioritizes patient's care. This will be achieved by gaining a deeper apprehension of the quality of health care services and measuring them to determine the aspects of quality in which hospitals are falling short. Only then will they be able to concentrate on creating better healthcare services. There has been an increasing consideration of the quality assurance activities, including the audits, patients' feedbacks and the quality improvement activities, especially among the accredited hospitals. However, these practices are not so common in all health care centers.

## 2 Background of the Study

Any excellent health care service must have the attributes like safety, security, cleanliness, comfortable, timeliness, continuity of care, equity, technical Competence, patient-represented, peaceful, respect for the patient's needs and preferences, values and the availability of clinical service options. Since the components of health care services are so complicated, the quality in health care service is a difficult notion to grasp.

A hospital's ability to measure and manage service quality depends on its ability to effectively assess stakeholder views, which include patients, doctors, Government and regulatory bodies, nurses, pharmaceutical and medical device companies, Non-Governmental Organizations (NGOs), advocacy groups, academic and research institutions, policy makers and legislators, technology providers and others. In order to develop a more comprehensive understanding of service quality this study has the intention to determine the key components of service quality as reported by patients. Additionally, SERVQUAL model is used to gauge the quality of service provided by private hospitals in Chennai.

## 3 Methodology

Chennai city, Tamil Nadu, India was taken as the site for the research titled, "Assessment of patients' perceived and expected health care service quality." Using the cluster sampling technique, 320 outpatients from 32 private hospitals in which ten outpatients from each hospital were chosen as samples. The primary data were gathered via SERVQUAL questionnaire. The survey is divided into two sections: PART A: the patient's expectations and views of the hospital service quality characteristics, PART B: the demographic profile of the respondents. On a 5-point Likert scale, from strongly disagree (1) to strongly agree (5), patients were asked to indicate how much they agreed with the statement. The survey was conducted between December 1, 2023, and December 31, 2023. Data were checked to ensure consistency and completeness and then analyzed using SPSS for windows version 21.0. Descriptive analysis was done by running simple frequencies and proportions.

## 4 Literature Review

Numerous researchers have evaluated service quality. Shabbir et al. used SERVQUAL to study the quality of care at the Combined Military Hospital (CMH), private and public hospitals. The findings revealed that both CMHs and private hospitals were able to fulfill patient expectations by providing punctual care as well as other amenities.

Using the gaps between expectations and perceptions score, Brown and Swartz assessed medical services from the standpoint of the healthcare service provider and the patients. From the research it is concluded that physician engagement was the key factor in patient satisfaction.

Patients' impressions of the quality of care in public and private hospitals in the United Arab Emirates were compared by Jabnoun and Chaker. With an objective of comparing service quality views in private and public hospitals, the researchers identified five dimensions: tangibles, empathy, responsiveness, supportive skills, and reliability. All factors, with the exception of the supportive skills component showed a substantial difference between these hospitals.

A service quality measuring methodology and quality dimensions, including nursing care, physician care, infrastructure, environment, social responsibility, outcome, allied services, administrative, operation and discharge mechanisms have been considered by a number of studies carried out in India. According to the research carried out by Aagja and Garg [1].

Donabedian suggested measuring the quality of healthcare services by combining three related items: structure, method and outcome. Therefore, in this case structure is defined as the environment, staff's qualification, and other organizational structures that make it possible to deliver health care. An outcome could be defined in context to health improvement where a patient has been restored to life or good health condition.

A method is the practice carried out in the healthcare sector. Health care facilities must have a holistic social approach in addition to a medical one. According to Donabedian [2], patients require performance in every service they receive not just correct diagnosis and treatment.

George [3] Economic Dimension of Primary Health Centres in Kerala: Understanding the Kerala Model, advocated an enhanced congruence and more people-centered health system. This paper emphasized the role of a healthy public policy in achieving good health at low cost for all its citizens. Using a modified version of SERVQUAL, Irfan et al. examined hospital service quality in public and private institutions, focusing on the attributes like empathy, tangibles, timeliness, responsiveness, and assurance. It is evident that public hospitals are not as good at providing services as private hospitals are Ladhari, [4] analysed over 150 studies using SERVQUAL across diverse service sectors including health care, education, banking, tourism, and public services. He emphasized that SERVQUAL remains a popular and influential tool, researchers often modify it to fit specific industries and cultural settings, raising concerns about its standardization. Minjoon et al. [5] explored the identification and measurement of service quality dimensions specifically within the healthcare sector. Using focus group interviews with healthcare professionals and patients, the study identified unique service quality attributes relevant to healthcare settings such as clinical competence, communication effectiveness, empathy, timeliness, and physical environment. Mostafa [6] conducted a study in Egyptian hospitals to examine the gap between patients' expectations and their actual experiences. Utilizing an adapted SERVQUAL model, the research uncovered notable deficiencies in service dimensions like reliability, responsiveness, and empathy. [7] Neupane and Gurung assessed service quality in Western Regional Hospital using the SERVQUAL model. The research highlighted the need for hospital management to focus on staff training, infrastructure improvement, and responsiveness to enhance overall service quality. Raposo, Alves, and Duarte [9] analyzed patient satisfaction in Portuguese primary healthcare centers using a satisfaction index. Their study identified key dimensions influencing satisfaction, including the patient-doctor relationship, facility quality, and administrative staff interactions. Talib et al. [10] reviewed existing studies on healthcare service quality, highlighting key dimensions such as reliability, empathy, and responsiveness. The paper emphasized the importance of continuous quality improvement for better patient satisfaction and healthcare outcomes. Zeithaml et al. [11] developed a key framework distinguishing desired and adequate service expectations, identifying factors like personal needs and past experiences as influencers. Their work is foundational in explaining gaps in service quality and shaping models like SERVQUAL.

# 5 The Servqual Model

Each of the 46 criteria that make up SERVQUAL's five service quality dimensions assesses the expectations and perceptions of the health care service quality. Reliability (09), responsiveness (11), assurance (10), empathy (08), and tangibility (08) are the five aspects. This approach records and quantifies the level of care that patients receive. The service quality is listed in this research is based on the discrepancy in scores or gaps between perception and expectation. The SERVQUAL instrument's validity and reliability have been demonstrated in a wide range of service contexts, which is a significant advantage.

# 6 Objectives of the Study

## *6.1 Primary Objective*

To examine and analyze patient's perceptions and expectations towards service quality attributes of private hospitals.

## *6.2 Secondary Objectives*

1. To examine the demographic variables impact on patient's perceptions and expectations.
2. To determine the discrepancies between patients' perceptions and expectations of each healthcare quality attributes.
3. To make suggestions for improvement of quality of services in private hospitals.

# 7 Data Analysis and Interpretation

## *7.1 Reliability and Validity*

Service quality attributes and service quality dimensions used in this study were tested with Cronbach's alpha values to ensure their reliability/ internal consistency.

Table 1 shows that, all the Cronbach's coefficients of alpha are more than 0.7 for all the variables. According to Fujun et al. (2007), a Cronbach's alpha value greater than 0.7 shows internal consistency. Thus, it may be inferred that the service quality features and variables are trustworthy.

**Table 1** Cronbach's alpha of all the variables

| Variables | Perception | Expectation |
|---|---|---|
| 46 service quality attributes | 0.814 | 0.881 |
| Reliability | 0.725 | 0.733 |
| Responsiveness | 0.784 | 0.791 |
| Assurance | 0.708 | 0.726 |
| Empathy | 0.752 | 0.762 |
| Tangibility | 0.711 | 0.753 |

## 7.2 *Demographic Profile of Respondents*

Table 2 shows that, 54% of respondents are male, 38% are between the ages of 36 and 50, and 39% have completed their undergrad. 32% of respondents have a higher secondary educational status, 35% have a monthly salary ranging from Rs. 0 to Rs. 10,000, and 29% are employed.

## 7.3 *Service Quality Ratings and Demographic Factors*

$H_0$: There is no significant difference between the health care service quality ratings and demographic factors.

$H_1$: There is a significant difference between the health care service quality ratings and demographic factors.

Table 3 shows that, the patient's service quality rating varies with demographic factors such as age, gender, educational level, monthly income in ₹ and occupation.

## 7.4 *Assessment of Service Quality Gaps*

Service quality gap between patient's expectations and patient's perceptions across five dimensions the reliability, responsiveness assurance, empathy, and tangible qualities of private hospitals are evaluated and analyzed.

Table 4 shows that, the five SERVQUAL gap scores calculated using the patient's expectations and perceptions mean score. The $P$ value of $< 0.05$ indicates that all five variables of the SERVQUAL scale differ significantly between exception and perception.

**Lowest Mean Expectation Score**

The mean expectation score for the reliability factor of health care services is low (4.27). The reason for this is that patients may believe they lack sufficient expertise to appraise the technical aspects of health care, and they may conclude that the capability

**Table 2** Demographic attributes

| Demographic attributes | No. of respondents | % |
|---|---|---|
| *Gender* | | |
| Male | 172 | 54 |
| Female | 148 | 46 |
| Total | 320 | 100 |
| *Age in years* | | |
| 19–25 | 43 | 13 |
| 26–35 | 47 | 15 |
| 36–50 | 120 | 38 |
| 51–60 | 78 | 24 |
| Above 61 | 32 | 10 |
| Total | 320 | 100 |
| *Education* | | |
| Secondary level | 65 | 20 |
| Higher secondary | 101 | 32 |
| UG degree | 126 | 39 |
| PG degree | 28 | 9 |
| Total | 320 | 100 |
| *Income in ₹* | | |
| 0–10,000 | 112 | 35 |
| 10,001–20,000 | 88 | 28 |
| 20,001–30,000 | 62 | 19 |
| Above 30,000 | 58 | 18 |
| Total | 320 | 100 |
| *Occupation* | | |
| Employed | 94 | 29 |
| Professional | 53 | 17 |
| Business | 45 | 14 |
| Student | 21 | 7 |
| Home maker | 49 | 15 |
| Retired | 52 | 16 |
| Others | 6 | 2 |
| Total | 320 | 100 |

to perform the service as per guarantee and in the correct way may sometimes be beyond the control of the service provider.

**Table 3** Hypothesis to test independence of service quality ratings and demographic factors

| Demographic variable | F statistic | *P* value | Significance at 5% | Acceptance of hypothesis |
|---|---|---|---|---|
| Age | 0.906 | 0.444 | Not significant | $H_1$ is accepted |
| Gender | 0.946 | 0.331 | Not significant | $H_1$ is accepted |
| Educational level | 2.764 | 0.515 | Not significant | $H_1$ is accepted |
| *Monthly Income in ₹* | 4.761 | 0.423 | Not significant | $H_1$ is accepted |
| Occupation | 9.509 | 0.365 | Not significant | $H_1$ is accepted |

**Table 4** Mean of patient's expectations and patient's perceptions

| Service quality dimensions | *EM | *PM | *Gap | Rank | **p* value |
|---|---|---|---|---|---|
| Reliability | 4.27 | 3.13 | 1.14 | 2 | < 0.05 |
| Responsiveness | 4.87 | 3.27 | 1.6 | 1 | < 0.05 |
| Assurance | 4.38 | 3.53 | 0.85 | 3 | < 0.05 |
| Empathy | 4.43 | 3.68 | 0.75 | 4 | < 0.05 |
| Tangibility | 4.72 | 4.87 | -0.15 | 5 | < 0.05 |

* *EM* expected mean; *PM* perceived mean; Gap-EM-PM; *p* value—*p* value for expectation and perception scores

### Highest Mean Expectation Scores

The responsiveness dimension of healthcare services obtained the highest expected score. This could be related to the fact that patients want the personnel to reply to all of their questions and concerns instantly. The second highest expectation score was for the tangibles dimension. The availability of current life-saving equipment may have a evident effect on patients' opinion towards the health care services quality.

### Lowest Mean Perception Score

The responsiveness dimension of health care services received the lowest perception scores. Because patients have the highest expectations for responsiveness, this dimension was perceived as low in relation to hospital services. Patients rated staff readiness and eagerness to assist patients by delivering prompt services as less favorable. Staff should be trained to meet the patient's needs cheerfully and swiftly as an urgent priority.

### Highest Mean Perception Scores

The tangibility dimension of healthcare received the highest perception score. Private hospitals provide physical amenities, equipment, staff appearance, visually appealing infrastructure, space in the hospital, cleanliness of the room/ward, water, electricity, parking facilities, and security services that meet the expectations of its patients.

**Reliability**

The average mean gap score for the attribute reliability is determined as 1.14. This demonstrates that private hospitals are unable to deliver the promised services consistently and accurately.

The reliability attribute of healthcare services is determined by the factors such as accuracy in diagnosis and treatment, patient safety, system reliability, furnishing certain information, providing services as promised, speedy & on time and error-free services, performing right on the bills and actively enforcing feedback mechanisms. It is clear that private hospitals have failed to meet this dimension.

**Responsiveness**

The average mean gap score for responsiveness is determined as 1.6, the highest among the health care service quality factors. Private hospitals are increasingly focusing on service quality features such as personnel willingness or readiness to deliver services, resulting in shorter procedure wait times.

**Assurance**

The average mean gap score for assurance is determined as 0.85. It is deduced that the staff's competency and efficiency, kindness and politeness, honesty, and seeking guidance from professional doctors are the visible service traits contributing to the average "Assurance" gap. It is observed that private hospitals have a moderate mean gap score for the dimension assurance. Hence, it is advisable that private hospitals should do continual efforts to increase the compassionate and politeness of their employees by training, soft skill development in interpersonal relationships and instilling trustworthiness to deliver better service quality in this dimension.

**Empathy**

The average mean gap score for empathy is determined as 0.75. From the analysis, it is observed that the service attributes that contributed to this gap score are, the hospital authority's maximum effort to cure patients, the staff's sincere concern to provide individualized attention to patients in need of care, suitable working hours, and the service facility's convenient location. Hospitals should make significant attempt to improve staff's genuine concern for patients, eagerness to help, and awareness of the patient's needs from their perspective so that they can meet the patient's expectations.

**Tangibility**

The average mean gap score for tangibility is determined as 0.15. Tangibility focuses on infrastructure facilities. It is observed that, the service quality attributes like visually attractive layout of the hospital, availability of essential equipment, easy access location of the hospital, the availability of attractive printed materials, staff's professional dressing, hygiene and healthy environment, parking facility for family members or visitors, personal cleanliness of patients, adequate and enhanced seating accommodations for the visitors, hygiene of restroom facilities, sanitation of visitor accommodations, laboratories and pharmacy facilities are enhances the tangibility component of the service quality.

## *7.5 Influence of Service Quality Variables on Expected Service Quality*

From Table 5 it is inferred that; the standardized estimate is taken for the regression weights to find out the influence of the independent variable over the dependents variable. In the present case, the impact of the independent variable like reliability, responsiveness, assurance, empathy, tangibility over the dependent variable expected service quality. If the independent variable reliability raises to one degree in the positive direction, then it can be seen that the dependent variable expected service quality rises to 0.064 in positive direction.

Similarly, if the independent variable responsiveness raises to one degree in the positive direction, then it can be seen that the dependent variable expected service quality rises to 0.18 in positive direction. If the independent variable assurance raises to one degree in the positive direction, then it can be seen that the dependent variable expected service quality rises to 0.15 in positive direction. Similarly, if the independent variable empathy raises to one degree in the positive direction, then it can be seen that the dependent variable expected service quality rises to 0.1 in positive direction. If the independent variable tangibility raises to one degree in the positive

**Table 5** Regression weight—influence of service quality variables on expected service quality

| Variable | Inf | Variable | UE | SE | S.E | C.R | P |
|---|---|---|---|---|---|---|---|
| Expected service quality | ← | Reliability | 0.064 | 0.121 | 0.019 | 3.43 | *** |
| Expected service quality | ← | Responsiveness | 0.18 | 0.152 | 0.042 | 4.317 | *** |
| Expected service quality | ← | Assurance | 0.15 | 0.196 | 0.027 | 5.595 | *** |
| Expected service quality | ← | Empathy | 0.1 | 0.103 | 0.034 | 2.949 | 0.003** |
| Expected service quality | ← | Tangibility | 0.16 | 0.132 | 0.042 | 4.215 | *** |

***Significance at 0.001% level; **Significance at 0.05% level

*UE* unstandardized estimate; *SE* standardised estimate; *S.E.* denotes the standard error; *C.R* denotes the critical ration; *P* denotes the probability value; *Inf.* represents influence

**Table 6** Squared multiple correlations

| Variable | Estimate |
|---|---|
| Perceived service quality | 0.049 |

direction, then it can be seen that the dependents variable expected service quality rises to 0.16 in positive direction.

### *7.6 Influence of Service Quality Variables on Perceived Service Quality*

Table 6 shows that, the extent to which, the dependent variable perceived service quality changes itself, when influenced by the independent variables. In this case, the dependent variable perceived service quality has 4.9% of its variance, when influenced by the independent variables like reliability, responsiveness, assurance, empathy, tangibility.

## 8 Limitations

Although this study was well conducted, several limitations should be noted. To begin, the poll was exclusively open to outpatients from private hospitals. Patients from various clinical departments are not given a particular priority. Furthermore, this study was conducted during the transition phase from pandemic to normal life, which may have influenced patient's perceptions and patient's expectations regarding health care service quality.

## 9 Findings, Suggestion and Conclusion

Patient care quality ratings vary according to demographic characteristics such as age, gender, educational level, monthly income in ₹ and occupation.

Responsiveness, reliability assurance, empathy, and tangibility were ranked first, second, third, fourth, and fifth in terms of service quality gap score. These gaps were statistically significant (values $\leq 0.05$), so hospital administrators should take these as a focal point.

In today's global market, giving high-quality health care services to patients can provide a competitive advantage, differentiation, and service excellence. Patients are key stakeholders in every firm, and their satisfaction is paramount. To satisfy and gain patient patronage, private hospitals should prioritize enhancing the responsiveness and reliability dimensions of service quality. Health care leaders, administrators, practitioners and policy makers can bring about significant change in hospitals by focusing on the identified areas for improvement.

Finally, there is a need for additional research to build enhanced conceptual reinforcement frameworks and formalized quantitative models. The conclusions derived from this research will help the parent healthcare practitioners, quality professionals and other gross and net stakeholders in the health care sector to understand the relative significance of the service quality dimensions and take up a step forward to adorning their respective organizations with hospital service quality dimensions where necessary.

## References

1. Aagja, J.P., Garg, R.: Measuring perceived service quality for public hospitals (PubHosQual) in the Indian context. Int. J. Pharm. Healthc. Mark. **4**(1), 60–83 (2010)
2. Donabedian, A.: Quality assessment and assurance: unity of purpose, diversity of mean. Inquiry **25**(1), 173–192 (1988)
3. George, S.: Economic dimension of primary health centres in Kerala: understanding the Kerala model. In: Aloysius Edward, J., Jaheer Mukthar, K.P., Asis, E.R., Sivasubramanian, K. (eds.) Current trends in economics, business and sustainability. ICEBS 2023. Contributions to environmental sciences & innovative business technology. Springer, Singapore (2023). https://doi.org/10.1007/978-981-99-3366-2_14
4. Ladhari, R.: A review of twenty years of SERVQUAL research. Int. J. Qual. Serv. Sci. **1**(2), 172–198 (2009)
5. Minjoon, J., Peterson, R.T., Zsidisin, G.A.: The identification and measurement of quality dimensions in health care: focus group interview results. Health Care Manag. Rev. **23**(4), 81–96 (1998)
6. Mostafa, M.M.: An empirical study of patients' expectations and satisfactions in Egyptian hospitals. Int. J. Health Care Qual. Assur. **18**(7), 516–532 (2005)
7. Neupane, B.P., Gurung, D.S.: An assessment of service quality in Western Regional Hospital. J. Bus. Financ. **02**(02), 65–72 (2014)
8. Parasuraman, A., Zeithaml, V.A., Berry, L.L.: A conceptual model of service quality and its implications for future research. J. Mark. **49**(3), 44 (1985)
9. Raposo, M., Alves, H., Duarte, P.: Dimensions of service quality and satisfaction in healthcare: a patient's satisfaction index. Serv. Bus. **3**(1), 85–100 (2009)
10. Talib, F., Azam, M., Rahman, Z.: Service quality in healthcare establishments: a literature review. Int. J. Behav. Healthc. Res. (IJBHR) **5**(1/2), 1–24 (2015)
11. Zeithaml, V.A., Berry, L.L., Parasuraman. A.: The nature and determinants of customer expectations of service. J. Acad. Mark. Sci. **12**(1), 1–12 (1993)

# Technological Innovation and Business, Environment, Economics Sustainability

# Consumer Perception and Behaviour Towards Organic Personal Care Products

**L. Jayanthi, J. Rajalakshmi, and K. P. Jaheer Mukthar**

**Abstract** Consumer interest in organic products has surged in recent years, driven by concerns over health, sustainability, and environmental impact. This paper examines the dynamic relationship between consumer perception and behaviour towards organic products. Drawing upon a review of existing literature, the study explores various factors influencing consumer attitudes and purchasing decisions in the organic market. These factors encompass product attributes, such as quality, price, and availability, as well as psychological factors, including trust, perceived benefits, and perceived risks associated with organic consumption. Furthermore, the paper explores the role of marketing strategies, certification, and labelling in shaping consumer perceptions of organic products. Additionally, it investigates demographic variables, such as age, income, and education, which influence consumer attitudes towards organics. It is essential to understand these factors for businesses and policymakers who seeks to effectively target and engage with consumers in the organic market. This research findings contribute valuable insights into the complexities of consumer behavior in the context of organic products, facilitating informed decision-making for stakeholders across the supply chain.

**Keywords** Consumer attitude · Product attributes · Marketing strategies · Organic products

L. Jayanthi · J. Rajalakshmi
Faculty of Management, SRM Institute of Science and Technology, Kattankulathur, India
e-mail: jayanthl@srmist.edu.in

J. Rajalakshmi
e-mail: rajalakj@srmist.edu.in

K. P. Jaheer Mukthar (✉)
Kristu Jayanti College, Bengaluru, India
e-mail: jaheermukthar@gmail.com

 
M. Al Mubarak (ed.), *Sustainable Digital Technology and Ethics in an Ever-Changing Environment*, Studies in Systems, Decision and Control 237,
https://doi.org/10.1007/978-3-031-86708-8_43

# 1 Introduction

This proves that customers' awareness of health consciousness indicates that customers would like to select the appropriate good quality product. The people will spend their money in purchasing quality products which are safe to use and the one that will lead to the intended outcome. However this awareness has led to awareness on the environment and hence, use of more environmental-friendly products [1]. Research studies show that the relationship between the consumer behaviour and the Organic and Personal Care Products is very weak. The prevailing situation of constructing the harmonized and consistent environment, the concern for environmental awareness was developed among the academicians, corporate, media, government and non-government organizations. The etiquette of these issues have raised awareness among the consumer. This means that they become more conscious of the way they consume and this leads to the movement promoting and practicing the green behaviour. Thus, a person who practices environment behaviour (a person who takes action to protect the environment) will encourage healthier communities and therefore, it is important to understand the concern with regard to environment among the consumers. This grasp a large impact on public health. People are becoming more conscious about the product they use daily in their life for good health they transform to various types of organic body products [2].

However, the attitude and buying trends of the consumers in the organic personal care products differ one way or the other. Quality sensitivity, price sensitivity, perceived brand image and perceived trust in the certification of the said products are some of the key factors that influence the purchase behaviour of the customers. Therefore, it is essential to understand these factors by marketers, manufactures and the decision makers to address the needs of this potential consumer segment [3].

In light of this, this research aims at evaluating the consumers' attitude and behaviour towards the OPCCs, the factors influencing the purchase intentions of the OPCCs, and factors that lead to the PDCA of the OPCCs. Thus, the understanding of the nature of consumption beliefs and behaviour, motives for the purchase of organic personal care products, and their purchasing decision will be helpful for deepening the existing knowledge of the market of organic personal care products and the improvement of the current state of marketing practices and products.

# 2 Review of Literature

Gurusanthosini and Gomathi [4] analyzes the Consumer Preference concerning to Patanjali Products. In this research study, they noted that there are a number of limited sources that affect the consumer buying behaviour. They also analyzed that there was a correlation between some demographic characteristics and tendency of the consumers of the products of Patanjali brand. They were able to discover that the factors such as hygiene and standard quality are relatively cheap in the direction of consumer preference.

Customer satisfaction is very crucial in the marketing concept [5]. One cannot run a business for a long time without customers being happy with the products that they are using. Besides, customers mind is a transitioning one. Concerning the different brands in this study, customers' overall satisfaction level to the brand Indulekha is relatively high much as most of them express contentment with the product. The entire customers are only concerned with the quality of the product rather than even the price.

The paper by Khanna [6] demonstrates there are many essential factors that help to create the buying decision of the product. Customer perception toward a brand has been mostly developed based on the satisfactory value of the user. The value proposition is also established, which is the benefit that a customer is likely to experience after cherishing a product of his or her money. It could be in ability of the product to solve the problem be it fully or partially. Retention of customer follows satisfaction which implies that a business organisation should aim to satisfy a customer to increase possibility of repeat patronage. To reduce customer churn rate to the company's advantage, the product brand must meet the promises the company makes about the brand before another brand emerges.

The research study by Lavanya and Velumani [7] was carried out to assess customer satisfaction towards Himalaya herbal and healthcare products and to explore the impact of the product dimensions on customer satisfaction and customer loyalty and to know about the special effect of the herbal product. The study finds out that in those areas where many of the traditional communities had scarce access to modern health facilities, or where the facilities that were available were substandard, the great preponderance of the populace embraced the culture.

## 3 Objective of the Study

To find out the influence of demographic factors of the respondents.

To identify the consumer awareness towards organic personal care product.

To recognize the factors affecting the consumer perception towards organic personal care product.

To determine out the most preferred organic personal care product by the consumers.

To find the relationship between factors influencing organic personal care products.

## 4 Research Methodology

The study titled consumer perception and behaviour towards organic personal care product in Chennai is an attempt to understand why people prefer organic products and what factors influence them to buy the organic personal care products. The awareness and the satisfaction level of the customers are measured using the subjective sampling method. The primary data is collected through a questionnaire from the 100 respondents using Organic personal care products.

## 5 Data Analysis and Interpretation

Table 1 shows that, Majority of 74% of the respondents are female are comparing to male is 26%. It is found that, the maximum 55% of the respondents belongs to 20–30 years age group the average 21% of the respondents are from the above 30–40 years age group. It is found that, the maximum 43% of the respondents are qualified UG/ diploma the average respondents 15% are qualified up to + 2. Among the respondents, it is recognized that maximum of 34% of the respondents are earning up to 20,000, the lower earning 17% of the respondents are earning above 30,000–40,000. Among the respondents, it is recognized that maximum of 53% of the respondents are occupation is others, the average 23% of the respondents are private employees.

Table 2 shows that, a Cronbach's Alpha of 0.915 for 12 items indicates a high level of internal consistency or reliability for the scale. Typically, Cronbach's Alpha values range between 0 and 1, with values above 0.7 considered acceptable, and values above 0.9 indicating excellent reliability.

In this case, the 12-item scale shows that the items measure the same underlying construct very consistently, which suggests that the scale is reliable for assessing consumer perception and behavior.

Table 3 shows the KMO value of 0.873 suggests that the sampling adequacy is excellent and Bartlett's test showed a significance of 0.000. Therefore, factor analysis could be applied to 12 variables such as 100% natural, No side effect, Excellent result, Chemical free, Reasonable price, Preferred by family members, Accessibility of product, Brand influence, Product packing, Health conscious, Product ingredient and Good quality in measuring consumer perception and behaviour towards organic personal care products.

Table 4 shows the Communalities from a Principal Component Analysis (PCA), which indicates how much of the variance in each variable is explained by the extracted components. Most variables are well represented by the extracted components, with communalities mostly ranging from 0.58 to 0.86. A few variables, such as "Good quality", have slightly lower communalities (0.583), indicating that less of their variance is explained by the factors, but they are still adequately represented. The high communalities for variables like "No side effect" and "100% natural" suggest that these factors are strongly explained by the components extracted through PCA.

**Table 1** Demographic profile of respondents

| Demographic factors | Responses | |
|---|---|---|
| | No of respondents | % |
| *Gender* | | |
| Male | 26 | 26 |
| Female | 74 | 74 |
| Total | 100 | 100 |
| *Age group* | | |
| Below 20 | 18 | 18 |
| 21–30 | 55 | 55 |
| 31–40 | 21 | 21 |
| 41–50 | 6 | 6 |
| Total | 100 | 100 |
| *Educational level* | | |
| Higher secondary | 15 | 15 |
| UG degree | 43 | 43 |
| PG degree | 55 | 55 |
| Professionals | 01 | 01 |
| Others | 06 | 06 |
| Total | 100 | 100 |
| *Monthly income in Rs.* | | |
| Up to 20,000 | 34 | 34 |
| 20,001–30,000 | 27 | 27 |
| 30,001–40,000 | 17 | 17 |
| Above 40,000 | 22 | 22 |
| Total | 100 | 100 |
| *Occupation* | | |
| Govt. employee | 12 | 12 |
| Private employee | 23 | 23 |
| Self employee | 12 | 12 |
| Others | 53 | 53 |
| Total | 100 | 100 |

**Table 2** Reliability statistics

| Cronbach's Alpha | No of items |
|---|---|
| 0.915 | 12 |

**Table 3** KMO and Bartlett's test

| KMO and Bartlett's test | | |
|---|---|---|
| Kaiser–Meyer–Olkin measure of sampling adequacy | | 0.873 |
| Bartlett's test of sphericity | Approx. chi-square | 764.922 |
| | Degrees of freedom | 66 |
| | Significance | 0.000 |

**Table 4** Communalities

| Communalities | | |
|---|---|---|
| Variables | Initial | Extraction |
| Health conscious | 1.000 | 0.775 |
| Product ingredient | 1.000 | 0.726 |
| Good quality | 1.000 | 0.583 |
| Reasonable price | 1.000 | 0.760 |
| Preferred by family members | 1.000 | 0.710 |
| Accessibility of product | 1.000 | 0.680 |
| Chemical free | 1.000 | 0.741 |
| No side effect | 1.000 | 0.861 |
| Excellent result | 1.000 | 0.749 |
| 100% natural | 1.000 | 0.849 |
| Brand influence | 1.000 | 0.684 |
| Product packing | 1.000 | 0.702 |

Extraction method: Principal Component Analysis

Table 5 shows the Eigenvalues and associated explained variance from a Principal Component Analysis (PCA), which helps to determine the number of significant components and how much variance they explain. The table lists 12 components, but only the first three have Eigenvalues greater than 1, suggesting that they are significant and explain most of the variance in the data. The total cumulative explained variance after rotation is 73.5%, meaning that the first three components still explain the same amount of total variance but more evenly distributed across components.

Table 6 shows the loadings of different variables (or items) on three components. Factor loadings close to $\pm$ 1 indicate a strong association between the item and the factor. The analysis has grouped the items into three distinct components: Safety Value, Motivating Value, and Hedonic Value. These components highlight the key dimensions along which the items are organized, and each item has been assigned to the component that it most strongly correlates with.

Table 7 shows that the correlation coefficient between Hedonic value and Safety value is 0.527 which indicate 52.7% positive relationship between them and is significant at 1% level. Correlation Coefficient between Motivational value and Safety value is 0.701 which indicate 70% relationship between them and is significant at

**Table 5** Total variance explained

| | Initial eigenvalues | | | Extraction sums of squared loadings | | | Rotation sums of squared loadings | | |
|---|---|---|---|---|---|---|---|---|---|
| | Total | % of variance | Cumulative % | Total | % of variance | Cumulative % | Total | % of variance | Cumulative % |
| 1 | 6.306 | 52.552 | 52.552 | 6.306 | 52.552 | 52.552 | 3.692 | 30.766 | 30.766 |
| 2 | 1.440 | 11.998 | 64.550 | 1.440 | 11.998 | 64.550 | 3.082 | 25.680 | 56.446 |
| 3 | 1.074 | 8.950 | 73.500 | 1.074 | 8.950 | 73.500 | 2.046 | 17.054 | 73.500 |
| 4 | 0.648 | 5.396 | 78.896 | | | | | | |
| 5 | 0.570 | 4.752 | 83.648 | | | | | | |
| 6 | 0.487 | 4.056 | 87.704 | | | | | | |
| 7 | 0.375 | 3.128 | 90.832 | | | | | | |
| 8 | 0.300 | 2.502 | 93.334 | | | | | | |
| 9 | 0.285 | 2.372 | 95.705 | | | | | | |
| 10 | 0.211 | 1.759 | 97.465 | | | | | | |
| 11 | 0.168 | 1.403 | 98.868 | | | | | | |
| 12 | 0.136 | 1.132 | 100.000 | | | | | | |

Extraction method: Principal Component Analysis

**Table 6** Rotated component matrix

| Rotated component matrix | | | |
|---|---|---|---|
| Particular | Component | | |
| | 1 | 2 | 3 |
| *Safety value* | | | |
| Natural | 0.861 | | |
| No side effect | 0.854 | | |
| Excellent result | 0.820 | | |
| Chemical free | 0.735 | | |
| *Motivating value* | | | |
| Reasonable price | | 0.831 | |
| Preferred by family members | | 0.785 | |
| Accessibility of product | | 0.699 | |
| Brand influence | | 0.649 | |
| Product packing | | 0.599 | |
| *Hedonic value* | | | |
| Health conscious | | | 0.874 |
| Product ingredient | | | 0.808 |
| Good quality | | | 0.538 |

Extraction method: Principal Component Analysis. Rotation method: Varimax with Kaiser Normalization
Rotation converged in 6 iterations

**Table 7** Karl Pearson correlation

| Particular | Hedonic value | Motivational value | Safety value |
|---|---|---|---|
| Hedonic value | 1 | 0.456** | **0.527**** |
| Motivational value | – | 1 | **0.701**** |
| Safety value | – | – | 1 |

** Correlation is significant at the 0.01 level (two-tailed)

1% level. The hedonic value factors considered health conscious, product ingredient and good quality which indicates positive relationship among the safety value. The motivational value factors considered reasonable price, preferred by family members, accessibility of product, brand influence and Product packing that indicates positive relationship among the safety value.

## 6 Findings and Conclusion

Consumer behaviour is central to understanding the organic personal care products segment. The organic personal care products marketers have many challenges and opportunities which call for flexibility and creativity come up with effective strategies. The findings suggest that the organic personal care product market is driven by a complex interplay of health consciousness, environmental awareness, price sensitivity, and the effectiveness of marketing strategies. As awareness continues to grow, the demand for these products is expected to increase, provided that the industry addresses the barriers to adoption. The results and analysis point to the fact that the consumers' perceptions of quality, value and brand trust are important factors in determining their attitudes and intentions towards organic personal care products. The high importance of organic personal care product was overlooked for some time as well, and as a result, the existing impact of information technology was identified as a crucial factor that was relatively ignored for a fairly long time period by scholars. Thus, due to the environmental sustainability, importance has been accorded they have shifted to buying organic personal care products. The study also reveals the fact that the people were aware of image and availability though not entirely loyal with the organic personal care products. It will be true in the sense that among the respondents some of them have indeed developed an inclination towards the idea of organic personal care products, However the marketers need to develop promotion that are realistic and uplifting of morale. The value chain has therefore been effective in selling personal care products that are organic in nature. From the various findings, it could be concluded that knowledge and awareness of the organic products change the attitude and perception of the consumers towards the product and thus their buying decision. Companies that tend to consider these factors as it relates to sensitivity of prices and social influence are likely to conquer this burgeoning market.

## References


1. Kim, M.J., Lee, C.K., Yoon, H.H.: How do consumers' perceptions of organic beauty products influence purchasing behavior? J. Consum. Behav. **15**(5), 437–450 (2016). https://doi.org/10.1002/cb.1603
2. Mastrogiacomo, T.: The purchase intention for organic beauty products: an application of the theory of planned behavior. J. Clean. Prod. **220**, 95–102 (2019). https://doi.org/10.1016/j.jclepro.2019.02.031
3. Kim, H., Lee, J., Lee, D.: Understanding consumers' repurchase intentions toward certified organic personal care products: an integration of TPB and perceived risk theory. Sustainability **11**(18), 5053 (2019). https://doi.org/10.3390/su11185053
4. Gurusanthosini, G., Gomathi, G.: A study on consumer preference towards Patanjali products. Int. J. Curr. Res. Mod. Educ. **2**(1), 163–166 (2017)
5. Kariappa, A.: A study on customer preference and brand awareness of Max Fashion. Int. J. Manag. IT Eng. **6**(1), 464–474 (2016). ISSN: 2249-0558. Available at SSRN: https://ssrn.com/abstract=2973181
6. Khanna, R.: E marketing customer perception towards brand: a study on 'Patanjali.' Glob. J. Manag. Bus. Res. **15**, 40–45 (2015)
7. Lavanya, R., Velumani, M.: A study on customer satisfaction towards herbal products with special reference to Himalaya. Indian J. Appl. Res. **4**(12) (2014)

# From Waste to Wellness: Circular Economy Strategies for the Indian Healthcare Industry

**Jean Maria George, U. S. Syama, Geril Scaria George, M. M. Nirmala, and K. Vinoda Devi**

**Abstract** The Indian healthcare industry is expanding quickly, creating a large number of jobs and bringing in a substantial amount of money. But this expansion has also resulted in a lot of medical waste being produced, which presents serious environmental problems. In order to solve these problems, the article highlights the necessity of circular economy techniques, with a particular emphasis on reduce, reuse, and recycle methods. Wastes from infectious, pathological, sharp, chemical, pharmacological, and radioactive sources are all produced by the healthcare industry. Depletion of resources, air pollution, and contamination of water and soil are examples of environmental consequences. The industry's current waste management situation is typified by a monotonous enforcement of regulations. Reducing waste production, effectively managing resources, raising awareness, and implementing sustainable technologies are some examples of prudent practices that could benefit the industry. The absence of infrastructure, financial resources, and awareness are obstacles in garbage management. There is also discussion of government initiatives and the possibility of the healthcare industry implementing circular economy concepts for sustainable development.

**Keywords** Circular economy · Healthcare · Medical waste management

J. M. George
Assistant Professor, Department of Economics, Christ College (Autonomous), Irinjalakuda, University of Calicut, Irinjalakuda, Kerala, India

U. S. Syama (✉) · M. M. Nirmala · K. Vinoda Devi
Kristu Jayanti College Autonomous, Bangalore, India
e-mail: syama.us@kristujayanti.com

G. S. George
Marian College Kuttikkanam Autonomous, Kuttikkanam, Kerala, India



M. Al Mubarak (ed.), *Sustainable Digital Technology and Ethics in an Ever-Changing Environment*, Studies in Systems, Decision and Control 237,
https://doi.org/10.1007/978-3-031-86708-8_44

# 1 Introduction

## *1.1 Overview of Indian Healthcare Sector*

India's healthcare sector is rapidly growing, becoming a significant revenue provider and employment generator [1]. With 7.5 million employees in 2024, the sector is expected to create 3 million new tech jobs. The Indian medical tourism sector, valued at $7.69 billion in 2024, attracts high foreign tourist spending, with 6.87% of international tourists visiting for medical treatment. However, the industry faces challenges in managing aftereffects and residuals, as the industry's advancements and achievements increase waste and impacts at a faster pace.

A sustainable healthcare sector and public health are crucial, but environmental challenges persist. India produced 774 tons of biomedical waste daily in 2020, and proper disposal methods are not widely available. Public awareness and professional training are lacking, and funds for waste management systems are insufficient. Addressing these issues is essential for a sustainable healthcare sector.

The Indian healthcare sector needs to adopt environmentally sustainable, affordable, and effective waste management strategies, such as the 'Reduce-Reuse-Recycle' (RRR) approach, to mitigate environmental damage caused by improper medical waste disposal. This circular economy model, which is not fully utilized in the sector, is a proven remedy for sustainable development.

# 2 Review of Literature

Matschewsky et al. [2] examined how the European Union is concentrating on public procurement as a means of transitioning from a linear to a circular economy. Nonetheless, there hasn't been much research done on circular public procurement; most of it has concentrated on IT and furnishings. In order to implement circular economy concepts, product design and procurement criteria need to be changed. Plastic medical goods have not gotten enough attention in the literature. The healthcare industry's use of circular indicators is assessed in the paper "Circular indicators for the design and procurement of plastic products in the healthcare sector—a review."

D'Alessandro et al. [3] in his paper "Exploring Circular Economy Practices in the Healthcare sector: A Systematic review and bibliometric analysis" presented a thorough analysis of effective circular processes in the healthcare industry, exposing current gaps and offering insightful information to stakeholders—despite the limits of the research criteria. Since the healthcare industry generates two gigatons of $CO_2$ a year, implementing circular economy (CE) principles should be a top priority. This paper offers a thorough analysis of the literature on CE procedures in the healthcare industry, highlighting important practices and assessing the current state of the field. 36 papers from reliable databases, including Web of Science, Scopus, and

VOSviewer, are used in the study's literature analysis. By highlighting important CE practices and their application domains, the findings support the larger goal of accomplishing SDGs 12 and 9.

Because of poor infrastructure, the healthcare sector has difficulties with energy use, waste management, and environmental protection. Circularity and sustainability are essential for mitigating these adverse effects. The purpose of this study is to use big data analytics to find barriers to the circular economy (CE) in the healthcare industry and suggest ways around them. The study "Big Data-Enabled Solutions Framework to overcoming the barriers to circular economy initiatives in the healthcare sector" [4] offers suggestions for theoretical, policy, and managerial implementations to support sustainable development activities in the healthcare sector. It also suggests a framework using fuzzy VIKOR and fuzzy best-worst Method (BWM).

The study by Fanta et al. [5] explores the potential benefits of Industry 4.0 digital technologies in the healthcare industry. The research focuses on the interactions between additive manufacturing, the Internet of Things, and big data analytics, which enable remanufacturing, reuse, and recycling of healthcare products. The study highlights the importance of digital technologies in promoting a circular economy in the healthcare sector.

This study "Circular economy as a COVID-19 cure?" [6] explores the link between COVID-19-related issues in public health and healthcare sectors and Circular Economy principles. It aims to drive sustainability and resilience in the sector, addressing challenges in implementing circular strategies. The framework outlines three resource loop strategies: closure, narrowing, and slowing, and a fourth strategy type: maintaining life.

Eswara Reddy and Tom [7] examines the economic aspects of solid waste management (SWM), emphasizing on the cost-benefit analysis and its funding mechanisms. The study points towards the need and importance of incorporating sustainable practices for ensuring economic efficiency by optimizing waste collection, encouraging recycling and reusing practices and thereby reducing the overall waste generation. The study also discusses the different challenges of developing countries in executing effective and sustainable waste management practices and also brings suggestions and recommendations to improve the economic feasibility.

Regarding the application of circular economy techniques in the Indian healthcare sector, researchers rarely agree. Concerns include those related to policy and regulation, economic viability, environmental impact, social and cultural acceptability, and logistical and infrastructural difficulties. While some support market-driven strategies and volunteer initiatives, others call for strict controls. In spite of these differences, an increasing amount of research highlights the potential advantages of circular economy approaches and emphasizes the need for more investigation and teamwork to overcome these obstacles.

The paper will examine the possibilities of circular economy model in the Indian healthcare system and how it will transform the sector in an environmentally sustainable manner, by reducing the wastes, disposing in a more scientific and sustainable way and also conserving the valuable resources. The paper goes through the urgent need for sustainable solutions in the Indian healthcare sector.

# 3 Research Objectives

- To understand the essential elements of the circular economy model which can be used to solve waste management problems in the healthcare sector of India.
- To understand what specific kinds of medical waste are produced in the Indian healthcare industry, and what are the potential environmental consequences of improper management of these wastes.
- To analyse which essential elements of the circular economy model can be used to solve waste management problems in the healthcare sector of India.

# 4 Results and Discussions

## *4.1 Circular Economy and Circular Healthcare Sector*

The circular economy model focuses on reusing and recycling old products to reduce waste generation and promote sustainable development. It encourages the reuse of waste from one sector as raw material in another, reducing resource demand and waste production. The life cycle of a product is extended through repair, reusing, and recycling [8]. This approach can reduce negative impacts on the environment and ecosystem. Circular healthcare is an example of a circular economy strategy that reduces harmful waste production in the healthcare industry. The reuse-repair-recycle approach encourages a shift towards sustainable supply chains, improving global health and sustainability while taking social and environmental responsibility. Developed products should integrate with the environment and can be reused or recycled, enhancing patient and worker experiences and reducing care costs. The Indian healthcare sector faces challenges in waste management and prevention, necessitating the implementation of a circular economy. This approach involves ecologically friendly practices in resource utilization, production, and waste generation, promoting sustainable practices and adhering to circular economy principles.

## *4.2 Types of Medical Wastes Generated in the Health Sector*

Bio-medical wastes, classified by WHO as hazardous or non-hazardous, are generated by healthcare activities. About 85% are non-hazardous, while 15% are hazardous, potentially infectious, chemicals, or radioactive [9]. The residuals and the by-products include a diverse range of hazardous waste materials which includes the following:

1. Infectious wastes: the waste contaminated with the patient's blood or body fluids or pathogens are considered to be infectious wastes. This includes wastes like those from discarded diagnostic samples, cultures and stocks of infectious agents

like wastes from autopsies and other infected agents from laboratory, and also the wastes from patients with infections and the things they used like swabs, bandages, etc.

2. Pathological waste: this is also a very serious type of waste which are the human tissues, organs or fluids, and other body parts of the infected patients.
3. Sharp wastes: this includes the medical devices like syringes, needles, disposable scalpels and blades, etc.
4. Chemical wastes: many chemicals that are used in the laboratories like solvents and reagents, disinfectants and sterilant produce hazardous chemical wastes. This also includes the heavy metals used in medical devices like mercury in broken thermometer and also the batteries.
5. Pharmaceutical wastes: these are another broad category of waste generated from drugs and vaccines which are unused or expired or sometimes contaminated. The unscientific disposal of these causes serious damages to the environment.
6. Radioactive wastes: these are the wastes which contain radioactive materials and is mostly generated from radiotherapy or other radioactive diagnostic material.

Thus, some of the major sources of these healthcare wastes includes the primary health providers and hospitals, laboratories, research centres, blood banks, collection centres, mortuary and autopsy centres and also the nursing homes for old aged. A serious damage happened, often in low-income countries is that these healthcare wastes are not separated into hazardous or non-hazardous wastes and this makes the real quantity of hazardous wastes much higher than actually generated. These improper management and carelessness can cause far reaching consequence and health risks to all related ones who may be the patients, health workers and also the general public. These adverse health outcomes include injuries due to sharps, exposure to toxic chemicals and pharmaceutical products, chemical burns in different contexts, air pollution mainly due to incineration of medical wastes, thermal injuries and also radiation burns.

### *4.3 Environmental Impacts of Indian Healthcare Sector*

The healthcare sector, which stands as a substantial element for public health in the country is also a major contributor of different wastes into the economy and also a significant environmental degrader. The key environmental concerns include the generation of substantial amounts of hazardous and non-hazardous medical wastes. These medical wastes, if not properly disposed can cause soil and water contamination, air pollution and ultimately it harms the public health itself.

- The wastes with chemical disinfectants are hazardous to the environment. The treatment of these wastes releases chemical substances into the environment which are mostly toxic if not properly handled and disposed in an environmentally sound manner.

- Incineration practice is the most common waste treatment method adopted in many health centres and hospitals. But inadequate incineration or the incineration of unsuitable materials can cause the release of harmful pollutants and ash residue into the air.
- The water contamination is another serious concern. The wastewater from hospitals often contains many chemicals, pharmaceutical products, microorganisms, heavy metals, chlorine, etc. The conventional methods of water treatment which is maintained in almost all health centres and hospitals may not be effective enough in removing many of these harmful products found in water.
- The energy consumption is also highly increasing with the advancements in the health industry. High dependence on electricity and fossil fuels can lead to the emission of greenhouse gases and climate change. There is also high consumption of valuable resources for the production of medical equipment and supplies which can lead to resource depletion.

### *4.4 Current Waste Management Scenario in the Indian Healthcare Sector*

The Indian healthcare sector is a rapidly evolving sector and as a result of its faster growth, a substantial amount of wastes is generated and the management is a very challenging task. There exist many regulations for the management of these medical wastes generated, including the Biomedical Waste Management and Handling Rules, 1998, but its proper implementation and monitoring is inefficient and uneven all throughout the nation [10]. All the rules and obligations are just formalities in papers only and the effectiveness and enforcement is still a distant vision to be achieved. The major limitations include the lack of concern, awareness and motivation among the common people as well as the health workers and also the cost factor associated with the management of these biomedical wastes [11].

Waste management and thereby effective utilisation of the available minimum resources is the most important phase of a circular economy. Handling wastes in one part of circular economy concept; the other and more sensible part is reducing the amount of wastes generated. This will bring environmentally sustainable practices and effective circular economy. Once generated, these wastes should be treated or disposed in some way and so the prime motive should be to control the generation of these wastes. Only then a sustainable economy can be formulated. It is a fact that world's population is expanding, leading to resource shortages and waste expansions, especially in urban areas. This necessitates the implementation of circular economy and its conservation practices. Some of the smart practices of circular economy for healthcare sector involves the different ways and methods for ensuring environmental sustainability in the sector. It includes the following smart practices:

- Minimizing the waste generation and hazardous chemicals—The practical solution to minimise these biomedical wastes is to switch to waste reuse and recycling policy. The possible types of biomedical wastes should be produced as such. Also, utmost attention should be taken to substitute hazardous chemicals with less hazardous alternatives.
- Efficient resource management—The resources, natural and other resources are to be efficiently managed. Proper utilisation of resources is a major step in circular economy. Some of the techniques includes implementing water safety plans, using automating water taps and censor, thereby reducing energy usage in public institutions or hospitals and increasing energy efficiency and its conservation.
- Assessing environmental impacts—This is important for ensuring sustainable development by making proper assessment of environmental impacts of products and services used in health sector. Environmentally sustainable, reusable, and less energy consuming equipment or services are preferred as maximum as possible.
- Increasing awareness of environmental and health aspects—Educating the healthcare workers and professionals regarding the importance of proper waste management and the serious risks associated with neglecting this is very essential.
- Incentives for sustainable practices—Encouraging health providers and health care institutions in practicing sustainable mode of operation by providing incentives can bring positive outcomes. This encourages them to adopt waste management practices and reduce biomedical waste generations. Financial support, low-interest credit supports, tax incentives, additional recognition or certification to such institutions or hospitals are all paves the way for a sustainable healthcare sector.
- Simplifying energy efficient procedures—Installing machines and equipment like solar panels, water reuse plants, waste treatment plants, recycling units in hospitals and health care institutions can reduce the energy consumption to a large extent. It is an important step to simplify the procedures and installation of these energy efficient techniques.
- Strict regulations for waste management—The handling and disposing of medical wastes needs the implementation of stricter regulations and also sufficient infrastructure.
- Waste segregation—In waste management, segregation of wastes is also a very important step. Separation of different types of waste is necessary for easy recycling and reusing.
- Environment friendly technology alternatives—Waste management techniques like incineration and autoclaving are already used in the health sector, but mostly not in an environment friendly manner. It is very important to explore more sustainable and environment friendly alternative technologies in this regard. Only modern incinerators operating at 850–1100 °C and connected with special gas cleaning equipment can comply with international standards for dioxins and furans.
- More preference to less emissions—Preference should be given to those alternative techniques and technologies which minimizes the formation of hazardous

emissions. Techniques like autoclaving methods, microwaving, steam treatment integrated with internal mixing are all considered as energy efficient and sustainable measures.
- Public–Private Partnership model—A collaborative system of management can bring more options and opportunities in effective waste management. A cooperative movement and teamwork by the government, healthcare providers and private agencies can implement circular economy objectives and ensure sustainable development approach in the healthcare sector.

## *4.5 Waste Management in Health Sector: Challenges Faced*

Waste management in healthcare sector often faces many challenges and difficulties that leads to a failure in many contexts. Lack of proper awareness, inadequate training in proper waste management, absence of proper waste management and disposal systems and also insufficient financial resources are some of the most common challenges and the reasons for the failure of the waste management of health care sector.

- The healthcare sector, especially in rural and remote areas faces the difficulty of inadequate infrastructure facilities for proper isolation, storage and disposal of medical wastes and its byproducts.
- The lack of proper awareness about the gravity of the situation is another major challenge faced. The public and even the healthcare workers are unaware or unbothered about the risks of improper waste management and are not concerned about the need and importance of its proper segregation.
- Another serious issue faced is the financial constraints related to setting up and operating the waste management facilities in the sector. The healthcare providers and the institutions, especially small clinics and hospitals feels it as a burden to set up waste management plants and other facilities.
- There are a number of informal agents involved in the collection and disposal of wastes from healthcare institutions. These informal agents are usually handling these wastes in an improper manner, leading to environmental degradations.
- As a developing economy, the incorporation of advanced technologies in waste treatments is still limited in India.

Despite all these challenges and limitations, Indian economy is still taking up all the necessary actions and steps to ensure a sustainable mode of development in all its sectors. In the interim budget 2024–25, the government allocated Rs. 90,659 crores to the Ministry of Health and Family Welfare (MoHFW). The government is also planning to introduce credit support and incentive programme worth Rs. 50,000 crores to boost the healthcare infrastructure of the country [12]. These investments can contribute significantly to the holistic development of the healthcare sector of the country, focusing primarily on the formulation of sustainable practices in the sector.

## 5 Conclusion

India has immense potential to become a best performer in circular healthcare system. With increasing competition and investments in the rapidly growing health industry, the health providers are eagerly looking forward to explore and invest in the latest technologies and modern dynamics. Nowadays, people are more conscious towards their healthcare upkeep. With the increasing demand and performance of the sector, the residuals and waste generation will also go ahead in a faster pace. Effectively controlling this waste generation by switching over to recycling and reusing processes is an urgent need of the hour, otherwise its far-reaching consequences will be serious and destructing, both for human existence and environmental sustainability. More research and development activities in this regard is highly encouraged for opening new options and sustainable alternatives in the healthcare sector. Transition to renewable sources of energy and promoting effective and scientific waste management techniques and energy efficient technologies are the essential requirements for a sustainable healthcare system.

## References

1. IBEF: Indian Healthcare Industry Analysis (2024). [Online]. Available: https://www.ibef.org/
2. Matschewsky, J., Lingegård, S., Martin, M.A.: Circular economy indicators for the design and procurement of plastic products in the healthcare sector—a review. Procedia CIRP **122**, 1095–1100 (2024)
3. D'Alessandro, C., Szopik-Depczyńska, K., Tarczyńska-Łuniewska, M., Silvestri, C., Ioppolo, G.: Exploring circular economy practices in the healthcare sector: a systematic review and bibliometric analysis. Sustainability **16**(1), 401 (2024). [Online]. Available: https://doi.org/10.3390/su16010401
4. Kazançoğlu, Y., Sağnak, M., Lafcı, Ç., Luthra, S., Kumar, A., Taçoğlu, C.: Big data-enabled solutions framework to overcoming the barriers to circular economy initiatives in healthcare sector. Int. J. Environ. Res. Public Health **18**(14), 7513 (2021). [Online]. Available: https://doi.org/10.3390/ijerph18147513
5. Fanta, G.B., Pretorius, L., Nunes, B.: Enabling circular economy in healthcare using industry 4.0 digital technologies. In: 30th Annual Conference of the International Association for Management of Technology (IAMOT 2021) (2021). [Online]. Available: https://doi.org/10.52202/060557-0085
6. Wuyts, W., Marin, J., Brusselaers, J., Vrancken, K.: Circular economy as a COVID-19 cure? Resour. Conserv. Recycl. **162**, 105016 (2020). [Online]. Available: https://doi.org/10.1016/j.resconrec.2020.105016
7. Eswara Reddy, E., Tom, T.: A review on economic assessment on solid waste management. In: Aloysius Edward, J., Jaheer Mukthar, K.P., Asis, E.R., Sivasubramanian, K. (eds.) Current Trends in Economics, Business and Sustainability. ICEBS 2023. Contributions to Environmental Sciences & Innovative Business Technology. Springer, Singapore (2023). https://doi.org/10.1007/978-981-99-3366-2_4
8. Javaid, M., Haleem, A., Khan, I.H., Singh, R.P., Khan, A.A.: Industry 4.0 and circular economy for bolstering healthcare sector: a comprehensive view on challenges, implementation, and futuristic aspects. Biomed. Anal. **1**(2), 174–198 (2024)
9. World Health Organization (WHO): Health-Care Waste. World Health Organization (2018)

10. The Economic Times: India to generate 775.5 tonnes of medical waste daily by 2020: study, 23 Mar 2018. [Online]. Accessed 23 Aug 2024. Available: https://economictimes.indiatimes.com
11. Henam, S., Shrivastav, R.: Biomedical Waste Management in India: Still a Looming Concern. Down To Earth (2019). [Online]. Available: https://www.indiabudget.gov.in/
12. Ministry of Finance, Government of India: Interim Union Budget 2024–25. Government of India (2024). [Online]. Available: https://www.indiabudget.gov.in/

# The Impact of Economically Weaker Section Reservation in India: A Study on the Syro Malabar Community in Kerala, India

**Subin Thomas, Smita Satapathy, and Sheen Thankalayam**

**Abstract** India is renowned for its multifaceted unity amid diversity, which proves the harmonious fusion of different communities. This harmony is rooted in the Indian Constitution by ensuring social justice to all citizens regardless of their individual differences and unique characteristics. It is crafted with empathy and foresight to benefit the most disadvantaged sections of society. It includes provisions for reservation programs under Articles 15 and 16 to combat discrimination and promote inclusiveness in the present Indian societal scenario. The 103rd amendment to the Indian constitution expanded the concept of backwardness to include more marginalized and disadvantaged sections of society. It reflects the United Nations' sustainable development goals (SDG 10) in India to reduce inequalities. The amendment introduced a 10% reservation in government employment and education for the Economically Weaker Sections (EWS) ensuring equitable development of all the sections of society. The study examines the impacts of the economically weaker sections (EWS) reservation in India by analyzing data collected from the Syro Malabar Community in Kerala, India.

**Keywords** Economically weaker section (EWS) · Reservation · Minorities · SDG 10 · Equitable society

S. Thomas (✉) · S. Satapathy
Alliance School of Law, Alliance University, Bangalore, India
e-mail: subinpolackal@gmail.com

S. Satapathy
e-mail: smita.satapathy@alliance.edu.in

S. Thankalayam
CHRIST (Deemed to be University), Bangalore, India
e-mail: sheen.tr@res.christuniversity.in



M. Al Mubarak (ed.), *Sustainable Digital Technology and Ethics in an Ever-Changing Environment*, Studies in Systems, Decision and Control 237,
https://doi.org/10.1007/978-3-031-86708-8_45

# 1 Introduction

India is united amidst diverse communities, religions, dialects, and cultures. This unity in diversity is deeply rooted in the Indian Constitution which establishes social justice and ensures equality for all. It is envisioned to lead the nation toward a more egalitarian society. The reservation policies in India are intended for this purpose. The reservation programs foster social harmony and national unity. It helps the backward and marginalized sections of the society to overcome decades of oppression which is rooted in the ancient caste system. The Sustainable Development Goal 10 (SDG 10) of the United Nations is focused on reducing inequalities. The Indian reservation programs are designed to align with the targets of SDG 10 and to uplift the marginalized and disadvantaged sections of society into the mainstream.

The Sinho Commission report on "Economically Backward Classes" recommended using economic criteria in the reservation programs [1]. The 103rd Constitutional Amendment Act of 2019 addressed this need by establishing a 10% quota for the economically weaker section (EWS) in higher education and government employment recruitment [2]. It caused a paradigm shift from caste-based reservations to economic-based criteria. The Supreme Court upheld the constitutional validity of the 103rd Amendment to the Indian Constitution and established a precedent for future initiatives [3]. The study emphasized the implications of the economically weaker section (EWS) reservation on the Syro-Malabar community in Kerala. The community represented India's diverse ethos. The study analyzes the implementation, perception, and consequences of the economically weaker section (EWS) reservation within the community.

## *1.1 Objectives*

The main objective of the research was to evaluate the necessity and relevance of the economically weaker section (EWS) reservation policy in India. The other objectives were as follows: (i) to comprehend the impact of judicial judgments on reservation policies within the democratic system of India; (ii) to examine the implications and requirements of the EWS reservation within the unique context of the Syro Malabar community in Kerala, India; (iii) to recommend future policymaking concerning reservation.

## *1.2 Methodology*

The participants of the research were the members of the Syro-Malabar community in Kerala, India. We collected data from 530 individuals comprising 240 males (45.3%) and 290 females (54.7%). Informed consent was obtained from the participants before data collection. This consent included their participation in the research and the use of their data for academic and publication purposes. A structured questionnaire was used to collect primary data from the participants. The responses were then statistically analyzed using the Jamovi software. We also employed a review of the literature to comprehend the impacts of the economically weaker sections (EWS) reservation in India.

### *1.3 Ethics Approval Statement*

Ethical approval for the study was obtained from the Ethics Committee of Alliance School of Law, Alliance University, Bangalore. It includes the validation of the survey questionnaire.

## 2 Review of Literature

### *2.1 Development of the Indian Reservation System*

The Indian reservation system originated in an attempt to reduce societal inequalities and promote fairness for marginalized communities, dating back to the time before independence [4]. In British India, certain communities and castes enjoyed preferential treatment under quota systems. It resulted in a growing demand for positive discrimination in 1882 and 1891 [4]. An early and noteworthy figure in this regard is Shahu, Maharaja of Kolhapur, who in 1902 introduced reservations for non-Brahmin and backward classes, ensuring their employment and advocating for a class-free India and the abolition of untouchability, creating a precedent with a 50% reservation for these communities [5].

The Government of India Act of 1909 was one of the pre-independence legislations that the British Raj used to further institutionalize the reservation system in India [6]. Major changes occurred in 1932 when British Prime Minister Ramsay MacDonald issued the Communal Award, which divided representation among Muslims, Sikhs, Indian Christians, Anglo-Indians, Europeans, and the Depressed Classes [7]. The latter group is roughly equivalent to the Scheduled Tribes (STs) and Scheduled Castes (SCs) of today. This led to the Poona Pact, which, after Gandhi's fast in protest, ensured a single Hindu electorate with reserved seats for Dalits. This measure served as the foundation for India's post-independence reservation policies [7].

After independence, the reservation system was codified in the Indian Constitution, with an initial emphasis on political and employment reservations. Reservations in educational institutions were subsequently incorporated [8]. Today, the system provides reservations in legislatures, government jobs, promotions, and educational scholarships for SCs, STs, Other Backward Classes (OBC), and Economically Backward Classes, with the aim of improving their socio-economic status [9].

The Indian government began adopting quota policy for SCs and STs in educational institutions in 1954, advocating a 20% reservation with optional qualifying mark relaxation. In 1978, the Mandal Commission recommended a 27% reservation for Other Backward Classes (OBCs) in accordance with the 1931 census to address educational and employment disparities among socially and educationally backward sections [7]. However, these recommendations were not implemented until the 1990s. In 2019, India implemented a 10% reservation in employment and education for economically weaker sections of the general category [7].

## 2.2 *Constitutional Provisions Related to the Reservation*

The Constitution of India provides specific provisions regarding reservations under its dedication to social justice and equality. They are described as follows:

**Educational Institutions**

Article 15(4) of the Indian Constitution empowers the state to establish special provisions for the advancement of any socially and educationally backward classes of citizens, as well as for the Scheduled Castes (SCs) and the Scheduled Tribes (STs). It enhances their access to educational facilities [10]. Article 15(5) enables the state to establish provisions for the reservation of seats for Scheduled Castes (SCs) and the Scheduled Tribes (STs), and Backward Classes in private educational institutions, whether aided or unaided, except minority educational institutions. This guarantees underprivileged populations to have access in all types of educational institutions [10]. The inclusion of Clause 6 to Article 15 of the Constitution of India is intended to grant reservations for economically weaker sections (EWS) for admission to educational institutions, including private educational institutions [10].

**Employment**

Article 16(4) of the Indian Constitution authorizes the state to set aside positions or appointments for disadvantaged groups if it determines that these groups are not adequately represented in the services provided by the State. This promotes inclusivity and diversity by guaranteeing fair representation of all groups in government employment [10]. Reservation for individuals from economically weaker sections (EWS) in government jobs is ensured by the addition of Clause 6 to Article 16 of the Indian Constitution [10].

**Political Representation**

The reservation of seats for Scheduled Castes (SCs) and the Scheduled Tribes (STs) in the House of the People (Lok Sabha), is described in Article 330 of the Constitution of India. It guarantees their participation in the national legislative process and decision-making [10]. Article 332 of the Indian Constitution ensures the reservation of seats for the Scheduled Tribes (STs) and the Scheduled Castes (SCs) in the Legislative Assemblies of the States. It facilitates their representation and participation in state-level governance and policymaking [10]. These constitutional provisions demonstrate the nation's commitment to redress historical injustices and promote social equality.

## *2.3 Reasons for the Implementation of Reservation Policy in India*

The reservation policies in India are meant to rectify the historical inequalities and social disparities among vulnerable communities [9]. Its primary objective is to improve the social, educational, and economic status of these communities to integrate them into the mainstream of Indian society. It guarantees a specific percentage of seats in educational institutions, government jobs, and political representation for individuals from these communities. It ensures an equal opportunity for them to contribute to the progress and development of the nation. It addresses the continuing disparities that have impeded the full participation of vulnerable communities in the sociopolitical and economic spheres of India and promotes social equity [9]. The reservation system guarantees a more equitable distribution of resources and opportunities, reduces the inequality between the rich and poor, and empowers backward classes [4]. It admits inequality and social exclusion caused by the institutional barriers that have kept these marginalized communities from getting respectable jobs and higher education. It aids these communities to get out of poverty and caste-based discrimination by providing reservations in educational institutions and government jobs. It improves their living standards, reduces inequalities, and creates a more inclusive and equitable society [3].

## *2.4 Judicial Perspectives on Reservation*

The judiciary has a vital role in interpreting and shaping the reservation policies in India. It gives judicial justification for the decisions that have a significant impact on social justice and equality in the Indian scenario. The Supreme Court of India has given its first judgment on the reservation in the case of 'the State of Madras v. Smt. Champakam Dorairajan'. It resulted in the ratification of the First Amendment of the Indian Constitution in 1951. The Apex Court held that Article 16(4) of the Indian Constitution permits the reservation of public employment for backward classes, but not in education [11]. As per the direction of the Court, Clause 4 was added to Article 15 of the Indian Constitution to grant educational reservations for backward classes [10]. The Supreme Court of India has given the key interpretations of Article

15(4) in the landmark case of 'M.R. Balaji and Ors. v. State of Mysore'. It was observed that reservations for backward classes should be reasonable and not exceed 50% to preserve the essence of equality as enshrined in Article 15(1) of the Indian Constitution. Also, it held that these reservations should be based on both social and educational backwardness rather than caste-based [12].

In the case of Indra Sawhney v. Union of India, the Supreme Court held that reservations for backward classes in government employment cannot exceed 50% to uphold the principles of equality enshrined in Articles 14 and 16 of the Indian Constitution. It acknowledged the discretion of the state to make reservations through executive orders within the constitutional limits [13]. It emphasized reservations as a tool for promoting equality and ensuring the representation of disadvantaged sections in higher education [14]. The inclusion of politically influential caste groups in OBC lists (Other Backward Classes List) without sufficient requirements was criticized by the Supreme Court in Ram Singh and Ors v. Union of India. It was observed that social deficiencies may exist beyond the concept of caste. It proposed a more nuanced approach to identify the most deserving backward classes [15].

The 103rd Amendment of the Indian Constitution, which guarantees 10% reservation for economically weaker sections (EWS) was held constitutionally valid by the Supreme Court in the case of Janhit Abhiyan v. Union of India [3]. Its main objective is to promote social and economic justice by offering equal chances in employment and education [16]. Thus, the judiciary safeguards the rights of the backward and marginalized classes and upholds the fundamental values of the Indian Constitution.

## *2.5 Economically Weaker Section (EWS)*

The recommendations of the Sinho Commission (Commission for Economically Backward Classes) have influenced the Indian government to amend the Constitution of India to introduce reservations for the Economically Weaker Sections [1]. The Economically Weaker Sections (EWS) are considered an unreserved category, which denotes that they are not a member of the Scheduled Castes (SC), Scheduled Tribes (ST), or Other Backward Classes (OBC) and have an annual family income of less than Rs. 8 lakhs [2]. Thus, economic backwardness is introduced as a criterion for reservation in employment opportunities and education rather than cast-based inequalities.

The 103rd Amendment of the Indian Constitution was a progressive legal move to introduce provisions granting 10% reservation for the economically weaker sections (EWS) in the general category by amending Articles 15 and 16 of the Constitution of India [2]. The addition of clause (6) to Article 15 authorizes the state to make special provisions, including reservation in educational institutions, whether assisted or unaided by the state, with the exception of minority educational institutions covered by Article 30 clause (1) of the Indian Constitution. Similarly, clause (6) of Article 16 allows for reservation in government jobs [2]. The reservations for EWS go over and beyond the existing quotas of 15% for SCs, 7.5% for STs, and 27% for the non-creamy layer of OBCs, bringing the total reservation in India to 59.5% [17]. This move represented a substantial policy change towards economic criteria for affirmative action, giving advantages to economically disadvantaged people outside of the conventional caste-based reservation system [18]. The numerous provisions of the Constitution of India acknowledge that economic criteria can be the sole basis for reservation [19].

## 2.6 The Eligibility Criteria to Claim EWS Reservation

In order to be eligible for an EWS (Economically Weaker Section) reservation, a person must meet a number of requirements. First and foremost, the applicant must fall into the 'general' category, which means they cannot be protected by the current reservations made for Other Backward Classes (OBC), Scheduled Tribes (ST), or Scheduled Castes (SC). Second, the individual's overall yearly income (before deductions) from all sources, including employment, business, farm, and profession, must be less than Rs. 8 lakhs. Moreover, the person cannot own more than five acres of agricultural land. The person should not possess a residential flat larger than 1000 $ft^2$ and a residential plot larger than 100 square yards in a sector of a notified municipality [20]. These requirements are meant to ensure the benefits of the EWS quota to economically disadvantaged individuals.

## 2.7 Road to EWS Reservation

The reservations for Economically Weaker Sections (EWS) have undergone legal scrutiny and judicial interpretation. The constitutional validity of the 103rd Amendment was challenged immediately after its enactment in 2019. The critics alleged that the amendment violated the basic structure of the Indian Constitution. The matter was brought before a three-judge bench in August 2020, which recognized the complexity of the matter and referred it to the Constitutional Bench [21]. It upheld the constitutional validity of the 103rd Amendment granting the reservation for the Economically Weaker Sections (EWS) [3]. The constitutional validity of the 103rd Amendment of the Indian Constitution has been confirmed by the Apex Court with a ratio of 3:2 majority. It upheld a 10% reservation for Economically Weaker Sections (EWS) in educational institutions and public employment. While Justices Maheshwari, Trivedi, and Pardiwala supported the amendment, Justice Bhat and Chief Justice Lalit, dissented. However, both the majority and Justice Bhat agreed that economic criteria alone can be considered for affirmative action in education as per Article 15 of the Constitution [22].

## 3 Analyzing the Effects of EWS Reservation on the Syro Malabar Community in Kerala, India

The Syro-Malabar Church, one of the prominent Christian denominations in Kerala, India, holds a significant historical legacy tracing back to St. Thomas, often hailed as 'the Apostle of India' [23]. According to tradition, St. Thomas embarked on missionary journeys across India between 42 and 72 AD, laying the foundations for what would become the Syro-Malabar Church, among other Churches of St. Thomas tradition in the region. Presently, the church possesses an estimated five million members from around the world, demonstrating its immense impact not only in Kerala but as well [23]. The community has garnered recognition for its significant impact on Kerala's educational environment. The government of Kerala has identified the Syro-Malabar Community as the Economically Weaker Section [2]. The study recognized the need for economically weaker section (EWS) reservations for the Syro-Malabar community in Kerala.

Table 1 describes a detailed analysis of the economic conditions of the Syro Malabar community in Kerala, India. It classifies the research population based on several economic indicators, such as economic status, possession of ration cards, ownership of property, and income levels, providing insight into the distribution of wealth and access to government reservations. The research population is characterized into two categories according to their economic status: individuals classified above poverty line (APL) and those classified below poverty line (BPL). 71.7% of the research population falls into the APL category, indicating that they have a higher socioeconomic status. However, 28.3% of the research population falls into the BPL category, indicating that a sizable proportion of the research population is economically disadvantaged.

The possession of ration card is a crucial sign of eligibility for government subsidies. The data show that the population has a broad distribution of ration card kinds. 32.5% of people have white ration cards, which frequently indicate a higher socioeconomic position or disqualification from some benefits. 38.9% of people have blue ration cards, whereas 20% have pink ration cards, which may indicate eligibility for different levels of subsidies. Yellow ration cards, which indicate a higher level of subsidy, are the least prevalent at 5.3%. 3.4% of the population do not have a ration card, possibly because they are unregistered.

Property ownership is another key component of the research. It is divided into four categories according to the amount of land possessed. A sizable proportion of people, 51.3%, possess between 10 cents and less than one acre of land. 27.5% of the population owns less than 10 cents. Individuals with 1 to less than 5 acres and those with 5 acres and above comprise 18.7% and 2.5% of the population, respectively. The figures show that the majority of the population meets the criterion for EWS reservation, with 97.5% owning less than 5 acres of land and 78.8% owning less than one acre.

**Table 1** Economic status of the sample

| Economic status | | Number | Percent (%) |
|---|---|---|---|
| Types | APL | 380 | 71.7 |
| | BPL | 150 | 28.3 |
| Total | | 530 | 100 |
| Ration card | White | 172 | 32.5 |
| | Blue | 206 | 38.9 |
| | Pink | 106 | 20.0 |
| | Yellow | 28 | 5.3 |
| | No ration card | 18 | 3.4 |
| Total | | 530 | 100 |
| Property | Below 10 cents | 146 | 27.5 |
| | 10 cents to below 1 acre | 272 | 51.3 |
| | 1 acre to below 5 acres | 99 | 18.7 |
| | 5 acres and above | 13 | 2.5 |
| Total | | 530 | 100 |
| Income | Less than 1 lakh | 275 | 52.1 |
| | Less than or equal to 2 lakhs | 162 | 30.6 |
| | Less than or equal to 5 lakhs | 48 | 9.1 |
| | Less than or equal to 8 lakhs | 21 | 4.0 |
| | 8 lakhs and above | 23 | 4.3 |
| Total | | 530 | 100 |

*Note* APL denotes above poverty line, BPL denotes below poverty line and ration card denotes an official document issued by state governments to get food grains at subsidized rates under the National Food Security Act (NFSA), 2013 [24]. Yellow card indicates most economically backward section of society. Pink card represents priority or below poverty line (BPL). Blue card designates non-priority subsidy or above poverty line (APL). White card represents non-priority [24]

The income distribution provides insight into the population's financial well-being. 52.1% of the individuals earn less than 1 lakh per annum, indicating a lower-income category. Individuals with annual incomes ranging from 2 to 5 lakhs make up 30.6 and 9.1% of the population, representing the lower-middle and middle-income categories. 8.3% of individuals earn between 5 lakhs to above 8 lakhs, indicating upper-middle to higher income levels. According to the data, a huge majority (82.7%) of the population falls into the lower to middle-income groups, potentially qualifying for EWS reservation based on an annual gross income of less than Rs. 8 lakhs [2].

According to Fig. 1, 73% (387 persons) of respondents are aware of Economically Weaker Sections (EWS), whereas 27% (143 individuals) are unaware, with a total of 530 surveyed individuals belonging to the Syro Malabar Church in Kerala, India [23]. Thus, the data indicates that EWS is widely known among the population.

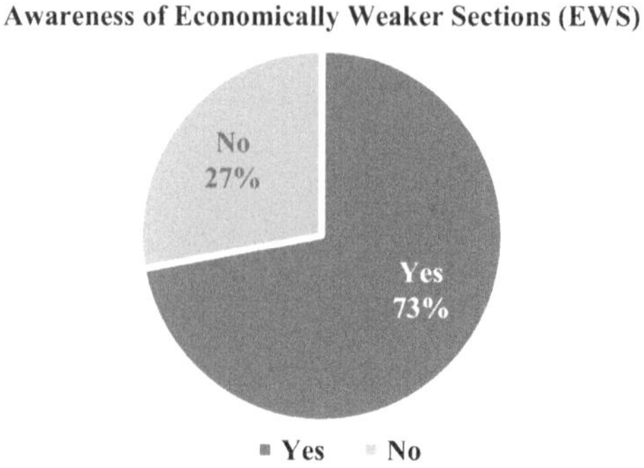


Fig. 1 Graphical representation of awareness of economically weaker sections (EWS)

Figure 2 indicates that 217 people (40.9%) of the 530 respondents in total have earned an Economically Weaker Sections (EWS) certificate. Conversely, 313 respondents (59.1%) of the sample reported not having an EWS certificate.

Figure 3 displays the distribution of Economically Weaker Sections (EWS) certificate submissions among different institutions by 217 respondents. 33.2% of people have not submitted their EWS certificate, while 29.0% have submitted it to government departments and 16.1% to private institutions. 9.7% of individuals submitted their EWS certificate to minority institutions, while 12.0% chose to submit theirs to various other places.

The various advantages that individuals have obtained by submitting an Economically Weaker Sections (EWS) certificate are depicted in Fig. 4. A notable proportion

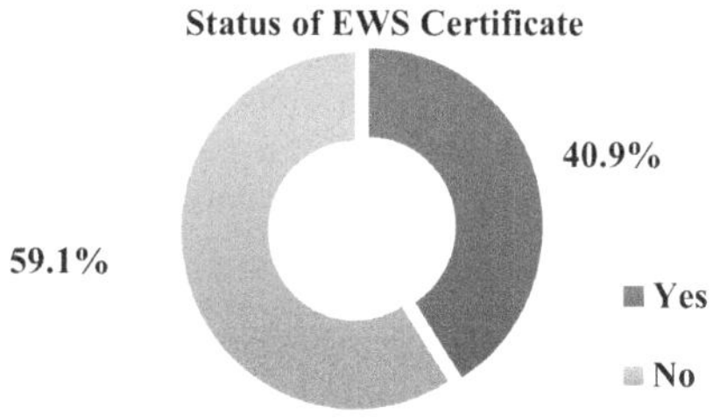


Fig. 2 Graphical representation of the status of economically weaker sections (EWS) certificate

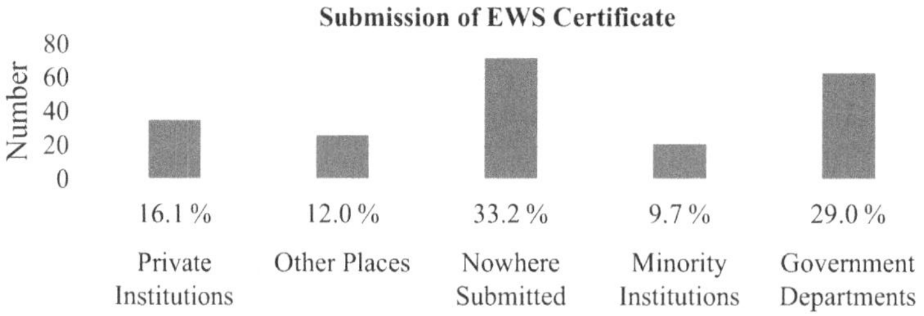


Fig. 3 Graphical representation of submission of EWS certificate

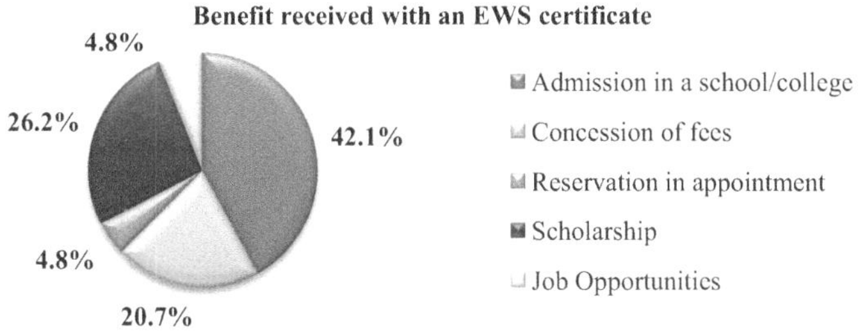


**Fig. 4** Graphical representation of the benefit received with an EWS certificate

of participants (42.1%) were granted admission to colleges or institutions, underscoring the certificate's capacity to facilitate educational opportunities for those who are economically disadvantaged. Concession of fees were the second most often mentioned advantage (20.7% of respondents), highlighting the financial assistance given to holders of EWS certificates. 26.2% of participants received scholarships to aid their study. However, the EWS certificate appears to have a less significant influence on career chances, with 6.2% of respondents reporting job opportunities and only 4.8% having appointment reservations. Overall, these results demonstrate the EWS certificate's important significance in providing educational possibilities and financial support, with less noticeable benefits in employment sectors.

Figure 5 offers clear insights to determine the necessity of Economically Weaker Sections (EWS) reservation for Syro Malabar Christians in Kerala. A substantial majority of the 530 respondents (85.7%) concur that EWS reservation is necessary for this community, with 58.3% strongly agreeing and 27.4% agreeing. This indicates a prevailing belief among the individuals surveyed that Syro-Malabar Christians, notwithstanding their religious minority status, encounter economic hardships that warrant their inclusion in EWS reservation policies. The necessity for such reservations is disagreed by a negligible percentage of respondents (1.5%), not sure (11.7%), or strongly disagree (1.1%). The community's economic vulnerabilities are underscored by the outpouring of support for EWS reservations, which indicates that many members believe these reservations could offer economically disadvantaged members essential assistance and opportunities. The data points out that the members of the Syro Malabar Christians in Kerala are aware of the need for policy interventions to rectify economic disparities.

## 4 Findings, Recommendations, and Conclusion

The majority of members of the Syro Malabar community in Kerala come under the category of above the poverty line (APL) and their economic status remains modest. The government must restructure the manner of identifying economically weaker

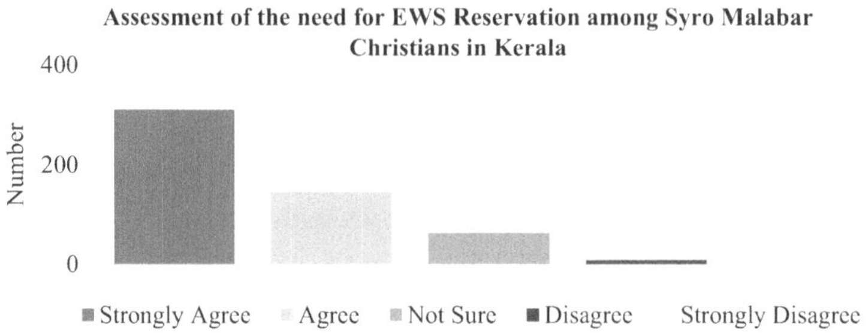


**Fig. 5** Graphical representation of assessment of the need for EWS reservation among Syro Malabar Christians in Kerala

households within the Syro-Malabar community in Kerala, India. They should guarantee the distribution of the EWS reservation benefits to the right beneficiaries. The issuance of EWS certificate should be simplified so that deserving persons can obtain it without encountering bureaucratic obstacles.

We observed that a significant portion of the community has not obtained or utilized the EWS certificate for educational or employment purposes. It is the role of the Syro-Malabar community administration to educate its people about the benefits of using government-aided programs such as the EWS reservation. The effective utilization of the EWS reservation can have a great impact on the upliftment of the community within a decade.

We noted that there is still space for further research on the impacts of the EWS reservation on different communities in India. It was also observed that the economic status of the family shall be considered as a criterion for reservations. Thus, the marginalized ones from all segments of society can be included in the reservation policy of India.

The EWS reservation policy serves as a positive mechanism to address economic inequalities to promote the socioeconomic advancement of weaker sections in India. The research identified the potential of the EWS reservation to reduce inequalities and attain sustainable development of the nation.

**Conflict of Interest** The authors declare that there are no conflicts of interest.

**Source of Funding** The authors received no external funding for this research.

## References

1. Sinho, S.R.: Report of the Commission for the Economically Backward Classes. Ministry of Social Justice and Empowerment, India (2010)
2. The Constitution (One Hundred and Third Amendment) Act 2019. The Gazette of India, New Delhi (2019)

3. Janhit Abhiyan v. Union of India, Livelaw (SC) 922 (2022)
4. Chopra, M.: Reservation system in India: advantages and disadvantages. Int. J. Perspect. **15**(1), 30–37 (2021)
5. Rana, M.S.: Reservation in India: Myths and Realities. Concept Publishing Company, New Delhi (2008)
6. Thampi, J., Girdhar, S., Kumar, S.: Does the Indian reservation system still make sense in the modern India? Satraachee **38**(26), 205–215 (2023)
7. Gopalan, K.R.: A study on reservations in the educational sector for the economically weaker sections. Parishodh J. **9**(3), 6175–6182 (2020)
8. Yoganandham, G., et al.: An evaluation of the reservation system in India. Int. J. All Res. Educ. Sci. Methods **11**(3), 218–229 (2023)
9. Raj, Y.S., Gokulraja, P.: An analysis of reservation system in India. Int. J. Res. **2**(10), 1038–1045 (2015)
10. India. The Constitution of India. Government of India, New Delhi (1950)
11. The State of Madras v. Smt. Champakam Dorairajan, AIR 1951 SC 226 (1951)
12. M. R. Balaji and Ors. v. State of Mysore, AIR 649 (1963)
13. Indra Sawhney v. Union of India, AIR 1993 SC 477 (1992)
14. Ashoka Kumar Thakur v. Union of India, 6 SCC 1138 (2008)
15. Ram Singh and Ors v. Union of India, AIR 2015 SCW 1718 (2015)
16. Kumari, B.: Janhit Abhiyan vs Union of India. Accessed 25 July 2024. https://corpbiz.io/learning/janhit-abhiyan-v-union-of-india (2023)
17. Jain, M.: EWS reservation case: Janhit Abhiyan vs Union of India. Just. Law Bull. **2**(2), 30–36 (2023)
18. Singh, G.: Constitutional validity in the economically weaker section reservation. Int. J. Law Manag. Humanit. **4**(2), 631–652 (2021)
19. Singh, L.M.R.: Reservation for economically weaker section (EWS): paradigm shift in policy to achieve equality. Think India J. **22**(17), 3064–3075 (2019)
20. Government of India, Ministry of Social Justice and Empowerment, Department of Social Justice and Empowerment: Office Memorandum: Reservation for Economically Weaker Sections (EWSs) in Civil Posts and Services in the Government of India and Admission in Educational Institutions (2019)
21. Janhit Abhiyan v. Union of India, AIRONLINE 2020 SC 670 (2020)
22. Chowdhury, S.: Reservation Solely Based on Economic Criteria Does Not Violate Basic Structure of Constitution: Supreme Court in EWS Quota Case. Accessed 30 July 2024. https://www.livelaw.in/top-stories/reservation-solely-based-on-economic-criteria-does-not-violate-basic-structure-of-constitution-supreme-court-in-ews-quota-case-213516
23. Syro-Malabar Church, Brief History. Accessed 05 Aug 2024. https://www.syromalabarchurch.in
24. India Filings, Kerala Ration Card. Accessed 10 Aug 2024. https://www.indiafilings.com/learn/kerala-ration-card/

# Sustainability in the Context of Remote Work in IT Sector

**Tisha Tomy and R. Sheeja**

**Abstract** Remote work effectively mitigates a substantial amount of expenses and energy consumption that are typically associated with physical office spaces. The sustainability benefits of working remotely are very tangible. The article presents the findings of a study concerning the IT sector's remote work employment in relation to the Sustainable Development Goals. Initially, a review of the literature is made. The research questions stated in the paper are: examining how remote work affects the achievement of Sustainable Development Goals; finding out the reasons why remote work is preferred by employees in the IT sector, and identifying the significant aspects of the environment concerning sustainability in the context of remote work. A survey was conducted among 150 IT sector employees working in Kerala. The data obtained in the survey underwent statistical analysis. The COVID-19 pandemic has perpetuated the need for remote work in the IT sector. Employees exhibit a preference for remote work on account of its time-saving benefits, work comfort, safety, and flexibility. The environmental factors that are most often cited in relation to sustainability are the reductions in hazardous substance emissions from automobile exhaust fumes and the decrease in electricity consumption at office buildings.

**Keywords** Remote work · Sustainability · IT sector · Environmental impact

## 1 Introduction

Remote working has been experiencing a meteoric rise in popularity due to the widespread adoption of flexible work arrangements prompted by the COVID-19 pandemic [1]. In this article an effort is made to identify the sustainability benefits of working remotely to understand why IT sector employers should allow more

T. Tomy (✉) · R. Sheeja
Department of Commerce and Management, Amrita Vishwa Vidyapeetham, Amritapuri, India
e-mail: tishatomy1@gmail.com

R. Sheeja
e-mail: sheejar@am.amrita.edu



M. Al Mubarak (ed.), *Sustainable Digital Technology and Ethics in an Ever-Changing Environment*, Studies in Systems, Decision and Control 237,
https://doi.org/10.1007/978-3-031-86708-8_46

flexibility when it comes to work-from-home opportunities moving forward. Remote work effectively mitigates a substantial amount of expenses and energy consumption that are typically associated with physical office spaces. Thus, the sustainability benefits of working remotely are very tangible [2].

Sustainable development goals pertaining to remote work may hold for the atmosphere, the weather, temperature, and the overall environment. In numerous businesses, digitalization, and telecommunication have already significantly reduced paper usage. Still, people continue to waste paper because of superfluous bureaucratic procedures at physical work locations. Remote work assists in eradicating all of that [3]. Employers have the potential to significantly reduce the enormous amount of paper used annually for business operations by instituting a work-from-home policy. When working remotely, every piece of information on the team can be shared online. Reduced paper consumption primarily results in paper manufacturers reducing the number of trees felled during the production process. Employees are found more conscious of energy conservation also [4]. But offices and structured work environments all require significant amounts of electricity to function.

The objective of the study was to seek the perspective of employees regarding remote work, to identify the category of employees who wish to work remotely, and to establish the reasons why they choose it as a better and cozy in the circumstances of sustainability. The following research questions were raised:

1. What are the reasons why employees in the IT sector prefer remote work?
2. What is the impact of remote work on sustainability?
3. What are the most significant environmental factors pertaining to sustainability in the context of remote work?

The paper is structured in such a manner that first it presents the literature on remote work with a focus on the motives behind IT sector employees to prefer or not prefer remote work. The next part of the review relates to sustainability in general and raises issues connected to sustainability because of working at offices over work from home. The review concludes by discussing the contribution of remote work in minimizing its impact upon environment. Then the article covers the methodology of the study and includes analysis and results. The paper ends up with a conclusion followed by a scope for future research studies.

## 2 Literature Review

With the COVID-19 outbreak, most organizations were forced to opt to Work from Home (WFH) in early 2020 for a major fraction of the workforce in various countries worldwide. Organizations in India have moved workplaces into the homes of their employees, with several companies providing facilities like hardware and Wi-Fi infrastructure [5] often without facilitating employees with the adequate skills necessary to work remotely [6].

## 2.1 Preference to Remote Work

Many people wish to work remotely. The introduction of these flexible ways of working appears to be a planned decision that requires the organization to develop, prepare, and adapt steps to effectively support employee output and achieve a good work-life balance [7]. Like all other organizational changes, remote work has several advantages and disadvantages [8]. Working at home tends to increase employees' level of autonomy when moving from the factory or office to the home. The employee at home is less exposed to control and surveillance measures, and close supervision and therefore can enjoy more flexibility in the choice of working hours and work patterns [9]. However, despite the obvious benefits of remote work for both individuals and employers, many challenges arise [10] as the effective installation of remote work initiatives poses significant challenges in human resource management [11]. Studies report both the benefits and disadvantages of working remotely. Its effects, therefore, have been quite explored. It is important to investigate how the 'new way of working', remote work, has impacted the performance and productivity of workers with no previous work experience of telecommuting, and to identify those specific work environments impacting remote work during the COVID-19 crisis [12].

Difficulties in reconciling personal and work responsibilities are included in studies as family work conflicts, a situation in which workers find it increasingly difficult to fulfill work responsibilities as they are involved in family activities [13]. Family–work conflict is very often perceived as a sex-dependent phenomenon. Very often, women are more penalized than men during times of adverse situations [14]. COVID-19 has left millions of people at home, making it harder to distinguish between one's personal and professional life, regardless of age and gender. For this reason, it may be argued that conflict between work and family can affect both men and women who work remotely. On the contrary, previous research has shown that teleworking can negate the benefits of Working from Home [15] by simultaneously reducing and increasing work-family conflicts [16].

At home, the presence of spouses and children who are busy with their jobs and schools, the disruption of child-care and education services during the pandemic, and having to perform domestic work all have a significant impact on teleworkers [17]. Being isolated at the workplace was a key feature of remote work during the pandemic. Even if researchers propounded that loneliness and being isolated at the workplace as the major demerits of remote work, [18] its negative impact and influence had greatly risen during this period. The pandemic has put people at higher levels of loneliness [19] which can lead to reduced job satisfaction and productivity, as well as increased stress between working from home and individual and team presentations.

Studies suggested that a positive full-time work-from-home experience is highly related with the quality of the workspace, such as adequate lighting, noise reduction, and acoustic isolation [20] and a works environment that is adequately delineated from the living space [21]. When such a separation is impossible, working in an atmosphere that is distracting and detracts from it may be an extra and reasonable job requirement and a corresponding job search. Based on available literature, remote work is distinguished based on two important characteristics. First of all, workers operate away from their typical work-stations. Secondly, there is a dominant link between the workplace and home. Information and communication technologies can be used to share information and communicate with colleagues [22]. Also, the studies put forward that remote workers should have an appropriate working atmosphere. Remote work is best appropriate for employees, who mainly have knowledge-based tasks to perform, maintain few personal contacts, and expects for a high degree of autonomy [23].

## 2.2 *Sustainability and Remote Work*

Sustainability in the workplace is crucial and a sustainable workplace is one that operates in a manner that protects the environment, fulfills the requirements of the present generation without sacrificing the capacity of future generations to satisfy their own requirements. The rise of remote work and the growing importance of sustainability are two key trends in the IT Sector. While the remote work model has gained attraction in recent years, sustainability practices and principles are still being adopted and established in IT Companies.

These two trends intersect in several ways, as the hybrid work model has the potential to be highly sustainable in several aspects. By allowing employees to work from home, it reduces the need for commuting and thus reduces carbon emissions. Additionally, remote work can reduce the amount of energy used in the workplace and many other sustainability effects [24]. When assessing the environmental impacts of remote work, it is crucial to bear in mind that numerous factors are involved, and their effects are very often interconnected [25].

## 2.3 *Remote Work and Environmental Effects*

The environment literature is still in its infancy; however, many of the existing studies concentrate on $CO_2$ emissions associated with reduced transportation, which typically involves commuting with private automobiles as opposed to sustainable public transportation [26]. In the context of remote work, there are several additional considerations that can be made, including consumption and non-work travel patterns, the energy efficacy of residences, and the carbon footprint associated with different modes of transportation [27].

By concentrating on home-based remote work to comprehend its environmental impacts, one can diminish the number of intricate aspects associated with remote work that are delegated along global value chains, thereby complicating environmental traceability [28]. In addition to six million barrels of diesel, the exhaustive global analysis conducted by the IEA [29] indicates that gasoline consumption decreased by over nine million barrels per day after the lockdowns implemented during the initial phase of COVID-19.

Barcelona's urban air quality improved substantially because of closure measures and homeworking implemented as an outcome of the COVID-19 pandemic. Average concentrations of $NO_2$, which is the primary pollutant produced by traffic emissions, decreased considerably. The modeling framework they have developed indicates a positive correlation between the quantity of remote work days per week and $NO_2$ emissions. Specifically, they find that two such days per week result in a 4% reduction in $NO_2$, whereas four days can cause a 10% decrease [30].

## *2.4 Remote Work and Energy Use*

A thorough international examination conducted during the initial stage of lockdowns unveiled that a single day of remote work was correlated with a 7–23% surge in residential energy usage. The exact magnitude of this increase varied based on several factors, such as the initial energy efficiency of the residence and the number of individuals conducting remote work from within the same household. Furthermore, research demonstrates that the proportion of energy used for teleworking during the week is comparable to the mean energy consumption on Sundays, the day when most individuals are at home [31].

Bachelet et al. propose that telework reduces annual transportation expenditures by 840 euros per employee, while increasing annual thermal energy expenditures by 110 euros per employee [32], using the German Microcensus and corresponding energy and carbon prices as data. Based on the research conducted in Canada, it can be inferred that an increase in productivity through remote working would likely result in a corresponding growth in income and production, as well as consumption patterns that extend beyond the bare necessities (e.g., increased leisure travel, consumption of goods beyond necessities, etc.) [33]. Consequently, the total energy demand will rise. In this scenario, although remote work could initially reduce emissions by eliminating the need for commuting, it may ultimately result in a more substantial ecological footprint due to heightened energy demands and increased consumption [34].

The energy consumption of data and cloud centers contributes to the carbon footprint associated with internet usage. This carbon footprint varies between 28 and 63 g of $CO_2$ equivalent per gigabyte [35]. When considering the range of internet services available, it is observed that frequently utilized tools like video conferencing consume the most energy. However, continuous technological advancements in this field are resulting in the introduction of more energy-efficient solutions [36].

### 2.5 Remote Work and Mobility

The primary prospective advantage of remote work primarily pertains to its capacity to mitigate mobility [38]. The discouragement surrounding the contribution of remote work to sustainable mobility has become more significant as its prevalence has grown. This is because remote or hybrid work is frequently cited as a significant method of mitigating carbon emissions on account of the reduced frequency of home-to-work commutes [39]. Brynjolfsson et al. present the results of a two-wave representative survey conducted in 2020 regarding working behavior in the US at times of COVID-19 pandemic. The author discovers that approximately half of those who were employed prior to the outbreak are now operating remotely, with 35.25 of those who previously commuted stating that they recently transitioned to working from home [40]. Several studies have demonstrated that remote work contributes positively to sustainable mobility primarily through a reduction in automobile usage; thus, it is possible to decrease the carbon footprint of transportation by eliminating the need for lengthy commutes [41–43].

This obviously indicates, nevertheless, that reduced mobility (not necessarily more sustainable mobility) is the only anticipated consequence of remote work in terms of mitigating emissions and pollution associated with transportation [44]. Conversely, alternative research raises doubts regarding this contribution and attributes any sustainability benefits to the longer travel distances undertaken by remote workers [26, 45]. During pandemics, environmental concerns receive considerable scientific attention [46]. Energy consumption increased in India following the re-establishment of social freedom and the lifting of lockdown measures, following a brief decline caused by the lockdowns. Electricity will be indispensable for technological advancement and progress in the near and distant future [47, 48].

In conclusion, sustainability and remote work are enduring trends that will continue to influence the contemporary workplace for the foreseeable future. While the correlation between the two is intricate and diverse, it is critical for organizations striving to establish sustainable and efficient hybrid work models to meticulously evaluate both the advantages and the obstacles [49].

## 3 Methods

Data collection was made using a well-structured questionnaire distributed among 150 IT sector employees working in Kerala. The statistical analysis-based section was intended to delineate two categories in accordance with the opinions of respondents about the reasons why they prefer remote work and its beneficial impact on sustainable development. Focus was also given to the other side through the replication of the study on a set of variables related to ill-effects of remote work. The combining of the outcomes of these to analyses resulted in the ascertainment of subsequent groups of factors: favoring remote work and its positive impact on sustainable development, and obstacles to remote work.

Six of the factors dealt with aspects favoring remote work and the other six dealt with obstacles to remote work. With respect to the sustainable development aspect, four variables were included.

*Favorable aspect variables*:

V1: Increased efficiency and quality of work
V2: More comfort and freedom at work
V3: Flexible working hours
V4: Reduced fuel and utility consumption
V5: Private financial savings
V6: Reduction in the emission of hazardous substances to the environment.

*Obstacle variables*:

V1: Lack of peaceful work environment and workspace at home
V2: Long working hours and increased workload
V3: Absence of direct interaction with clients/colleagues affected the excellency
V4: Chances of ambiguities and misunderstandings during electronic communication
V5: High control or surveillance measures from the side of superiors
V6: Lack of technical support.

*Sustainable development aspect variables*:

V1: Lower fuel consumption
V2: Energy efficiency and other utility consumption in offices
V3: Reduction in the emission of hazardous substances to the environment
V4: Private financial savings.

## 4 Analysis and Results

The calculations were based on survey data obtained from 150 IT sector employees. Table 1 presents the indicator value and Standard Deviation for each of the factors favoring remote work in IT sector.

From Table 1, it can be inferred that the major factor why employees favored remote work is the flexible working hours that it offers followed by more comfort and freedom at work. The same was done for obstacle factors in remote work also. Table 2 shows the indicator value and Standard Deviation for each of the obstacle factors to remote work in IT sector.

From Table 2, it can be inferred that the biggest obstacle that IT sector employees face in remote work is the absence of direct interaction with clients/colleagues affected the excellency. Another major barrier is the long working hours and increased work load. The smallest problem was lack of technical support.

**Table 1** Factors favoring remote work

| Factors favoring remote work (variables) | Indicator value | Standard deviation |
|---|---|---|
| V1: Increased efficiency and quality of work | 2.00 | 5.93 |
| V2: More comfort and freedom at work | 3.33 | 13.81 |
| V3: Flexible working hours | 4.00 | 14.10 |
| V4: Reduced fuel and utility consumption | 2.86 | 8.60 |
| V5: Private financial savings | 2.50 | 7.80 |
| V6: Reduction in the emission of hazardous substances to environment | 2.00 | 6.32 |

*Source* Author's own work

**Table 2** Obstacle factors to remote work

| Obstacle factors to remote work (variables) | Indicator value | Standard deviation |
|---|---|---|
| V1: Lack of peaceful work environment and workspace at home | 2.22 | 8.17 |
| V2: Long working hours and increased work load | 3.33 | 11.24 |
| V3: Absence of direct interaction with clients/colleagues affected the excellency | 5.40 | 8.35 |
| V4: Chances of ambiguities and misunderstandings during electronic communication | 1.53 | 5.95 |
| V5: High control or surveillance measures from the side of superiors | 2.00 | 7.90 |
| V6: Lack of technical support | 1.00 | 7.46 |

*Source* Author's own work

Tables 1 and 2 are graphically presented in Figs. 1 and 2 respectively.

Using the same principle, an attempt was made to identify the significant environmental factors of sustainability in the context of remote work. Table 3 shows the indicator value and standard deviation of various aspects of sustainable development.

Table 3 is graphically presented in Fig. 3.

Table 4 summarizes the results of the analysis made above regarding the major and weaker factors—reasons for IT Sector employees working in Kerala, to prefer remote work and do not prefer remote work and also the sustainable development contributions of remote work.

## 5 Conclusion and Scope for Future Research Studies

Given the findings of the undertaken investigation, it is pertinent to highlight the theoretical ramifications. We were able to discern the core of employees' favorable sentiments concerning remote work by utilizing the employed methodology. It is

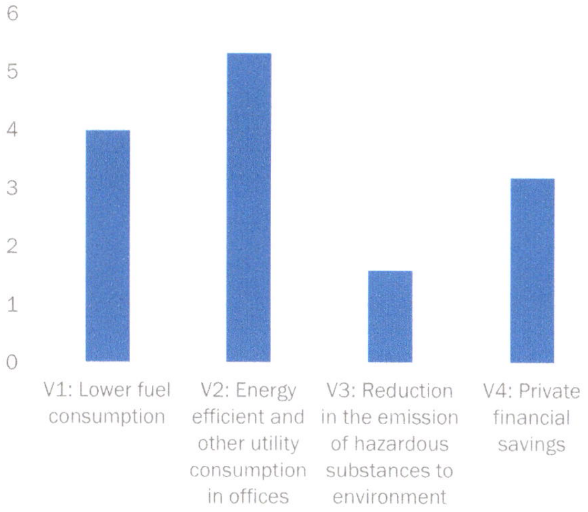


**Fig. 1** Factors favoring remote work

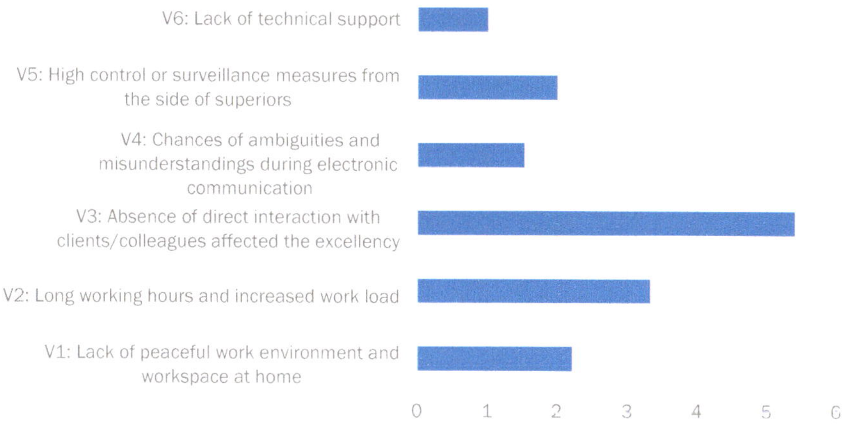


**Fig. 2** Obstacle factors to remote work

**Table 3** Various aspects of sustainable development

| Sustainable development aspects | Indicator value | Standard deviation |
|---|---|---|
| V1: Lower fuel consumption | 4.00 | 12.07 |
| V2: Energy efficiency and other utility consumption in offices | 5.32 | 11.07 |
| V3: Reduction in the emission of hazardous substances to the environment | 1.61 | 5.11 |
| V4: Private financial savings | 3.20 | 13.55 |

*Source* Author's own work

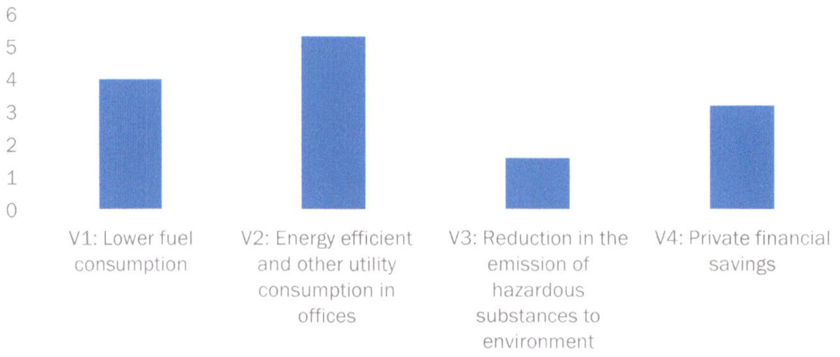


**Fig. 3** Sustainable development aspects

valuable to emphasize the effects of outsourcing work to employees and human resource management from a managerial standpoint, and not just in the context of a pandemic. This publication contributes to the existing body of literature and infuses IT sector-related research with new insights regarding employees' and business owners' perceptions of remote work. It also addresses the various ramifications of the COVID-19 pandemic and establishes connections between the domains in relation to environmental sustainability.

Future research is warranted to examine novel technological approaches that augment sustainability. This includes devising environmentally friendly solutions and conducting research on the environmental impact of WFH technologies. Furthermore, research examining the incorporation of nascent technologies such as virtual reality and augmented reality into remote work may yield valuable insights regarding prospective work environments. Future research may also be subject to developing and assessing policy frameworks that encourage sustainable practices in remote work infrastructure and incentives for environmentally favorable practices. Finally, as crucial components of sustainability, the health and well-being of remote workers should be considered in future research endeavors.

Forecasting the potential occurrence of future pandemics is not pessimistic, but rather pragmatic. The discourse ought to center on the temporal occurrence of forthcoming epidemics rather than the question of their potential occurrence [50]. Hence, in pursuit of a sustainable future, it is critical to comprehend how a novel paradigm can be established regarding sustainable work systems.

It is important to acknowledge that the study and findings presented in this publication were applied exclusively to a subset of individuals who advocated for remote work within their respective industries and regarded it as the most advantageous model of business operation in terms of convenience, productivity, and attention to detail. This article accurately asserts that the psychological well-being and motives of IT sector employees are also influenced by external factors, including ambiguity

**Table 4** Summary of results

| | | | |
|---|---|---|---|
| *Favorable factors to remote factor* | | | |
| Flexible working hours | Major attractions | Weaker attractions | Private financial savings |
| More comfort and freedom at work | | | Increased efficiency and quality of work |
| Reduced fuel and utility consumption | | | Reduction in the emission of hazardous substances to environment |
| *Obstacle factors to remote work* | | | |
| Absence of direct interaction with clients/colleagues affected the excellency | Major obstacles | Weaker obstacles | High control or surveillance measures from the side of superiors |
| Long working hours and increased work load | | | Chances of ambiguities and misunderstandings during electronic communication |
| Lack of peaceful work environment and workspace at home | | | Lack of technical support |
| *Sustainable development factors* | | | |
| Energy efficient and other utility consumption in offices | Major contributions to sustainability | Weaker contributions to sustainability | Private financial savings |
| Lower fuel consumption | | | Reduction in the emission of hazardous substances to environment |

*Source* Author's own work

in role, organizational culture, work environment and colleagues. These are supplementary factors that ought to be considered when formulating a claim regarding the remote work model's superiority to the conventional one.

## References

1. Moglia, M., Hopkins, J., Bardoel, A.: Telework, hybrid work and the United Nation's sustainable development goals: towards policy coherence. Sustainability **13**(16), 9222 (2021)

2. Hestad, D., Tàbara, J.D., Thornton, T.F.: The three logics of sustainability-oriented hybrid organisations: a multi-disciplinary review. Sustain. Sci. **16**, 647–661 (2021)
3. Ewert, C., Maggetti, M.: Regulating side by side: the role of hybrid organisations in transnational environmental sustainability. Policy Soc. **35**(1), 91–102 (2016)
4. Babapour Chafi, M., Hultberg, A., Bozic Yams, N.: Post-pandemic office work: perceived challenges and opportunities for a sustainable work environment. Sustainability **14**(1), 294 (2021)
5. Gartner: Gartner HR survey reveals 41% of employees likely to work remotely at least some of the time post coronavirus pandemic (press releases) (2020)
6. Molino, M., Ingusci, E., Signore, F.: Wellbeing costs of technology use during COVID-19 remote working: an investigation using the Italian translation of the technostress creators scale. Sustainability **12**(15), 5911 (2020)
7. Allen, T.D., Golden, T.D., Shockley, K.M.: How effective is telecommuting? Assessing the status of our scientific findings. Psychol. Sci. Public Interest **16**, 40–68 (2015)
8. Konradt, U., Hertel, G., Schmook, R.: Quality of management by objectives, task-related stressors, and non-task-related stressors as predictors of stress and job satisfaction among teleworkers. Eur. J. Work Organ. Psychol. **12**, 61–79 (2003)
9. Schaufeli, W.B., Bakker, A.B.: Job demands, job resources, and their relationship with burnout and engagement: a multi-sample study. J. Organ. Behav. **25**, 293–315 (2004)
10. Crandall, W., Longge, G.: An update on telecommuting: review and prospects for emerging issues. S.A.M. Adv. Manage. J. **70**(3), 30–311 (2005)
11. Hall, L., Liddicoat, L.: Challenges to developing effective family friendly work practices: findings from New Zealand. Res. Pract. Hum. Resour. Manag. **13**(1), 1–17 (2005)
12. Vander Elst, T., Verhoogen, R., Godderis, L.: Teleworking and employee well-being in corona times. J. Occup. Environ. Med. **62**, e776–e777 (2020)
13. Weer, C., Greenhaus, J.H.: Family-to-work conflict. In: Encyclopedia of Quality of Life and Well-Being Research, pp. 2210–2211 (2014)
14. Galanti, T., Cortini, M.: Work as a recovery factor after earthquake: a mixed-method study on female workers. Disaster Prev. Manag. **28**, 487–500 (2019)
15. Hilbrecht, M., Shaw, S.M., Johnson, L.C., Andrey, J.: "I'm home for the kids": contradictory implications for work–life balance of teleworking mothers. Gend. Work Organ. **15**, 454–476 (2008)
16. Toscano, F., Zappalà, S.: Smart working in Italia: origine, diffusione e possibili esiti. Psicol. Soc. **15**(2), 203–223 (2020)
17. Xiao, Y., Becerik-Gerber, B., Lucas, G., Roll, S.C.: Impacts of working from home during COVID-19 pandemic on physical and mental well-being of office workstation users. J. Occup. Environ. Med. **63**, 181–190 (2014)
18. Cooper, C.D., Kurland, N.B.: Telecommuting, professional isolation, and employee development in public and private organizations. J. Organ. Behav. **23**, 511–532 (2002)
19. Ellis, W.E., Dumas, T.M., Forbes, L.M.: Physically isolated but socially connected: psychological adjustment and stress among adolescents during the initial COVID-19 crisis. Can. J. Behav. Sci. **52**, 177–187 (2020)
20. Raguseo, E., Gastaldi, L., Neirotti, P.: Smart work: supporting employees' flexibility through ICT, HR practices and office layout. Evid. Based HRM **4**, 240–256 (2016)
21. Mello, J.A.: Managing teleworks programs effectively. Empl. Responsib. Rights J. **19**, 247–261 (2007)
22. Bélanger, F.: Workers' propensity to telecommute: an empirical study. Inf. Manag. **35**(3), 139–153 (1999)
23. Bailey, D.E., Kurland, N.B.: A review of telework research: findings, new directions, and lessons for the study of modern work. J. Organ. Behav. **23**(spec issue), 383–400 (2002)
24. Giovanis, E.: The relationship between teleworking, traffic and air pollution. Atmos. Pollut. Res. **9**(1), 1–14 (2018)
25. Eurofound: Is Telework Really 'Greener'? An Overview and Assessment of Its Climate Impacts. WPEF 22031 (2022)

26. Cerqueira, E.V.C., Motte-Baumvol, B., Chevallier, L.B., Bonin, O.: Does working from home reduce $CO_2$ emissions? An analysis of travel patterns as dictated by workplaces. Transp. Res. Part D Transp. Environ. **83**, 102338 (2020)
27. Carbon Trust: Homeworking report: an assessment of the impact of teleworking on carbon savings and the longer-term effects on infrastructure services (2021)
28. EcoAct: Homeworking emissions whitepaper, whitepaper-2020 (2020)
29. IEA: Working from home can save energy and reduce emissions. But how much? Commentary, 12 June 2020. International Energy Agency (2020)
30. Badia, A., Langemeyer, J., Codina, X., Gilaber, J., Guilera, N., Vidal, V., Segura, R., Vives, M., Villalba, G.: A take-home message from COVID-19 on urban air pollution reduction through mobility limitations and teleworking. npj Urban Sustain. **1**, 35 (2021)
31. Kylili, A., Afxentiou, N., Georgiou, L., Panteli, C., Morsink-Georgalli, P.-Z., Panayidou, A., Papouis, C., Fokaides, P.A.: The role of remote working in smart cities: lessons learnt from COVID-19 pandemic. Energy Sources Part A Recov. Util. Environ. Eff. (2020)
32. Bachelet, M., Kalkuhl, M., Koch, N.: What if working from home will stick? Distributional and climate impacts for Germany. IZA discussion paper 14642. Institute of Labor Economics (2021)
33. Lachapelle, U., Tanguay, G.A., Neumark-Gaudet, L.: Telecommuting and sustainable travel: reduction of overall travel time, increases in non-motorised travel and congestion relief? Urban Stud. **55**(10), 2226–2244 (2018)
34. Efoui-Hess, M.: Climate crisis: the unsustainable use of online video: the practical case for digital sobriety. The Shift Project (2019). https://theshiftproject.org/wp-content/uploads/2019/07/2019-02.pdf
35. Obringer, R., Rachunok, B., Maia-Silva, D., Arbabzadeh, M., Nateghi, R., Madani, K.: The overlooked environmental footprint of increasing internet use. Resour. Conserv. Recycl. **167**, 105389 (2021)
36. Sovacool, B.K., Hook, A., Martiskainen, M., Baker, L.: The whole systems energy injustice of four European low-carbon transitions. Glob. Environ. Change **58**, 101958 (2019)
37. EEA: Greenhouse Gas Emission Efficiency of Different Transport Modes. European Environment Agency (2022). https://www.eea.europa.eu/data-and-maps/figures/ghgefficiency-of-different-transport
38. Schupak, A.: Is remote working better for the environment? Not necessarily. The Guardian, 2 Aug 2021
39. Aguiléra, A., Pigalle, E.: The future and sustainability of carpooling practices. An identification of research challenges. Sustainability **13**(21), 11824 (2021)
40. Brynjolfsson, E., Horton, J.J., Ozimek, A., Rock, D., Sharma, G., TuYe, H.Y.: COVID-19 and remote work: an early look at US data. Working paper 27344. National Bureau of Economic Research (2020)
41. Elldér, E.: Telework and daily travel: new evidence from Sweden. J. Transp. Geogr. **86**, 102777 (2020)
42. López Soler, J.R., Christidis, P., Vassallo, J.M.: Teleworking and online shopping: socioeconomic factors affecting their impact on transport demand. Sustainability **13**(13), 7211 (2021)
43. Yum, S.: Differences between telecommuters and commuters: the case of the Twin Cities metropolitan area. Transp. Plan. Technol. **44**(3), 303–318 (2021)
44. Clausen, J., Schramm, S.: Klimaschutzpotenziale der Nutzung von Videokonferenzen und Homeoffice: Ergebnisse einer repräsentativen Befragung von Geschäftsreisenden. Borderstep Institut. https://www.borderstep.de/wp-content/uploads/2021/02/AP3-5_Repraesentativbefragung-11-02-2021.pdf (2021)
45. Budnitz, H., Tranos, E., Chapman, L.: Telecommuting and other trips: an English case study. J. Transp. Geogr. **85**, 102713 (2020)
46. Severo, E.A., De Guimarães, J.C.F., Dellarmelin, M.L.: Impact of the COVID-19 pandemic on environmental awareness, sustainable consumption and social responsibility: evidence from generations in Brazil and Portugal. J. Clean. Prod. **286**, 124947 (2021)

47. Pradhan, S., Ghose, D., Shabbiruddin: Present and future impact of COVID-19 in the renewable energy sector: a case study on India. Energy Sources Part A Recov. Util. Environ. Eff. 1–11 (2020)
48. Steffen, B., Egli, F., Pahle, M., Schmidt, T.S.: Navigating the clean energy transition in the COVID-19 crisis. Joule **4**, 1137–1141 (2020)
49. Weitz, N., Carlsen, H., Nilsson, M., Skånberg, K.: Towards systemic and contextual priority setting for implementing the 2030 agenda. Sustain. Sci. **13**, 531–548 (2018)
50. Donthu, N., Gustafsson, A.: Effects of COVID-19 on business and research. J. Bus. Res. **117**, 284–289 (2020)

# Evaluating the Impact of Financial Inclusion on the Effectiveness of the Pradhan Mantri Fasal Bima Yojana

**Ashwitha, K. A. Thanuja, K. Vinodhadevi, and Priyanka Ghosh**

**Abstract** One such program that aims to give farmers in India financial security is the Pradhan Mantri Fasal Bima Yojana (PMFBY). This study explores a number of PMFBY topics, with a focus on the contribution of financial inclusion to the facilitation of direct benefit transfer to farmers. A mixed-methods approach will be used in terms of methodology, including quantitative surveys and qualitative interviews. Farmers, public servants, and other pertinent parties participating in PMFBY administration and implementation shall provide data. While qualitative data will be thematically evaluated to extract nuanced insights, statistical analysis will be carried out to identify patterns and trends. With regard to farmer awareness, enrolment facilities, benefits distribution, managerial consequences, and stakeholder expectations, this research attempts to offer thorough insights into PMFBY.

**Keywords** Pradhan Mantri Fasal Bima Yojana · PMFBY · Financial inclusion · Direct benefit transfer · Farmers · Awareness · Enrolment facilities

## 1 Introduction

Farming in India is heavily dependent on nature, and the changing climate, along with natural phenomena such as rain, drought, flood, and storms, directly impact crop cultivation [1]. Numerous governments globally have introduced the concept of group insurance. In 2016, the Government of India initiated its ambitious program known as the Pradhan Mantri Fasal Bima Yojana (PMFBY). This comprehensive program aims to provide insurance coverage for all crops, farmers, and regions, ensuring that every crop is insured against various losses [2]. The PMFBY is not only a crop

Ashwitha (✉) · K. A. Thanuja · P. Ghosh
Department of Commerce (UG), Kristu Jayanti College (Autonomous), Bengaluru, Karnataka, India
e-mail: ashwitha@kristujayanti.com

K. Vinodhadevi
Department of Economics, Kristu Jayanti College (Autonomous), Bengaluru, Karnataka, India



M. Al Mubarak (ed.), *Sustainable Digital Technology and Ethics in an Ever-Changing Environment*, Studies in Systems, Decision and Control 237,
https://doi.org/10.1007/978-3-031-86708-8_47

insurance scheme but also an integrated IT solution and web-based platform designed to expedite service delivery, consolidate fragmented databases, achieve a unified view of data, and streamline manual processes [3]. Through the PMFBY, insurance services are provided to farmers more efficiently than ever before. Administered by the Ministry of Agriculture and Farmers Welfare, Government of India, the PMFBY represents a significant initiative to enable farmers to insure their crops and address queries and concerns through an online interface (3). This portal also facilitates the dissemination of information regarding various schemes and programs aimed at serving farmers, across different branches of the Government of India.

This paper examines the impact of financial inclusion on the effectiveness of the Pradhan Mantri Fasal Bima Yojana (PMFBY) by analysing how access to financial services influences farmer participation and insurance outcomes. Through a review of existing literature, data analysis, and case studies, the research highlights the correlation between financial inclusion and the success of PMFBY in protecting farmers against crop losses. The findings offer policy recommendations to enhance both financial access and insurance effectiveness in India's agricultural sector [4].

## 2 Implementation Agency (IA)

The Scheme will be carried out through a multi-office structure by chosen insurance agency under the general direction and control of the Department of Agriculture, Cooperation and Farmers Welfare (DAC&FW), Ministry of Agriculture and Farmers Welfare (MoA&FW), Government of India (GOI) and the concerned State in co-appointment with different organizations; viz. Financial Institutions like Commercial Banks, Co-usable Banks, Regional Rural Banks and their administrative bodies, Government Departments viz. Farming, Co-activity, Horticulture, Statistics, Revenue, Information/Science and Technology, Panchayati Raj and so forth. DAC&FW has assigned/empanelled Agriculture Insurance Company of India (AIC) and some private insurance agency by and by to take an interest in the Government supported agribusiness/crop protection plans in light of their monetary strength, foundation, labor and skill and so forth.

The concerned State Government would select the insurance agency to act as an IA from among the empanelled insurance agencies in order to carry out the plan in their State. This kind of IA determination will be made from among the designated/empanelled firms, which will initially be thoroughly pre-qualified based on

- Presence of framework nearby
- Nature of administrations like inclusion of farmers and region
- Pay-outs with regards to quantum and ideal settlement thereof
- Eagerness to do exposure and mindfulness crusades and so forth.

The last choice of IA from among the pre-qualified insurance agency will be done in view of the least weighted premium cited by a pre-qualified organization for all informed harvests inside the group of areas

**The Board of the Scheme**: Appropriate administration of the Scheme will be the responsibility of the State Level Co-appointment Committee on Crop Insurance (SLCCCI), Subcommittee to SLCCCI, District Level Monitoring Committee (DLMC), which was previously in charge of overseeing the implementation and verification of continuous harvest protection plans. IA will be an operational member of the plan's District Level Monitoring Committee (DLMC) and SLCCCI.

**Unit of Insurance**: The Plan will be implemented based on the "Region Approach" tenets, which state that each guaranteed farmer in an insurance unit to be designated as a "Notified Area" for a yield will face comparable risk opportunities, bring about a generally indistinguishable cost of creation per 3 ha, receive equivalent homestead pay per hectare, and suffer a comparative degree of yield misfortune due to the activity of a protected risk, in the advised region. Characterized Areas, also known as unit areas of protection, are at the Village/Village Panchayat level. These regions may be referred to as such for notable yields, or they may be a unit of size larger than the Village/Village Panchayat level for different harvests.

**Farmers to Be Covered**: All farmers developing informed crops in a told region 4 during the season who have insurable interest in the yield are qualified.

**Insurance Coverage**: The following categories of farmers will need to enroll in the plan, subject to possessing an insurable interest in the growth of the informed crop in the informed region.

Throughout the harvest season, farmers in the informed region with a Crop Loan account or KCC account—referred to as Loaned Farmers—have their credit limit allowed or refilled based on the informed yield.

These diverse farmers, whom the government may from time to time decide to include

**Voluntary Coverage**: Voluntary inclusion might be acquired by all farmers not shrouded in 7.1 above, counting Crop KCC/Crop Loan Account holders whose credit limit isn't re-established.

**Risks to Be Covered and Exclusions**: 8.1. Chances: Following dangers prompting crop misfortune are to be covered under the plan.

**Yield Losses (Standing Harvests, on Recommended Region Premise)**: Full gamble protection is provided to cover yield losses due to unavoidable risks, such as (i) Lightning and Natural Fires, (ii) Storms, Hailstorms, Cyclones, Typhoons, Tempests, Hurricanes, Tornadoes, and so forth, (iii) Landslides, floods, and inundations, (iv) Droughts and dry spells and (v) Insects and diseases, and so on [5].

**Forecast in Sowing (On Advised Region Premise)**: In situations where larger part of the guaranteed farmers of an informed region, having goal to plant/plant and caused use for the object, are kept from planting/establishing the safeguarded crop because of unfriendly weather patterns, will be qualified for repayment claims upto a limit of 25% of the total protected.

**Post-harvest Losses (Individual Ranch Premise)**: Coverage is accessible upto a greatest time of 14 days from gathering for those yields which are kept in "cut and spread" condition to dry in the field in the wake of reaping, against explicit risks of tornado/cyclonic downpours, unseasonal downpours all through the country.

**Restricted Calamities (Individual Farm Premise)**: Loss/harm coming about because of event of distinguished restricted gambles for example hailstorm, avalanche, and Inundation influencing segregated ranches in the informed region.

**Exclusions**: Risks and Losses emerging out of following dangers will be rejected: War and fellow hazards, atomic dangers, riots, malevolent harm, burglary, demonstration of hatred, touched and additionally annihilated by homegrown or potentially wild creatures, if there should be an occurrence of Post-Harvest misfortunes the gathered yield packaged and piled at a spot prior to sifting, other preventable dangers.

**Aggregate Insured/Limit of Coverage**: In the event that Loanee farmers fall under the Compulsory Component, the Sum Insured would match the Harvest Scale of Finance set by the District Level Technical Committee (DLTC). This could go up to the guaranteed rancher's preferred maximum amount, which is the edge yield of the crop. The Sum Insured will be greater if the edge yield's value is less than the Scale of Finance. A rise in the Notional Threshold Yield in relation to the current year's Minimum Support Price (MSP) indicates the overall protected value. Any location The MSP for this year is not available; the MSP for previous years will be accepted. The yields for which the ranch door cost is specified by MSP but not declared.

## 3 Risk Management Process

Aris et al., ISSN: 1790-0832

In certain instances, though, the scenario is so unusual that it is challenging to quantify and, hence, evaluate the risk. Three critical variables are information, control, and time in all risky situations.

A common theme in crisis management is the lack of time to evaluate the situation and make a choice. In a crisis management scenario, a lack of control and/or ignorance of the risk are also typically the norm. Risk assessment is intrinsically subjective, even when it appears objective. According to Fischhoff et al. [6], a risk's acceptability is a relative rather than an absolute concept, hence it is impossible to define an exact figure that accurately represents the acceptable risk for a community or a group.

Kahan [7] explained the risks associated with farming has classified the farmers into three categories, risk aversion, risk takers and risk neutral. The risks are classified into five areas, production, marketing, financial, institutional and human. The production totally depends on weather changes. Unexpected rain, flood, cyclone etc., affect productivity and it is the first risk. The collection of crop by the institutions like food corporations and other bodies that collect the crops.

## 4 Review of Literature

Prasad and Mohapatra's [1] work, this literature review explores the implications of climate change on agriculture and sustainable development in India. Climate change exacerbates extreme weather events in India, posing a threat to agricultural productivity and food security. India's agricultural productivity and food security are threatened by global climate change.

Sharma [2] findings suggest that the PMFBY has fallen short of anticipated success. Politics is important since the states foot the bill while the Centre introduces the policy as PMFBY. Opposition-led states are unable to fully adopt the program due to competitive populism. Furthermore, online registration is necessary for the Government of India's Digital India effort. This presents difficulties for the PMFBY, including digital illiteracy, a lack of cellphones, and inconsistent internet connectivity.

Asravor [8], the word 'risk' is a typical and broadly utilized piece of the present jargon, yet to some degree shockingly, there is still no expansive agreement on the importance of this term. Knight was quick to recognize risk from vulnerability. He recognize quantifiable vulnerability and un-quantifiable vulnerability, we might utilize the term 'risk' to assign the previous and the term 'vulnerability' for the last option.

Holton [9] most creators track down a more helpful differentiation between vulnerability as blemished information and hazard as openness to unsure troublesome monetary outcomes. Information on farmers' mentality towards risk is significant in deciding how farmers act for new rural rehearses.

Danso-Abbeam et al. [5] as a result of the study, it is recommended that the design and implementation of a cocoa insurance scheme by Ghana COCOBOD or the government of Ghana should carefully consider the factors influencing farmers' willingness to pay (WTP) for cocoa insurance. Additionally, it is important to ensure that farmers are adequately educated about the cocoa insurance scheme.

Golait and Pradhan [10] expressed that however various elements influence the Indian farming creation as well as efficiency, the fundamental gamble being its extreme reliance on 'climate' which is past the control of people. The yield protection as a gamble the executives instrument has been practically speaking very for quite a while in Indian horticulture in various variations. They referenced that data on weather conditions isn't liberated from cost and, accordingly, protection becomes costly to farmers. However farmers are seen to understand their dangers better than the Government or different establishments, it is amazing that the two farmers and leaders tend to underrate the gamble of horticulture particularly because of unconventionality of the nature's misfortune.

Ashan et al. [11], in this paper, we construct a theory surrounding crop insurance. Initially, we emphasize the crucial function of competitive crop insurance markets in distributing risk. However, we contend that when confronted with the challenges of imperfect information, the existence of a truly competitive crop insurance market may be questionable. We propose two alternative approaches. The first involves market insurance wherein the public sector assumes the role of gathering and disseminating (albeit costly) information. The second approach suggests the direct provision of crop insurance by the public sector.

Vijay and Olivier [12], the paper presents an evaluation of the crop insurance scheme in Karnataka, a state in southern India known for its dry climate. Despite being the second driest state in the country, the analysis reveals several shortcomings in the current scheme regarding its design, implementation, and operational efficiency. These issues result in the scheme failing to fulfill its intended objectives, which include both explicit goals such as risk management and implicit objectives like providing a safety net and containing government liabilities.

## 5 Research Gap

The Government of Karnataka has already introduced a portal called 'Samrakshana' in their official website of Department of Agriculture. The farmers have to register their credentials in the scheme and the claim will be distributed as Direct Benefit Transfer to the farmer's account. The data in Samrakshana itself shows that the registration of farmers in only four districts are noticeable and in other districts, the registration is negligible. The districts where registration is noticeable are, by Kopala, Havery, Gadag, Davangare, Mandya etc. This study analyses the reasons for poor involvement of farmers in crop insurance scheme.

Problem Statement

The awareness on the PMFBY is the prime factor in increasing enrolment. The Karnataka government has its own website for enrolment called 'Samrakshana' and found that the distribution of enrolment. More than 60% of the enrolment is done through common service centres (CSC). This shows that there is a facility utilization gap and awareness gap in the implementation of the project.

## 6 Need of the Study

The success of implementation of any scheme, especially government level, depends on the system involved in it. The system should develop a process to increase both inclusion of farmers in the scheme through awareness development. Hence, this study is essential to find why the enrolment is low in many districts.

Objectives

- To analyse the awareness on PMFBY among farmers
- To analyse the available facilities for the enrolment of farmers in the schemes
- To assess the information on benefits of PMFBY
- To analyse the managerial implications of the scheme
- To analyse the expectations from scheme.

Hypothesis Definition

Hypothesis 1

$H_0$: There is no significant effect of awareness in increasing enrolment of farmers in PMFBY.

Hypothesis 2

$H_0$: There is no significant effect of available sources of for enrolment of farmers in PMFBY.

Hypothesis 3

$H_0$: There is no significant effect of benefits available in PMFBY on farming productivity.

Hypothesis 4

$H_0$: There is no significant effect of benefits given in PMFBY on solving the present issues of farmers.

Scope of Study

The scope of the study is the farmers in districts (name) where the enrolment is very low. The reasons for low reenrolment in Samrakshanam Scheme. This study also analyses the risks faced by the farmers and seek solutions introduced by the Government to make the farmers aware on it. Information asymmetry is a challenge in low enrolment of the scheme. The benefit is transferred as Direct Benefit Transfer system.

Research Methodology

The farmers who participated in this study are from various districts. The selection of responders is based on their willingness to provide personal information. The research shall be primarily exploratory and descriptive in nature, followed by an empirical and analytical survey intending to expand knowledge.

## 7 PMFBY Data Sources

The research is proposed to be based on both Primary data as well as Secondary data.

Secondary data shall be collected from books, journals, periodicals, magazines, newspapers, internet, regulatory authorities, Websites etc.

The research shall attempt to focus on effectiveness of KKSAN, Samrakshana.

Primary data shall be collected through structured questionnaires using Expert opinion poll as Pilot Study.

After modification of Questionnaire, final questionnaire will be framed after improving reliability and validity.

Selection of Respondents

Stratified sampling is employed in response to the respondent selection criteria, which are designed to guarantee sufficient samples for analysis. This usage of non-probabilistic sampling is due to the size of the population. Stratification is done using demographic information. Land area, crop count, production volume, and yearly income and expenses are additional strata.

Data Collection

Reliability of Questionnaire

A pilot study with workers from various industries is planned. Three aspects of the pilot study are analysed: comprehension of the questions, response clarity, and answer consistency. The responses are employed to verify the coherence.

Repeated testing helps to increase the tools' reliability. The dependability and consistency are analysed using kappa and Cronbach alpha.

In factor analysis, the KMO factor is used to analyse the data's validity.

Sampling Size

The population size is not defined in this research as it is high. Hence the Cochran's sample size formula.

$$N_0 - Z^2pq/e^2$$

p is the taken as 50 and 95% confidence level and hence, Z is taken as 1.96. 'e' is the error level and percentage of error is taken as 5%. The sample size is 384 will be from each district selected for survey.

Stratified sampling used in this research. The primary level of strata are, farmers who enrolled in PMFBY and not enrolled in PMFBY.

Data was collected from those who are ready to share the information.

Being the population is large and considering it as infinite, minimum 384 samples to be collected.

Based on response in pilot study, the ratio of farmers of enrolled and not enrolled will be estimated.

The sample size will be increased if the number of cases in any strata is less and inadequate for doing data analysis.

Scale of Measurement

The questions are 'subjective' and hence a qualitative scale is used. To get more precision, 11 point scale is used (0–5 scale).

Statistical Tools

Descriptive statistics is used in this research as the data is from Pilot study and the main intention is response analysis and response orientation.

Table 1 shows that 41.88% of the respondents have land less than 3 acres while 27.23% have land in between 3 and 6 acres. Awareness level of farmers on insurance (15%), Prime Minister Fisala Beema Yojana (19%), Samrakshana (12%), and Local administration Schemes (19%).

Table 2 shows that, No Land (23%), No efforts to develop awareness (29%) and lack of involvement in awareness (21%) have no awareness on government schemes.

**Table 1** Demographic variables

| Categorising variable | Sub variables | Interview method | |
|---|---|---|---|
| | | (Qualitative data) | |
| Gender | Male | 81 | 42.41% |
| | Female | 110 | 57.59% |
| Age in years | < 20 | 37 | 19.37% |
| | 20–40 | 63 | 32.98% |
| | 40–60 | 45 | 23.56% |
| | < 60 | 46 | 24.08% |
| Land | < 3 acres | 80 | 41.88% |
| | 3–6 acres | 52 | 27.23% |
| | 6–9 acres | 39 | 20.42% |
| | 9–12 acres | 12 | 6.28% |
| | < 12 acres | 8 | 4.19% |
| Awareness on government facilities for farmers | Insurance | 29 | 15% |
| | PMFBY | 36 | 19% |
| | Enrolment in Samrakshana | 23 | 12% |
| | Local administration schemes | 36 | 19% |
| Awareness on government facilities for farmers | Not known | 56 | 29% |
| | Awareness | 28 | 15% |
| | Registered | 61 | 32% |
| | Availed | 56 | 29% |

**Table 2** Distribution of respondents who do not know about government schemes

| | Illiteracy (%) | No contact with government offices (%) | Not own land (%) | No efforts to develop awareness (%) | Lack of involvement in awareness programs (%) | Total (%) |
|---|---|---|---|---|---|---|
| Illiteracy | 5 | 4 | 0 | 0 | 5 | 14 |
| No contact with government offices | 0 | 5 | 0 | 7 | 0 | 13 |
| Not own land | 0 | 0 | 23 | 0 | 0 | 23 |
| No efforts to develop awareness | 7 | 4 | 0 | 11 | 7 | 29 |
| Lack of involvement in awareness programs | 2 | 0 | 0 | 11 | 9 | 21 |
| Total | 14 | 13 | 23 | 29 | 21 | 100 |

In Table 3, the rural areas, common service centres acts as the main information point and then TV and Radios. Agriculture office also play an important role in developing awareness in farmers.

Future of Research

The study was implemented by the ITC in UP with both input and output benefits to the farmers. The research can be extended to the possibility to include output end in the model. The output means, how to find market for the farmer's products at a genuine price.

**Table 3** Awareness on government projects

| | From agriculture office (%) | NGOs (%) | Common service centre (%) | TVs and radio (%) | Other farmers (%) | Total (%) |
|---|---|---|---|---|---|---|
| Agriculture office | 0 | 7 | 4 | 0 | 7 | 18 |
| NGOs | 0 | 4 | 0 | 7 | 4 | 14 |
| Common service centre | 0 | 0 | 29 | 0 | 0 | 29 |
| TV and video | 4 | 7 | 0 | 11 | 4 | 25 |
| Other farmers | 4 | 0 | 0 | 7 | 4 | 14 |
| Total | 7 | 18 | 32 | 25 | 18 | 100 |

Limitations

The limitation of this study to find the farmers who real avail benefit of DBT to compare the perception of farmers who availed the DBT and non-registered. The geographical coverage and shrinking farming process are other limitations.

## 8 Conclusion

The paper analysed the distribution of information on facilities and benefits offered to farmers and how it reaches the farmers. Information asymmetry is the main reason for the poor registration in PMFBY or Samrakshana. Illiteracy, non-access of farmers to agriculture office, inadequate interaction by the Government employees with farmers in rural areas are a few challenges in developing awareness. Table 1 shows that the knowledge on different schemes are less and only 61% respondents registered and availed the benefits. The registration is permitted for the head of the family and others have to move away from the parental house and should have ration card and forms to avail benefits. It is a barriers for those dependent farmers. Main source of information is common service centres and the radio and TVs. Even the study done by Bhushan and Kumar [3] concluded the same thing in their particular study.

## References

1. Prasad, R.S., Mohapatra, T.: Climate Change and Indian Agriculture: Impacts, Coping Strategies, Programmes and Policy. Indian Council of Agricultural Research (2019)
2. Sharma, S.: A critical appraisal of Pradhan Mantri Fasal Bima Yojana. Int. J. Soc. Sci. Arts Humanit. (IJSSAH) (2021)
3. Bhushan, C., Kumar, V.: Pradhan Mantri Fasal Bima Yojana: An Assessment. Centre for Science and Environment, New Delhi (2017)
4. Ellis, F.: The determinants of rural livelihood diversification in developing countries. J. Agric. Econ. **51**(2), 289–302 (2000)
5. Danso-Abbeam, G., Addai, K.N., Ehiakpor, D.: Willingness to pay for farm insurance by smallholder cocoa farmers in Ghana. J. Soc. Sci. Policy Implicat. **2**(1), 163–183 (2014)
6. Slovic, P., Fischhoff, B., Lichtenstein, S.: Perceived risk: psychological factors and social implications. Proc. R. Soc. Lond. A Math. Phys. Sci. **376**(1764), 17–34 (1981)
7. Kahan, D.: Managing Risk in Farming. Food and Agriculture Organization of the United Nations, Rome (2008)
8. Asravor, R.K.: Farmers' risk preference and the adoption of risk management strategies in Northern Ghana. J. Environ. Plann. Manage. **62**(5), 881–900 (2019)
9. Holton, G.A.: Defining risk. Financ. Anal. J. **60**(6), 19–25 (2004)
10. Golait, R.B., Pradhan, N.C.: Relevance of weather insurance in Indian agriculture. CAB Calling 37–41 (2008)
11. Ashan, Syed, M., Ali, A.A.G., Kurian, N.J.: Towards a theory of agricultural crop insurance. Am. J. Agric. Econ. **64**(3), 520–529 (1982)
12. Vijay, K., Olivier, M.: Crop insurance in Karnataka. Policy Research Working Paper Series (3654) (2005)

13. Jodha, N.S.: Role of credit in farmers' adjustment against risk in arid and semi-arid tropical areas of India. Economic and Political Weekly, pp. 1696–1709 (1981)
14. European Commission: Risk Management Tools for EU Agriculture with a Special Focus on Insurance. Working Document. Accessed 20 Mar 2018. http://ec.europa.eu/agriculture/publi/insurance/text_en.pdf (2001)

# Harnessing the Potential of Artificial Intelligence: Transforming Industries and Workforce Dynamics

**Bijin Philip, Geethu Anna Mathew, Jinu Mathew, Roshen Therese Sebastian, Mary Jacob, and M. Jeeva**

**Abstract** The world is evolving significantly due to artificial intelligence (AI). Numerous industries, economies, and our way of life are all changing as a result. This study investigates the effects of AI. We'll discuss its impact on the economy, how it affects employment, and the policies that are required. We gathered information from a variety of sources, including research conducted by investment banks and reports from the McKinsey Global Institute. Our objective is to examine how AI may boost economic growth while also examining how it impacts employment and causes disparities in some sectors. Our discovery is fascinating! The GDP might increase significantly with AI. To put it another way, it has the potential to significantly expand economies. There is a drawback, though. Many individuals are concerned about employment losses and rising inequality, particularly in certain places that don't have good infrastructure or enough skilled workers. Another factor is that generative AI is already available! It presents fresh possibilities as well as new challenges. Both job functions and employee work styles are evolving. It is critical that businesses, workers, and politicians collaborate to manage all of this change. They can work together to develop an ethical AI strategy. This entails preparing individuals for the workforce of the future by teaching them new skills and considering ethical applications of AI. Great things can happen if we foster an environment that is conducive to using AI! We can ensure that everyone wins, encourage creativity, and increase our team's productivity. This study emphasises the significance of adopting proactive policies and ethical guidelines. We can reap the benefits of the AI revolution while

B. Philip (✉) · G. A. Mathew · R. T. Sebastian
Department of Management, Kristu Jayanti College Autonomous, Bengaluru, India
e-mail: bijin.p@kristujayanti.com

J. Mathew
Department of Professional Accounting and Finance, Kristu Jayanti College Autonomous, Bengaluru, India

M. Jacob
Department of Computer Science, Kristu Jayanti College Autonomous, Bengaluru, India

M. Jeeva
Department of English, Kristu Jayanti College Autonomous, Bengaluru, India

 
M. Al Mubarak (ed.), *Sustainable Digital Technology and Ethics in an Ever-Changing Environment*, Studies in Systems, Decision and Control 237,
https://doi.org/10.1007/978-3-031-86708-8_48

also being aware of its potential drawbacks if we manage it well. In this new AI era, we can work towards a better and brighter future for everyone by addressing the socioeconomic challenges related to AI use.

**Keywords** Artificial intelligence (AI) · Job displacement · Upskilling and reskilling · Workforce dynamics · Economic impact and responsible AI adoption

## 1 Introduction

Artificial Intelligence, or AI, is changing everything! It's really something amazing. You see, AI can do jobs that we usually think only humans can handle. It's smart software that analyzes a ton of data to give us great insights. Because of this, we are entering a whole new era of working smarter and being more creative. Just think about it—self-driving cars, helpful virtual assistants, diagnosing health problems, and even analyzing finances. The ways we use AI are huge! It's set to change just about every part of our lives.

The job market feels AI's impact too. Many simple tasks can now be done automatically. This helps people make better decisions and create new job options! According to the folks at the McKinsey Global Institute, by 2030, we might see an extra $13 trillion in the global economy thanks to AI—a big jump of around 16% in the total GDP! But with all these benefits come some challenges. We need to think about retraining workers and how to deal with job changes.

Goldman Sachs has more thoughts on this. They predict that up to 300 million full-time jobs worldwide could be at risk due to AI taking over tasks. Yes, this might really boost productivity and get the economy moving. But it also means many people may have to change their jobs. In fact, about two-thirds of jobs in places like the U.S. and Europe could see some part affected by AI automation. This raises important questions about the future of work.

And let's not forget about other parts of the world! Emerging markets and poorer countries will also feel these changes. They may not face as many disruptions right away, but without strong infrastructure or enough skilled workers, they could miss out on the benefits of AI. This might widen the gap between rich and poor countries.

This research aims to look closely at how AI influences jobs around the world. We want to find out what trends are happening and what challenges lie ahead. By exploring technology, economics, and society together, we hope to guide smarter policymaking that takes advantage of all the good things about AI while reducing its negative effects.

## 2 Theoretical Outline and Literature Indications

The advent of AI technologies is poised to reshape the job landscape in the foreseeable future by automating various job tasks. AI is anticipated to replace both routine and nonroutine tasks, leading to a transformation in how humans and robots collaborate, potentially in unprecedented ways. This shift is expected to necessitate changes in the skills required for employability. Given the significant impact on jobs, reports from consulted experts emphasize the need for collaborative solutions that extend beyond the confines of individual organizations and industries to involve relevant stakeholders. Furthermore, organizations will need to reconsider their human resource (HR) function, ensuring it is adeptly aligned with the realities of AI integration [1]. The progress in artificial intelligence (AI) has sparked renewed discussions regarding its implications for the future of work, prompting concerns over potential widespread job displacement. This presentation will delve into descriptive findings that explore the relationship between the implementation of AI and its effects on job creation and elimination within organizations [2]. The rise of artificial intelligence (AI) has profoundly influenced both management practices and society at large. This paper endeavors to investigate its potential ramifications on employees in the hospitality industry, providing insights for both employees and managers. The findings reveal a positive correlation between awareness of AI and job burnout. However, no significant direct correlation was observed between AI awareness and career competencies [3]. Discovering the importance of collective models between humans and AI, this research highlights successful instances where AI systems bolster human productivity and decision-making capabilities. It also ventures into the future trajectory of AI, discussing emerging technologies and anticipating both challenges and opportunities on the horizon [4]. The way AI affects workers is by fostering a sense of pessimism about the future, which ultimately leads to burnout and diminishes their motivation for learning on the job. Furthermore, AI's effects such as replacement, mismatch, and deskilling contribute to a decrease in people's income while also extending their working hours. This, in turn, reduces their available financial resources and time for additional learning endeavors. Research indicates that AI's impact on on-the-job learning is particularly noticeable among older individuals, females, and those with lower levels of education. Additionally, it affects those without labor contracts and with limited job autonomy and work experience. In regions where there is heightened competition between humans and AI, as well as more labor-management conflicts and inadequate labor protection, the inhibitory effect of AI on further learning is even more pronounced [5]. Although automation can result in job displacement, the key to the future of work is the collaboration between humans and AI systems. Through upskilling and reskilling efforts, individuals can obtain the essential competencies needed to excel in the AI-driven era. Governments and businesses must invest in policies and initiatives that facilitate a seamless transition, generate fresh job prospects, and guarantee inclusivity within the evolving job landscape. By approaching the future of work with foresight and adaptability, we can leverage the complete potential of AI for the betterment of society [4].

AI satisfaction with service quality acted as a mediator in the relationship between AI service quality and employees' job satisfaction. Additionally, job level served as a moderator in the impact of AI service quality on both AI satisfaction and job satisfaction. Specifically, AI service quality influenced AI satisfaction solely within the non-supervisory group, while it affected job satisfaction in both non-supervisory and supervisor/managerial groups. Furthermore, AI service quality impacted AI satisfaction across job roles (frontline and back-end employee roles), but its influence on job satisfaction was observed only within the back-end employee role [6].

## 3 Objectives

- To investigate the multifaceted impact of Artificial Intelligence (AI) adoption on industries, economies, and societal structures, with a focus on assessing its economic implications, workforce dynamics, and policy considerations.
- To assess the influence of artificial intelligence (AI) on diverse job roles within identified prospective companies.
- To evaluate which sectors are most and least affected by the implementation of AI.

## 4 Results and Discussion

For workers, this change means there are some good things but also some tough ones. Some people might get to take new programs. This will help them learn to do AI-related tasks. But for others, it could mean layoffs or even losing their jobs since their work can be done by machines now.

Start-ups, especially under pressure from investors, really want to make money fast. This focus on profit can make things harder for workers. It leads to tricky situations about how to treat employees well.

It's super important companies to be open and honest while adding AI into how they work. For example, when some staff at Grapevine shared their thoughts about Saarthi.ai, it showed just how vital clear talk and good decisions are in businesses using AI.

Also, there's a big need for skilled workers in AI these days. Look at Netflix—they're hiring lots of professionals! This shows just how important it is for everyone to keep learning and growing their skills. That way, we can help those who might lose jobs find new opportunities in this changing world.

**SAP**: SAP is investing $2 billion in AI and changing up 8000 roles. This shows big move towards automation and digital change. Some employees might learn new skills for AI tasks. But, there are others who could lose their jobs. It's a strategic choice by SAP that highlights how AI can affect traditional jobs in the company. Sadly, some roles might be seen as unnecessary.

**Google**: Google is letting some of its ad sales team go while pushing forward with AI tech. This shows how they want to use automation to work more efficiently. These layoffs aren't all about AI, but they do suggest that jobs might disappear because of it. As Google aims to stay strong in the digital world, more automation could affect many workers.

**Salesforce, Duolingo and Humane**: Recently, Salesforce, Duolingo, and Humane have made cuts and put hiring on hold. This trend points toward using AI to make things run smoother and boost productivity. Still, there are worries about job losses in these companies. As they focus on bringing in more AI tools, layoffs or fewer new hires in other areas might happen, leading to unemployment for those impacted.

**Dukaan**: Dukaan has decided to replace 90% of its staff with an AI chatbot! That really raises worries about job security and what all this means for workers. The CEO says it's all about making money, but there are real concerns about what this means for jobs down the line. Many workers losing their positions reflects how disruptive AI can be.

**Dunzo**: Dunzo has delayed employee salaries and made layoffs while changing their business plans. It shows how hard it is to keep up with new tech changes. As innovations push companies like Dunzo to adjust, some may see job cuts or adjustments that lead to unemployment among their staff.

**Netflix**: Netflix may not have directly laid off anyone due to AI yet, but they're offering high-paying jobs for skilled AI engineers. This shows there's a rising need for people with AI knowledge! Still, there's a shortage of qualified candidates which creates challenges for businesses looking toward an AI future. While these roles create chances for skilled workers, they highlight a growing gap and the risk of job loss for those without expertise in AI.

**Policy Considerations**

- Regulatory Frameworks: Analyze existing and proposed regulations governing AI adoption, including measures to ensure safety, fairness, and responsible use.
- Investment and Innovation Policies: Evaluate government initiatives to support AI research, development, and deployment, including funding mechanisms, tax incentives, and intellectual property rights.
- Education and Workforce Policies: Assess policies aimed at promoting AI education, training, and workforce development, as well as initiatives to address the social impacts of AI-driven job displacement.

To thoroughly investigate these dimensions, interdisciplinary research methods incorporating economic analysis, sociological studies, policy evaluations, and technological assessments would be necessary. Additionally, case studies from diverse industries and regions can provide valuable insights into the real-world implications of AI adoption.

The nature of work will change as generative AI becomes more widely used, with tools replacing some activities and changing typical work profiles across a range of industries. Nonetheless, a lot of vocations are still "uniquely human"; the effects of AI will vary depending on the industry. The power of generative AI to find patterns in massive data sets and produce original content is "an ability that has often been considered uniquely human," according to consultancy firm McKinsey. GenAI can "write code, design products, create marketing content and strategies, streamline operations, analyze legal documents, provide customer service via chatbots, and even accelerate scientific discovery" thanks to its creative abilities.

**AI Job Loss Statistics: How GenAI May Affect Sectors**

Hiring Lab determined thematic skill sets that assist in defining the professions that are most and least exposed to possible change brought on by AI by examining the number of skills stated in different job advertisements.

Figures 1 and 2 highlights careers most likely to be affected by AI and High Demand Jobs Least Exposed to GenAI Change respectively.

These abilities are badly scored by AI competency; therefore, job security may be secure for some time to come:

- Operating
- Well-being and beauty
- Personal hygiene and at-home medical care
- Nursing
- Child Support
- Food preparation and serving
- Veterinary Architecture.

Similarly, McKinsey projects that by 2030, there will be a rise in the labor need for STEM (science, technology, engineering, and math) experts as well as health

**Fig. 1** The careers most likely to be affected by AI. *Source* Techopedia

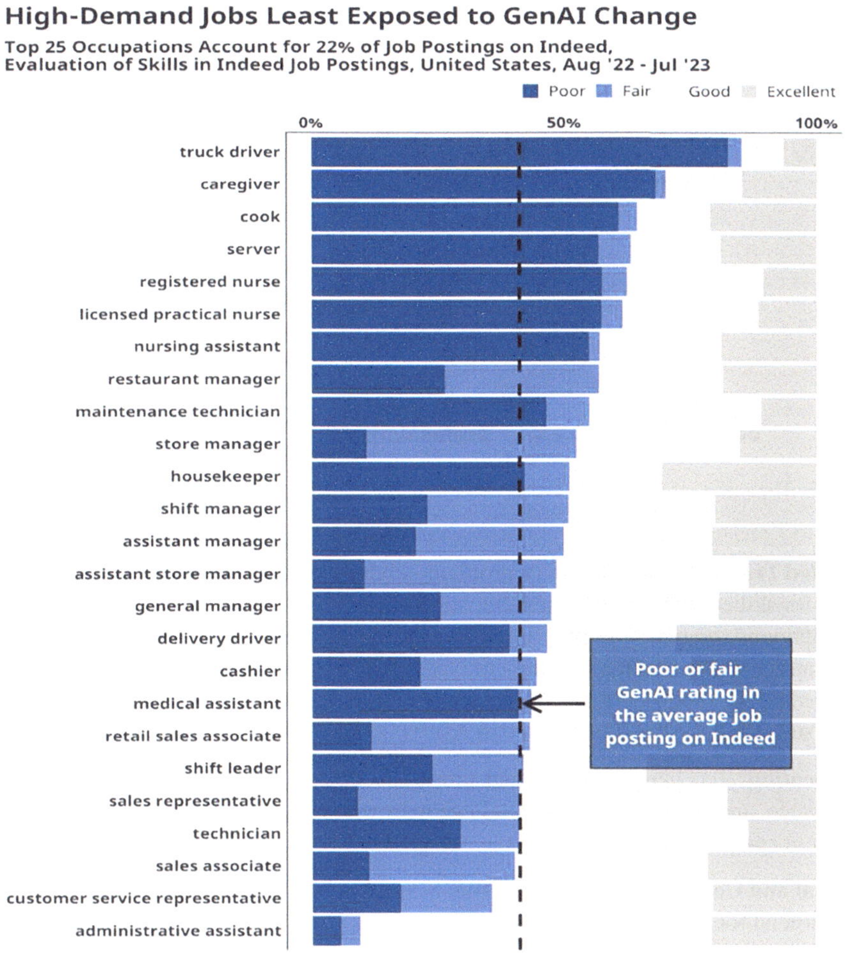


**Fig. 2** High demand jobs least exposed to GenAI change. *Source* Indeed

professionals, assistants, and technicians due to the adoption of automation made possible by the acceleration of generative AI. The likelihood of a job loss due to AI increases with the likelihood that a job can be completed remotely, nevertheless, as the global labor market continues to adapt to the expansion of remote working.

Figure 3 highlights the Careers least likely to be affected by AI.

**Healthcare**: Artificial intelligence (AI) has a lot of potential in the field of healthcare, but it is not likely to take the position of medical experts. It can help with drug discovery, diagnosis, therapy, and remote surgery, among other things. Due to their strong emphasis on interpersonal skills and emotional intelligence, many jobs held by doctors, nurses, therapists, and counselors are resistant to automation.

**Fig. 3** The careers least likely to be affected by AI. *Source* Techopedia

**Skilled Trades**: Manual dexterity, problem-solving abilities, and flexibility are necessary for skilled tradespeople including mechanics, electricians, and plumbers. These occupations frequently require labor in erratic settings, which AI finds difficult to operate in. Furthermore, talented craftspeople like glassblowers, blacksmiths, and carpenters frequently create one-of-a-kind, personalized products that are unlikely to result in employment losses due to AI.

**Education**: Teachers, coaches, and other educators offer invaluable mentorship, guidance, and emotional support. AI can help with the development of educational content and provide tailored learning and grading.

**Social and Community Services**: Human connections, empathy, and understanding are crucial for careers in social work, community outreach, and advocacy.

Overall, as AI continues to permeate various industries, companies need to consider the broader societal implications and prioritize ethical considerations while navigating the evolving tech landscape.

## 5 Conclusion

The current AI revolution is reinventing customer experiences, improving operational efficiencies, and drastically changing traditional businesses. A few useful suggestions stand out for organisations hoping to successfully negotiate this change. First and first, it is imperative to invest in the training and development of employees, as the human element is still critical to the effective use of AI technologies. Businesses like Siemens have upskilled their employees through continuous learning efforts, guaranteeing alignment with AI projects. Furthermore, the creation of cross-functional teams can facilitate cooperation and creative problem-solving. Ford is one example of this, having included artificial intelligence (AI) into its supply chain management, which led to a 25% cost reduction. Finally, before implementing AI applications widely, companies should think about running pilot programs to evaluate them on a smaller scale. This will enable iterative learning and adaptation. Traditional sectors can prosper in the AI era by adopting these tactics, opening up new development opportunities and preserving their competitive edge.

## References


1. Poba-Nzaou, P., Galani, M., Uwizeyemungu, S., Ceric, A.: The impacts of artificial intelligence (AI) on jobs: an industry perspective. Strateg. HR Rev. **20**(2), 60–65 (2021)
2. Hunt, W., Sarkar, S., Warhurst, C.: Measuring the impact of AI on jobs at the organization level: lessons from a survey of UK business leaders. Res. Policy **51**(2), 104425 (2022)
3. Kong, H., Yuan, Y., Baruch, Y., Bu, N., Jiang, X., Wang, K.: Influences of artificial intelligence (AI) awareness on career competency and job burnout. Int. J. Contemp. Hosp. Manag. **33**(2), 717–734 (2021). https://doi.org/10.1108/IJCHM-07-2020-0789
4. Jumaev, G.: The impact of AI on job market: adapting to the future of work. Mod. Sci. Res. **3**(1) (2024)
5. Li, C., Zhang, Y., Niu, X., Chen, F., Zhou, H.: Does artificial intelligence promote or inhibit on-the-job learning? Human reactions to AI at work. Systems **11**(3), 114 (2023)
6. Nguyen, T.M., Malik, A.: A two-wave cross-lagged study on AI service quality: the moderating effects of the job level and job role. Br. J. Manag. **33**(3), 1221–1237 (2022)

# Navigating Corporate Governance's Role in Health Insurance Performance: A Bibliometric Approach

**N. Rakesh and Vishal Srivastava**

**Abstract** This study delves into the intricate relationship between corporate governance and performance within the health insurance sector spanning from 2019 to 2024. Employing bibliometric analysis via the Dimension database, it scrutinizes publication trends, citation behaviors, authorship patterns, and global cooperation dynamics. The aim is to discern the evolution of this field and pinpoint seminal works and influential contributors. Results illuminate a dynamic landscape marked by a waning growth in publications yet enduring impact, underscoring continual scholarly contributions and international partnerships. Key revelations include insights into prominent sources, authors, and globally referenced documents, alongside thematic portrayals spotlighting pivotal topics and stakeholder interrelations. This study accentuates the enduring relevance of corporate governance in shaping financial performance within health insurance and lays groundwork for forthcoming research endeavours in this critical arena.

**Keywords** Corporate governance · Performance · Health insurance companies · Bibliometrics · R · Biblioshiny

## 1 Introduction

The exploration of Corporate Governance has become a compelling subject for both scholars and professionals in recent times. Issues associated with corporate governance are not novel, having arisen concurrently with the establishment of corporations. The significant attention directed towards corporate governance is largely influenced by the substantial bankruptcies of Enron and WorldCom in the early 2000s.

N. Rakesh (✉) · V. Srivastava
School of Commerce, Jain (Deemed to be University), Bengaluru, Karnataka, India
e-mail: sairakeshnrao@gmail.com

V. Srivastava
e-mail: s.vishal@jainuniversity.ac.in



M. Al Mubarak (ed.), *Sustainable Digital Technology and Ethics in an Ever-Changing Environment*, Studies in Systems, Decision and Control 237,
https://doi.org/10.1007/978-3-031-86708-8_49

Corporate governance (CG) comprises a collection of practices, procedures, and policies that function as directives for the management and oversight of a company. The significance of this subject has surged in popularity across both developed and developing nations, primarily driven by the prolonged failures of major corporations on a global scale [1]. Highlight that corporate governance has become a focal point of discourse in response to numerous corporate breakdown. Consequently, nearly every country has established corporate governance institutions, delineated optimal practices, and endeavoured to enforce suitable governance frameworks [2].

Corporate Governance is described as 'the processes and structures utilized to guide and oversee business and corporate affairs, with the overarching objective of enhancing business prosperity and corporate accountability.' From an economic standpoint, corporate governance assumes a pivotal role in achieving efficiency by channelling scarce funds toward investment projects with the highest returns. Moreover, it serves as a crucial determinant of institutional investment [3] and is integral to initiatives aimed at reforming institutional investors [4]. Corporate governance mechanisms fall into two categories: internal mechanisms (such as board size and independence) and external mechanisms (including competitive market conditions, markets for managerial labor and talent, and markets for corporate control). Research on corporate governance has predominantly concentrated on countries like the United States and the United Kingdom. Recognized as indispensable, corporate governance plays a pivotal role in shaping corporate performance and fostering enduring macroeconomic growth. The significance of wealth is underscored in a literature review, emphasizing concerns related to conflicts of interest between shareholders and management [5]. Numerous scholars contend that corporate governance holds a crucial position in overseeing the operations of a firm [6, 7]. Nevertheless, earlier investigations have yielded diverse results regarding the correlation between corporate governance practices and firm performances.

The core objective of effective corporate governance is to fundamentally safeguard shareholder value by ensuring the prudent utilization of company resources, facilitating access to capital, and bolstering investor confidence [8]. This pertains to both internal organizational aspects and external market dynamics. The company's adaptability to external conditions relies significantly on its management practices and the efficacy of its corporate governance framework [9, 10]. Certain scholars posit that robust corporate governance acts as a safeguard against managerial misuse of company resources, fostering sound decision-making and efficient management [11]. This, in turn, leads to a more effective allocation of company resources and, ultimately, an enhancement in overall performance.

The paper consists of Introduction, underscores the importance of corporate governance, particularly within the health insurance sector, and lays the groundwork for the research. The Research Questions and Objectives delineate the study's focus on trends, collaboration, and influential publications. The Methodology section outlines the bibliometric analysis performed using the Dimensions database, detailing the data collection process and the tools employed. The Results section presents the key findings, including publication trends, leading sources, and prominent authors, supported by tables and figures. The Conclusion then summarizes the study's implications, highlighting the enduring significance of corporate governance in the context of health insurance.

## 2 Research Questions

1. What are the trends in scholarly publications concerning corporate governance and financial performance in health insurance from 2019 to 2024?
2. How does collaboration, both domestically and internationally, impact research in the field of corporate governance and financial performance in health insurance?
3. What are the most cited and influential papers in the realm of corporate governance and financial performance within health insurance companies?
4. How has the scholarly landscape evolved over time in terms of publication output, citation impact, and international collaboration?

## 3 Research Objective

1. To examine the bibliometric characteristics of scholarly publications related to corporate governance and financial performance in health insurance, including publication trends, citation patterns, and authorship dynamics.
2. To assess the extent and impact of collaboration among researchers in the field, both domestically and internationally, and its implications for knowledge production.
3. To evaluate the most cited and influential papers in the domain of corporate governance and financial performance within health insurance companies, highlighting notable contributions to the literature.
4. To analyse the evolution of the scholarly landscape over time, including changes in publication output, citation impact, and international collaboration, to understand the field's development.

## 4 Methodology

This study utilized data from the Dimension database as of February 08, 2024, chose for its extensive coverage in research. A bibliometric analysis was performed, focusing on documents related to corporate governance, financial performance, and health insurance based on their titles. The dataset spans five years (2019–2024), providing a contemporary snapshot of the subject matter. Out of 358 documents, 204 sources contributed, indicating a diverse literature pool. Predominantly, articles (275) formed the bulk, supplemented by chapters (3), edited books (77), and monographs

(3). Analysis was conducted using R software, specifically with the "Biblioshiny" extension in R-studio, enabling comprehensive examination of various parameters such as access type, year, authorship, subject field, document type, and geographic distribution, among others.

Table 1 specifies the impact of the research, measured by mean total citations (TC) per article and mean TC per year, varied from 2019 to 2024. In 2019, it was substantial with a mean TC per article of 2.87 over six years. The peak was in 2020, sustaining over five years. However, 2021 saw a decline, and in 2022, both metrics decreased further over three years. Subsequently, 2023 and 2024 witnessed significant drops, indicating a declining trend overall.

Table 2 displays the local impact of the top 10 sources in corporate governance and financial performance in health insurance. Notably, journals such as Sustainability and Journal of Corporate Finance show strong citation metrics and consistent impact, while Lecture Notes in Networks and Systems has a robust publication history. Despite variations, each source contributes significantly to the field, offering valuable insights and aiding knowledge dissemination. These findings highlight the diversity and richness of scholarly contributions in understanding corporate governance and financial performance dynamics in health insurance.

Table 3 depicts Authors' bibliometric profiles vary significantly. Kweh and Ting share similar attributes, with each having an h-index, g-index, and m-index of 3. Both have three publications and garnered 16 citations since 2021. Alhassan AL has an h-index of 2, g-index of 2, and m-index of 0.5, with 13 citations from 2 publications since 2021. Banker RD, starting in 2019, accumulated 30 citations from 2 publications. Bhattacharyya SS, starting in 2021, earned 58 citations from 3 publications. DE Zoysa A, initiating in 2019, accrued 20 citations from 2 publications. Gupta A, starting in 2023, gathered 4 citations from 2 publications. Huang J, Konstantios D, and Li W exhibit varied profiles, reflecting diverse contributions within the field.

Table 4 presents key citation metrics for various papers across journals. Barnett (2020) paper in the Journal of Management garnered 178 total citations with a high TC per Year of 35.6. Ahmed (2020) work in Corporate Social Responsibility and Environmental Management received 141 citations, displaying a TC per Year of 28.2. Renneboog (2019) paper in the Journal of Corporate Finance accumulated 138 total

**Table 1** Data showing average citation per year

| Year | Mean TC per Art | Number | Mean TC per Year | Citable Years |
|---|---|---|---|---|
| 2019 | 17.21 | 42 | 2.87 | 6 |
| 2020 | 20.65 | 55 | 4.13 | 5 |
| 2021 | 12.01 | 85 | 3 | 4 |
| 2022 | 6.14 | 74 | 2.05 | 3 |
| 2023 | 1.2 | 94 | 0.6 | 2 |
| 2024 | 0.12 | 8 | 0.12 | 1 |

*Source* Biblioshiny output

**Table 2** Top 10 sources of local impact

| Element | h_index | g_index | m_index | TC | NP | PY_start |
|---|---|---|---|---|---|---|
| Sustainability | 7 | 12 | 1.4 | 176 | 26 | 2020 |
| Journal of Corporate Finance | 5 | 6 | 0.833 | 205 | 6 | 2019 |
| Lecture Notes in Networks and Systems | 5 | 11 | 0.833 | 128 | 18 | 2019 |
| Journal of Cleaner Production | 4 | 5 | 0.667 | 47 | 5 | 2019 |
| The Geneva Papers on Risk and Insurance—Issues and Practice | 4 | 6 | 1 | 37 | 6 | 2021 |
| Contributions to Management Science | 3 | 8 | 0.5 | 68 | 14 | 2019 |
| Journal of Business Research | 3 | 3 | 0.5 | 151 | 3 | 2019 |
| Journal of Intellectual Capital | 3 | 3 | 0.5 | 72 | 3 | 2019 |
| Springer Proceedings in Business and Economics | 3 | 4 | 0.5 | 22 | 19 | 2019 |
| Studies in Computational Intelligence | 3 | 3 | 0.75 | 61 | 3 | 2021 |

*Source* Biblioshiny output

**Table 3** List of top 10 authors

| Element | h_index | g_index | m_index | TC | NP | PY_start |
|---|---|---|---|---|---|---|
| Kweh | 3 | 3 | 0.75 | 16 | 3 | 2021 |
| Ting | 3 | 3 | 0.75 | 16 | 3 | 2021 |
| Alhassan | 2 | 2 | 0.5 | 13 | 2 | 2021 |
| Banker | 2 | 2 | 0.333 | 30 | 2 | 2019 |
| Bhattacharyya | 2 | 3 | 0.5 | 58 | 3 | 2021 |
| de Zoysa | 2 | 2 | 0.333 | 20 | 2 | 2019 |
| Gupta | 2 | 2 | 1 | 4 | 2 | 2023 |
| Huang | 2 | 2 | 0.333 | 26 | 2 | 2019 |
| Konstantios | 2 | 2 | 0.667 | 43 | 2 | 2022 |
| Li | 2 | 2 | 0.333 | 19 | 2 | 2019 |

*Source* Biblioshiny output

citations with a TC per Year of 23. Flammer (2021) paper in the Strategic Management Journal and Sood (2022) in Contemporary Studies in Economic and Financial Analysis also showed significant impact with 77 and 64 citations, respectively.

Figure 1 indicates a notable rise in scientific output over three years. Articles surged from 25 in 2019 to 50 in 2021, further increasing to 75 in 2023, marking a remarkable 200% growth. The period from 2019 to 2021 observed a 100% increase, while 2021 to 2023 saw a 50% rise, suggesting potential acceleration in growth. However, drawing specific conclusions necessitates additional context, such as the field or domain of study and data sources. Without such information, the analysis is confined to numerical changes in scientific production, lacking a comprehensive understanding of underlying trends.

Figure 2 data illustrate a significant surge in publications across all six sources in 2023 compared to the previous four years. Contributions to Management Science led with 20,191 publications, while The Geneva Papers on Risk and Insurance Issues and Practice had only one in 2023. Overall, total publications increased from 12,439 in 2019 to 25,240 in 2023, marking a remarkable 102.9% increase over five years.

**Table 4** Most globally cited documents

| Paper | Year | Journal/Source | Total citations | TC per year | Normalized TC |
|---|---|---|---|---|---|
| Barnett | 2020 | Journal of Management | 178 | 35.6 | 8.62 |
| Ahmed | 2020 | Corporate Social Responsibility and Environmental Management | 141 | 28.2 | 6.83 |
| Renneboog | 2019 | Journal of Corporate Finance | 138 | 23 | 8.02 |
| Kaffash | 2020 | European Journal of Operational Research | 91 | 18.2 | 4.41 |
| Harris | 2019 | Journal of Business Research | 88 | 14.67 | 5.11 |
| Flammer | 2021 | Strategic Management Journal | 77 | 19.25 | 6.41 |
| Sood | 2022 | Contemporary Studies in Economic and Financial Analysis | 64 | 21.33 | 10.43 |
| Visconti | 2020 | International Journal of Environmental Research and Public Health | 63 | 12.6 | 3.05 |
| Kim | 2020 | Strategic Management Journal | 59 | 11.8 | 2.86 |
| Singhania | 2021 | Journal of Sustainable Finance and Investment | 54 | 13.5 | 4.5 |

*Source* Biblioshiny output

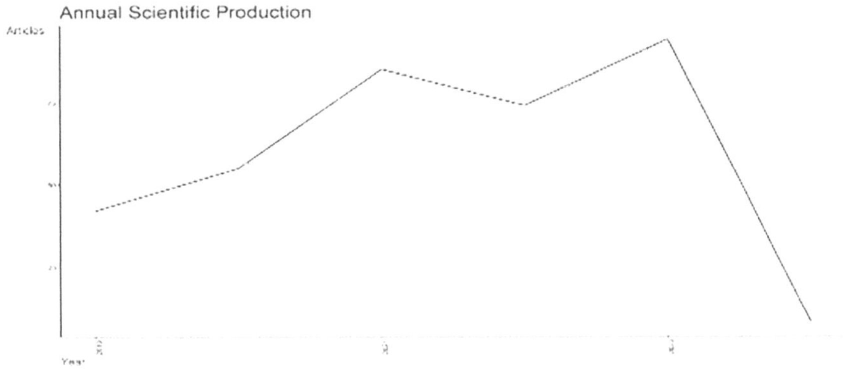


**Fig. 1** Annual scientific production. *Source* Authors compilation using R Studio: Biblioshiny library

Factors such as heightened research demand, expanded researcher presence, and increased funding may have contributed. However, the lack of insights into publication quality or impact highlights the need for a comprehensive evaluation of their significance.

Figure 3 signifies the past five years, the publication output in health insurance corporate governance and financial performance has steadily decreased by an annual growth rate of – 28.23%. Factors such as reduced research funding, evolving priorities, or topic saturation may contribute to this trend. Figure 4 despite the decline, the field remains active, with an average document age of 2.59 years and high average citations of 9.631 per document. Collaboration among 870 authors is increasing,

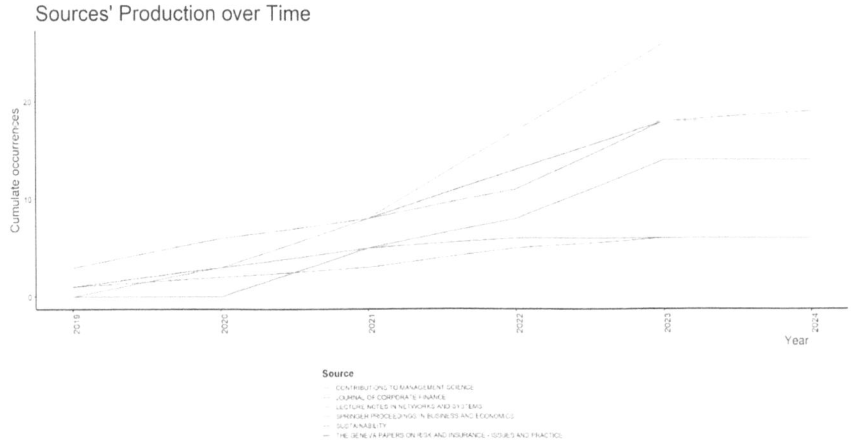


**Fig. 2** Sources' production over the time of corporate governance in health insurance sector. *Source* Authors compilation using R Studio: Biblioshiny library

with 18.72% international collaboration, indicating sustained scholarly engagement despite declining publication rates.

Figure 4 word cloud prominently centers around the theme of Corporate Governance and Financial Performance within the healthcare sector. Notable terms include "hospitals," "public," "private," "leadership," "administration," "financial management," "CSR," "economics," and "access to information." The predominant language is English, with occasional inclusion of terms like "sement regalastan" (potentially Catalan) and "coats and cost analysis" (likely Spanish). References to sources like

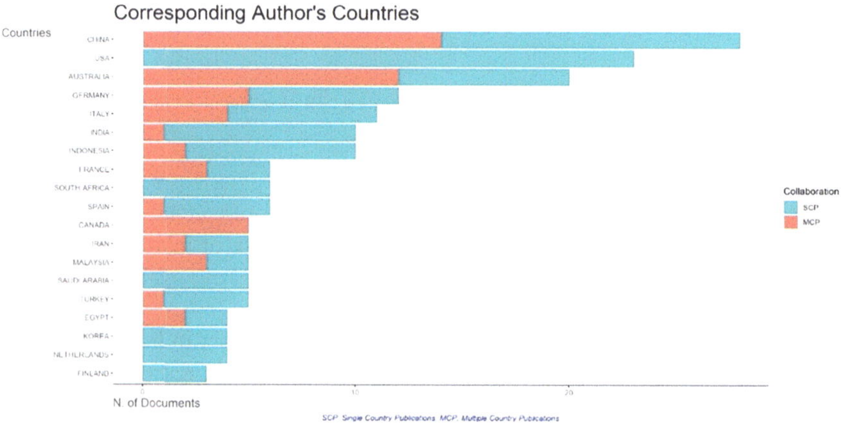


**Fig. 3** Corresponding author's countries. *Source* Authors compilation using R Studio: Biblioshiny library

**Fig. 4** Word cloud. *Source* Authors compilation using R Studio: Biblioshiny library

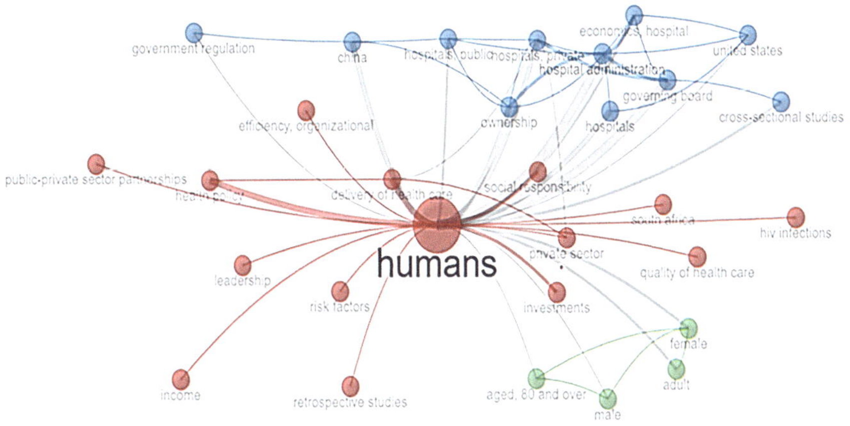


**Fig. 5** Thematic map. *Source* Authors compilation using R Studio: Biblioshiny library

"hospitals publicat," "hospitals private," and "united states" suggest a focus on the US healthcare system.

While the word cloud offers insight into key issues related to CSR in healthcare, further context is needed to grasp its specific purpose, likely serving as an illustrative tool for a research paper or report.

Figure 5 indicates healthcare system's network diagram reveals intricate connections between stakeholders, including Humans, the Government, and Hospitals (public/private). Government plays a dual role in regulating and collaborating with hospitals. Public and private hospitals, governed by boards of directors, impact healthcare quality through leadership, risk factors, and investments. Despite lacking specific data, the diagram underscores the need to understand these relationships. Key questions arise, addressing the impact of government regulation, funding mechanisms, dynamics of public–private partnerships, and strategies for healthcare investments. Answering these questions is crucial for informed decision-making and improving the healthcare system.

## 5 Conclusions

In conclusion, our comprehensive exploration spanning from 2019 to 2024 illuminates the dynamic interplay between corporate governance and performance within the health insurance sector. Despite a dwindling annual growth rate in publications, our analysis unveils the resilience of this field, characterized by ongoing scholarly contributions and substantial impact as evidenced by high citation rates. Notably, international collaboration emerges as a crucial element, reflecting the global nature of research efforts in this sphere.

By delving into top sources, authors, and globally cited documents, our study provides valuable insights into influential works and the collective scholarly endeavor to comprehend corporate governance dynamics in health insurance. Visual aids such as word clouds and thematic maps enrich our narrative, offering nuanced perspectives on critical themes and stakeholder relationships within the healthcare system.

While grappling with challenges such as fluctuating publication trends, our research underscores the enduring importance of corporate governance in shaping financial performance within the health insurance realm. It serves as a springboard for further inquiry, laying the groundwork for advancing understanding in this pivotal area. Despite the fewer publications, our impactful findings offer actionable insights for policymakers, industry professionals, and researchers alike. Emphasizing the imperative of international collaboration, our study highlights both global challenges and opportunities for knowledge exchange and the adoption of best practices.

## References

1. Antwi, I.F., Carvalho, C., Carmo, C.: Corporate governance and firm performance in the emerging market: a review of the empirical literature. J. Gov. Regul. **10**(1) (2021)
2. Prusty, T.: Surveillance of corporate governance compliance through a prolific advance. Indian J. Corp. Gov. **2**(2), 117–138 (2009)
3. Bushee, B.J., Carter, M.E., Gerakos, J.: Institutional investor preferences for corporate governance mechanisms. J. Manag. Account. Res. **26**(2), 123–149 (2014)
4. Karpoff, J.M.: The impact of shareholder activism on target companies: a survey of empirical findings. Available at SSRN 885365 (2001)
5. Jensen, M.C., Meckling, W.H.: Theory of the firm: managerial behavior, agency costs and ownership structure. In: Corporate Governance. Gower, pp. 77–132 (2019)
6. Fama, E.F.: Agency problems and the theory of the firm. J. Polit. Econ. **88**(2), 288–307 (1980)
7. Fama, E.F., Jensen, M.C.: Separation of ownership and control. J. Law Econ. **26**(2), 301–325 (1983)
8. Denis, D.K., McConnell, J.J.: International corporate governance. J. Fin. Quant. Anal. **38**(1), 1–36 (2003)
9. Gregory, H.J., Simms, M.E.: Corporate governance: what it is and why it matters. In: 9th International anti-corruption conference, Kuala Lumpur, pp. 10–15 (1999)
10. Mohamad, S.: Why effective corporate governance matters. Taylor's Bus. Rev. **7**(2), 1–17 (2018)
11. Rwegasira, K.: Corporate governance in emerging capital markets: whither Africa? Corp. Gov.: Int. Rev. **8**(3), 258–267 (2000)

# A Systematic Review of the Impact of Artificial Intelligence on Digital Art

**Diego Pozú-López, Johan Peralta-Rodríguez, and Orlando Iparraguirre-Villanueva**

**Abstract** The systematic review aimed to analyse Artificial Intelligence (AI)'s impact on the different areas of its use in digital art. Four databases from reliable sources were used for the analysis: IEEE, Springer, Scopus and ScienceDirect. The PRISMA methodology was used to regulate the article collection process and a total of 50 articles associated with the research topic were counted the results were then sectioned into 3 key questions that helped to compare the observations of the related works. The results expose the different advantages of using AI in art, such as fostering creativity and novelty, creating new art forms, as well as the use of techniques used by AI for image generation such as deep learning algorithms or convolutional neural networks. On the other hand, the various limitations of AI in the generation of images were identified, such as errors in distributional displacements or generation of faces with noticeable distortions. Finally, it was concluded that AI is a very helpful tool that enhances traditional art, but, given that it still has limitations, it is not entirely perfect.

**Keywords** Art · Artificial intelligence · Generative · Social

## 1 Introduction

At the end of 2022, AI systems that automatically translate text into original images became an Internet sensation, marking an innovation by improving the creation of digital art [1]. Text-to-image conversion technologies such as GLIDE, DALLE-2,

D. Pozú-López · J. Peralta-Rodríguez · O. Iparraguirre-Villanueva (✉)
Universidad Tecnológica del Perú, Chimbote, Perú
e-mail: oiparraguirre@ieee.org

D. Pozú-López
e-mail: U20215706@utp.edu.pe

J. Peralta-Rodríguez
e-mail: U20215774@utp.edu.pe

 
M. Al Mubarak (ed.), *Sustainable Digital Technology and Ethics in an Ever-Changing Environment*, Studies in Systems, Decision and Control 237,
https://doi.org/10.1007/978-3-031-86708-8_50

Midjourney or Stable Diffusion have established a gap between the traditional and the innovative, making the classic paper and pencil be replaced by a more up-to-date input object such as keyboard and mouse [2], therefore, the use of Machine Learning (ML) technology opens endless possibilities for artistic creation and degrades the traditional lines between authorship and creativity [3, 4]. In contrast, other artists use AI to produce more avant-garde artworks [5].

The impact of artificial learning generates an ethical problem by allowing anyone with sufficient knowledge to access these tools, which allows them to explore and automate techniques that were difficult for other artists to achieve, such as creativity [6]. At the same time, it generates complications in the social issue, discussing the veracity of the creation of media content by the media [7]. Likewise, the use of this technology can have a great impact by being a threat to the field of graphic design, future artists will have to have the incorporation of AI as another tool [8], improving processes, time and effort [9, 10]. The importance of addressing the impact of AI in digital art is to contribute to the different situations where it has been used, as well as the different opinions that go against its use due to the ethics of image generation, to enrich the knowledge regarding the management of ML, being perhaps a benefit as a useful tool rather than a mistake for artists [11].

The exploration of AI-generated digital art has been a topic of growing interest in academia and science in recent years. For example, in [12] they analysed the future of Fine Arts students, to understand their vision and position on the subject. To do so, they used an interview methodology to find out their opinions. Finally, it was concluded with negative responses that the impact that AI can have will generate distrust and concerns in the art industry.

The study [13] addressed the issue of social responsibility in the face of an increase in artificial creations, its methodology being the presentation of two debates in which firstly copyright 2.0 laws and intellectual property on AI-generated content were discussed, and then the proposals of anti-AI. To finally conclude AI can change the way works are made today. On the other hand, we have the article by [14] that aimed to evaluate the human–machine interaction between AI and children to conclude their activity. Their methodology was based on long-term research on the incorporation of AI in school education, concluding that the generation of artificial digital art depends on the human hand to be processed. At the same time, research [15] stated that for better or worse ML is already present in our lives, giving examples such as the daily queries made to Siri, ChatGPT, DALLE-2, etc. To clarify the controversies surrounding the use of AI in digital art, the research methodology used was a literature review, concluding that AI should not be seen as a threat to work, creations or the development of future competencies, but rather as another tool in the field of graphic design. Finally, we have the contribution of [16], which talks about how AI innovated the process of creating images by being established through texts entered from a keyboard to the computer, it aimed to provide information on the great ethical gap that the situation of artificial learning is going through, to achieve it using as a methodology interviews with leading professionals, in addition to using a literature review, and then concluding that ML drives human creativity in the artistic field, enhancing traditional techniques.

The aim of this systematic literature review (SLR) is to provide an extended view of AI in its use in digital design, as there is much debate about the implementation of AI in art.

This paper is organized in 5 sections. In this section contextualizes the topic of AI in digital art. Section 2 presents the process of the method used for SLR. Section 3 presents the results obtained. Finally, in the last sections, we discuss our findings and conclude this study.

## 2 Methodology

This section contains the research elements that were necessary to answer the purpose of the research, such as the prism diagram, the research questions and the search strategy.

Based on the literature review, the study seeks to investigate the impact that AI has had on the different fields of digital art. Therefore, the following research questions were posed:

RQ1. What are the benefits of using AI in digital art? RQ2. What are the AI techniques for digital art? RQ3. What are the limitations of current drawing techniques in comparison to the use of AI in digital art?

The RSL was chosen as a procedure, which provides a procedure for collecting articles to analyze them and answer specific questions [17]. To produce a document based on quality standards, the PRISMA method was used, being a highly recognized tool for its use in SLR, as it minimizes the search for information by standardizing the process. This procedure was essential to analyze the impact of AI on digital art [18].

The following inclusion and exclusion criteria shown in Table 1 were applied for the SLR.

The search strategy employed for the SLR was based on applying a search sequence composed of the keywords, expressed in the following string: "art and artificial intelligence and generative and social" in different reliable databases, such as Scopus, ScienceDirect, IEEE and Springer. A total of 684 articles were found in

**Table 1** Inclusion and exclusion criteria

| Inclusion | Exclusion |
|---|---|
| Documents that are between the years 2015–2024 | Exclusive or paid access documents |
| Documents with relevance to the research topic | Documents out of time range |
| Open Access Documents | Documents with a lack of relevance to the proposed theme |
| Documents in English language | Documents with non-specific methodology |

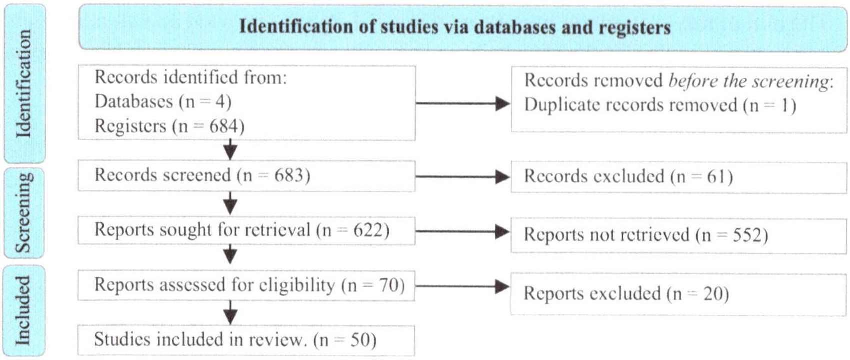


**Fig. 1** Prism diagram

the research, of which only 50 were important for the research based on inclusion and exclusion criteria. 4 stages were used for the selection of articles. In the first stage, duplicate articles were discarded, leaving 683 articles, in the second stage, articles were discarded if their titles were not related to the research topic, leaving 622 articles, in the third stage, articles were analysed by re-summary, leaving 70 research documents, and finally, in the fourth stage, 50 articles were selected that responded to the questions proposed in this research. Figure 1 shows the selection process.

## 3 Results

In this section of the SLR, the different findings discovered in the databases were analysed, pointing out sections such as the bibliometric analysis, the percentage of documents used by each database and the focus of the articles reviewed, among others.

Figure 2 shows the bibliometric analysis of the most frequent keywords among the articles reviewed. The most recurrent keyword among the articles reviewed was "artificial intelligence", given that it is one of the pillar themes for the basis of this article, while the Deep learning technique is another of the keywords that most make up the documents reviewed, as it is one of the techniques that collaborates with the advance of AI in digital art. Finally, the keyword Digital Art is the third most present keyword due to the reason for the inclusion and exclusion of the articles.

As can be seen in the bibliometric analysis, the related topics of the reviewed papers emphasise study focuses such as Deep learning algorithms, having a niche in digital art automation since it is used not only for the creation of images with AI but also for their restoration. In turn, issues related to image generation and digital art are addressed. Other significant topics include the use of AI in art, authenticity in digital art through blockchain technology and non-fungible tokens (NFTs), as well

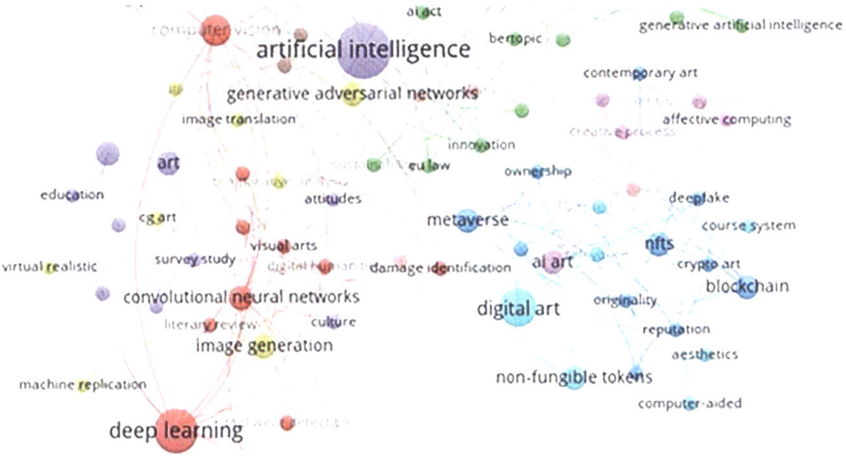


Fig. 2 Bibliometric analysis of the keywords

Fig. 3 Number of items per database

as the regulation of AI, with special attention to the AI Act and its implementation in various sectors. This mapping of diverse topics reflects in digital art the breadth and diversity of current discussions in the field of AI and its applications. Figure 3 shows the number of papers found per database.

Figure 3 shows how the number of articles is divided by databases used in the SLR, each belonging to each database used. The predominant database is ScienceDirect, occupying more than 50% of the articles chosen for the review, with an exact number of 28.

In second place, with 32% of the articles used, is Scopus, which shows that all databases were of great relevance in the review. On the other hand, the literature reviewed was composed of three types of approaches: qualitative, quantitative and mixed. The quantitative approach was the most representative, followed by the qualitative approach. It is also evident that 99% of the literature reviewed came from original articles and only one manuscript came from a Proceeding Conference. On the other hand, the distribution of the number of articles on digital art by country was also analysed. China stands out as the main contributor with a total of 13 articles, indicating a strong focus and development of interest in this country. The United States and Finland contribute 4 articles each, reflecting considerable research activity in the field of digital art. Canada and Germany are also present with 3 articles each,

**Table 2** Benefits of AI in digital art

| Benefits | Quantity | References |
|---|---|---|
| Encouraging creativity | 14 | [19–32] |
| Encourage novelty | 15 | [33–47] |
| Increasing the number of users | 6 | [48–53] |
| Creation of new art forms | 5 | [54–58] |
| Generating new and different styles | 1 | [59] |
| Increasing sales of digital art | 1 | [26] |

demonstrating their commitment to the field, showing a global diversity in digital art research, albeit at an earlier stage.

## 4 Discussions

### *4.1 What Are the Benefits of Using AI in Digital Art?*

AI as an enhancer towards the creation of digital images has been a success for this field, given that unlike the study [12] where we see negative responses towards generative AI, according to the RSL, we found several benefits that it can provide towards digital art, as shown in Table 2. Of the 50 articles reviewed, 10 confirm that the use of AI in digital art encourages creativity in new users, allowing more people to get involved in the creation of art, while 1 article mentions how positive it can be in the digital art sales market for users. This can be related to a study where AI is seen as the innovator in the process of image creation [16]. On the other hand, it is noted that another of the most recognised benefits is that it allows users who are not familiar with the concept of art to synthesise their ideas clearly and concisely.

### *4.2 What Are the AI Techniques for Digital Art?*

The techniques mostly used for AI are the use of DALLE -2, deep learning algorithms, use of DeepDream algorithms and the use of Stable Diffusion. Table 3 shows the different techniques that are used for AI in digital art. Also, in Table 3 it can be evidenced that a higher use of generative applications such as DALL E-2 and Stable Diffusion is a total of 11 articles. This finding is related to the work [15] where they used generative AI for daily queries made by users, given that DALL E-2 and Stable Diffusion stand out for their ability to generate images from text written by users.

Table 3 AI techniques in digital art

| Limitations | Quantity | References |
|---|---|---|
| DALL E-2 | 6 | [37, 47, 58, 60–62] |
| Deep learning algorithms | 4 | [30, 35, 38, 41] |
| Deep dream algorithms | 2 | [39, 63] |
| Stable diffusion | 5 | [37, 46, 57, 58, 62] |
| Generative adversarial Networks | 4 | [29, 36, 42, 64] |
| Convolutional neural networks | 3 | [20, 65, 66] |

### 4.3 RQ3: What Are the Limitations of Current Drawing Techniques Compared to the Use of AI in Digital Art?

Although AI offers very powerful advantages to the creation of generative images not only for users who are steeped in the subject of art but also for users who are completely unfamiliar with the subject, for now, AI is not perfect, because although it creates images that may go unnoticed to the uninformed human eye or to those who do not have a certain knowledge of art, AI generates certain deformities that are not visible to the naked eye. Table 4 shows some of the limitations of AI in image creation.

As seen in Table 4, generative AI still has many limitations in different areas, one of the most prominent is the reliance on human creativity. Also, it highlights the lack of human touch, leaving aside traditional art, this is related to the work of [59], where they evidenced the limitation of creating realistic images for fashion, also highlights the generation of deformed or notorious faces in their use of AI, then distributional shifts is also a limitation to highlight, since AI still cannot distribute objects in a way that does not make its use noticeable. All the above is supported by the article [15]

Table 4 Limitations of AI in digital art

| Limitations | Quantity | References |
|---|---|---|
| Distributional shifts | 3 | [21, 22, 31] |
| Reproduction of existing patterns | 3 | [23, 37, 64] |
| Creation of photorealistic images | 1 | [59] |
| Parameter complexity | 2 | [38, 39] |
| Generation of faces with noticeable distortions | 2 | [30, 41] |
| Dependence on human creativity | 5 | [40, 47, 49, 50, 67] |
| Grammatical flaws | 2 | [52, 67] |
| Lack of human touch | 5 | [42, 45, 51–53] |
| Controversial question posing | 1 | [43] |
| Noticeable textural and color features | 2 | [64, 68] |

which tells us that for better or worse, generative AI is already incorporated into the daily lives of different users despite the limitations it may have.

## 5 Conclusions

After performing all the procedures of an RSL, with a total of 50 articles reviewed for the methodology that was applied to the 4 phases of the prism method, a total of 49 original articles and 1 conference were obtained, of which 28 were from ScienceDirect (56%), 1 from IEE (2%), 16 from Scopus (32%) and 5 from Springer (10%). It is concluded that generative AI has had a high impact in different areas of digital art, whether for fashion, image generation, or even education. But not everything is perfect, given that, despite generating an abysmal advantage against traditional art, for now, AI has many limitations that can be improved in the near or distant future. On the other hand, we found different tools or techniques that make possible the impact on digital art such as DALLE-2, Stable Diffusion, Generative Adversarial Networks (GAN), as well as Deep learning or dream algorithms. Likewise, we recommend further research and development of improvements in generative AI, to continue promoting technological progress in the generation of images in the different areas of digital art.

## References

1. Hutson, J., Cotroneo, P.: Praxis and augmented areativity: a case study in the use of generative artificial intelligence (AI) art in the digital art classroom. Int. J. Technol. Learn. **31**(1), 113–132 (2023). https://doi.org/10.18848/2327-0144/CGP/v31i01/113-132
2. Pan, Z., Zhou, X., Tian, H.: Arbitrary style guidance for enhanced diffusion-based text-to-image generation. In: Proceedings—2023 IEEE Winter Conference on Applications of Computer Vision, WACV 2023, pp. 4450–4460 (2023). https://doi.org/10.1109/WACV56688.2023.00444
3. Cousins, S.: The rapid rise of AI art. Eng. Technol. **18**(2), 20–25 (2023). https://doi.org/10.1049/ET.2023.0208
4. Mathew, D., Brintha, N.C., Jappes, J.T.W.: Artificial Intelligence Powered Automation for Industry 4.0, pp. 1–28 (2023). https://doi.org/10.1007/978-3-031-20443-2_1
5. Notaro, A.: State-of-the-art: AI through the (artificial) artist's eye, pp. 322–328 (2020). https://doi.org/10.14236/EWIC/EVA2020.58
6. Rani, S., Jining, D., Shah, D., Xaba, S., Singh, P.R.: Revolutionizing the creative process: exploring the benefits and challenges of AI-driven art. In: Lecture Notes in Networks and Systems, vol. 854, pp. 234–243. Springer Science and Business Media Deutschland GmbH (2023). https://doi.org/10.1007/978-3-031-50151-7_23
7. Franganillo, J.: La inteligencia artificial generativa y su impacto en la creación de contenidos mediáticos. methaodos revista de ciencias sociales **11**(2), m231102a10 (2023). https://doi.org/10.17502/mrcs.v11i2.710
8. Prada, J.M.: ENG Visual Artistic Creation in the Face of the Challenges of Artificial Intelligence. Creative Automation and Ethical Questions [La creación artística visual frente a los

retos de la inteligencia artificial. Automatización creativa y cuestionamientos éticos1]. Eikon Imago **13**, e90081 (2024). https://doi.org/10.5209/eiko.90081

9. Liu, M.: Overview of artificial intelligence painting development and some related model application. SHS Web of Conferences, vol. 167, no. 01004 (2023). https://doi.org/10.1051/SHSCONF/202316701004
10. Iparraguirre-Villanueva, O., Obregon-Palomino, L., Pujay-Iglesias, W., Sierra-Liñan, F., Cabanillas-Carbonell, M.: Productivity of incident management with conversational bots-a review. IAES Int. J. Artif. Intell. (IJ-AI) **12**(4), 1543–1556 (2023). https://doi.org/10.11591/ijai.v12.i4.pp1543-1556
11. Then, C., Soewandi, E.J., Danial, M.F., Achmad, S., Sutoyo, R.: The impact of artificial intelligence on art—a systematic literature review. In: Proceeding—IEEE 9th Information Technology International Seminar, ITIS 2023, Malang (2023). https://doi.org/10.1109/ITIS59651.2023.10420208
12. Albar Mansoa, P.J.: Artificial intelligence for image generation in art: how does it impact on the future of Fine Art students? Encuentros (Maracaibo) **20**, 145–164 (2024). https://doi.org/10.5281/zenodo.10052355
13. Bernaschina, D.: Artes mediales e inteligencia artificial: la crisis de la ética y la precariedad laboral en el campo artístico-digital Media arts and artificial intelligence: the crisis of ethics and the precarious work in the artistic-digital field. Available at www.fundacionavenir.com.co/avenir [Online]
14. Vartiainen, H., Tedre, M., Jormanainen, I.: Co-creating digital art with generative AI in K-9 education: socio-material insights. Int. J. Educ. Art **19**(3), 405–423 (2023). https://doi.org/10.1386/eta_00143_1
15. Sesé, J.R.: El diseñador gráfico en la era de la Inteligencia Artificial. EME Exp. Illustration Art Des. **8**(8), 66–73 (2020). https://doi.org/10.4995/EME.2020.13210
16. Rani, S., Jining, D., Shah, D., Xaba, S., Singh, P.R.: Revolutionizing the creative process: exploring the benefits and challenges of AI-driven art. In: Lecture Notes in Networks and Systems, vol. 854, pp. 234–243 (2023). https://doi.org/10.1007/978-3-031-50151-7_23
17. Producir una revisión sistemática. Accessed 20 June 2024. Available at https://psycnet.apa.org/record/2010-00924-039 [Online]
18. Humana Dietética, N., et al.: Ítems de referencia para publicar Protocolos de Revisiones Sistemáticas y Metaanálisis: declaración PRISMA-P 2015. Revista Española de Nutrición Humana y Dietética **20**(2), 148–160 (2016). https://doi.org/10.14306/RENHYD.20.2.223
19. Cui, Z.: Application of line integral convolution techniques in the generation of fine art painting effects. Mob. Inf. Syst. **2022**, 6872214 (2022). https://doi.org/10.1155/2022/6872214
20. Wadinambiarachchi, S., Kelly, R.M., Pareek, S., Zhou, Q., Velloso, E.: The effects of generative AI on design fixation and divergent thinking. In: Conference on Human Factors in Computing Systems—Proceedings (2024). https://doi.org/10.1145/3613904.3642919
21. Hurst, W., Spyrou, O., Tekinerdogan, B., Krampe, C.: Digital art and the metaverse: benefits and challenges. Future Internet **15**(6), 188 (2023). https://doi.org/10.3390/fi15060188
22. Shen, D., Ye, X.: Robot collaborative interactive artistic design for digital art using embedded systems. Comput. Aided Des. Appl. **21**(S8), 1–12 (2024). https://doi.org/10.14733/cadaps.2024.S8.1-12
23. Notaro, A.: All that is solid melts in the Ethereum: the brave new (art) world of NFTs. J. Vis. Art Pract. **21**(4), 359–382 (2022). https://doi.org/10.1080/14702029.2022.2129204
24. Shi, Y., Liu, Y.: Research on the application of virtual reality technology in digital media art teaching. Comput. Aided Des. Appl. **21**(S17), 171–186 (2024). https://doi.org/10.14733/cadaps.2024.S17.171-186
25. Meng, X., Li, H.: Intelligent design and application of traditional cultural and creative products based on digital art elements. Appl. Math. Nonlinear Sci. **9**(1), 20231684 (2024). https://doi.org/10.2478/amns.2023.2.01684
26. He, X.: Computer-aided relevance analysis of digital art course systems informed by medical research. Comput. Aided Des. Appl. **21**(S24), 157–168 (2024). https://doi.org/10.14733/cadaps.2024.S24.157-168

27. Yao, Y., Liu, Y.: Digital public system of urban art: navigating human-computer interaction in artistic design for innovative urban expressions. Int. J. Adv. Comput. Sci. Appl. **15**(5), 369–376 (2024). https://doi.org/10.14569/IJACSA.2024.0150536
28. Alsultan, S., Kourtis, A., Markellos, R.N.: Can we price beauty? Aesthetics and digital art markets. Econ. Lett. **235**, 111572 (2024). https://doi.org/10.1016/j.econlet.2024.111572
29. Castellano, G., Vessio, G.: Deep learning approaches to pattern extraction and recognition in paintings and drawings: an overview. Neural Comput. Appl. **33**(19), 12263–12282 (2021). https://doi.org/10.1007/S00521-021-05893-Z/TABLES/2
30. Di Zio, S., Calleo, Y., Bolzan, M.: Delphi-based visual scenarios: an innovative use of generative adversarial networks. Futures **154**, 103280 (2023). https://doi.org/10.1016/j.futures.2023.103280
31. Lee, J., Park, J.: AI as 'another I': journey map of working with artificial intelligence from AI-phobia to AI-preparedness. Organ. Dyn. **52**(3) (2023). https://doi.org/10.1016/j.orgdyn.2023.100994
32. Zhu, S., et al.: Exploring the impact of ChatGPT on art creation and collaboration: Benefits, challenges and ethical implications. Telematics Inf. Rep. **14**, 100138 (2024). https://doi.org/10.1016/j.teler.2024.100138
33. Crothers, E.N., Japkowicz, N., Viktor, H.L.: Machine-generated text: a comprehensive survey of threat models and detection methods. IEEE Access **11**, 70977–71002 (2023). https://doi.org/10.1109/ACCESS.2023.3294090
34. Mubarak, R., Alsboui, T., Alshaikh, O., Inuwa-Dutse, I., Khan, S., Parkinson, S.: A survey on the detection and impacts of deepfakes in visual, audio, and textual formats. IEEE Access **11**, 144497–144529 (2023). https://doi.org/10.1109/ACCESS.2023.3344653
35. Li, L., Zhang, X., Chen, K., Feng, G., Wu, D., Zhang, W.: Image steganography and style transformation based on generative adversarial network. Mathematics **12**(4), 615 (2024). https://doi.org/10.3390/math12040615
36. Rosenberg, H., Ahmed, S., Ramesh, G., Fawaz, K., Vinayak, R.K.: Limitations of face image generation. In: Proceedings of the AAAI Conference on Artificial Intelligence, pp. 14838–14846 (2024). https://doi.org/10.1609/aaai.v38i13.29403
37. Vartiainen, H., Tedre, M.: How text-to-image generative AI Is transforming mediated action. IEEE Comput. Graph. Appl. **44**(2), 12–22 (2024). https://doi.org/10.1109/MCG.2024.3355808
38. Fu, X.: Digital image art style transfer algorithm based on CycleGAN. Comput. Intell. Neurosci. **2022**, 6075398 (2022). https://doi.org/10.1155/2022/6075398
39. Messer, U.: Co-creating art with generative artificial intelligence: implications for artworks and artists. Comput. Human Behav.: Artif. Humans **2**(1), 100056 (2024). https://doi.org/10.1016/J.CHBAH.2024.100056
40. Tao, F., Ma, X., Liu, W., Zhang, C.: Digital Engineering: state-of-the-art and perspectives. Digital Eng. **1**, 100007 (2024). https://doi.org/10.1016/J.DTE.2024.100007
41. Hasan, H.R., Salah, K., Jayaraman, R., Yaqoob, I., Omar, M.: NFTs for combating deepfakes and fake metaverse digital contents. Internet of Things **25**, 101133 (2024). https://doi.org/10.1016/J.IOT.2024.101133
42. López-Varela Azcárate, A.: Intermedial semiotics in the age of artificial intelligence: challenges and opportunities for the arts. New Techno Humanit. **3** (2024). https://doi.org/10.1016/J.TECHUM.2024.04.001
43. Ye, Y., et al.: Generative AI for visualization: state of the art and future directions. Visual Inf. **8**(2), 43–66 (2024). https://doi.org/10.1016/J.VISINF.2024.04.003
44. Xiang, J.: On generated artistic styles: image generation experiments with GAN algorithms. Visual Inf. **7**(4), 36–40 (2023). https://doi.org/10.1016/J.VISINF.2023.10.005
45. Oksanen, A., et al.: Artificial intelligence in fine arts: a systematic review of empirical research. Comput. Human Behav.: Artif. Humans **1**(2), 100004 (2023). https://doi.org/10.1016/J.CHBAH.2023.100004
46. Mall, P.K., et al.: A comprehensive review of deep neural networks for medical image processing: recent developments and future opportunities. Healthc. Analytics **4**, 100216 (2023). https://doi.org/10.1016/J.HEALTH.2023.100216

47. Bergs, T., Holst, C., Gupta, P., Augspurger, T.: Digital image processing with deep learning for automated cutting tool wear detection. Procedia Manuf. **48**, 947–958 (2020). https://doi.org/10.1016/J.PROMFG.2020.05.134
48. Dehouche, N., Dehouche, K.: What's in a text-to-image prompt? The potential of stable diffusion in visual arts education. Heliyon **9**(6), e16757 (2023). https://doi.org/10.1016/J.HELIYON.2023.E16757
49. Choi, S.K., DiPaola, S., Gabora, L.: Art and the artificial. J. Creativity **33**(3), 100069 (2023). https://doi.org/10.1016/J.YJOC.2023.100069
50. Mishra, M., Lourenço, P.B.: Artificial intelligence-assisted visual inspection for cultural heritage: State-of-the-art review. J. Cult. Herit. **66**, 536–550 (2024). https://doi.org/10.1016/J.CULHER.2024.01.005
51. Weigand, H., Johannesson, P., Andersson, B.: An artifact ontology for design science research. Data Knowl. Eng. **133**, 101878 (2021). https://doi.org/10.1016/j.datak.2021.101878
52. Shedko, A.Y.: Semantic-map-based assistant for creative text generation. Procedia Comput. Sci **123**, 446–450 (2018). https://doi.org/10.1016/j.procs.2018.01.068
53. Demmer, T.R., Kühnapfel, C., Fingerhut, J., Pelowski, M.: Does an emotional connection to art really require a human artist? emotion and intentionality responses to AI-versus human-created art and impact on aesthetic experience. Comput Human Behav **148**, 107875 (2023). https://doi.org/10.1016/j.chb.2023.107875
54. Leivada, E., Murphy, E., Marcus, G.: DALL·E 2 fails to reliably capture common syntactic processes. Soc. Sci. Humanit. Open **8**(1), 100648 (2023). https://doi.org/10.1016/j.ssaho.2023.100648
55. Antunes, R.F., Correia, L.: Virtual simulations of ancient sites inhabited by autonomous characters: lessons from the development of Easy-population. Digit. Appl. Archaeol. Cult. Heritage **26** (2022). https://doi.org/10.1016/j.daach.2022.e00237
56. Hassan, M.U., Alaliyat, S., Hameed, I.A.: Image generation models from scene graphs and layouts: a comparative analysis. J. King Saud Univ.—Comput. Inf. Sci. **35**(5), 101543 (2023). https://doi.org/10.1016/j.jksuci.2023.03.021
57. Hautala, J., Heino, H.: Spectrum of AI futures imaginaries by AI practitioners in Finland and Singapore: the unimagined speed of AI progress. Futures **153** (2023). https://doi.org/10.1016/j.futures.2023.103247
58. Barman, D., Guo, Z., Conlan, O.: The dark side of language models: exploring the potential of LLMs in multimedia disinformation generation and dissemination. Mach. Learn. Appl. **16**, 100545 (2024). https://doi.org/10.1016/j.mlwa.2024.100545
59. Hashmi, M.F., Ashish, B.K.K., Keskar, A.G., Bokde, N.D., Geem, Z.W.: FashionFit: analysis of mapping 3D pose and neural body fit for custom virtual try-on. IEEE Access **8**, 91603–91615 (2020). https://doi.org/10.1109/ACCESS.2020.2993574
60. Latikka, R., Bergdahl, J., Savela, N., Oksanen, A.: AI as an Artist? A two-wave survey study on attitudes toward using artificial intelligence in art. Poetics **101**, 101839 (2023). https://doi.org/10.1016/j.poetic.2023.101839
61. Gupta, P., Ding, B., Guan, C., Ding, D.: Generative AI: A systematic review using topic modelling techniques. Data Inf. Manag. **8**(2), 100066 (2024). https://doi.org/10.1016/j.dim.2024.100066
62. Hacker, P.: The European AI liability directives—critique of a half-hearted approach and lessons for the future. Comput. Law Secur. Rev. **51**, 105871 (2023). https://doi.org/10.1016/j.clsr.2023.105871
63. Dipaola, S., Gabora, L., McCaig, G.: Informing artificial intelligence generative techniques using cognitive theories of human creativity. Procedia Comput. Sci. **145**, 158–168 (2018). https://doi.org/10.1016/j.procs.2018.11.024
64. Sun, R., Huang, C., Zhu, H., Ma, L.: Mask-aware photorealistic facial attribute manipulation. Comput. Vis. Media (Beijing) **7**(3), 363–374 (2021). https://doi.org/10.1007/S41095-021-0219-7/METRICS
65. Rosin, P.L., et al.: NPRportrait 1.0: a three-level benchmark for non-photorealistic rendering of portraits. Comput. Vis. Media (Beijing) **8**(3), 445–465 (2022). https://doi.org/10.1007/S41095-021-0255-3/METRICS

66. Sheng, K., et al.: Learning to assess visual aesthetics of food images. Comput. Vis. Media (Beijing) **7**(1), 139–152 (2021). https://doi.org/10.1007/S41095-020-0193-5/METRICS
67. Yang, J., Zhang, W., Liu, J., Wu, J., Yang, J.: Generating De-identification facial images based on the attention models and adversarial examples. Alexandria Eng. J. **61**(11), 8417–8429 (2022). https://doi.org/10.1016/j.aej.2022.02.007
68. Spratt, E.L., Elgammal, A.: Computational beauty: aesthetic judgment at the intersection of art and science. Lecture Notes in Computer Science. Lecture Notes in Artificial Intelligence and Lecture Notes in Bioinformatics, vol. 8925, pp. 35–53 (2015). https://doi.org/10.1007/978-3-319-16178-5_3

# Social Entrepreneurship as Catalyst for Social Change: A Mediation Approach

**Debolina Gupta, R. Mallieswari, Jayashree Kowtal, Niharika Mishra, and S. Jayashree**

**Abstract** Social entrepreneurship in India has grown rapidly over the past decade. Access to knowledge, better financing, and top-notch training and seminars for social entrepreneurs and leaders have boosted the quantity and quality of creative ideas and business strategies. Social enterprises have grown due to financial, advisory, direct, and indirect support from aid groups. India lacks social enterprise ideas, which is unexpected given its dynamic climate and huge industry potential. Furthermore, the social business sector lacks knowledge about the policies and courses offered by Indian universities. This research seeks to explore social entrepreneurship's impact on Bengaluru's positive social change. This study focuses on descriptive analysis. The descriptive analysis may be considered an assertion since the researcher has no influence on the component under study. We gather information from both primary and secondary sources. We examined social entrepreneurs' autobiographies and biographies to enhance our findings. For this census survey, we contacted 30 social entrepreneurs based in Bengaluru. Despite a moderately direct correlation, social enterprise led to societal transformation. Social entrepreneurship appears to have a major impact on society. They combine entrepreneurial, commercial, and social work skills and techniques with social support to mobilize financial, human, and infrastructural resources to sustain their social entrepreneurship company and achieve their social objectives.

**Keywords** Social entrepreneurship · Social enterprise · Social change · Mediation approach

---

D. Gupta · R. Mallieswari · J. Kowtal · N. Mishra
M S Ramaiah Institute of Management, Bangalore, India

S. Jayashree (✉)
Department of Professional Accounting and Finance, Kristu Jayanti College (Autonomous), Bangalore, India
e-mail: jayashree.s@kristujayanti.com



M. Al Mubarak (ed.), *Sustainable Digital Technology and Ethics in an Ever-Changing Environment*, Studies in Systems, Decision and Control 237,
https://doi.org/10.1007/978-3-031-86708-8_51

## 1 Introduction

As a concept in political economy, "entrepreneurship" refers to the trait of becoming into a person who "businesses." The word emphasizes the dedication and risk-taking required to start and maintain a firm [1], as well as the innovative thinking that results from this pursuit of economic progress [2]. In response to an opportunity in the market, "entrepreneurship" may help launch new businesses or breathe fresh life into old ones. One definition of a start-up is a business that an entrepreneur creates from the ground up. Over the past few years, the idea has evolved to include the political and social dimensions of entrepreneurship [3]. There is a wide range of entrepreneurial practices based on the type of company and level of innovation involved. From small, one-person operations to massive, multi-employee behemoths, or even just part-time endeavors, there is a wide spectrum of entrepreneurship. Aspiring company owners can find support from a variety of sources, including government agencies with specialized knowledge, business incubators, research parks, and non-governmental organizations (NGOs) [4]. Social entrepreneurship, political entrepreneurship, and expertise-based entrepreneurship are some examples of the more recent developments in the definition of entrepreneurship as a mindset that yields entrepreneurial endeavors [5].

Recognizing when a segment of society is mired in inertia and coming up with innovative solutions to free it is the responsibility of a social entrepreneur [6]. To identify and address systemic societal issues, we need a committed leader who can see the big picture and keep going even when things get tough. Social entrepreneurs are driven to make a real difference by creating new opportunities for marginalized and disadvantaged people and by releasing society's full potential for positive change [7]. Therefore, social entrepreneurs consider a beneficial return to society. Social entrepreneurs are often associated with the nonprofit sector and work to advance wide-ranging social, cultural, and environmental causes [8].

The number of social entrepreneurs starting and running businesses is growing [9]. For quite some time, they have been actively involved in a wide range of fields, including healthcare, renewable energy, garbage management, housing, issues related to the expansion of slum areas, sanitation and water purification, domestic violence against women, and other pressing concerns affecting children, teenagers, and the elderly. This pattern shows how social entrepreneurs help the least fortunate while simultaneously employing less qualified people and creating jobs for themselves and others [10]. More than 22% of Indians live in poverty and lack access to healthcare, adequate nutrition, and formal education. Quality healthcare, education, and other necessities are essential for these people. Social entrepreneurs are currently focusing on developing innovative, practical, and long-term strategies in this area [11]. People in India are coming to understand that social entrepreneurs play a crucial role in bringing about direct social change and are altering the power dynamics within society [12]. Several government programs and contemporary tactics in the corporate sector that recognize Social Enterprises as powerful tools for long-term social and economic change prove this [13].

With this background, the current study entitled Social Entrepreneurship as Catalyst for Social Change—A Mediation Approach is presented under 5 sections. In this section gives the background and the introduction to the study. The review of literature, which enables identifying the gap, is discussed in the Sect. 2. The research methodology is presented in the Sect. 3. The results and Discussion from statistical analysis of data are presented in the Sect. 4. The research article concludes with recommendations and scope for further research.

## 2 Review of Literature

The study used a systematic literature analysis to analyze background reviews and independent social entrepreneurship research [14–16]. The analysis includes 54 English-language articles published within 15 years. After collecting the full study articles, 44 pieces of literature were selected to evaluate their quality. Four papers were excluded from the analysis due to iterative methods and quality issues. This analysis was based on a detailed review of 40 relevant scientific articles. Social enterprise uses business principles to improve social, environmental, and humanitarian challenges [17]. Such as providing jobs in organizations that solve social or environmental issues and innovate to improve their primary business [18]. Even while many social entrepreneurs value effect over profit [19], they seek to improve the world in sustainable ways [20]. They can have a big impact on the lives and communities they serve and show commercial influence for social good [21]. By questioning existing business paradigms and providing fresh solutions to social concerns, social entrepreneurs help transform society [22]. Social entrepreneurs can solve some of the world's most difficult problems creatively and effectively by producing value for people and the planet [23]. Social entrepreneurs demonstrate that profit and purpose can coexist for the greater good through their dedication and inventive ideas [24]. Social entrepreneurs impact the world and inspire future generations [25–27].

### *2.1 Research Gap*

Research toward the goal of producing lasting communal wealth, as opposed to merely focusing on private affluence and firm performance, can be advanced by studying the principles of Social Entrepreneurship and clarifying the unique behaviors, typologies, and characteristics of Social Enterprises. The idea of social entrepreneurship has the potential to be a feature of social expansion through models that are both realistic and sustainable. Making a positive difference in people's lives is the primary goal of social entrepreneurs.

## 2.2 Statement of the Problem

The field of social entrepreneurship in India has come a long way in the last ten years [3]. Thanks to more information, better funding, and high-quality training and seminars available to social entrepreneurs and social enterprise leaders, the amount and quality of creative ideas and business strategies have improved [11]. An uptick in social enterprises can be traced to the proliferation of aid groups that offer direct, indirect, monetary, and advisory support. In light of the evolving environment and the industry's potential, there is a dearth of social enterprise concepts in India. It is also not apparent who the heads of social enterprises are, how many there are in India, or how they compare to the country's gross domestic product or labor force. In addition, the policies and courses provided by universities in India that are supposed to help the social enterprise sector are not well understood. This study's overarching goal is to learn how social entrepreneurship in Bengaluru has contributed to positive societal development.

## 2.3 Research Methodology

The current study is focused on techniques of descriptive analysis [27, 28]. The Biographies and Auto biographies of the Social Entrepreneurs are also accessed to enhance the study. The study relied on primary data collected from respondents about their perceptions and beliefs regarding their relationship with their employers. We contacted all 30 social entrepreneurs in Bangalore to collect the data. The study employs a Likert scale with convenience sampling. SPSS Software is used for Data analysis [29, 30].

# 3 Result and Discussion

This section analyzes and interprets study social entrepreneur data. Seven (23.3%) of the 30 entrepreneurs in the sample are women and 23 (76.7%) are men. 75% (23 responders) of Social Entrepreneurs are over 38. Three (10%) Social Entrepreneurs are 33–37 years old and four (13.3%) are 28–32. Ninety-three percent of 28 respondents graduated in their specialties. Three Post Graduates have Ph.Ds. 6.7% of few responders graduated recently. 7 (23.33%) of the 30 Social Entrepreneurs started their careers under 30. Ten responders (33.33%) started between 30 and 35. Nine respondents (30%) started their Social Enterprise late in their 30s. 4 13.33% of Social Entrepreneurs joined ventures after 40. About 10 respondents (33.33%) had their careers in top-level management of different businesses; 5 (16.67%) were engineering organization administrators, 3 (10%) were managing directors, and 2 (6.67%) were CEOs. Seven (23.33%) Social Entrepreneurs were managers, five (16.67%)

were engineers, and one (3.33%) were financiers. Three respondents (10%) have experience founding businesses before Social Entrepreneurship projects. One female social entrepreneur started ventures without job experience. Seven (23.33%) of the 30 Social Entrepreneurs started their firms in health care, hospitals, jobs, and training and development. 8 founders (26.67%) were in finance and microfinance. 5 Social Entrepreneurs promote clean energy and climate sustainability. Only 10% of Social Entrepreneurs promote fruit, livestock, and horticulture.

## *3.1 Social Enterprise Profile*

Almost 14 (46.66%) of Social Enterprises created in 2005–2010, 4 (13.33%) started before 2000. Around 2000–2005, 5 (16.66%) Social Enterprises formed. The report found 7 (23.33%) Social Entrepreneurs started projects after 2010. The thesis's last initiative began in 2017. Most 20 (66.7%) Social Entrepreneurs registered their businesses via the Companies Act, 1956, while 3 (10%) created them under Section 25 of the Indian Companies Act and the Indian Trusts Act, 1882. The Second Schedule to the Reserve Bank of India Act, 1934 registers 2 (6.7%) financial technology social enterprises and 1 (3.3%). 1 Social Enterprise started in Bengaluru but moved to the UK under the 501(c)(3) Social Enterprises for Profit category. South Bengaluru hosts 46.7% of the 14 Social Entrepreneurship enterprises. Two Social Entrepreneurs started in Bengaluru but moved elsewhere. Seven (23.3%) Social Entrepreneurs started projects in North Bengaluru and three (10%) in Central Bengaluru. Social Entrepreneurs are concentrated in West Bengaluru (3.3%) and East Bengaluru (6.7%). Only six Social Entrepreneurs (20%) served the public. Other social pioneers focused on one demography and societal purpose and vision. Both Social Enterprises targeted 2 (6.67%) commercial, retail, and domestic energy consumers, manufacturers who need financial assistance, fishermen, the general public looking for organic toys and home decor, and rural electricity suppliers. 1(3.3%) of the Social Entrepreneurs targeted businesses, philanthropists, charities and social organizations, food stalls and other commercial food preparation units, micro-entrepreneurs, clinics, special disease patients, pregnant mothers, social investors, and tier II and III city residents. 3 (10%) of Social Entrepreneurs target micro-finance and other financing institutions, prospective employees, and working employees. Predicting the quantity of clients at the start is difficult. 22 (73.3%) Social Entrepreneurs reported no consumer awareness and no clientele. When they started their social enterprises, founders with a proven client base stated they had customers.

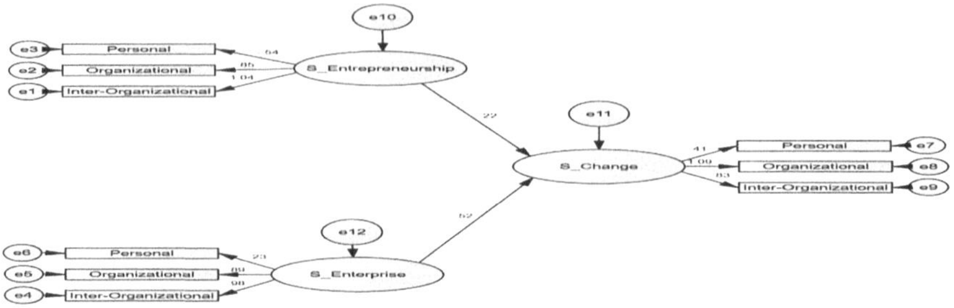


Fig. 1 Structural relationship between social entrepreneurship and social enterprise on social change

## 3.2 Testing of Hypothesis 1

***Hypothesis: There is an impact of Social Entrepreneurship and Social Enterprise on Social change***

From the output of CFA, the model fit the data well because Chi-Square value of 74.535 with 2.891 degree of freedom is statistically significant since the p value is 0.00. It Goodness of Fit Index (GFI) = 0.809 which is below the expected Criteria. The Model shown below is the pictorial output of the evidence that the model has run and the standardized estimates for relationships are shown in the Model.

From the Fig. 1 it can be indicated that when Social Entrepreneurship goes up by 1, social change increases by 0.224 or 22.4%. When Social Enterprise activities go up by 1, social change increases by 0.524 or 52.4%. This shows that both Social Entrepreneurship and Social Enterprise activities have a significant positive impact on driving social change. The higher percentage increase in social change associated with Social Enterprise activities indicates that these initiatives may have a more direct and immediate impact on addressing social issues compared to Social Entrepreneurship. Overall, these findings highlight the importance and effectiveness of both approaches in creating meaningful and sustainable social impact. Therefore Alternate Hypothesis—There is an impact of Social Entrepreneurship and Social Enterprise on Social change is accepted.

## 3.3 Testing of Hypothesis 2

***Hypothesis: Social Enterprise has a mediating effect on the relationship between Social Entrepreneurship and social change***

Using Fig. 2, the following information can be interpreted: (a) At the lower level of confidence ($p = 0.0130$), the relationship between social enrepreneurship and social enterprise is statistically significant ($t(28) = 2.6514$, $p = 0.0000$).There is no change from 1985 to the higher level confidence interval of 1.5473, which remains

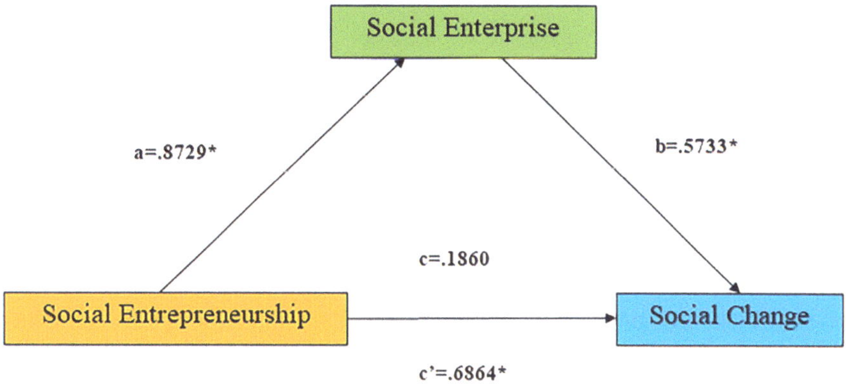


**Fig. 2** Social enterprise has a mediating effect on the relationship between social entrepreneurship and social change

at 0. Because of this, the connection is noteworthy. A positive change in social enterprises of 87.29% is caused by social entrepreneurship, as indicated by the coefficient of 0.8729. (b) At t(28) = 0.3.5568, p = 0.000, p = 0.014, the lower level confidence interval is.2426, and the upper level confidence interval is 0.9040, none of which moves beyond 0. The relationship between social enterprise and societal change is statistically significant. Because of this, the connection is noteworthy. Social entrepreneurship results in a favorable change in social enterprises by 57.33%, as indicated by the coefficient of 0.5733.

(C) How X influences Y directly At t(28) = 0.1860, p = 0.0000, p = 0.5586, the lower level confidence interval is – 0.4584, and the higher level confidence interval is 0.8305, both of which go through 0. The relationship between social entrepreneurship and societal change is not significant. Consequently, there is no significance in this relationship. (C1) All-Inclusive X–Y Effects A statistically significant relationship exists between social entrepreneurship and social change achieved through social enterprises (t(28) = 0.6864, p = 0.0000, p = 0.0494). Neither the lower nor the upper confidence intervals (0.6949 and 0.3619, respectively) cross the threshold of 0. Because of this, the connection is noteworthy. The total impact association between social entrepreneurship and social change is considerable, indicating that social entrepreneurship leads to social change through social enterprise, albeit the direct relationship between the two is minor. Therefore Alternate Hypothesis—Social Enterprise has a mediating effect on the relationship between Social Entrepreneurship and social change is accepted.

## 4 Conclusion

According to the literature on social entrepreneurship, the term arose when non-profit leaders adopted innovative earned income strategies due to a lack of government and philanthropic support. Later, the Ashoka Foundation, the pioneer organization for assisting social entrepreneurs, defined social entrepreneurship with five requirements. They are fresh ideas, approaches, or answers to social problems; social effect and explicability; resourcefulness; entrepreneurial qualities; and ethical fiber. The foundation applies these five criteria in seventy countries including India. Notably, the earned income strategy is not found in the criteria. In this study, the researcher could also find out that the Social Entrepreneurs are depending on different sources of income for their organization. Most of the Social Entrepreneurs depend on mixed sources of incomes including earned income, donation and government grants. There are also some of the Social Entrepreneurs who purely depend either on earned income or donation and government grants. Therefore in India, the earned income strategy cannot be one of the criteria for considering individuals as Social Entrepreneurs.

Case studies and primary data show that Indian social entrepreneurs have found innovative solutions to social issues like poverty, child rights violations, a lack of infrastructure in rural and tribal areas, violence against women, caste discrimination, deteriorating agricultural production, a lack of quality education and health care, and rural problems. To address these challenges, social entrepreneurs have formed a limited corporation, nongovernmental organization, or trust. Based on the aforementioned findings, the researcher can characterize Indian social entrepreneurs as follows: Social entrepreneurs are socially conscious individuals or organizations that use creative models to solve social and environmental issues. Through the application of entrepreneurial, business, and social work skills and strategies, along with social support, social entrepreneurs mobilize financial, human, and infrastructural resources to sustain their social entrepreneurial venture and fulfill their social mission.

**Implications of the study**—Implications of the study include recognizing the importance of incorporating social entrepreneurship into existing frameworks for social change. By utilizing a mediation approach, organizations and individuals can work together to address complex social issues in a collaborative and innovative manner. This study highlights the potential for social entrepreneurship to serve as a catalyst for positive social change, ultimately leading to a more inclusive and equitable society.

## References

1. Gupta, P., Chauhan, S., Paul, J., Jaiswal, M.P.: Social entrepreneurship research: a review and future research agenda. J. Bus. Res. **113**, 209–229 (2020)
2. Canestrino, R., Ćwiklicki, M., Magliocca, P., Pawełek, B.: Understanding social entrepreneurship: a cultural perspective in business research. J. Bus. Res. **110**, 132–143 (2020)

3. Gupta, P., Srivastava, R.: Research on social enterprises from an emerging economy—systematic literature review and future research directions. J. Soc. Entrepreneurship **15**(2), 458–493 (2024)
4. Maksum, I.R., Rahayu, A.Y.S., Kusumawardhani, D.: A social enterprise approach to empowering micro, small and medium enterprises (SMEs) in Indonesia. J. Open Innov.: Technol. Market Complexity **6**(3), 50 (2020)
5. Kruse, P., Wach, D., Wegge, J.: What motivates social entrepreneurs? A meta-analysis on predictors of the intention to found a social enterprise. J. Small Bus. Manage. **59**(3), 477–508 (2021)
6. Pinheiro, P., Daniel, A., Moreira, A.: Social enterprise performance: the role of market and social entrepreneurship orientations. VOLUNTAS: Int. J. Voluntary Nonprofit Organ. **32**(1), 45–60 (2021)
7. Diaz Gonzalez, A., Dentchev, N.A.: Ecosystems in support of social entrepreneurs: A literature review. Soc. Enterp. J. **17**(3), 329–360 (2021)
8. Bruder, I.: A social mission is not enough: reflecting the normative foundations of social entrepreneurship. J. Bus. Ethics **174**(3), 487–505 (2021)
9. Xiao, Y., Liu, Z., Ahlstrom, D., Bai, Y.: Entrepreneurs' personality traits and social enterprise: a legitimation perspective. Entrepreneurship Res. J. **14**(3) (2022)
10. Kickul, J., Lyons, T.S.: Understanding social entrepreneurship: the relentless pursuit of mission in an ever changing world. Routledge (2020)
11. Haugh, H.M., Talwar, A.: Linking social entrepreneurship and social change: the mediating role of empowerment. J. Bus. Ethics **133**, 643–658 (2016)
12. Perrini, F., Vurro, C.: Social entrepreneurship: Innovation and social change across theory and practice. In: Social Entrepreneurship, pp. 57–85. Palgrave Macmillan UK, London (2006)
13. Ney, S., Beckmann, M., Graebnitz, D., Mirkovic, R.: Social entrepreneurs and social change: tracing impacts of social entrepreneurship through ideas, structures and practices. Int. J. Entrep. Ventur. **6**(1), 51–68 (2014)
14. Stone, K.B.: Four decades of lean: a systematic literature review. Int. J. Lean Six Sigma **3**(2), 112–132 (2012)
15. Okoli, C.: A guide to conducting a standalone systematic literature review. Commun. Assoc. Inf. Syst. **37** (2015)
16. Xiao, Y., Watson, M.: Guidance on conducting a systematic literature review. J. Plan. Educ. Res. **39**(1), 93–112 (2019)
17. Luke, B., Chu, V.: Social enterprise versus social entrepreneurship: an examination of the 'why' and 'how' in pursuing social change. Int. Small Bus. J. **31**(7), 764–784 (2013)
18. Daru, M.U., Gaur, A.: Social entrepreneurship-a way to bring social change. Innov. J. Bus. Manag. **2**(1), 26–29 (2013)
19. Bansal, S., Garg, I., Sharma, G.D.: Social entrepreneurship as a path for social change and driver of sustainable development: a systematic review and research agenda. Sustainability **11**(4), 1091 (2019)
20. Steyaert, C., Hjorth, D. (eds.).: Entrepreneurship as Social Change: A Third New Movements in Entrepreneurship Book, vol. 3. Edward Elgar Publishing (2008)
21. Hietschold, N., Voegtlin, C., Scherer, A.G., Gehman, J.: Pathways to social value and social change: an integrative review of the social entrepreneurship literature. Int. J. Manag. Rev. **25**(3), 564–586 (2023)
22. Kury, K.W.: Sustainability meets social entrepreneurship: a path to social change through institutional entrepreneurship. Int. J. Bus. Insights Transformation **4** (2012)
23. Auvinet, C., Lloret, A.: Understanding social change through catalytic innovation: empirical findings in Mexican social entrepreneurship. Can. J. Adm. Sci./Revue Canadienne des Sciences de l'Administration **32**(4), 238–251 (2015)
24. Lumpkin, G.T., Bacq, S., Pidduck, R.J.: Where change happens: community-level phenomena in social entrepreneurship research. J. Small Bus. Manage. **56**(1), 24–50 (2018)
25. Abdul Ghani Azmi, I., Che Hashim, R., Mohamed, H.A.B., Sahol Hamid, N.: A study of Muslim women entrepreneurs' SMEs challenges and motivation in the Asia Pacific region. In: Rafiki,

A., Dana, L.P., Nasution, M.D.T.P. (eds) Open Innovation in Small Business. Contributions to Environmental Sciences & Innovative Business Technology. Springer, Singapore (2023). https://doi.org/10.1007/978-981-99-5142-0_13

26. Gundry, L.K., Kickul, J.R., Griffiths, M.D., Bacq, S.C.: Creating social change out of nothing: the role of entrepreneurial bricolage in social entrepreneurs' catalytic innovations. In: Social and Sustainable Entrepreneurship, pp. 1–24. Emerald Group Publishing Limited (2011)
27. Constantin, P.N., Stanescu, R., Stanescu, M.: Social entrepreneurship and sport in Romania: how can former athletes contribute to sustainable social change? Sustainability **12**(11), 4688 (2020)
28. Lans, W., Van Der Voordt, T.: Descriptive research. In: Ways to Study and Research Urban, Architectural and Technical Design (2002)
29. Verma, J.P.: Data analysis in management with SPSS software. Springer Science & Business Media (2012)
30. Barnidge, M., De Zúñiga, H.G.: Amos (software). In: The International Encyclopedia of Communication Research Methods, pp. 1–3 (2017)

# Analysing the Status of Agricultural Finance in Rural India in Promoting Agriculture

**Roopa Adarsh, E. Jothi, K. Sivasubramanian, and R. Madhusudhanan**

**Abstract** Agriculture is an integral sector of the global economies and is the backbone of many developing economies, particularly in India. This sector is contributing significantly to the gross domestic product and employment generation. According to the Economic Survey 2022–23, the gross value addition of the agricultural sector to national income has been recorded as 18.3% at current prices. It is also observed that there was a huge decline in the share of gross domestic product by agricultural production from 55.41% to 17.8% in 1950–51 and 2020–21 respectively. The agricultural output has declined over the last few decades as the industrial and service sectors have seen a robust improvement. The lack of agrarian finance is one of the various problems of the farm sector in discouraging agricultural activities. This study aims to identify the significance of agrarian finance on rural development in India and analyse the impact of agricultural finance on farmers in India.

**Keywords** Agricultural · Finance · Farmers · Economic and empowerment

## 1 Introduction

Agricultural finances are prominent in improving technology and productivity to empower the rural poor. The main purpose of boosting agriculture is to facilitate people worldwide to ensure food security [1]. Finance for agriculture is an important

R. Adarsh
Department of Economics, Mount Carmel College, Bangalore, India

E. Jothi
Department of Commerce (CA), Sona College of Arts and Science, Salem, India

K. Sivasubramanian (✉)
Department of Economics, Kristu Jayanti College, Bangalore, India
e-mail: sivasubramanian@kristujayanti.com

R. Madhusudhanan
Department of Management, Kristu Jayanti College, Bangalore, India



M. Al Mubarak (ed.), *Sustainable Digital Technology and Ethics in an Ever-Changing Environment*, Studies in Systems, Decision and Control 237,
https://doi.org/10.1007/978-3-031-86708-8_52

element in seeding the agricultural inputs and deriving effective output. Technological inputs can be procured and utilized by the farmer through the availability of funds in hand or borrowed from other sources. But the own funds are always not enough to meet costs and the farmers need financial support through credits. In India, the local lenders supply farm loans for agricultural activities and they impose a high rate of interest [2]. Poverty and hunger become the greatest challenges of the world countries. The developing countries taking major initiatives to arrest this problem [3].

### 1.1 Paper Structure

This research work is prepared with certain specific objectives such as to identify the significance of agricultural finance on rural development in India and analyse the impact of agricultural finance on farmers in India. The study is articulated as follows, the introduction of the chapter, objectives, methodology, review of literature, analysis, findings and conclusion.

### 1.2 Objectives

To identify the significance of agricultural finance on rural development in India.
To analyse the impact of agricultural finance on farmers in India.

## 2 Methodology

A simple analytical method has been adopted to proceed with this research paper. Various existing works which are closely related to the selected topic have been collected and reviewed. The reviewed literature has been critically evaluated to embrace and emphasise various new findings to provide suggestions. Along with the descriptive analysis, this study applied certain available secondary data to exhibit the significance and impact of agricultural finances in India. Data was collected from different published sources such as RBI and NABARD.

## 3 Literature Review

As stated in the introduction segment, agricultural finance or farm loans are crucial elements of the agrarian sector as a whole. This section examines and compares the related works of agricultural finance in India in terms of its definition, scope, problems, and future opportunities. The term agricultural finance emphasises the examination and analysis of financial components in the farm sector. It is very significant due to the objective of enhancing the agricultural and [4]. The nature of the farm sector: It is very tough to forecast risks and problems encountered by the farm sector across the globe. The landlord or the farmer engaging in agriculture is vulnerable towards numerous uncertain conditions such as cyclones, monsoon failure, water shortage, labour issues, and non-availability of formal agricultural finances [5]. Because of these various circumstances, the farm sectors cannot meet the expected development. The farmer's income was recorded as low compared with other sectors in informal employment. Predominant farmers are carrying out their agricultural activities with their resources without cooperation [6]. To achieve sustainable development in rural India, formal agricultural finance creates an avenue for growth and development [7].

### *3.1 Agricultural Credit in Developing Economy*

It is identified that there are a huge number of small and marginal farmers. They used to face the problem of access to timely and sufficient institutional finance [8, 9]. According to the Economic Survey 2022–23, the gross value addition of the agricultural sector to national income has been recorded as 18.3% at current prices. It is also observed that there was a huge decline in the share of gross domestic product by agricultural production from 55.41% to 17.8% in 1950–51 and 2020–21 respectively [10].

### *3.2 Problems of Agricultural Finance*

Poor communication and lack of transportation technology is the major issue for the agricultural production and marketing in India. However, our country is a prominent developing nation and the villages do not have connected roads and non-availability of sufficient infrastructure facilities led to a decline in production [11]. The credit facilities are very limited to farmers in rural India. They are freely accessible to moneylenders in rural areas and the local financiers charge an excessive amount of interest for the agricultural loans [12]. Along with these problems, the accessibility of bank loans makes the farmers vulnerable to the production and productivity of agricultural commodities [13].

### 3.3 Opportunities, Government Support and Role of Banks

Timely and enough credit at reasonable rates is critical to the growth of agriculture, as is any other productive activity. Agricultural credit was originally provided in India by indigenous loan providers such as landowners, agricultural and professional financiers, traders, commission agents, and family members. While these avenues had certain advantages, such as ease of access and availability and the lack of loan procedures, they also had some drawbacks, such as excessive and usurious interest rates, the exploitative nature of moneylenders, and insufficiency [14]. To overcome these constraints and provide adequate credit to agriculturists, modern financial institutions have been created in a multi-agency structure, with credit cooperative societies and banks, regional rural banks, and commercial banks serving as all three parts of this multiagency strategy [15, 16].

## 4 Analysis and Interpretation

The agricultural sector has to be well developed through various initiatives of both union and state governments in India. In that scenario, support through technological advancement, and providing subsidised raw materials and irrigation facilities would yield the results. In this context, agricultural finance plays a key role in progressing the agricultural output.

Table 1 depicts the status of agricultural credit flow from 2016–17 to 2019–20. Data shows that there is consistent financial support from the banks. But while we compare the amount of loan in line with the size of farming activities in India, it is on the lesser side.

Figure 1 emphasises the grant for the agricultural sector relating to the purchase of seeds, fertilizers, electricity and solar panels. During the year 2020–21, the grant by the government was given at Rs. 21,175 rupees and it has been declined to Rs. 19,468 crores in the year 2021–22. There was a drastic decline in the grant and the same amount has been followed in the next financial year also with a grant of Rs. 19,500 crores.

**Table 1** Status of agricultural credit

| Year | 2016–17 | 2017–18 | 2018–19 | 2019–20 |
|---|---|---|---|---|
| Total agricultural credit | 1,065,755 | 1,168,503 | 1,256,830 | 1,392,729 |

*Source* NABARD, 2020

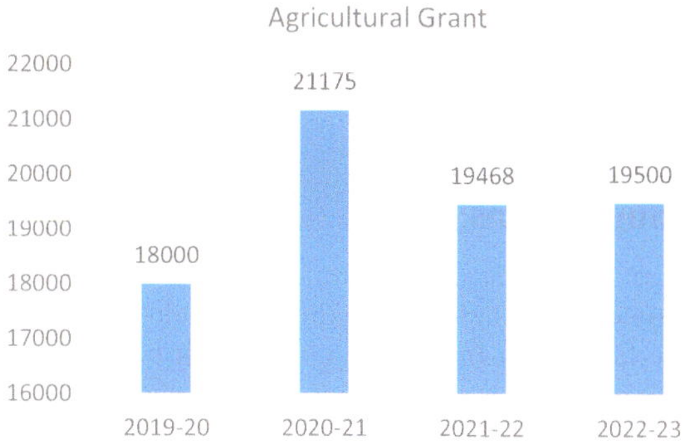


**Fig. 1** Agricultural grant from 2019–20 to 2022–23 (in crore). *Source* GOI, Ministry of Agriculture

## 5 Key Findings

According to the Economic Survey 2022–23, the gross value addition of the agricultural sector to national income has been recorded as 18.3% at current prices. It is also observed that there was a huge decline in the share of gross domestic product by agricultural production from 55.41% to 17.8% in 1950–51 and 2020–21 respectively.

Technological inputs can be procured and utilized by the farmer through the availability of funds in hand or borrowed from other sources. Predominant farmers are carrying out their agricultural activities with their resources without cooperation. Agricultural credit flow from 2016–17 to 2019–20. Data shows that there is consistent financial support from the banks. But while we compare the amount of loan in line with the size of farming activities in India, it is on the lesser side.

The credit facilities are very limited to farmers in rural India. They are freely accessible to moneylenders in rural areas and the local financiers charge an excessive amount of interest for the agricultural loans.

## 6 Conclusion

The policymakers and government are taking several necessary initiatives to enhance the financial accessibility of farmers. It helps to promote timely and adequate financial support to widen the agricultural activities [17]. Banks play a crucial role in credit creation and circulate the money to the farmers by way of providing affordable loans. Under this characteristic, banks serve as a vehicle for mobilising the savings of a vast number of people who have only little sums to deposit. The amount so mobilised has a high turnover, which is then given as credit to various industries, small, medium, and large, in order to enhance industrial activity in the national economy and so raise the general economic level of society. This in turn contributes to more savings and further economic expansion. As a result, a cycle is formed, which leads to overall economic growth. To conclude, the availability of formal agricultural loans is sufficient and efficient to promote farm production in India. It requires a robust supply of bank loans to increase the contribution of this sector to the nation's total output.

## 7 Suggestions

The state and the central government should consider more collateral-free farm loans to the farmers in India. The government has to create awareness among the rural people about the availability of various schemes for the farmers.

## 8 Limitations and Future Direction

This part of the research is limited and related to financial resources only. There is a huge amount of opportunity to improve this study further with other factors. The other influential factors are the educational status of the farmers, technological application in agriculture, marketing-related aspects and government schemes.

## References

1. Khan, F.U., Nouman, M., Negrut, L., Abban, J., Cismas, L.M., Siddiqi, M.F.: Constraints to agricultural finance in underdeveloped and developing countries: a systematic literature review. Int. J. Agric. Sustain. **22**(1) (2024). https://doi.org/10.1080/14735903.2024.2329388
2. Damodaran, A.: Financial inclusion: issues and challenges. AKGEC Int. J. Technol. **4**(2), 54–59 (2013)
3. De Janvry, A., Sadoulet, E.: Using agriculture for development: supply-and demand-side approaches. World Dev. **133**, 105003 (2020). https://doi.org/10.1016/j.worlddev.2020.105003
4. Kishore, N.K.: Agriculture credit in India: an integrated rural credit approach. Int. J. Sci. Res. Publ. **2**(3), 1–5 (2012)
5. Kambali, U., Niyaz.: An overview of agriculture finance in India. Int. J. Case Stud. Bus. IT Educ. **5**(2), 197–210 (2021)
6. Kesarwani, S.K., Tiwari, S.C., Jee, K., Kumar, A.: Advancement of agricultural finance in India: a chronicled point of view. Empirical Econ. Lett. **21**(3), 339–365 (2022)
7. Kumar, C.S., Turvey, C.G., Kropp, J.D.: The impact of credit constraints on farm households: survey results from India and China. Appl. Econ. Perspect. Policy **35**(3), 508–527 (2013). https://doi.org/10.1093/aepp/ppt002
8. Yadav, P., Sharma, A.K.: Agriculture credit in developing economies: a review of relevant literature. Int. J. Econ. Financ. **7**(12), 219–244 (2015). https://doi.org/10.5539/ijef.v7n12p219
9. Weng, C., Wang, H., Heerink, N., van den Berg, M.: Credit constraints and rural households' entrepreneurial performance in China. Emerg. Markets Finance Trade **58**(2), 570–583 (2020)
10. Kumar, A., Singh, M.K., Sinha, S.: Institutional credit to agriculture sector in India: status, performance and determinants. Agric. Econ. Res. Rev. **23**(2), 253–264 (2010)
11. Sukanya, N., Vishwanatha.: Commercial bank finance to agriculture in India—the recent trends and issues. J. Rural Dev. **33**(4), 493–508 (2014). Retrieved from https://www.nirdprojms.in/index.php/jrd/article/view/93284

12. Kumar, M.: Problems of agriculture marketing in India. Int. J. Res. Anal. Rev. **5**(4), 296–299 (2018)
13. Jatav, S.S., Nayak, S.: Access and determinants of formal agriculture credit in Uttar Pradesh, India. J. Rural Dev. **41**(2), 185–197 (2023). https://doi.org/10.25175/jrd/2022/v41/i2/172905
14. Kumar, V.: Innovative loan products and agricultural credit: a case study of KCC scheme with special reference to Alwar District of Rajasthan. Indian J. Agric. Econ. **66**(3), 477–488 (2011)
15. Agarwal, A.K.: Banking penetration in rural areas and villages: trends and Challenges. Int. J. Res. Commer. Manage. **6**(2)
16. Kumar, A., Singh, R.K.P., Jee, S., Chand, S., Tripathi, G., Saroj, S.: Dynamics of access to rural credit in India: patterns and determinants. Agric. Econ. Res. Rev. **28**, 151–166 (2015)
17. Bedru, B.B., Motunrayo, A.O.: Agricultural credit constrains in smallholder farming in developing countries: evidence from Nigeria. World Dev. Sustain. **1**, 100012 (2022). https://doi.org/10.1016/j.wds.2022.100012

# Challenges on Wheels: A Study of Food Delivery Executives and Influencing Factors in Bangalore

**Bijin Philip and Soumya Simon**

**Abstract** Technological advancement in the industry has accelerated as customers' inclination towards home delivery services has replaced eating out. Right now, food delivery services are dominating the internet industry. The overall revenue of the online meal ordering marketplace is predicted to increase by 8.8% until 2024, based on statistics. Many sit-down restaurants become Ghost Kitchens or delivery-only businesses as a result of COVID-19. According to a Business Wire analysis, it is anticipated that the global trend of online meal delivery services would continue, with this market growing at an annual pace of 11% to reach $192 billion by 2025. There has been a steady rise in unemployment in India. Many engineers and graduates are either unemployed or underemployed. This has led to an increase in the no. of engineers and graduates opting for the job of delivery executives as it has good income and incentives. Youth are the building blocks of the nation understanding the challenges and stress levels faced at their workplace is of primary concern. The current study will play a pivotal role in understanding various challenges faced by the online food delivery executives and provide valuable suggestions to overcome the day to day challenges faced, and in improving the work culture by taking necessary measures to reduce the stress levels. Some of the major issues that meal delivery companies are currently facing are covered in this article. Based on the identification of noteworthy obstacles such disobedient clients, delays in logistics, and problems with payments, our results emphasise the urgent requirement for focused interventions to assist these employees.

**Keywords** Food delivery executives · Delivery obstacles · Employee well-being · Customer service

B. Philip (✉)
Department of Management, Kristu Jayanti College Autonomous, Bengaluru, India
e-mail: bijin.p@kristujayanti.com

S. Simon
Department of Psychology, Kristu Jayanti College Autonomous, Bengaluru, India



M. Al Mubarak (ed.), *Sustainable Digital Technology and Ethics in an Ever-Changing Environment*, Studies in Systems, Decision and Control 237,
https://doi.org/10.1007/978-3-031-86708-8_53

# 1 Introduction

India has seen a boom in the e-commerce industry in the past few years and it has led to a rise in the demand for delivery executives. The latest technological advancements have played a pivotal role in the food delivery services, it has revolutionized consumer behaviour. Consumers have numerous options at their disposal and are motivated to have everything online, including cooked meals delivered at their door steps. Availability of products and various services are a matter of just few clicks on any mobile device and has made errands effortless. The excessive reliance on technology, comfort and fast delivery of food have become the key motivational factors for the users to opt for online food ordering and delivering services. The expectations of consumers have increased with the increase in the popularity of online food ordering and delivery services.

A substantial increase in the earnings and incentives of food delivery executives has inflated the number of graduates and engineers opting for this Job. These delivery executives become the ambassadors of the company and the business to the customers. Approximately 3 Lakh plus delivery executives are working with various food delivery application across India.

There has been a steady rise in unemployment in India. Many engineers and graduates are either unemployed or underemployed. This has led to an increase in the no. of engineers and graduates opting for the job of delivery executives as it has good income and incentives.

Youth are the building blocks of the nation understanding the challenges and stress levels faced at their workplace is of primary concern. The current study will play a pivotal role in understanding various challenges faced by the online food delivery executives and provide valuable suggestions to overcome the day to day challenges faced, and in improving the work culture by taking necessary measures to reduce the stress levels.

# 2 Theorectical Outline and Literature Indications

Singh [1] This research investigates the complex issues faced by food delivery executives—also known as delivery boys—in the ever-changing food delivery sector. This research paper identified diffent challeges that can be impact the effectiveness of the food delivery boys.

Aggarwal and Kapoor [2] The researcher's list of challenges includes things like not knowing about mobile apps, managing delivery times, keeping customers, controlling food quality, managing large quantities, having not enough personnel for delivery and logistics, setting up online food ordering gateways, and the complexities of ordering and delivery within the online food delivery industry.

Hao [3] In addition to examining the working conditions of food delivery workers in China's gig economy, this study also looks at public opinions and impressions of this workforce using a mixed qualitative and quantitative methodology. We found positive correlations between the residential area and the frequency of ordering meal delivery services through correlation and regression analyses. Additionally, we saw a strong correlation between people's opinions of delivery workers and their readiness to help this group as well as their assessment of the difficulties associated with the job.

Lal [4] Several publications draw attention to the risks posed by fast delivery riders who put their own and other people's lives in jeopardy by driving too fast and too safely. In their quest for quick service, start-up businesses like Zepto, Zomato, Domino's, and Swiggy have been accused of supporting dangerous practices and unsafe situations, which has resulted in numerous near-misses. This article, which draws conclusions from 201 industry professionals who thought about the problems and possible solutions, promotes the use of qualitative research to solve these concerns. It is crucial to uphold safety regulations to shield riders from harm and death if the delivery sector is to remain viable.

Nasreen [5] This essay explores the problems with social security and working conditions that delivery workers in Delhi, India experience. It does so by applying a Marxian perspective, paying special attention to the ideas of "alienation" and "precariousness." Delivery personnel have become indispensable to the e-commerce sector, particularly in the last phase of product delivery, as a result of the industry's explosive growth and the rising demand for doorstep delivery services. According to the paper, delivery workers who perform services related to consumption are especially vulnerable to precariousness and feel alienated from the delivery process. They experience uncertainty about social protections and employment stability, which increases their susceptibility and alienation from larger social systems.

## 3 Statement of the Problem

To analyse the influence of various factors like challenges faced, stress levels, coping styles and personality on job satisfaction among online food delivery executives. Despite the overwhelming rise in the ecommerce industry and the use of online food delivery apps and the increase in the demand for food delivery executives there has been a dearth of research in understanding the challenges and the various factors related to the wellbeing and optimal functioning of these individuals.

## 4 Objectives

- To identify the various challenges faced by the delivery executives while delivering food.
- To determine the correlation between the obstacles experienced by food dlivery executives and their age, education, and daily delivery count.

# 5 Methodology

**Source of Data:** The participants of this study were food delivery executives employed by various food delivery platforms Zomato and Swiggy and were operating in Bangalore. A total of 147 participants were selected using convenience sampling. The inclusion criteria were as follows:

1. Must be actively working as a food delivery executive.
2. Must be able to comprehend and respond to the questionnaire in English.

**Sampling:** The study was exploratory and used purposive sampling. The sample consisted of 147 food delivery executives in Bangalore.

**Inclusion Criteria:**

1. Online food delivery executives working with Swiggy and Zomato
2. Age 18–60
3. Food delivery executives working in Bangalore.

**Exclusion Criteria:**

1. Food delivery executives working with restaurants.
2. Online food delivery executives working with online applications other than Swiggy and Zomato
3. Food delivery executives working outside Bangalore.

**Analysis:** The data will be analyzed by using Descriptive statistics, ANOVA, correlation and regression analysis.

# 6 Results and Discussion

The modern workplace presents a myriad of challenges, particularly for gig economy workers such as food delivery executives. This segment of the workforce faces unique stressors including irregular work hours, job insecurity, and constant exposure to environmental hazards.

Table 1 shows the socio economic background of the respondents. The study surveyed 147 food delivery personnel, revealing a diverse demographic profile. The majority of respondents are aged between 18 and 25 years (48.3%) and 26–40 years (45.6%), with smaller percentages under 18 (1.4%) and above 40 (4.8%). The workforce is predominantly male (92.5%), with females representing 7.5%. In terms of education, most have completed SSLC (36.7%) or PU (38.1%), while fewer have attained Diplomas (8.2%), Degrees (12.9%), or Postgraduate qualifications (4.1%). Employment is almost evenly split between Swiggy (52.4%) and Zomato (47.6%). Monthly incomes through food delivery vary, with 40.1% earning 10,000–20,000 INR, 44.9% earning 20,000–40,000 INR, 10.2% earning less than 10,000 INR, and

4.8% earning over 40,000 INR. Regarding daily deliveries, 46.9% complete 20–30 deliveries, 23.8% complete 10–20, 19% complete 30–40, 6.8% complete fewer than 10, and 3.4% complete more than 40 deliveries.

Table 2 descriptive statistics reveal the array of challenges encountered by food delivery boys in Bangalore. The most significant hurdles appear to be uncooperative customers, late orders from restaurants, and inadequate navigation instructions, with mean ratings of 7.796, 7.143, and 7.027 respectively, showcasing their prevalence and impact. Payment-related issues follow closely behind with a mean rating of 6.333. Moderately challenging factors include order modifications or cancellations, vehicle parking, traffic congestion, environmental conditions, carrying heavy packages, and the pressure for swift deliveries, each exhibiting mean ratings between 4.204 and 4.973. These statistics underscore the complexity of the food delivery ecosystem, encompassing not only logistical and environmental obstacles but also the dynamics of customer interactions and time constraints that delivery personnel navigate daily in the bustling city of Bangalore.

**Table 1** Socio-demographic details

| | Categories | Percentage (%) |
|---|---|---|
| Age group | Less than 18 | 1.4 |
| | 18–25 | 48.3 |
| | 26–40 | 45.6 |
| | Above 40 | 4.8 |
| Gender | Male | 92.5 |
| | Female | 7.5 |
| Education | SSLC | 36.7 |
| | PU | 38.1 |
| | Diploma | 8.2 |
| | Degree | 12.9 |
| | Post graduate | 4.1 |
| Food delivery app | Swiggy | 52.4 |
| | Zomato | 47.6 |
| Monthly Income through food delivery | Less than 10,000 | 10.2 |
| | 10,000–20,000 | 40.1 |
| | 20,000–40,000 | 44.9 |
| | Above 40,000 | 4.8 |
| How many food delivery per day | Less than 10 | 6.8 |
| | 10–20 | 23.8 |
| | 20–30 | 46.9 |
| | 30–40 | 19 |
| | 40–50 | 3.4 |

**Table 2** Descriptive analysis of challenges faced by food delivery executives

| Challenges faced by food delivery executives | N | Mean | Std. deviation |
|---|---|---|---|
| Uncooperative customer | 147 | 7.796 | 2.9212 |
| Late orders to delivery boys by restaurants | 147 | 7.143 | 2.7548 |
| The customer was not given proper navigation | 147 | 7.027 | 2.8089 |
| Payment related issues | 147 | 6.333 | 2.6310 |
| Order modification or cancellation | 147 | 4.973 | 2.3230 |
| Vehicle parking | 147 | 4.463 | 2.4108 |
| Traffic | 147 | 4.442 | 2.6331 |
| Environment condition | 147 | 4.388 | 2.4591 |
| Carrying heavy packages | 147 | 4.231 | 2.3381 |
| Hard and fast delivery | 147 | 4.204 | 2.0869 |
| Valid N (listwise) | 147 | | |

Table 3 ANOVA analysis results indicate varied significance levels across different factors impacting delivery operations. The analysis reveals that for most factors, such as customer navigation issues ($F = 0.901$, $p = 0.443$), late orders to delivery boys ($F = 0.408$, $p = 0.748$), uncooperative customers ($F = 0.717$, $p = 0.543$), traffic ($F = 0.627$, $p = 0.599$), order modification or cancellation ($F = 0.087$, $p = 0.967$), payment-related issues ($F = 1.436$, $p = 0.235$), environmental conditions ($F = 0.806$, $p = 0.493$), and hard and fast delivery requirements ($F = 0.324$, $p = 0.808$), there are no statistically significant differences between groups, suggesting these issues are consistently problematic across different age group of the respondents. However, vehicle parking ($F = 3.270$, $p = 0.023$) and carrying heavy packages ($F = 2.888$, $p = 0.038$) show statistically significant differences between groups, indicating that these issues vary significantly among the different age groups of the respondent. This analysis highlights that while most operational challenges are uniformly experienced, specific issues like vehicle parking and carrying heavy packages exhibit significant variability and may require targeted interventions

Table 4 ANOVA analysis reveals significant differences between groups for several factors impacting delivery operations. Specifically, issues such as customers not being given proper navigation ($F = 4.005$, $p = 0.004$), late orders to delivery boys by restaurants ($F = 4.432$, $p = 0.002$), uncooperative customers ($F = 4.969$, $p = 0.001$), environmental conditions ($F = 5.708$, $p = 0.000$), hard and fast delivery requirements ($F = 4.432$, $p = 0.002$), and carrying heavy packages ($F = 4.516$, $p = 0.002$) exhibit statistically significant differences between groups. These results suggest that these issues vary considerably across different educational backgrounds and may require targeted interventions to address them effectively.

Conversely, factors such as traffic ($F = 1.238$, $p = 0.298$), order modification or cancellation ($F = 0.829$, $p = 0.509$), vehicle parking ($F = 1.035$, $p = 0.391$), and payment-related issues ($F = 0.461$, $p = 0.764$) do not show statistically significant

**Table 3** ANOVA—age and challenges faced by the food delivery executives

| | | Sum of squares | df | Mean square | F | Sig |
|---|---|---|---|---|---|---|
| The customer was not given proper navigation | Between groups | 16.370 | 3 | 5.457 | 0.901 | 0.443 |
| | Within groups | 866.528 | 143 | 6.060 | | |
| | Total | 882.898 | 146 | | | |
| Late orders to delivery boys by restaurants | Between groups | 6.767 | 3 | 2.256 | 0.408 | 0.748 |
| | Within groups | 791.369 | 143 | 5.534 | | |
| | Total | 798.136 | 146 | | | |
| Uncooperative customer | Between groups | 9.426 | 3 | 3.142 | 0.717 | 0.543 |
| | Within groups | 626.452 | 143 | 4.381 | | |
| | Total | 635.878 | 146 | | | |
| Traffic | Between groups | 13.111 | 3 | 4.370 | 0.627 | 0.599 |
| | Within groups | 997.555 | 143 | 6.976 | | |
| | Total | 1010.667 | 146 | | | |
| Order modification or cancellation | Between groups | 1.543 | 3 | 0.514 | 0.087 | 0.967 |
| | Within groups | 847.001 | 143 | 5.923 | | |
| | Total | 848.544 | 146 | | | |
| Vehicle parking | Between groups | 50.576 | 3 | 16.859 | 3.270 | 0.023 |
| | Within groups | 737.315 | 143 | 5.156 | | |
| | Total | 787.891 | 146 | | | |
| Payment related issues | Between groups | 29.606 | 3 | 9.869 | 1.436 | 0.235 |
| | Within groups | 982.652 | 143 | 6.872 | | |
| | Total | 1012.259 | 146 | | | |
| Environment condition | Between groups | 19.153 | 3 | 6.384 | 0.806 | 0.493 |
| | Within groups | 1132.738 | 143 | 7.921 | | |
| | Total | 1151.891 | 146 | | | |
| Hard and fast delivery | Between groups | 8.418 | 3 | 2.806 | 0.324 | 0.808 |
| | Within groups | 1237.459 | 143 | 8.654 | | |
| | Total | 1245.878 | 146 | | | |
| Carrying heavy packages | Between groups | 63.296 | 3 | 21.099 | 2.888 | 0.038 |
| | Within groups | 1044.704 | 143 | 7.306 | | |
| | Total | 1108.000 | 146 | | | |

differences between groups. This indicates that these issues are consistently experienced across different educational backgrounds of the respondents and may require more generalized solutions.

Overall, the analysis highlights specific areas that need more focused attention, such as navigation, timely order delivery, customer cooperation, environmental conditions, delivery demands, and the challenge of carrying heavy packages, while other issues appear uniformly problematic.

**Table 4** ANOVA—education and challenges faced by the food delivery executives

| | | Sum of squares | df | Mean square | F | Sig |
|---|---|---|---|---|---|---|
| The customer was not given proper navigation | Between groups | 89.507 | 4 | 22.377 | 4.005 | 0.004 |
| | Within groups | 793.391 | 142 | 5.587 | | |
| | Total | 882.898 | 146 | | | |
| Late orders to delivery boys by restaurants | Between groups | 88.577 | 4 | 22.144 | 4.432 | 0.002 |
| | Within groups | 709.559 | 142 | 4.997 | | |
| | Total | 798.136 | 146 | | | |
| Uncooperative customer | Between groups | 78.082 | 4 | 19.520 | 4.969 | 0.001 |
| | Within groups | 557.796 | 142 | 3.928 | | |
| | Total | 635.878 | 146 | | | |
| Traffic | Between groups | 34.055 | 4 | 8.514 | 1.238 | 0.298 |
| | Within groups | 976.612 | 142 | 6.878 | | |
| | Total | 1010.667 | 146 | | | |
| Order modification or cancellation | Between groups | 19.370 | 4 | 4.842 | 0.829 | 0.509 |
| | Within groups | 829.175 | 142 | 5.839 | | |
| | Total | 848.544 | 146 | | | |
| Vehicle parking | Between groups | 22.327 | 4 | 5.582 | 1.035 | 0.391 |
| | Within groups | 765.565 | 142 | 5.391 | | |
| | Total | 787.891 | 146 | | | |
| Payment related issues | Between groups | 12.978 | 4 | 3.244 | 0.461 | 0.764 |
| | Within groups | 999.281 | 142 | 7.037 | | |
| | Total | 1012.259 | 146 | | | |
| Environment condition | Between groups | 159.567 | 4 | 39.892 | 5.708 | 0.000 |
| | Within groups | 992.324 | 142 | 6.988 | | |
| | Total | 1151.891 | 146 | | | |
| Hard and fast delivery | Between groups | 138.284 | 4 | 34.571 | 4.432 | 0.002 |
| | Within groups | 1107.594 | 142 | 7.800 | | |
| | Total | 1245.878 | 146 | | | |
| Carrying heavy packages | Between groups | 125.043 | 4 | 31.261 | 4.516 | 0.002 |
| | Within groups | 982.957 | 142 | 6.922 | | |
| | Total | 1108.000 | 146 | | | |

Table 5 ANOVA analysis reveals several factors with statistically significant differences between groups, indicating varying impacts across different groups. For "Customer not given proper navigation," the analysis shows a significant difference ($F = 5.244$, $p = 0.001$), suggesting this issue is perceived differently among the groups. Similarly, "Late orders to delivery boys by restaurants" ($F = 2.785$, $p = 0.029$), "Traffic" ($F = 3.007$, $p = 0.020$), and "Carrying Heavy Packages" ($F = 3.843$, $p$

= 0.005) also exhibit statistically significant differences, indicating these factors are significant concerns that vary based on the number of deliveries payday.

Conversely, some factors did not show statistically significant differences between groups. These include "Uncooperative Customer" ($F = 2.223$, $p = 0.070$), "Order modification or cancellation" ($F = 1.332$, $p = 0.261$), "Vehicle Parking" ($F = 2.175$, $p = 0.075$), "Payment Related issues" ($F = 0.831$, $p = 0.508$), "Environment condition" ($F = 0.967$, $p = 0.428$), and "Hard and Fast Delivery" ($F = 1.393$, $p = 0.239$).

**Table 5** ANOVA—number of delivery and challenges faced by the food delivery executives

| | | Sum of squares | df | Mean square | F | Sig |
|---|---|---|---|---|---|---|
| The customer was not given proper navigation | Between groups | 113.631 | 4 | 28.408 | 5.244 | 0.001 |
| | Within groups | 769.267 | 142 | 5.417 | | |
| | Total | 882.898 | 146 | | | |
| Late orders to delivery boys by restaurants | Between groups | 58.062 | 4 | 14.515 | 2.785 | 0.029 |
| | Within groups | 740.074 | 142 | 5.212 | | |
| | Total | 798.136 | 146 | | | |
| Uncooperative customer | Between groups | 37.467 | 4 | 9.367 | 2.223 | 0.070 |
| | Within groups | 598.411 | 142 | 4.214 | | |
| | Total | 635.878 | 146 | | | |
| Traffic | Between groups | 78.920 | 4 | 19.730 | 3.007 | 0.020 |
| | Within groups | 931.747 | 142 | 6.562 | | |
| | Total | 1010.667 | 146 | | | |
| Order modification or cancellation | Between groups | 30.686 | 4 | 7.672 | 1.332 | 0.261 |
| | Within groups | 817.858 | 142 | 5.760 | | |
| | Total | 848.544 | 146 | | | |
| Vehicle parking | Between groups | 45.494 | 4 | 11.374 | 2.175 | 0.075 |
| | Within groups | 742.397 | 142 | 5.228 | | |
| | Total | 787.891 | 146 | | | |
| Payment related issues | Between groups | 23.155 | 4 | 5.789 | 0.831 | 0.508 |
| | Within groups | 989.104 | 142 | 6.966 | | |
| | Total | 1012.259 | 146 | | | |
| Environment condition | Between groups | 30.550 | 4 | 7.637 | 0.967 | 0.428 |
| | Within groups | 1121.342 | 142 | 7.897 | | |
| | Total | 1151.891 | 146 | | | |
| Hard and fast delivery | Between groups | 47.040 | 4 | 11.760 | 1.393 | 0.239 |
| | Within groups | 1198.837 | 142 | 8.443 | | |
| | Total | 1245.878 | 146 | | | |
| Carrying heavy packages | Between groups | 108.239 | 4 | 27.060 | 3.843 | 0.005 |
| | Within groups | 999.761 | 142 | 7.041 | | |
| | Total | 1108.000 | 146 | | | |

These results indicate that these issues are consistently experienced across different groups, with no significant variation in their impact.

In summary, while issues such as customer navigation, late orders, traffic, and carrying heavy packages show significant variability based on the number of orders per day and may require targeted interventions, other issues like uncooperative customers, order modifications, vehicle parking, payment issues, environmental conditions, and hard delivery requirements are uniformly perceived and might benefit from more generalized solutions.

## 7 Summary of Findings

The present study sheds light on the challenges faced by food delivery executives in Bangalore. This discussion compares our findings with recent research from the past five years to provide a comprehensive understanding of these dynamics.

### *7.1 Challenges Faced by Food Delivery Executives*

Our data indicates that there are some genuine difficulties in the world. First, a large number of clients simply don't cooperate, with an average score of 7.796. Then, there are order deliveries that are delayed and are now at 7.143. With a score of 7.027, the navigation directions are also subpar. Similar findings were made in other studies, which also highlight these problems in the workplace. For instance, a 2019 study by Van Dorn and colleagues highlighted the difficulties delivery workers may have with customers and with logistical delays. Performance and job satisfaction are severely harmed by these difficulties.. Furthermore, research by Shankar and Gopinath [6] confirmed that navigation problems and delays in order preparation significantly impact the efficiency and stress levels of food delivery personnel.

Another significant worry that surfaced was payment-related (mean = 6.333). This is consistent with research by de Souza et al. [7], which found that gig workers frequently complain about inconsistent and inadequate income, which increases stress and financial instability. The literature also mentions somewhat difficult elements that we found in our study, like environmental concerns, road congestion, and the need for quick deliveries. According to D'Cruz and Noronha [8], time constraints and urban environmental stresses make it more difficult for gig workers to succeed, which emphasises the need for policies and infrastructure that support them.

## 8 Conclusion

This study shows how different personality traits mix with job stress and coping methods for delivery workers in Bangalore. It out some tough challenges like dealing with unco customers, late deliveries, and payment. Our findings really highlight the need for strategics to help these workers out. There's a strong link between CISM and ineffective coping ways., we see how aversion plays a subtle role in stress and coping. These insights can help create better stress management programs.

Organizations should think about these personality traits when they're planning training and support systems. This way, it'll help enhance flexibility in coping and boost resilience among their teams! Moving ahead, future studies should look at how these ideas work in other gig economy jobs and different cultures. It's also important to explore how the long-term effects of personality traits and coping methods influence job stress and overall happiness.

In short, this study helps us better understand what's happening in the gig economy. It opens doors for better support and well-being for food delivery workers!

## References


1. Singh, K.: A study on challenges faced by food delivery executives or delivery boys. IJSREM (2024). https://ijsrem.com/download/a-study-on-challenges-faced-by-food-delivery-executives-or-delivery-boys/
2. Aggarwal, B., Kapoor, D.: A study on managing online food delivery business in vuca environment: challenges and strategies. VUCA **108** (2020)
3. Hao, X.: An investigation on the food-delivery men group. In: 2020 3rd International Conference on Humanities Education and Social Sciences (ICHESS 2020), pp. 453–462. Atlantis Press, December 2020
4. Lal, H.: Fast Food delivery boys, at-risk behaviours, fatalities and solutions-qualitative perspective. Indian J. Health Well Being **13**(2), 270–273 (2022)
5. Nasreen, A., Purohit, S.K.: Working conditions and social security issues of e-commerce delivery workers in India: understanding from Marxian perspectives. Int. J. Soc. Sci. Econ. Res. **3**(3), 1049–1067 (2018)
6. Shankar, A., Gopinath, S.: Urban environmental stressors and food delivery. J. Urban Manag. **9**(2), 135–147 (2020)
7. de Souza, K.R., de Almeida, J.A., Moreira, S.: Payment and financial insecurity among gig workers. Int. J. Employ. Stud. **29**(1), 123–145 (2021)
8. D'Cruz, P., Noronha, E.: The interface between technology and customer worker relations in India. Ind. Labor Relat. Rev. **72**(3), 767–794 (2019)

# The Devastating Impact of Greenhouse Gas Emissions: A Looming Threat to Sustainable Development in India

**P. Remmiya Rajan and K. B. Rishikesh**

**Abstract** This study investigates the trends in greenhouse gas emissions across various sectors in India, focusing on carbon dioxide ($CO_2$), methane ($CH_4$), and nitrous oxide ($N_2O$). By estimating and evaluating the emissions and removals from key sectors, including industrial, agricultural, and other significant subsectors, this research aims to comprehensively analyse the types and sources of greenhouse gases emitted by India's industries. The study conducts a detailed trend analysis within multiple subsectors, identifying significant patterns and discrepancies over different time periods. A comparative analysis between the industrial and agricultural sectors highlights the differences in emission sources and levels. Furthermore, this research delves into the measures and initiatives undertaken by the Indian government to mitigate greenhouse gas emissions and promote sustainable development. This includes policies, regulations, and programs to reduce emissions and enhance environmental sustainability. The findings of this study are expected to contribute to a deeper understanding of sector-specific emissions in India, offering valuable insights into the effectiveness of current policies and suggesting areas for improvement. This research aims to inform and influence policy-making for enhanced environmental sustainability in India by comprehensively evaluating emissions trends and government actions.

**Keywords** Green house gases · Looming threat · Sustainable development

P. Remmiya Rajan (✉)
Department of Economics, Zamorin's Guruvayurappan College, Calicut University, Malappuram, India
e-mail: remmiyadinesh@gmail.com

K. B. Rishikesh
Department of Commerce, School of Commerce, Finance and Accountancy, CHRIST (Deemed to Be University), Bangalore, India
e-mail: rishikeshkb@christuniversity.in



M. Al Mubarak (ed.), *Sustainable Digital Technology and Ethics in an Ever-Changing Environment*, Studies in Systems, Decision and Control 237,
https://doi.org/10.1007/978-3-031-86708-8_54

## 1 Introduction

The greenhouse effect, a natural phenomenon where certain gases in the Earth's atmosphere trap heat from the sun, is the basis for several Sustainable Development Goals (SDGs). It is directly related to SDG 13 (Climate Action) because human actions, particularly the combustion of fossil fuels, have increased greenhouse gas concentrations, exacerbated the greenhouse effect, and caused global warming. It also impacts SDGs 14 (Life Below Water) and 15 (Life on Land) due to the altered habitats and threatened biodiversity brought on by the ensuing climatic changes. Additionally, because crop yields and water availability are impacted by climate change, it has an effect on SDGs 6 (Clean Water and Sanitation) and SDG 2 (Zero Hunger). Last but not least, the greenhouse effect is closely related to is closely related to the advocacy of switching to renewable and low-carbon energy sources in order to reduce greenhouse gas emissions. According to the World Resources Institute Climate Analysis Indicators Tool, 68.7% of India's total emissions in 2014 came from a variety of sectors. Manufacturing and construction accounted for 24% of emissions within the varied sector, while energy and heat production provided 49%. Agriculture was the second-highest source of emissions (19.6% of total emissions), with enteric fermentation accounting for 45% of emissions. India is attempting to improve energy security, affordability, and accessibility while shifting its energy production toward more renewable sources. Its economy is expected to expand at one of the fastest rates in the coming years, which would significantly increase the need for energy. It has the ability to affect the environment whether it meets those criteria by using fossil fuels or ecologically beneficial alternatives. The government has made great strides toward fulfilling its emissions reduction targets under the Paris Agreement, but under its existing policies, India's overall greenhouse gas emissions will still increase by more than 40% by 2030. While meeting targets for energy security and poverty reduction may require a slight rise in short-term emissions, accelerated implementation of current policies could significantly reduce emissions over the medium term and move India closer to achieving net zero emissions by 2070.

## 2 The Structure of the Paper

Including the structure in your paper as you described would likely mean presenting it as an outline or a roadmap within the document such as

- **Introduction:** Start by introducing the research topic and providing relevant background information to set the stage for the study.
- **Rationale of the Study:** Explain why this study is necessary and relevant, linking it to the gaps identified in existing literature.
- **Objectives:** State the overall aim of your research

- **Methodology:** The study employed a mixed-methods approach, combining quantitative surveys with qualitative interviews to comprehensively analyze the population, using stratified random sampling to ensure diverse representation and employing statistical analysis tools for data interpretation.
- **Analysis and Interpretation**: Provide an in-depth interpretation of the data, explaining what the results mean in the context of your research objectives.
- **Conclusion** Discuss the implications of your findings for theory, practice, and future research.
- **References** Provide a complete list of all sources cited in your research, formatted according to your chosen citation style

## 3 Rationale of the Study

One of the most important ways to slow down climate change is to stabilize greenhouse gas (GHG) emissions from croplands as the demand for agricultural goods rises. Depending on how they are run, the energy, transportation, industrial, and agricultural sectors might all be net carbon sinks or generators. At the moment, it contributes 25% of the world's anthropogenic carbon emissions. Despite accounting for only 8% of India's total greenhouse gas emissions, this industry might play a major role in the country's efforts to reach net-zero emissions by 2070. In this analysis, we assess the carbon footprints of India's agriculture, forestry, and other land-use sectors, with particular attention to enteric fermentation, rice paddies, burning crop residues, forest fires, fertilizer and manure management, and food waste. Additionally, employing the conventional autoregressive. The article primarily focuses on proactive strategies for utilizing the energy, transportation, industrial, and agricultural sectors to help achieve net-zero emission targets. The key to lowering India's carbon footprint in the agriculture, forestry, and other land-use sectors is to increase the amount of forest cover, agroforestry, and other tree-based land-use systems; enhance soil health through soil management; lower emissions from rice ecosystems; and reduce food waste.

## 4 Objectives of the Study

This study examines trends in greenhouse gas emissions measured over a range of time periods. This exercise covers carbon dioxide (CO), methane (CH), and nitrous oxide (NO) as greenhouse gases. The following sectors' greenhouse gas emissions and removals are estimated and evaluated by various Indian sectors.

- To analyse the many types of greenhouse gases that India's industries emit.
- To examine trend analysis conducted by several Indian subsectors
- To contrast the emissions from the industrial and agricultural sectors

- To research the actions the government has made to reduce greenhouse gas emissions and promote sustainable development

## 5 Methodology

The study's secondary sources were the JRC Highlights Report 2023 and the Primary Analysis of Cumulative Emission and Energy Statistics (2018). Access to Clean Cooking Energy and Electricity: Survey of States 2018, India's transport sector: state-wise synthesis, Atmospheric Environment [1], Ministry of Statistics, Government of India, IMF ENV model Jean, JRC Highlights Report 2023 [2], and Solid waste management in India [3].

**Statistical technique**

**Industrial waste calculation**

$$\text{Mean} = \frac{\Sigma\ \text{Industrial Emission}}{\text{Number of Industries}}$$

$$\text{Variance} = \frac{\Sigma(\text{Emission} - \text{Mean})2}{\text{Number of Industries} - 1}$$

**Agricultural waste calculation**

$$\text{Mean} = \frac{\Sigma\ \text{Agricultural Emission}}{\text{Number of Industries}}$$

$$\text{Variance} = \frac{\Sigma(\text{Emission} - \text{Mean})2}{\text{Number of Agricultural} - 1}$$

**Percentage Growth** = ((Value in Current Year − Value in Previous Year)/Value in Previous Year) * 100.

**The Pareto analysis:** The chart and the bar chart show the emissions by each country and the cumulative percentage of emissions, highlighting the major contributors to the analysis.

## 6 Analysis and Interpretation of Data

**Pareto Analysis**

The Pareto analysis chart and the bar chart show the emissions by each country and the cumulative percentage of emissions, highlighting the major contributors. The use of modern technology in the future will help achieve the successful completion of the United Nations' Sustainable Development Goals (SDGs), but it always leads to gas emissions 10) Joseph [4].

**Largest contributors to greenhouse gas emissions worldwide in 2023, by country** (in billion metric tons of carbon dioxide equivalent) (Table 1).

**Interpretation of the data:** The Fig. 1 shows that China is the largest emitter, contributing 15.7 units and accounting for 47% of cumulative global emissions. The United States follows with 6.0 units, bringing the cumulative total to 65%. India, the EU 27, and Russia contribute significantly as well, with cumulative percentages reaching 76.7%, 87.5%, and 95.2%, respectively. Other notable contributors include Brazil, Indonesia, Japan, Iran, and Mexico. The top five emitters account for over 95% of global emissions, highlighting their major environmental impact.

**Interpretation of the data:** As shown in the Fig. 2, there was a slight increase in 2018–19 to 0.9648 tCO2/MWh, followed by a small decrease in 2019–20 to 0.9603 tCO2/MWh. However, the emissions intensity increased again in 2021–22, rising to 0.9675 tCO2/MWh. This fluctuation indicates variable progress in efforts to reduce the carbon footprint of electricity generation over these years.

1. **To analyse various greenhouse gases emitted by various sectors**

**Interpretation of the data:** As shown between 2006 and 2018, emissions in India's energy sectors have shown significant growth. Public electricity generation emissions nearly doubled, reflecting increasing energy demand and industrialisation (Table 2).

As shown in Table 3, Public Electricity Generation has the steepest positive slope (43.64), indicating a significant increase in emissions over the years, with a very high R-squared value (0.9964), signifying a strong linear relationship.

**Interpretation of the data:** As shown in Table 4, it indicates that the Waste sector is the largest contributor to emissions, with a significant value of 267, overshadowing

**Table 1** Table showing the largest contributors of greenhouse gas emissions worldwide in 2023

| Country | Emission | Cumulative emission | Cumulative percentage |
|---|---|---|---|
| China | 15.7 | 15.7 | 47.0 |
| United States | 6.0 | 21.7 | 65.0 |
| India | 3.9 | 25.6 | 76.7 |
| EU 27 | 3.6 | 29.2 | 87.5 |
| Russia | 2.6 | 31.8 | 95.2 |
| Brazil | 1.3 | 33.1 | 99.0 |
| Indonesia | 1.2 | 34.3 | 102.6 |
| Japan | 1.2 | 35.5 | 106.3 |
| Iran | 1.0 | 36.5 | 109.2 |
| Mexico | 0.8 | 37.3 | 111.6 |

*Source* Simpson, C., JRC Highlights Report 2023 and Primary analysis of cumulative Emission [2]

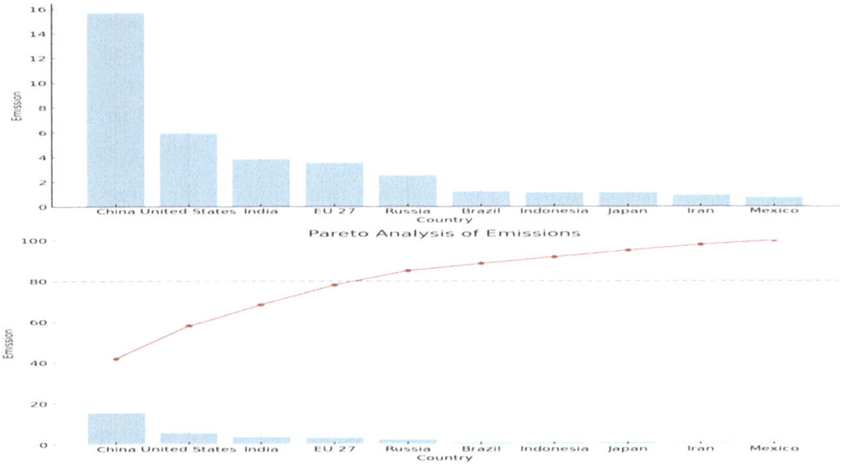


**Fig. 1** Pareto analysis of emission of greenhouse gases. *Source* Primary analysis

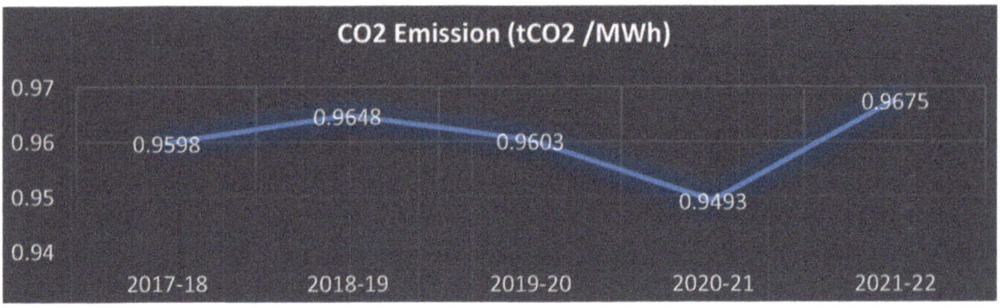


**Fig. 2** Year-wise trend of $CO_2$ emission. Source Primary analysis

**Table 2** Year-wise emissions by different sectors in India

| Year | Fugitive growth (%) | Transport growth (%) | Other sectors growth (%) | Electricity growth (%) |
|---|---|---|---|---|
| 2008 | 5.71 | 18.60 | 7.94 | 14.31 |
| 2010 | 27.03 | 21.57 | 3.68 | 18.15 |
| 2012 | – 10.64 | 15.59 | 2.84 | 7.81 |
| 2014 | – 9.52 | 9.30 | 3.45 | 13.39 |
| 2016 | – 5.26 | 11.91 | 9.33 | 8.34 |
| 2018 | 8.33 | 13.69 | 7.93 | 7.70 |

*Source* Primary Source

**Table 3** Trend analysis of the emission under different sectors

| Trend analysis | Fugitive emissions | Transport | Other sectors | Public electricity generation |
|---|---|---|---|---|
| Slope | 0.01785 | 13.9107 | 3.89285 | 43.642857 |
| Intercept: | 3.2142 | 27,776.92 | 7684.0 | 86,985.571 |
| R-squared:1 | 0.000347 | 0.996376 | 0.95048 | 0.9964 |
| P-value: | 0.96837 | 2.68731 | 0.000188 | 2.64207 |
| Standard error: | 0.428497 | 0.37517 | 0.39737 | 1.173038 |

*Source* Primary Source

**Table 4** Emission by different sectors in India

| Sectors | Emission |
|---|---|
| Agriculture | 23.18 |
| Manufacturing | 15 |
| Electricity | 35 |
| Waste | 267 |
| Fugitive | 2.78 |
| Building | 5.08 |
| Transportation | 8.64 |
| Other fuel combustion | 2.03 |
| Industrial processed | 4.67 |

*Source* Primary Source

all other sectors. Electricity generation also plays a major role, contributing 35 to emissions, while sectors like Agriculture and Manufacturing have moderate impacts with emissions of 23.18 and 15, respectively.

2. **To analyse the trend analyses by different sub-sectors in India**

A. **Emissions from Energy sector**

**Interpretation of the data:** Table 5 shows a significant increase in coal (97.95%) and an extraordinary rise in naphtha consumption (1291.66%) from 2005 to 2014, indicating a shift towards these energy sources. Conversely, there is a notable decrease in the use of furnace oil, LDO/HSD, LSHS/HHS, natural gas, and diesel, reflecting a reduction in reliance on these fuels.

B. **Emission from Vehicles**

**Interpretation of the data in a paragraph:** Table 6 indicates that different vehicle types contribute varying levels of pollutants. Comparative study of agricultural and Industrial waste**,** India's transport sector: state wise synthesis, Atmospheric Environment [1].

The Table 7 shows that industry emissions are higher when compared to the Agricultural sector.

**Table 5** Energy consumption

| Energy | 2005 | 2014 | Percentage change |
|---|---|---|---|
| Coal | 583.3 | 1154.7 | 97.95 |
| Furnace oil | 3.0 | 1.4 | – 53.33 |
| LDO/HSD | 1.6 | 0.7 | – 56.25 |
| LSHS/HHS | 2.3 | 0.7 | – 69.56 |
| Natural gas | 36.3 | 30.6 | – 15.70 |
| Lignite | 24.1 | 33.4 | 38.58 |
| Naphtha | 2.4 | 33.4 | 1291.66 |
| Diesel | 12.8 | 10.0 | – 21.875 |

*Source* Primary Source

**Table 6** Emission from different vehicle types in India. *Source* Primary Source

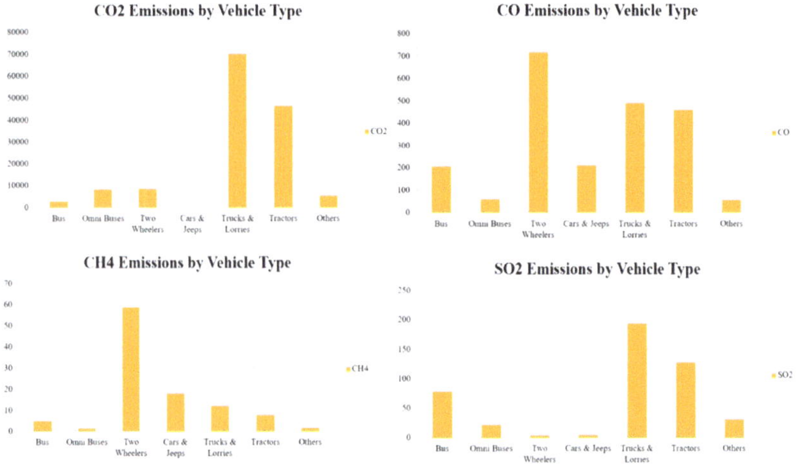


**Table 7** Comparison of industry and agricultural emission (A)

| Industry (type) waste | Mt CO2e (2018) | Agricultural type (waste) | Emission |
|---|---|---|---|
| Glass production | 12.27 | Enteric fermentation | 50.5 |
| Ammonia production | 25.24 | Crop residues | 3.7 |
| Chemical industry | 46.62 | Rice cultivation | 17.4 |
| Metal industry | 30.1 | Cultivation using organic soil | 0.6 |
| Iron and steel industry | 23.64 | Synthetic fertilizer | 13.4 |
| Mineral industry | 130.89 | Manure | 14.2 |
| Cement industry | 117.22 | Total emission | 99.8 |
| Total emission | 385.98 | | |

*Source* Primary Source

Table 8 shown above reveals that industrial emission is more when compared to the agricultural sector.

**Interpretation of data:** Fig. 3 shows provides a breakdown of emissions from various industrial and agricultural sources for the year 2018. The total industrial emissions amount to 385.98 Mt CO2e, with the mineral industry being the largest contributor. Agricultural emissions total 99.8 Mt CO2e, with enteric fermentation contributing the most. This highlights significant emission sources that could be targeted for reduction efforts. **Source:** ***Greenhouse Gas Emissions in the Agricultural and Industrial Sectors—Change Trends, Economic Conditions*** (Fig. 4).

**Steps taken by government to mitigate the emission of greenhouse gases**

- India revealed its aim to reach 450 GW by 2030 at the UN Climate Summit.
- Updated statistics to reach 500 GW by 2030, as announced at COP26.
- By 2030, the nation hopes to have installed 50% non-fossil power sources.
- A hydrogen energy mission for grey and green hydrogen has also been announced.
- During its first and second cycles, the market-based PAT (Perform, Achieve and Trade) program has already avoided up to 92 million tons of $CO_2$ emissions.

**Table 8** Comparison of industry and agricultural emission (B)

| Sector | Mean emissions (Mt CO2e) | Variance (Mt CO2e) |
|---|---|---|
| Industrial waste | 55.14 | 2335.9463 |
| Agricultural waste | 16.633333 | 317.6506666 |

*Source* Primary Source

**Fig. 3** Distribution of emission by two sectors (box diagram). *Source* Primary analysis

**Fig. 4** Distribution of emission from different waste. *Source* Primary analysis

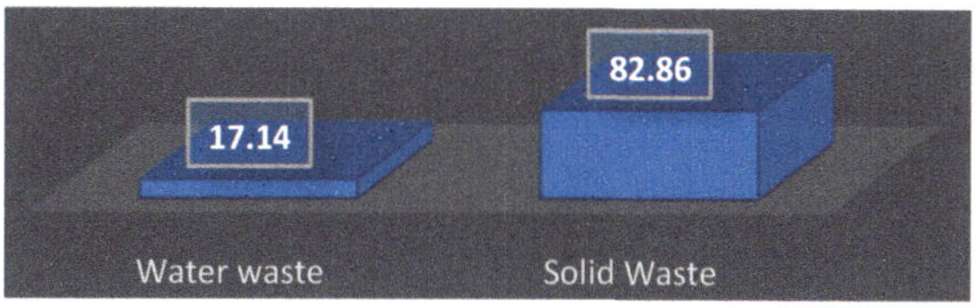

- The Government of India stands committed to combating climate change through its several programmes and schemes, including the National Action Plan on Climate Change (NAPCC), which comprises missions in specific areas of solar energy, energy efficiency, and water.

  **Source:** Jean Chateau [5] A Framework for Climate Change Mitigation in India.

## 7 Conclusion

The climate maintains equilibrium through the intricate interplay of solar radiation and the greenhouse effect. The greenhouse effect effectively captures solar heat, thereby sustaining life by ensuring stable temperatures. Nevertheless, the equilibrium is jeopardised by anthropogenic greenhouse gas emissions resulting from the combustion of fossil fuels, deforestation, and industrial operations, which consequently contribute to the phenomenon of global warming. As per this study, the main sectors responsible for these emissions are energy production, transportation, industrial operations, and agriculture, each emitting certain pollutants that worsen the greenhouse effect. The consequences are substantial, especially in nations such as India, which, in 2019, was classified as one of the most impacted by climate-related extreme weather events. These occurrences lead to acute fatalities, property damage, and long-lasting impacts on agriculture and biodiversity. This exacerbates susceptibility, particularly in developing areas. This circumstance highlights the pressing need for comprehensive climate policies tackling emission reduction and enhancing resilience and adaptation techniques to protect future generations and preserve ecological equilibrium.

## References


1. Ramachandran.: Emissions from India' s transport sector: state wise synthesis. Atmos. Environ. **43**, 5510–5517 (2019), November 2009. https://doi.org/10.1016/j.atmosenv.2009.07.015
2. Simpson, C.: JRC Highlights Report 2023. In: Bonjean, I., Fornara, M., Jonkers, K., Klaassen, T., Lehto, S., Soldatova, L., Thielen Del Pozo, J., Westra Van Holthe, M. (eds.) Publications Office of the European Union, Luxembourg (2024). https://doi.org/10.2760/947345, JRC136884
3. Patel, J.: Solid waste management in India—current status, management practices, models, impacts, limitations, and challenges in future June 2023. Adv. Environ. Res. **12**(2), 95–111 (2023). https://doi.org/10.12989/aer.2023.12.2.095
4. Joseph, A.: An exploratory study on metaverse and SDGs. In: El Khoury, R., Alareeni, B. (eds.) How the Metaverse Will Reshape Business and Sustainability. Contributions to Environmental Sciences & Innovative Business Technology. Springer, Singapore (2023). https://doi.org/10.1007/978-981-99-5126-0_9
5. Chateau J.: A framework for climate change mitigation in India. IMF Working Paper Asia and Pacific and Research Departments (2023)

6. Garg, A., Shukla, P.R., Kankal, B., Mahapatra, D.: $CO_2$ emission in India: trends and management at sectoral, sub-regional and plant levels. Carbon Manag. **8**(2), 111–123 (2021).https://doi.org/10.1080/17583004.2017.1306406
7. Leip, A., Caldeira, C., Corrado, S., Hutchings, N.J., Lesschen, J.P., Schaap, M., de Vries, W., Westhoek, H., van Grinsven, H.J.: Halving nitrogen waste in the European Union food systems requires both dietary shifts and farm level actions. Glob. Food Secur. **35**, 100648 (2022). https://doi.org/10.1016/j.gfs.2022.100648
8. Energy statistics.: Government of India Ministry of Statistics and Programme Implementation National Statistical Office (2023). https://www.mospi.gov.in/
9. Murawska, A., Goryńska-Goldmann, E.: Greenhouse gas emissions in the agricultural and industrial sectors—change trends, economic conditions, and country classification: evidence from the European Union. Agriculture **13**(7), 1354 (2023). https://doi.org/10.3390/agriculture13071354
10. Chateau, J., Jaumotte F.: Climate policy options: a comparison of economic performance international monetary fund, IMF working papers describe research in progress by the author(s) and are authorized for distribution by Antonio Spilimbergo (2023)
11. Jain, A., Tripathi, S., Mani, S., Patnaik, S., Shahidi, T., Ganesan, K.: Access to clean cooking energy and electricity: survey of states 2018. CEEW, New Delhi (2018)

# Factors Affecting the Performance of Women Entrepreneurs, Kothanallur Taluk of Thiruvarur District, Tamilnadu

**M. Rani Therasa and K. V. R. Rajandran**

**Abstract** Women entrepreneurs contribute to job creation and the socio-economic progress of the country. Thus, flagging the way for future peers of women entrepreneurs. They also face numerous difficulties in starting and successively a business with a lack of managerial skills, inadequate tutoring, less risk-taking skills, limited mobility and household obligations. This investigator attempted to establish the glitches women entrepreneurs face and their impact on business performance. The data were collected from 150 women entrepreneurs in questionnaire format, and it was collected using simple random sampling methods. The statistical tools of Simple percentage, Factor analysis and Regression statistical tools were followed to find the result of the study. From the study, it is ascertained that women entrepreneurs face more difficulty in managing family and financial problems, which automatically impact the business's overall performance.

**Keywords** Business performance · Family problem · Financial · Managerial problem · Marketing problem · Production problem · Women entrepreneur · Tamil Nadu

## 1 Introduction

Kothanallur Taluk is located in the Thiruvarur district of Tamil Nadu and is known for its agricultural activities, primarily in paddy cultivation. The region has a mix of rural and semi-urban setups that foster entrepreneurship opportunities. However, women entrepreneurs in Kothanallur Taluk, like many in similar regions, face challenges related to access to finance, limited market exposure, societal constraints, and a lack of necessary skill development opportunities. Access to credit and resources remains a significant hurdle for women who want to start or expand their businesses. Social

M. Rani Therasa (✉) · K. V. R. Rajandran
Department of Management Studies, Periyar Maniammai Institute of Science and Technology, (Deemed to be University), Thanjavur, India
e-mail: sivasubramanian16@gmail.com

 
M. Al Mubarak (ed.), *Sustainable Digital Technology and Ethics in an Ever-Changing Environment*, Studies in Systems, Decision and Control 237,
https://doi.org/10.1007/978-3-031-86708-8_55

norms and cultural expectations may also impact their ability to fully engage in entrepreneurial activities. The purpose of the study is to investigate the problems faced by women entrepreneurs and their impact on the business performance of women entrepreneurs [6, 12].

### *1.1 Structure of the Paper*

This paper aims to demonstrate the problems faced by women entrepreneurs in India, specifically in Tamil Nadu. This paper is designed and categorised as three different sections. The first segment of the paper depicts the introduction and literature review. The second section involves objectives and data analysis. The third portion deals with results, conclusion and future scope in the relevant areas.

## 2 Literature Review

Dima [4] and Swati [16] researched the restrictions and possibilities confronting female entrepreneurs. The primary data were gathered from ten female entrepreneurs through in-depth interviews. The percentage and frequencies were utilized to determine the outcome. The researcher concluded that the findings of this study demonstrate the importance of micro and macro-level elements in entrepreneurship research, as well as the importance of combining different lenses and units of analysis to fully comprehend the complexities of women entrepreneurs in any given setting [15].

Gargi and Sumen [5] investigated the problems and difficulties faced by female small-business owners. The data for the study were gathered from 68 female entrepreneurs using descriptive analysis. Engaged in fabric selling, artificial jewellery and craft creating, food catering and home delivery, and bakery product sales, with secondary data gathered from papers, journals, books, and the internet. The outcome was determined using simple percentage analysis and chi-square techniques. The researcher determined that the biggest hurdles experienced during their introduction into business were age, educational level, and marital status.

Mamadeep and Jainendru [7] investigated the demographics of female entrepreneurs, as well as the motivators and challenges they face. The study relies on primary data collected from four typical districts in Malawi. The information was gathered from 511 female entrepreneurs. ANOVA, percentage, frequency, simple mean, and ranking revealed that when starting and running a business, women entrepreneurs face challenges. The majority of women struggle with having to juggle their concurrent roles as mother, wife, and sister, which makes it difficult for them to operate their businesses. Other factors mentioned by some respondents include male domination and financial difficulties.

Najma [8] and Charulakhm [2] conducted study on the hurdles that female entrepreneurs experience in their businesses. The study collected data through several methodologies, including individual interviews, surveys, and observation investigations. Data were collected from 397 female entrepreneurs. The reliability and validity of additional information obtained from publications, relevant books, and periodicals were tested before data analysis using t-test, chi-square, factor analyses of mean or regression, ANOVA, and correlations to determine the outcome. The researchers discovered that the most significant problems faced by female entrepreneurs include a lack of educational and training opportunities, regulatory constraints, cultural and societal support, access to money, and work-family integration.

Palaniappan and Ramgopal [9] discovered that the primary source of primary data is obtained from women entrepreneurs via a well-structured questionnaire. The study includes 90 respondents from various Taluks in the erode districts. The primary data collection tool is the personal interview. Women entrepreneurs will be interviewed using a specific technique. Statistical methods used include percentage analysis, average, weightage score, and the Chi-Square test. Other obstacles that women commonly experience during the execution of their enterprises include a lack of competent leadership, poor planning, and insufficient financial resource allocation [10]. Women home-based business entrepreneurs are also facing several socio-economic and health issues such as isolation, formal credit from the government, capital formation, marketing issues and lack of branding for their commodities produced [3, 13, 14].

## 2.1 Problems of Women Entrepreneurs Identified

Stiff rivalry, non-cooperation of agents, a lack of marketing knowledge, insufficient packaging, and a poor warehousing facility all contribute to proper product pricing. Liquidity issues, a lack of permanent capital, a lack of credit, a high cost of investment, an absence of working capital, a high rate of interest, and a delay in the sanction of funds. Labour employee turnover, partially skilled labourers, absenteeism, and the demand for high wages. High production costs, a lack of development and research facilities, a scarcity of raw materials, a lack of knowledge about modern technology, and a power shortage. Lack of support from relatives, limitations due to religion, and family responsibility [1].

## 2.2 Objectives

To find the challenges faced by women entrepreneur in the study area.

To find the impact of challenges on the performance of women entrepreneur.

## 3 Analysis

To achieve the objectives of this study, the researcher has framed a specific model for depicted blow in Fig. 1. The major variable has been taken for this models such as financial problem, managerial issues, production related challenges, and marketing related issues.

### 3.1 *Result Findings and Discussion*

The age-wise circulation of sample respondents is presented in Table 1, which reveals significantly that the highest percentage of 36.7% of respondents belong to the age group of 30–40. It is also identified from the data that 23.3% of them are age.

Table 2 emphasises the category-wise businesses of sample respondents. It is identified that 46.7% of women entrepreneurs are engaging in production activities and 38% of them are carrying out service and service-enabled businesses. The rest of the 15.3% are in the retail business.

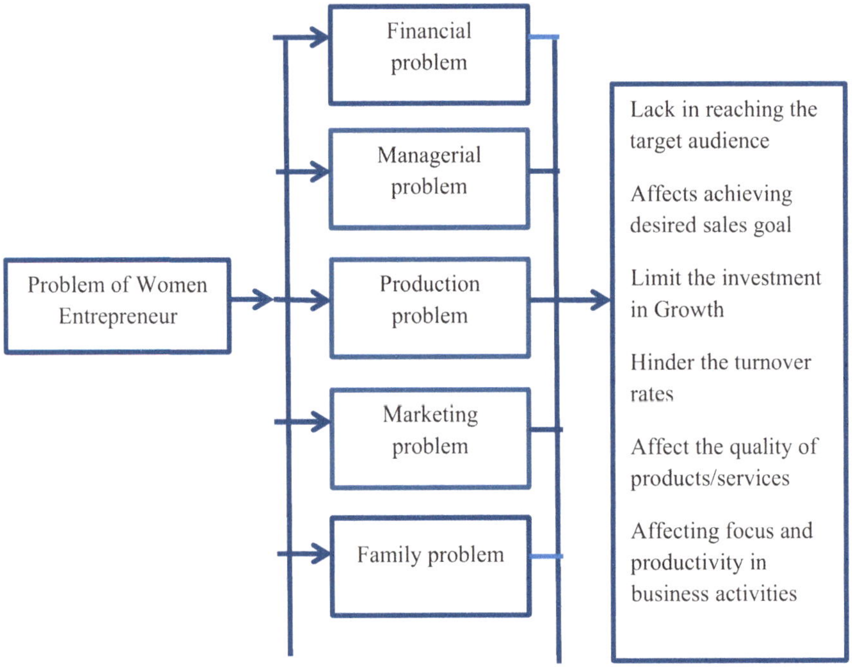


**Fig. 1** Research frame work model. *Source* Primary survey

**Table 1** Age-wise data distribution

| Age | | |
|---|---|---|
| | | Percent |
| Valid | 21–30 | 23.3 |
| | 31–40 | 36.7 |
| | 41–50 | 25.3 |
| | 51 and above | 14.7 |
| | Total | 100.0 |

*Source* Primary data

**Table 2** Business category of respondents

| Business category | | |
|---|---|---|
| | | Percent |
| Valid | Retailing | 15.3 |
| | Production | 46.7 |
| | service | 38.0 |
| | Total | 100.0 |

*Source* Primary data

The entrepreneurial experience of the sample is portrayed in below provided Table 3. This table provides clear evidence of the rich experience of women entrepreneurs.

The KMO test value is calculated at 0.699 which is more than 0.5 of significance, it can be considered acceptable and valid to conduct data reduction techniques. Bartlett's Test of Sphericity of significance value is measured as $< 0.001$, which shows a high correlation level between the analysed variables (Table 4).

**Table 3** Experience of the respondents

| Experience | | |
|---|---|---|
| | | Percent |
| Valid | less than 5 years | 18.0 |
| | 5–10 years | 37.3 |
| | 10–15 years | 27.3 |
| | 15 and above | 17.3 |
| | Total | 100.0 |

*Source* Primary data

**Table 4** KMO and Bartlett's test—major problem faced by women entrepreneur

| KMO and Bartlett's Test | | |
|---|---|---|
| Kaiser–Meyer–Olkin measure of sampling adequacy | | 0.699 |
| Bartlett's test of sphericity | Approx. Chi-square | 950.410 |
| | df | 300 |
| | Sig. | 0.000 |

*Source* Author's calculation based on the primary survey

**Table 5** Communalities—major problem faced by women entrepreneur

| Communalities | Initial | Extraction |
|---|---|---|
| Stiff competition | 1.000 | 0.636 |
| Non-cooperation of agents | 1.000 | 0.701 |
| Lack of promotion knowledge | 1.000 | 0.742 |
| Poor packaging | 1.000 | 0.505 |
| Poor warehousing facility | 1.000 | 0.698 |
| Problem of liquidity | 1.000 | 0.613 |
| Lack of fixed capital | 1.000 | 0.682 |
| Non availability of credit | 1.000 | 0.486 |
| High cost of capital | 1.000 | 0.603 |
| Shortage of working capital | 1.000 | 0.670 |
| High rate of interest | 1.000 | 0.629 |
| Delay in sanction of funds | 1.000 | 0.621 |
| Labour turnover | 1.000 | 0.668 |
| Semi-skilled labourers | 1.000 | 0.530 |
| Absenteeism | 1.000 | 0.709 |
| Demand for high wages | 1.000 | 0.638 |
| High cost of production | 1.000 | 0.536 |
| Lack of research and development facilities | 1.000 | 0.725 |
| Shortage of raw materials | 1.000 | 0.779 |
| Lack of awareness of modern technology | 1.000 | 0.649 |
| Power shortage | 1.000 | 0.731 |

*Source* Author's calculation based on the primary survey

Every variable in the commonality initially is expected to share 100% variance hence every item has a value of 1.00 which is 100% variance and the extraction value is ranking from 0.486 to 779 (Table 5).

**Table 6** Model summary-impact of the problem on business performance

| Model summary | | | | |
|---|---|---|---|---|
| Model | R | $R^2$ | Adjusted $R^2$ | Std. error of the estimate |
| 1 | 0.750[a] | 0.562 | 0.544 | 0.67556874 |

[a] Predictors: (Constant), Affecting focus and productivity in business activities, Limiting the investment in growth /expansion plan opportunities, Affecting the quality of products/services, Lack of reaching the target audience, Hinder the turnover rates, Affects achieving desired sales goal
[b] Dependent Variable: F1

According to Total Variance Explained Table 5, the Eigen-value for a given factor measures the variance in all the variables which are accounted for by those factors, it's also clear that nine components have having Eigen-value greater than 1 from the given set of variables.

The R-value 0.750 (Table 6) clearly shows in table there is a correlation between dependent and independent variables and the $R^2$ value 0.562 indicates that 54% of F1 has impacted the dependent variable.

According to the ANOVA table the sig value = 0.000, which indicates that there is a significant relationship between the dependent and independent variable (Table 7).

The coefficient table indicates that the dependent variable has an impact on all the independent variables with a sig value less than 0.05 (Table 8).

**Table 7** Table ANOVA—impact of the problem on business performance

| ANOVA | | | | | | |
|---|---|---|---|---|---|---|
| Model | | Sum of squares | df | Mean square | F | Sig. |
| 1 | Regression | 83.736 | 6 | 13.956 | 30.579 | 0.000[a] |
| | Residual | 65.264 | 143 | 0.456 | | |
| | Total | 149.000 | 149 | | | |

[a] Predictors: (Constant), Affecting focus and productivity in business activities, Limiting the investment in growth /expansion plan opportunities, Affecting the quality of products/services, Lack of reaching the target audience, Hinder the turnover rates, Affects achieving desired sales goal

**Table 8** Coefficients—Impact of the problem on business performance

Coefficients

| Model | | Unstandardized coefficients | | Standardized coefficients | t | Sig |
|---|---|---|---|---|---|---|
| | | B | Std. error | Beta | | |
| 1 | (Constant) | − 2.159 | 0.181 | | − 11.920 | 0.000 |
| | Lack in reaching the target audience | 0.225 | 0.062 | 0.258 | 3.637 | 0.000 |
| | Affects achieving the desired sales goal | − 0.143 | 0.07001 | − 0.153 | − 2.007 | 0.047 |
| | Limit the investment in growth / expansion plan opportunities | 0.159 | 0.062 | 0.184 | 2.585 | 0.011 |
| | Hinder the turnover rates | 0.207 | 0.066 | 0.233 | 3.131 | 0.002 |
| | Affect the quality of products/services | 0.146 | 0.058 | 0.174 | 2.528 | 0.013 |
| | Affecting focus and productivity in business activities | 0.256 | 0.061 | 0.284 | 4.229 | 0.000 |

Dependent Variable: F1

## 4 Conclusion

The study by the researcher shows that religious Restrictions, lack of fixed capital, family responsibility, non-availability of formal credit, lack of family support, and shortage of working capital were the problems faced by women entrepreneurs and it automatically affected the business performance of women entrepreneurs which includes lack in reaching the target audience, affects achieving desired sales goal, limit the investment in growth and expansion plan opportunities, hinder the turnover rates, affect the quality of products and services and also affecting focus and productivity in business activities. Therefore, effective financial support and family support could significantly help to overcome the various challenges and issues of women entrepreneurs. The report illustrates various problems that women entrepreneurs face, including a lack of financial resources, trouble locating suitable labour, work-home conflict, and low profit margins. The findings indicate that the most influential component is industry (SMEs) features, followed by entrepreneurial aims and motivations, and legal issues. Furthermore, opportunity recognition has been shown to moderate the relationship between business objectives and motivations, as well as industry features.

## 5 Suggestions

Based on the findings of the study the following suggestions are made for women entrepreneurs to succeed in their business.

Know Your Audience: Research the target market thoroughly.

Strategic Marketing: Use various channels effectively.

Invest in Marketing: Allocate budget for outreach.

Network and Partner: Expand reach through collaborations.

Feedback and Improvement: Regularly refine products/services.

Streamline Operations: Enhance productivity internally.

Continuous Learning: Stay updated and improve skills.

Financial Planning: Carefully manage resources for growth.

## 6 Limitations and Future Directions

This study is limited to a small taluk of a district and the sample size was relatively limited for a broad area of research. There is a wider scope for the expansion of the sample size to draw even better results accurately.

## References

1. Abdul Ghani Azmi, I., Che Hashim, R., Mohamed, H.A.B., Sahol Hamid, N.: A Study of Muslim women entrepreneurs' SMEs challenges and motivation in the Asia Pacific region. In: Rafiki, A., Dana, L.P., Nasution, M.D.T.P. (eds.) Open Innovation in Small Business. Contributions to Environmental Sciences & Innovative Business Technology. Springer, Singapore (2023). https://doi.org/10.1007/978-981-99-5142-0_13
2. Charulakhm, T.: A study on problems faced by the women entrepreneurs in Kancheepuram district. Int. J. Recent. Technol. Eng. **7**(6), 1191–1194 (2019)
3. Deepa, V., Ramesh, K., Sivasubramanian, K.: A comparative analysis of the business performance of selected public and private sector banks in India. In: Khamis, R., Buallay, A. (eds.) AI in Business: Opportunities and Limitations, Studies in Systems, Decision and Control 516, pp. 129–138. Springer (2024)
4. Dima, J.: Constraints and opportunities facing women entrepreneurs in developing countries. Gend. Manag.: Int. J. **24**(4). 232–251 (2009). https://doi.org/10.1108/17542410910961532
5. Gargi, L., Sumen, N.: A study on challenges faced by women entrepreneurs of small businesses in Kolkata. Mizoram Univ. J. Humanities Soc. Sci. **8**(1), 25–35 (2022)
6. Kasthuri, R., Rajeswari, S., Kumarasamy, P., Thamilmani, R., Sivasubramanian, K.: Performance of the agricultural sector in india through farm credit accessibility: a statistical analysis. In: Alareeni, B., Elgedawy, I. (eds.) AI and Business, and Innovation Research: Understanding

the Potential and Risks of AI for Modern Enterprises. Studies in Systems, Decision and Control, vol. 440. Springer, Cham (2024). https://doi.org/10.1007/978-3-031-42085-6_69
7. Mamadeep, K., Jainendru, K.: Problems and motivations for women entrepreneurs in Punjab, India. Turk. Online J. Qual. Inq. **12**(3), 762–777 (2021)
8. Najma, T.: The challenges facing by women entrepreneurs: A study on Gulf Cooperation Council (GCC) Countries. Unpublished Thesis, Brunel University. London (2016)
9. Palaniappan, G., Ramanigopal, C.S.: A study on problems and prospects of women entrepreneurs with special reference to Erode district. Int. J. Phys. Soc. Sci. **2**(3), 219–220 (2012)
10. Shilp, G.: Rural women entrepreneurship in India: challenges, opportunities and motivational factors. Int. J. Res. Anal. Rev. **5**(3), 91–96 (2018)
11. Sathiyabama, Velmuruga.: Challenges faced by women entrepreneurs in Coimbatore district, J. Adv. Res. Dyn. Control. Syst. **11**(9). Special Issue. 1091–1099 (2019). https://doi.org/10.5373/Jardcs/V11/20192676
12. Sathyadevi and Fasla (2016). A study on problems faced by women entrepreneurs in Pattambi block Panchayath, Palakkad District. Int. Conf. Res. Ave. Soc. Sci. **1**(3), 166–170
13. Sivasubramanian, K., Pushpa, A., Raju, V., Kumar, M.D.: Women and informal employment: an analysis of socio-economic and health conditions of women home-based workers in Chennai India. J. Int. Women Stud. **21**(5), 97–106 (2020)
14. Sivasubramanian, K., Adarsh, R., and Krishnamurthy, A.: Economic empowerment of women through household dairy farming in rural India. In: Khamis Hamdan, R. et al. (eds.) Information and Communication Technology in Technical and Vocational Education and Training for Sustainable and Equal Opportunity, Technical and Vocational Education and Training: Issues, Concerns and Prospects 38. Springer (2024). https://doi.org/10.1007/978-981-99-6909-8_9
15. Shakila, Tony, M.: Problems faced by women entrepreneurs in Thoothukudi district. Turk. J. Comput. Math. Educ. **12**(3), 4014–4018 (2021)
16. Swati, P.: Constraints faced by women entrepreneurs in developing countries. Rev. Rank.: Int. J. (2018). https://doi.org/10.1108/gm-01-2017-0003

The manufacturer's authorised representative in the EU is Springer
Nature Customer Service Centre GmbH, Europaplatz 3, 69115 Heidelberg,
Germany. If you have any concerns regarding our products, please
contact ProductSafety@springernature.com

Printed and bound by CPI Group (UK) Ltd, Croydon, CR0 4YY
01/10/2026
02225971-0019